National Audubon Society®
Field Guide to
North American Mammals

A Chanticleer Press Edition

National Audubon Society®
Field Guide to
North American Mammals

Revised Edition

John O. Whitaker, Jr.
Professor of Life Sciences,
Indiana State University

Alfred A. Knopf, New York

This is a Borzoi Book.
Published by Alfred A. Knopf, Inc.

Copyright © 1996 by Chanticleer
Press, Inc. All rights reserved under
International and Pan-American
Copyright Conventions. Published in
the United States by Alfred A. Knopf,
Inc., New York, and simultaneously in
Canada by Random House of Canada,
Limited, Toronto. Distributed by
Random House, Inc., New York.
Originally published in the United
States in somewhat different form
by Alfred A. Knopf, Inc., New York,
in 1980.

www.randomhouse.com

Knopf, Borzoi Books, and the
colophon are registered trademarks
of Random House, Inc.

Prepared and produced by
Chanticleer Press, Inc., New York.

Printed and bound by Toppan
Printing Co., Ltd., Tokyo, Japan.

Published December 1980
Second edition, fully revised March 1996
Seventh printing, July 2002

Library of Congress Cataloging-in-
Publication Number: 95-81456
ISBN: 0-679-44631-1

National Audubon Society® is a registered
trademark of National Audubon Society,
Inc., all rights reserved.

CONTENTS

Contents

NATIONAL AUDUBON SOCIETY

The mission of NATIONAL AUDUBON SOCIETY, *founded in 1905, is to conserve and restore natural ecosystems, focusing on birds, other wildlife, and their habitats for the benefit of humanity and the earth's biological diversity.*

One of the largest, most effective environmental organizations, AUDUBON has nearly 550,000 members, numerous state offices and nature centers, and 500+ chapters in the United States and Latin America, plus a professional staff of scientists, educators, and policy analysts. Through its nationwide sanctuary system AUDUBON manages 160,000 acres of critical wildlife habitat and unique natural areas for birds, wild animals, and rare plant life.

The award-winning *Audubon* magazine, which is sent to all members, carries outstanding articles and color photography on wildlife, nature, environmental issues, and conservation news. AUDUBON also publishes *Audubon Adventures,* a children's newsletter reaching 450,000 students. Through its ecology camps and workshops in Maine, Connecticut, and Wyoming, AUDUBON offers nature education for teachers, families, and children; through *Audubon Expedition Institute* in Belfast, Maine, AUDUBON offers unique, traveling undergraduate and graduate degree programs in Environmental Education.

AUDUBON sponsors books and on-line nature activities, plus travel programs to exotic places like Antarctica, Africa, Baja California, the Galápagos Islands, and Patagonia. For information about how to become an AUDUBON member, subscribe to *Audubon Adventures,* or to learn more about any of our programs, please contact:

NATIONAL AUDUBON SOCIETY
Membership Dept.
700 Broadway
New York, NY 10003
(800) 274-4201
(212) 979-3000
http://www.audubon.org/

THE AUTHOR

John O. Whitaker, Jr., is Professor of
Life Sciences at Indiana State University,
where he teaches courses in animal
ecology, vertebrate zoology, mammalogy,
and herpetology. During the past 30
years he has written more than 245
technical and popular articles on the
food, habits, reproduction, and parasites
of, chiefly, our native North American
fauna. He is co-author, with W. J.
Hamilton, Jr., of *Mammals of the Eastern
United States* (Cornell University Press,
revised edition due 1996), and co-author,
with R. E. Mumford, of *Mammals of
Indiana* (Indiana University Press, 1982).

ACKNOWLEDGMENTS

Since the appearance of the first edition of the *National Audubon Society Field Guide to North American Mammals* in 1980, a number of taxonomic changes have been introduced in the scientific literature, and much more has been learned about many species. For these reasons, we felt it was time to update this guide.

I would like to thank all the scientists and students of mammalogy who have contributed to the scientific literature. For about 25 years, the American Society of Mammalogists has been compiling the Mammalian Species accounts, which summarize information from the literature. Information from these accounts has been used in the updating of this guide.

For the first edition, Robert Elman served as text consultant, and Angus Cameron and Sydney Anderson also made valuable contributions. My wife, Royce, read and greatly improved many of the species accounts. Laura Bakken made many contributions. I owe a great debt to William J. Hamilton, Jr., for his advice and encouragement throughout graduate school and the ensuing years, as well as to my parents and my family for their help, support, and patience throughout. I also wish to thank Paul Steiner, the founding publisher of

Chanticleer Press, and his staff for the first edition: Gudrun Buettner, Milton Rugoff, Anne Knight, Susan Rayfield, Richard Christopher, Susan Linder, Carol Nehring, Helga Lose, Ray Patient, and Dean Gibson.

For this revised edition, I am again indebted to all the scientists and mammalogists who have contributed to the scientific literature on North American mammals. I owe a special debt to Phillip Frank and James D. Lazell for allowing us to include information they have compiled about the Little Mastiff Bat in the Florida Keys. Fiona Reid and Mark D. Engstrom, who served as consultants, scrutinized the text and the photographs and offered many helpful comments and suggestions. It has been a pleasure to work with Andrew Stewart, the publisher of Chanticleer Press, and his staff. Edie Locke, managing editor; Amy K. Hughes, senior editor; Drew Stevens, art director; Kristina Lucenko, typesetting and editorial assistant; and Sam Shaw, office intern, skillfully ushered the book through the editorial and design processes, and Alicia Mills saw it through production and printing. Giema Tsakuginow, photo editor, and Consuelo Tiffany Lee, photo assistant, patiently sifted through thousands of submissions to find the stunning photographs used in the guide. The book has benefited greatly from the editorial skills of Patricia Fogarty, Beth Greenfeld, Marian Appellof, Cathy Peck, and Karen Dubno. Sy Barlowe produced the fine animal illustrations, Dot Barlowe the animal tracks, and Paul Singer the original maps and silhouettes. Finally, I would like to thank the photographers, whose pictures make this guide to North American mammals not only unique, but a thing of beauty.

John O. Whitaker, Jr.

INTRODUCTION

The mammals of North America are a diverse and fascinating group. However, because most mammals are nocturnal, secretive, and make few sounds audible to our ears, they tend to be elusive. In a morning's walk on a tract of woods in the midwestern U.S., for instance, we can expect to see perhaps only five or six species of wild mammals: Eastern Fox Squirrels, Eastern Chipmunks, Eastern Cottontails, Woodchucks, pond-dwelling Common Muskrats, and occasionally a White-tailed Deer or two. We may also see or hear as many as 30 species of birds and 10 species of amphibians and reptiles, creatures that are much more easily seen though not necessarily more numerous. However, those woods contain many other kinds of mammals that are not immediately apparent. If we knock with a stick on the bases of enough dead upright trees with old woodpecker holes we are likely to rouse a Southern Flying Squirrel. If we examine the mud around small streams and pools we will see an abundance of tracks made by the Common Raccoon. In the dusty road along the edge of the woods, Virginia Opossum tracks are often discernible. Rabbit and deer droppings are common throughout the area. Ridges of earth pushed up by the Eastern Mole meander

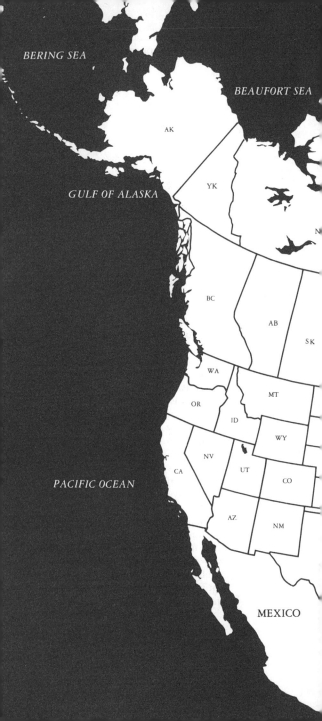

across grassy areas and some woodland clearings. If we look closely, we may find discarded black cherry pits, caches of stored food, or burrows under tree roots—all signs of the White-footed Mouse. If we peer into the leaf mold on the forest floor we may discover the rounded burrows of the Woodland Vole. Near moist logs we may find burrows of a short-tailed shrew, and under the lips of banks along streams or roadways the tiny pathways made by other shrews.

Mammal-watching often involves looking for the evidence of mammals rather than the mammals themselves. Patience, alertness, and careful observation are the keys to identifying the species in your area. This guide is designed to introduce the reader to the ways of mammals: their feeding and social behaviors, breeding habits, chosen habitats, and geographical range, as well as the tracks and other evidence that mark their presence when they cannot be seen. The detailed descriptive text is supplemented by hundreds of color photographs, black-and-white drawings, track illustrations, range maps, and unique range charts.

About the Introduction

The first section of the Introduction explains how this field guide works, and gives a breakdown of what the user will find in the color plates section, the species accounts, and the appendices. An essay on mammals explains what a mammal is and how mammals are classified scientifically. The second essay, "Habitats of North America," which describes the major habitat types in our region, will help the reader in identifying mammals by their habitats. "How to Find Mammals" offers a roundup of helpful tips for tracking these fascinating creatures. "How to Use This Guide" offers examples to help the reader employ the guide in the field to identify mammals.

The Mammals Because this guide is intended primarily for field use, we have included all the native land-dwelling or land-breeding species of wild mammals found in North America north of Mexico. While we include seals and sea lions, which breed on land, we do not include whales and dolphins and their kin, as they are exclusively ocean-dwelling.

The classification of mammals in this guide is based on the "Revised Checklist of North American Mammals North of Mexico, 1991," by J. Knox Jones, Jr., Robert S. Hoffmann, Dale W. Rice, Clyde Jones, Robert J. Baker, and Mark D. Engstrom (*Occasional Papers, The Museum, Texas Tech University* 146 [February 7, 1992]:1–23). In a few instances, we have differed from the Checklist in our acceptance of a species or subspecies; such cases are noted.

Introduced and Domesticated Species Some introduced mammals—animals that are not native to our region but have been brought here—have been included, as have some domesticated species that are also represented in the wild. The introduced species chosen for this guide have been present in North America north of Mexico for many years; they reproduce and maintain themselves under natural conditions, and interact with and influence native species. Among the introduced mammals that fulfill these criteria are the Black Rat and the House Mouse—which usually are found in North American field guides—and the Rhesus Monkey, which is not often represented. We have not included every species that has been brought to these shores, but have attempted to include those that are most established or influential. The feral horse, the feral pig, and the domestic cat and dog are examples of domesticated species that either have self-sustaining feral populations (the former two) or that have a notable influence on native fauna (the latter two).

Photographs as a
Guide to
Identification

This field guide presents the mammals in full-color photographs rather than in the more traditional paintings or drawings. An artist's rendering of an animal is his or her interpretation, whereas a good photograph captures the natural color and stance of the animals. The photographs are not only beautiful but are true to nature, presenting the mammals in their natural settings and often engaged in characteristic behavior. We have obtained color photographs of 265 species for this guide. Other animals—often species for which a good photograph was unobtainable—are represented in a black-and-white drawing, which accompanies the species description. Still other species, most of which are virtually identical to another species (often one from which they have recently been separated taxonomically), are not illustrated.

The Color-Plates
Section

Part I of this guide is the color-plates section, which includes two tables—the Silhouette and Thumb Tab Guide and the Track Guide—as well as the photographs of the animals.

*Organization of
the Color Plates*

While in the text descriptions the mammals are presented in the scientific (taxonomic) sequence (with the species within genera listed alphabetically), in the color plates they are arranged by type, shape, and size—features an observer notices in the field. In many cases the animals still fall into a taxonomic sequence, as taxonomy is based in part on an animal's appearance; this is the case with shrews and moles, which are of similar size and shape, and are closely related taxonomically. We have placed all the mice and rats together, grouping them mainly from small to large, and have put the voles, lemmings, and pocket gophers, all small, blunt-shaped rodents, with the pikas, which belong to the rabbit and hare family, but differ from them in appearance. The marmots and the Woodchuck, which are in the squirrel family, have been grouped with other large rodents, including the American Beaver and the Common Muskrat. The unrelated Common Porcupine and Nine-banded Armadillo are grouped together in a section called Armored Mammals because of their unique adaptations for defense against predators. The larger mammals— wolves, foxes, and the Coyote, cats, and hoofed mammals—fall into a more traditional scientific sequence, which correlates with their physical attributes.

Silhouette and Thumb Tab Guide The organization of the color plates is explained in a table preceding them called the Silhouette and Thumb Tab Guide. A silhouette of a typical member of each group appears on the left side of the table. Silhouettes of the mammals within that group are shown on the right. For example, the group Voles, Lemmings, Pikas, and Pocket Gophers is represented by a silhouette of a Brown Lemming. The representative silhouette is repeated in a thumb tab on the left edge of each double page of color plates devoted to that group of mammals.

Track Guide A guide to mammal tracks follows the Silhouette and Thumb Tab Guide in the color section. The tracks are arranged according to their general appearance. The group of hand- and foot-like tracks is divided into those with four-toed foreprints and five-toed hindprints, such as the Norway Rat; and those with foreprints and hindprints both having five toes, such as the Common Muskrat. Tracks that show as pads with four toes are divided into those with and without claw marks; canids, or dog-like mammals, leave tracks with claw marks, while cats do not. A familiarity with the basic track patterns will enable you to identify many mammals by the tracks they have left behind. Above each track drawing is the name of the species and the page number in the text on which that animal is discussed.

Captions   The captions under the photographs give the plate number, the common name of the species, total body length from tip of nose to tip of tail, and the page number of the species description in the text (Part II of the guide). For larger animals, the shoulder height may also be included. The animal's sex (♀ female, ♂ male) is given for those species for which it is known.

Organization of   Following the color plates is Part II of
the Text   the guide, the species accounts, which are text descriptions of each species in our region. The text discusses the larger groupings of mammals (orders, families, and genera) in phylogenetic groups— that is, according to their scientific classification. The orders of mammals are not arranged in a linear fashion from "primitive" to "advanced," as is often thought, but in a tree-like branching sequence. We have followed the standard sequencing for the orders, families, genera, and species, as outlined in the "Revised Checklist of North American Mammals"; species within genera are arranged alphabetically by scientific name. Aquatic carnivores (sea lions, seals, and the Walrus) were previously classed in a separate order, Pinnipedia, but are now considered part of the order Carnivora.

This guide contains brief descriptions of the orders and families represented; these provide basic information about the species the larger groups contain. Species accounts follow the family descriptions in the text sequence. When reading about a particular animal, readers should review the order and family accounts that precede the species account. The species accounts include the following information:

Names   Each species account begins with the name currently accepted as the species' proper common name. If a species has alternate common names, those are also given, within quotation marks. The scientific species name, shown below the common name, is italicized.

Description   This section presents the mammal's physical characteristics, including color, distinctive markings, and anatomical features. Key identification characteristics are italicized. For some of the smaller mammals, such as shrews and rodents, species are identified on the basis of dental patterns, so there is often a description of the teeth (see the illustrations of dental patterns on page 21). In bats, the salient features include the interfemoral (or tail) membrane, the calcar (a bony projection near the hindfoot), and the tragus (a fleshy lobe at the base of the ear); the parts of a bat are illustrated on page 22. Each species description ends with measurements: total length (L) from tip of nose to end of tail, length of tail (T), length of hindfoot (HF) from heel to tip of longest toe, and weight (Wt). Ear measurement (E), when given, is from notch to tip. For bats, the length of the forearm (FA) is given, as this is the best feature for determining a bat's overall size. For larger mammals, we also give height (Ht), measured from ground to shoulder level. The drawings on page 20 show how different features are measured.

Similar Species   Here we briefly identify other species that may be confused with the subject, especially those that have overlapping geographic ranges. Each similar species is described in a full species account elsewhere in the book.

Breeding   The general breeding habits of a family are usually described in the family introduction. Breeding information

How Mammals Are Measured

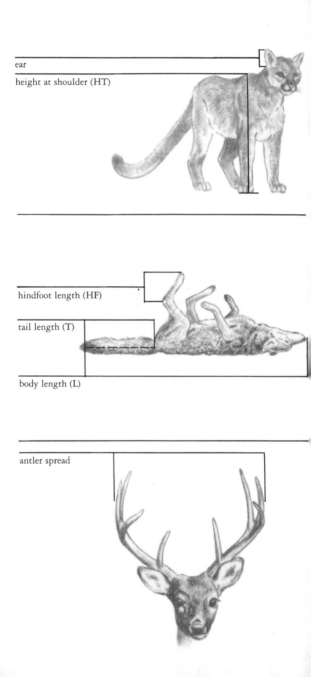

ear

height at shoulder (HT)

hindfoot length (HF)

tail length (T)

body length (L)

antler spread

Dental Patterns

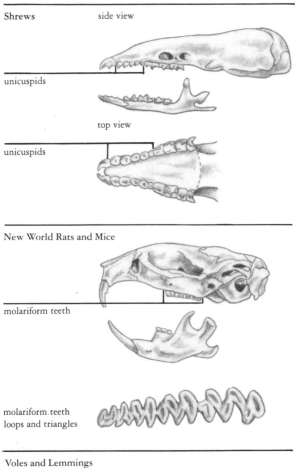

Shrews

side view

unicuspids

top view

unicuspids

New World Rats and Mice

molariform teeth

molariform teeth
loops and triangles

Voles and Lemmings

molariform teeth

molariform teeth
cusps

Parts of a Bat

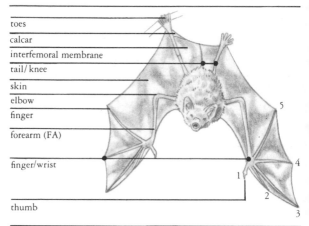

toes
calcar
interfemoral membrane
tail/ knee
skin
elbow
finger
forearm (FA)

finger/wrist

thumb

5

4

1

2

3

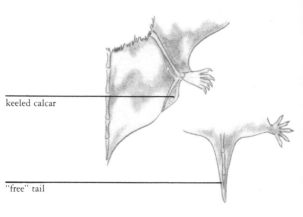

keeled calcar

"free" tail

noseleaf/tragus

chinleaf

unique to a particular mammal—such as reproductive season, number of litters per year and young per litter, and gestation period—is included in this section. For many species, especially some of the smaller ones, breeding habits are little known.

Sign The evidence that a particular animal is or has been in a particular area is called its sign. An animal's sign can take many forms, including tracks, scat (fecal matter), burrows and burrow openings, food remnants and caches, mud slides, and nests. Some smaller mammals leave such scanty evidence that their sign is not useful in identifying them and so is not described. The Sign section may include a description of the scat (its color, shape, consistency, contents, and where most likely to be found) and of the tracks, including foreprint and hindprint size and shape; straddle (width between prints); walking or running stride (distance between sets of prints); and distinguishing marks, such as claw impressions or tail drag. Many track descriptions are accompanied by drawings of foreprint (at left), hindprint (at right), and typical walking track. Feeding signs, den and nest areas, scentposts, and other evidence may also be described.

Habitat All mammals are restricted to certain natural areas called habitats. An animal's requirements for type of vegetation, soil, air, rainfall, light intensity, and food are met by its habitat. Some species, such as the Deer Mouse, have adapted to many habitats; others, such as the Water Shrew, are confined to a narrow range. Knowing the habitat can help the observer to identify a mammal.

Range Each species account includes a description of the animal's geographical range. In this section abbreviations are used for northern (n), southern (s), eastern (e), western (w), central (c), and any compounds of these words, such as northeast (ne) or south-central (sc). A range map is provided for all species except those that have an extremely limited range north of Mexico. A species is not distributed evenly throughout its range, and the various areas it occupies may be separated by many miles. The range information is often most accurate if used in conjunction with the habitat description. As an additional aid to identifying an animal's geographical location, range charts (in the Appendices) give range by state or province for many species of smaller mammals that are difficult to identify by sight alone.

Other Information Each account concludes with a section about the mammal's feeding, nesting, and social behaviors, its activity patterns, its life history, and other information of interest. Some mammals have been well studied and documented, while others, especially tiny secretive animals from remote areas, are little known; for these, this concluding paragraph contains whatever information is available.

The Appendices The Appendices section of the book, Part III, includes a glossary of terms used throughout the book, range charts, photo credits, and an index of the mammals. The unique range charts, prepared exclusively for this guide, help the user in identifying many smaller mammals that are similar in appearance. Since many mammals occupy specific geographic areas, the charts often make it possible to verify an identification based on where the mammal was seen. A list of the abbreviations used for states and provinces and a map of North America precede the charts.

Classification of Biologists divide plants and animals
Mammals into major groups called phyla. The phylum Chordata includes all mammals, as well as fishes, birds, amphibians, and

reptiles. Phyla are divided into classes, classes into orders, orders into families, families into genera (singular: genus), and genera into species. The species is the basic unit of classification and is generally what we have in mind when we talk about a "kind" of animal. The scientific name of a species consists of two Latin or Latinized words. The first word is the generic name; the second is called the trivial or species name. The species name of the Meadow Jumping Mouse is *Zapus hudsonius*. *Zapus* is the genus name, and *hudsonius* is the trivial name.

Mammals are in the class Mammalia. A mammal is a warm-blooded animal with hair; female mammals have mammary glands that secrete milk for the nourishment of the young. The class Mammalia includes 4,629 species in 25 orders. Twelve orders occur in North America north of Mexico, 10 native and two introduced (primates and odd-toed hoofed mammals). The order Cetacea, the whales and dolphins, is not included in this book as its members are not land mammals.

What Is a Species? Technically, species are populations or groups of populations that are able to interbreed and produce fertile offspring like themselves. They usually are not able to breed successfully or produce fertile offspring with members of other species. Although species occasionally crossbreed, the hybrid offspring are often sterile. Subspecies are populations of a species that differ from one another in some way and that may be developing into new species. The first step in speciation is the isolation of one or more of the populations from the rest of the populations within the species. The isolated population, having stopped interbreeding with other populations, may then evolve in a different direction from its counterparts. With the passage of time, enough differences may evolve that individuals from the separate populations are no longer able to interbreed. The differences might be any factors—genetic, morphological, behavioral, or ecological—that keep members of the different populations from interbreeding. For example, the populations might breed at different times of the year, or they might be genetically incompatible.

Habitats of
North America

The North American continent can be broken down by habitat type. These habitats—all of which are diminishing in some way—are often influenced by climate and elevation. The following are the major North American habitat types and some of the mammals found in them.

Woodlands

There are essentially two major types of woodlands in North America: coniferous and deciduous. The coniferous forest, composed of fir, spruce, and pine, is home to Moose, Mule Deer, and Snowshoe Hares, which browse the needles; squirrels, American Martens, and Fishers, which live in the trees; and Lynx, wolves, and Wolverines, all predators. The deciduous woods, made up of maple, oak, beech, and other trees that shed their leaves, often occur in areas with less severe winters. Many mammals thrive in these woods, where a greater variety of food is available, such as Black Bears, Bobcats, foxes, Common Raccoons, skunks, chipmunks, and mice, which occupy tunnels in the leaf mold of the forest floor.

Grasslands

The center of the North American continent, where rainfall is too scarce for trees to grow but sufficient to prevent deserts from forming, is covered by grasslands. Grasses and associated forbs (herbaceous plants) provide the basic food for many mammals, ranging from pocket gophers and jackrabbits to the larger grazing animals, such as Pronghorns and American Bison.

Arid Areas   Although the great arid areas of North America—the shortgrass prairies and deserts of the West—might be expected to harbor few mammals, a large variety of species have adapted to conditions of limited vegetation and little free water, and the diversity here actually is higher than in forests. Many of these animals collect food when it is available, storing the surplus for later. They obtain water from vegetation either directly or by metabolizing it. Many are pale in color, which helps them blend in with the landscape. Most are nocturnal, becoming active when the air is cooler and when there may be dew on the ground. Kangaroo mice, pocket mice, jackrabbits, and ground squirrels dwell in arid areas. Pronghorns and deer are common in the shortgrass prairies. Predators include the Bobcat, the American Badger, foxes, and the Coyote.

Wetlands   Ponds, marshes, and swamps support a variety of aquatic mammals, including the American Beaver, Common Muskrat, Nutria, and rice rats. Most eat the leaves and roots of wetland vegetation; rice rats also eat crayfish, clams, snails, and other small invertebrates. The Star-nosed Mole can be found in mucky areas in the northeastern U.S. The most common wetland predators are the Mink and the Common Raccoon, but foxes and Coyotes often stalk the edges of these habitats.

Streams and Lakes   American Beavers build their dams along small streams and may live in the banks of larger ones. Both American Beavers and Common Muskrats are found on lakes. Water Shrews live along tumbling mountain brooks and sometimes along other streams, and Northern River Otters play in rivers and lakes. Lakes and streams provide food and drinking water for a variety of other animals, such as Grizzly Bears, Common Raccoons, Moose, and deer.

Oceans Billions of organisms live in the oceans, providing food for ocean-dwelling mammals such as seals, sea lions, and the Walrus. Through the aeons, these mammals have evolved streamlined shapes and flippers that enable them to swim swiftly through the water in pursuit of fish and other food. The chief predators of these animals in North American waters are killer whales, large sharks, and the Polar Bear.

Tundra The tundra, a vast, treeless plain, stretches from the timberline north to the polar ice cap. Vegetation is limited to lichens and other low-growing plants, which form the principal foods of such mammals as Caribou, Muskoxen, lemmings, and Arctic Hares; shrews feed on the invertebrates common in this habitat. These animals are preyed upon by wolves and the Arctic Fox.

Other Habitats Other North American habitats include talus slopes (slopes formed mainly by an accumulation of rock debris), where long-tailed shrews and pikas live; mountains, where nimble Bighorn Sheep and Mountain Goats make their home; and cultivated areas, where some rodent species, such as the Deer Mouse and House Mouse, forage among corn, wheat, and soybean crops.

How to Find Mammals Because most mammals are extremely secretive, the best way to find them and learn about them is to look for and examine their sign: tracks, scat, feeding debris, burrows, and the like.

When to Go Most mammals are nocturnal, but many can be seen emerging at dusk or heading back home at dawn. Daytime animals can be seen by the quiet and patient observer, as can nocturnal animals in bright moonlight. Sit near a wooded trail or a stream, under a tree, or in the shelter of a rock. Before long, the mammals will forget you are there and begin to resume their normal activities. The best time to go looking for tracks and other signs is early in the morning, when they will be fresh and the animals that made them may still be nearby.

Tracks Animals can leave clear tracks in mud, fine dirt, snow, and sand. Mud often records tracks in fine detail, sometimes showing the claws or webbing. Because animals often come to ponds or streams to drink or feed, tracks are likely to be found on their muddy shores. Snow may leave a less clear impression, but the tracks can often be followed for a long distance, and the pattern of the animal's stride can sometimes be discerned. When you have encountered tracks, study the Track Guide in the color-plates section. If you suspect that they belong to a specific animal, look up that animal in the species accounts and study the sign description and the track drawing, if there is one. Follow the tracks to determine what the mammal might have been doing: for example, chasing prey or digging for insects. Discovering that an animal climbed a tree or descended into a burrow will help you to identify it.

Burrows and Runways Many small mammals that live close to or under the ground leave evidence of their burrowing activity, or they may wear paths, known as runways, among the vegetation, especially in thick grass or under leaf litter. If you find a burrow entrance, make note of its diameter and whether or not excavated dirt is piled up or fanned out around it. Perhaps piles of fecal pellets, plant cuttings, or feeding debris are strewn about. Such signs are described in the species accounts of many mammals.

Lodges and Nests Many other types of mammal homes can be seen by the careful observer. Nests of small mammals can be found by looking under logs or stones. The tree nests of squirrels are often evident in winter, when obscuring leaves have fallen away. American Beavers and Common Muskrats use logs and sticks to build large and often elaborate lodges in water. A depression in leaves or high grass may be the bed of a deer.

Droppings   The scat of mammals, and where it is typically deposited (often described in the species accounts), can also offer clues to identity. Some animals employ a latrine area, where they habitually eliminate. Others, such as rabbits and deer, make deposits as they travel about. Often the scat offers clues about what the mammal has been eating.

Food Remnants   The food remnants and caches of many
and Caches   mammals can be found and safely approached. It is never safe, however, to approach the food cache of a bear, which may be a shallow pit roughly covered with dirt, branches, or other debris. Bears also overturn rocks and logs, and tear apart stumps and trees in their search for grubs and insects. Other animals may leave grass or twig cuttings, parts of seeds or nuts, opened shellfish, parts of other animals, pieces of cattails or other aquatic plants, and piles of black cherry pits, other seeds, pinecones, or snails. The "haystacks" of the American Pika are easily identifiable in conjunction with the animal's rocky-slope habitat. The main foods of the different mammals are described in the final paragraph of the species account; feeding debris may be described in the Sign section.

Other Signs   Bears, cats, and porcupines may scratch or gnaw at the bark of a tree, leaving evidence of their passing. Beavers can do substantial damage to a tree, gnawing it all the way around and eventually felling it. Bears rub against trees, often wearing the bark smooth, and deer polish their antlers on saplings, wearing away the bark.

HOW TO USE THIS GUIDE

When you see a mammal in the field, there are several ways you can use the guide to determine its identity. You can look at the animal shapes in the Silhouette and Thumb Tab Guide for the shape that most resembles it, and then turn to the corresponding color plates. Once you find the animal's photograph or the photograph of one that seems to resemble it, you can flip to the species account (page numbers are given in the captions). There you can read about that species or look up other animals mentioned in the Similar Species section. If you find tracks, refer to the Track Guide in the color-plates section as well as individual track drawings in the species accounts. If you know what type of animal you have seen, you can look it up in the index or use the Range Charts to identify it by habitat. Following are two examples of how to use the guide to identify a mammal.

Example 1: In a damp wooded area in the western
Gray shrew-like part of Virginia, you observe a uniformly
mammal with short gray, shrew-like mammal with a short
tail, in woods tail. It is foraging on the forest floor.

1. Turn to the Silhouette and Thumb Tab Guide preceding the color plates. Among the silhouettes for shrews, you

see that the one for short-tailed shrews (color plates 11–14) most closely resembles the specimen you observed.

2. Turning to the color plates, you find that the Northern Short-tailed Shrew (plates 11 and 12) seems to match your specimen. The captions give you the length of the animal and the page number of the species description.

3. Reading the species description, you note that there are two similar species, the Southern Short-tailed Shrew and the Least Shrew. You flip to the description of the Least Shrew and are able to eliminate it because it is too small.

4. A comparison of the Northern and Southern short-tailed shrews' range maps, or a glance at the Eastern Range Chart in the Appendices, shows you that only the Northern occurs in western Virginia, confirming the identification.

Example 2: *Tan-colored deer with antlers, in brushy area in Southwest* In a brushy area in west Texas, you observe a tan-colored deer grazing in a meadow. As it darts away, you note that its antlers have one main beam that juts forward, and you see a flash of white as the tail is raised.

1. In the Silhouette and Thumb Tab Guide, you flip to the antler illustrations in the Hoofed Mammals section, and scan the antler patterns. The White-tailed Deer (plates 304–306) most closely resembles what you saw.

2. Turning to the color plates, you see that the male White-tailed Deer indeed matches the animal you observed. The photograph of the similar Mule Deer shows that it has equally branching antlers rather than a main forward tine.

3. Reading the species account helps to confirm your identification.

STAFF

Prepared and produced by
Chanticleer Press, Inc.

Founding Publisher: Paul Steiner
Publisher: Andrew Stewart

Staff for this book:
Managing Editor: Edie Locke
Senior Editors: Amy K. Hughes and
Lisa Leventer
Copyeditor: Patricia Fogarty
Editorial Assistant: Kristina Lucenko
Art Director: Drew Stevens
Production: Alicia Mills
Photo Editor: Giema Tsakuginow
Photo Assistant: Consuelo Tiffany Lee
Maps and Silhouettes: Paul Singer and
Sam Shaw
Drawings: Sy Barlowe
Tracks: Dot Barlowe

Original series design by
Massimo Vignelli

All editorial inquiries should be
addressed to:
Chanticleer Press
665 Broadway, Suite 1001
New York, NY 10012

To purchase this book, or other
National Audubon Society illustrated
nature books, please contact:
Alfred A. Knopf
299 Park Avenue
New York, NY 10171
(800) 733-3000
www.randomhouse.com

NATIONAL AUDUBON SOCIETY
FIELD GUIDE SERIES

Also available in this unique all-color, all-photographic format:

African Wildlife

Birds *(Eastern Region)*

Birds *(Western Region)*

Butterflies

Fishes

Fossils

Insects and Spiders

Mushrooms

Night Sky

Reptiles and Amphibians

Rocks and Minerals

Seashells

Seashore Creatures

Trees *(Eastern Region)*

Trees *(Western Region)*

Weather

Wildflowers *(Eastern Region)*

Wildflowers *(Western Region)*

NOTES

Key to the Color Plates

The color plates on the following pages are divided into 17 groups.

Shrews and Moles
Bats
Mice and Rats
Voles, Lemmings, Pikas,
    and Pocket Gophers
Chipmunks, Squirrels,
    and Prairie Dogs
Rabbits and Hares
Large Rodents: Muskrats, Nutria,
    Beavers, and Marmots
Armored Mammals
Opossum
Raccoons and Their Kin
Weasels, Skunks, and Their Kin
Otters
Cats
Wolves, Foxes, and Coyote
Hoofed Mammals
Bears
Marine Mammals

Silhouette and
Thumb Tab
Guide

Silhouettes of the different types of mammals within each group appear on the following pages. The group silhouettes have been inset as thumb tabs at the left-hand edge of each double page of plates, thus providing a quick and convenient index to the color section.

Track Guide

Since most mammals are nocturnal and reclusive, a knowledge of animal signs, such as tracks, scat, and runways, is of great value in identifying mammal species. The Track Guide, which follows the Silhouette and Thumb Tab Guide, will familiarize you with various types of animal tracks and the mammals that make them.

Thumb Tab	Group
	Shrews and Moles
	Bats

Mammals		Plate Numbers
	long-tailed shrews	1–10, 15
	short-tailed shrews	11–14
	Shrew-mole	16
	moles	17–22
	leaf-chinned bats and New World fruit bats	23–27, 31
	evening bats and myotises	28–30, 32–54
	big-eared bats	55–60
	free-tailed bats	61–64

Mice and Rats

Mammals		Plate Numbers
	kangaroo rats and pocket mice	65–74, 97, 98, 102–110
	harvest mice	75–78
	House Mouse	79
	deer mice, pygmy mice, and kin	80–89
	jumping mice	90–92
	grasshopper mice and woodrats	93–96, 111–118
	rice and cotton rats	99–101
	Old World rats	119, 120

Thumb Tab	Group
	Voles, Lemmings, Pikas, and Pocket Gophers
	Chipmunks, Squirrels, and Prairie Dogs

Mammals		Plate Numbers
	voles	121–134
	lemmings	135–137
	pikas	138–140
	pocket gophers	141–147
	chipmunks	148–159
	antelope and ground squirrels	160–180
	tree squirrels	181–191
	flying squirrels	192–195
	prairie dogs	196–199

Thumb Tab	Group
	Rabbits and Hares
	Large Rodents
	Armored Mammals

Mammals		Plate Numbers
	rabbits and cottontails	200–207
	hares and jackrabbits	208–215
	Common Muskrat	216, 217
	Nutria	218, 219
	American Beaver	220–223
	Mountain Beaver	224
	marmots and Woodchuck	225–229
	Common Porcupine	230, 231
	Nine-banded Armadillo	232, 233

Thumb Tab	Group
	Opossum
	Raccoons and Their Kin
	Weasels, Skunks, and Their Kin
	Otters

Mammals		Plate Numbers
	Virginia Opossum	234–238
	Common Raccoon, Ringtail, and White-nosed Coati	239–244
	weasels	245–250
	American Marten and Fisher	251, 252
	skunks	253–256
	Wolverine and American Badger	257, 258
	Northern River Otter	259, 260
	Sea Otter	261, 262

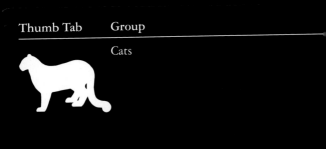

Cats

Wolves, Foxes, and Coyote

Hoofed Mammals

Mammals		Plate Numbers
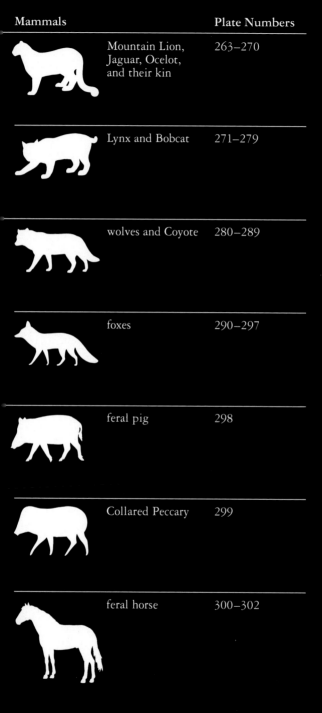	Mountain Lion, Jaguar, Ocelot, and their kin	263–270
	Lynx and Bobcat	271–279
	wolves and Coyote	280–289
	foxes	290–297
	feral pig	298
	Collared Peccary	299
	feral horse	300–302

Hoofed mammals
(with antlers)

Mammals		Plate Numbers
	Fallow Deer	303
	White-tailed Deer	304–306
	Mule, Sika, and Sambar deer	307–310
	Caribou	311, 312
	Moose	313, 314
	Elk	315–317

Hoofed mammals
(with horns)

Mammals		Plate Numbers
 Pronghorn		318, 319
Mountain Goat		320–323
Dall's Sheep		324
Barbary Sheep		325
Bighorn Sheep		326–328
American Bison		329–333
Muskox		334, 335

Bears

Marine Mammals

Mammals		Plate Numbers
	bears	336–349
	eared seals	350–355
	Walrus	356, 357
	hair seals	358–362, 364
	Hooded Seal	363
	Northern Elephant Seal	365
	Manatee	366, 367

Hand- and Foot-like
(4-toed foreprints, 5-toed hindprints)

Hand- and Foot-like
(5-toed foreprints and hindprints)

White-footed Mouse, *p. 590*

Norway Rat, *p. 659*

Round-tailed Muskrat, *p. 648*

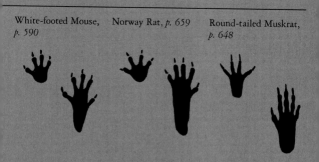

Eastern Gray Squirrel, *p. 487*

Eastern Chipmunk, *p. 434*

Long-tailed Weasel, *p. 761*

Common Muskrat, *p. 649*

Common Raccoon, *p. 747*

Striped Skunk, *p. 778*

Black-footed Ferret, *p. 764*

American Beaver, *p. 568*

Nutria, *p. 677*

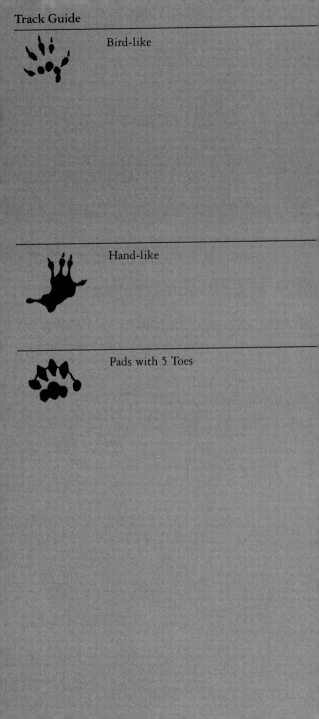

Bird-like

Hand-like

Pads with 5 Toes

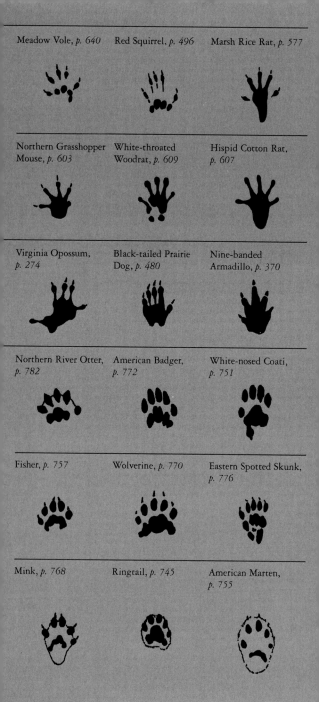

Meadow Vole, *p. 640*

Red Squirrel, *p. 496*

Marsh Rice Rat, *p. 577*

Northern Grasshopper Mouse, *p. 603*

White-throated Woodrat, *p. 609*

Hispid Cotton Rat, *p. 607*

Virginia Opossum, *p. 274*

Black-tailed Prairie Dog, *p. 480*

Nine-banded Armadillo, *p. 370*

Northern River Otter, *p. 782*

American Badger, *p. 772*

White-nosed Coati, *p. 751*

Fisher, *p. 757*

Wolverine, *p. 770*

Eastern Spotted Skunk, *p. 776*

Mink, *p. 768*

Ringtail, *p. 745*

American Marten, *p. 755*

Pads with 4 Toes
(without claws)

Pads with 4 Toes
(with claws)

Foot-like

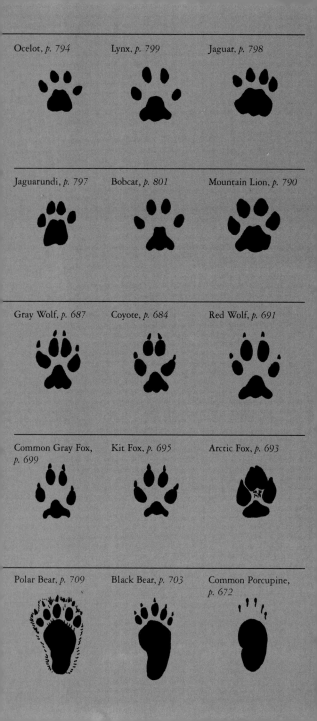

Ocelot, *p.* 794

Lynx, *p.* 799

Jaguar, *p.* 798

Jaguarundi, *p.* 797

Bobcat, *p.* 801

Mountain Lion, *p.* 790

Gray Wolf, *p.* 687

Coyote, *p.* 684

Red Wolf, *p.* 691

Common Gray Fox, *p.* 699

Kit Fox, *p.* 695

Arctic Fox, *p.* 693

Polar Bear, *p.* 709

Black Bear, *p.* 703

Common Porcupine, *p.* 672

Hoofed

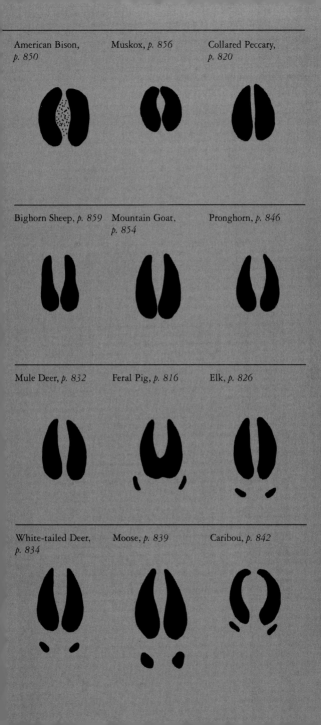

American Bison, *p. 850*

Muskox, *p. 856*

Collared Peccary, *p. 820*

Bighorn Sheep, *p. 859*

Mountain Goat, *p. 854*

Pronghorn, *p. 846*

Mule Deer, *p. 832*

Feral Pig, *p. 816*

Elk, *p. 826*

White-tailed Deer, *p. 834*

Moose, *p. 839*

Caribou, *p. 842*

The color plates on the following pages are numbered to correspond with the description of each species in the text. The caption under each color plate gives the mammal's common name, size (body and tail length, as well as height at shoulder if applicable), and the number of the page on which it is described. The sex of the species (♀ female, ♂ male) is also included in the caption if it is known. If the female and male of a species differ in size, the size given in the caption is the full range for both, from the smallest female to the largest male.

Shrews and Moles

Shrews and moles are small mammals with short, dense fur, five clawed toes on each foot, small eyes, and ears hidden beneath their fur. Shrews are very active animals with soft brownish or grayish fur, pointed snouts, and sharp teeth with chestnut-colored tips. Seldom seen and less often recognized, they often can be distinguished only by differences in their unicuspid teeth. Shrews live in a variety of habitats, but most species prefer moist areas. Moles, adapted for burrowing, have well-developed foreclaws and soft fur that can lie either backward or forward. Spending most of their time underground, they eat mainly earthworms.

1 Pacific Water Shrew, 5¾–6⅞", *p. 281*

2 Masked Shrew, 2¾–4⅜", *p. 282*

4 Smoky Shrew, 4¼–5″, *p. 284*

5 Pygmy Shrew, 3⅛–3⅞″, *p. 286*

6 Southeastern Shrew, 2⅞–4¼″, *p. 288*

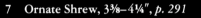

7 Ornate Shrew, 3⅜–4¼", *p. 291*

8 Water Shrew, 5⅝–6¼", *p. 293*

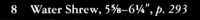

9 Trowbridge's Shrew, 4¼–5¼", *p. 295*

10 Vagrant Shrew, 3¾–4⅝″, *p. 297*

11 Northern Short-tailed Shrew, 3¾–5″, *p. 298*

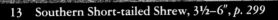
13 Southern Short-tailed Shrew, 3½–6″, *p. 299*

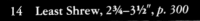
14 Least Shrew, 2¾–3½″, *p. 300*

16   Shrew-mole, *4–5″, p. 304*

17   Coast Mole, *5¾–6⅞″, p. 305*

18   Hairy-tailed Mole, *5½–6¾″, p. 307*

19   Star-nosed Mole, tentacles and foreclaws, *p. 309*

20   Star-nosed Mole, 6–8¼", *p. 309*

21 Eastern Mole, 3¼–8¾", *p. 308*

22 Eastern Mole, face and foreclaws, *p. 308*

Bats

The only mammals that fly, bats soar through the night air on wings that consist of a membrane thinly stretched across enlarged forearm bones and greatly elongated fingers. Although most bats have excellent vision, they navigate and locate prey by means of echolocation, which is similar to radar or sonar. Nearly all species in our range are insectivorous. In winter, when food supplies dwindle, most North American bats hibernate hanging upside down, often in vast numbers, while others migrate to warmer climes.

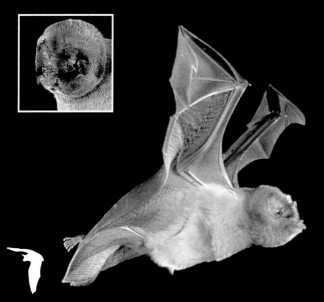

23   Ghost-faced Bat, 2⅜–2⅝″, *p. 316*

24   California Leaf-nosed Bat, 3¼–3⅝″, *p. 317*

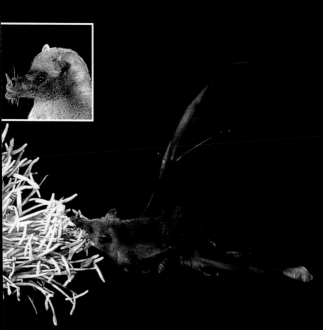

25 Mexican Long-tongued Bat, 2⅛–3⅛″, *p. 318*

26 Jamaican Fruit-eating Bat, 3⅛″, *p. 320*

27   Mexican Long-nosed Bat, 3–3½″, *p. 319*

28   Southeastern Myotis, 3¼–3⅞″, *p. 326*

29　California Myotis, 2⅞–3⅜″, *p. 327*

30　Fringed Myotis, 3⅛–3¾″, *p. 336*

31 Southern Long-nosed Bat, 2¾–3¼", p. 319

32 Long-eared Myotis, 3–3¾", p. 329

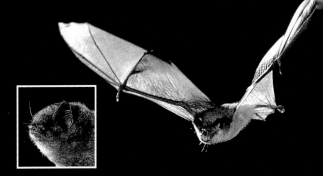

33 Gray Myotis, 3⅛–3¾", p. 330

34 Eastern Small-footed Myotis, 2¾–3¼″, *p. 332*

35 Northern Myotis, 3⅛–3½″, *p. 334*

36 Indiana Myotises, 2¾–3⅝″, *p. 335*

37 Little Brown Myotis, 3⅛–3⅝″, *p. 332*

38 Little Brown Myotis, *p. 332*

9 Cave Myotis, 3½–4½", *p.* 337

41 Yuma Myotis with prey, *p. 338*

43 Eastern Red Bat ♀ and young, *p. 340*

44 Eastern Red Bat ♂, 3¾–5″, *p. 340*

45 Hoary Bat, 4–6″, *p. 341*

46 Hoary Bat with prey, *p. 341*

47 Northern Yellow Bat, 4½″, *p. 343*

49 Silver-haired Bat, 3⅝–4¼", p. 345

51 Eastern Pipistrelle, 3⅛–3½″, *p. 347*

52 Eastern Pipistrelle hibernating, *p. 347*

53 Big Brown Bat, 4⅛–5″, *p. 348*

54 Evening Bat, 3⅛–3⅝″, *p. 350*

55 Spotted Bat, 4¼–4½″, *p. 351*

57 Townsend's Big-eared Bat, 3½–4¼", *p. 353*

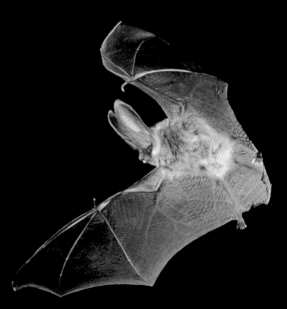

59 Allen's Big-eared Bat, 4–4⅝", *p. 354*

60 Pallid Bat eating scorpion, 4¼–5⅛", *p. 354*

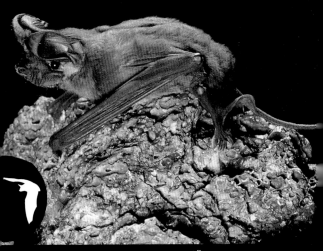

61   Big Free-tailed Bat, 5⅛–5⅝″, *p. 359*

62   Brazilian Free-tailed Bat, 3½–4¼″, *p. 356*

63 Western Mastiff Bat, 5½–7¼″, p. 361

64 Underwood's Mastiff Bat, 6⅜–6⅝″, p. 362

Mice and Rats

This huge group of small rodents, many of which are quite common, includes pocket mice, harvest mice, deer mice, cotton and rice rats, jumping mice, grasshopper mice, kangaroo mice and rats, woodrats, and Old World rats and mice; they are arranged in the color plates from small to large. Although all are rodents, these animals show numerous variations in form and habits. Most have large ears, moderately large eyes, and long tails. Primarily omnivorous and nocturnal, they live in nearly all habitats. Most use burrows, and only a few hibernate.

65   Arizona Pocket Mouse, 4⅞–6¾", *p. 527*

67 Silky Pocket Mouse molting, 3⅞–4¾", *p. 529*

68 Great Basin Pocket Mouse, 5¾–7¾", *p. 531*

69 Bailey's Pocket Mouse, 7⅞–9″, p. 536

70 California Pocket Mouse, 7½–9¼″, p. 537

71 San Diego Pocket Mouse, 6⅞–7⅞″, p. 537

72 Desert Pocket Mouse, 6⅜–8½″, p. 542

73 Spiny Pocket Mouse, 6½–8⅞″, p. 543

74 Little Pocket Mouse, 4¼–6″, p. 531

75 Fulvous Harvest Mouse, 5¼–7⅞″, *p. 579*

76 Eastern Harvest Mouse, 4¼–5⅞″, *p. 580*

77 Western Harvest Mouse, 4½–6¾″, *p. 580*

78 Salt-marsh Harvest Mouse, 4⅝–6⅞″, *p. 582*

79 House Mouse, 5⅛–7¾″, *p. 663*

80 Brush Mouse, 7⅛–9⅜″, *p. 584*

81 California Mouse, 8⅝–11⅛″, *p. 586*

82 Cactus Mouse, 6¼–8⅝″, *p. 587*

83 Cotton Mouse, 5⅝–8⅛″, *p. 588*

84 White-footed Mouse, 5⅛–8⅛″, *p. 590*

85 Deer Mouse ♀ and young, 4⅝–8¾″, *p. 591*

86 Oldfield Mouse, 4¾–6″, *p. 597*

87   Piñon Mouse, 6¾–9⅛″, p. 598

88   Golden Mouse, 5–7½″, p. 600

89   Northern Pygmy Mouse, 3⅜–4⅞″, p. 601

10   Meadow Jumping Mouse, 7⅜–10⅛", p. 666

11   Western Jumping Mouse, 8½–10¼", p. 668

93 Mearns' Grasshopper Mouse, 5½", *p. 602*

94 Northern Grasshopper Mouse juvenile, *p. 603*

96 Southern Grasshopper Mouse, 4⅝–6⅜", *p. 605*

97 Dark Kangaroo Mouse, 5⅞–7", *p. 543*

99   Hispid Cotton Rat, 8⅛–14⅜″, *p.* 607

100   Marsh Rice Rat, 7⅜–12″, *p.* 577

102   Desert Kangaroo Rat, 12–14¾″, *p. 548*

103   Heermann's Kangaroo Rat, 9⅞–13⅜″, *p. 551*

105   Merriam's Kangaroo Rat, 8⅝–10¼", p. 555

106   Fresno Kangaroo Rat, 8¼–10", p. 557

08 Panamint Kangaroo Rat, 11¼–13⅛″, p. 561

09 Banner-tailed Kangaroo Rat, 12¼–14⅜″, p. 562

111 White-throated Woodrat, 11⅛–15¾", p. 609

113   Florida Woodrat, 12¼–17⅜″, *p. 612*

114   Dusky-footed Woodrat, 13¼–18¾″, *p. 614*

115 Desert Woodrat, 8⅞–15⅛", *p. 615*

116 Desert Woodrat in stick house, *p. 615*

117 Mexican Woodrat, 11⅜–16⅝″, *p. 617*

119   Norway Rat, 12⅜–18⅛″, *p. 659*

120   Black Rat, 12¾–17⅞″, *p. 662*

Voles, Lemmings, Pikas, and Pocket Gophers

All of these similarly shaped mammals are rodents except pikas, which are in the same order as the rabbits and hares. Voles and lemmings, many of which live in open grassy areas, have small pudgy bodies, blunt faces with beady eyes, tiny, often inconspicuous ears, and short legs and tails. Pocket gophers are thick-set burrowers, with large foreclaws, small eyes and ears, yellow incisors, and external fur-lined cheek pockets, which they cram with vegetation for transport. Pikas are also small and stout, but have rounded ears and no visible tail. Extremely social creatures, pikas inhabit rocky mountain slopes, where they spend their days cutting grasses, which they dry and cure like hay.

121 Southern Red-backed Vole, 4¾–6¼″, *p. 625*

122 Mexican Vole, 4¾–6″, *p. 634*

123 California Vole, 6¼–8⅜″, *p. 630*

125 Long-tailed Vole, 6⅛–8¾", *p. 633*

27 Montane Vole, 5½–7½", p. 636

129 Rock Vole, 5⅜–7¼", *p. 632*

130 Tundra Vole, 6–8⅞", *p. 638*

132   Woodland Vole, 4⅛–5¾″, p. 642

133   Yellow-cheeked Vole, 7⅜–8⅞″, p. 646

134   Sagebrush Vole, 4–5⅝″, p. 646

135   Brown Lemming, 4¾–6⅝″, *p. 652*

136   Southern Bog Lemming, 4⅝–6⅛″, *p. 654*

137   Labrador Collared Lemming, 5⅞″, *p. 656*

138   Collared Pika, 7–7¾", p. 375

139   American Pika, 6⅜–8½", p. 376

140   American Pika grass cuttings, p. 376

141   Botta's Pocket Gopher in burrow, to 10¾", *p. 505*

142   Camas Pocket Gopher, 11¾", *p. 506*

43 Northern Pocket Gopher, 6½–9⅛", p. 510

145   Southeastern Pocket Gopher, 9–13¼", *p. 519*

146   Plains Pocket Gopher, 7⅜–14", *p. 515*

Chipmunks, Squirrels, and Prairie Dogs

These small to medium-size rodents have moderately to very bushy tails. Most are active by day, and many hibernate in winter. Chipmunks generally have light and dark stripes on the face and body. Most antelope and ground squirrels are uniformly colored, and a few are spotted, speckled, or striped, though never with stripes on the face. All are burrowers, and many live in colonies. The tree squirrels, mostly forest dwellers, have long, bushy tails and prominent eyes and ears. Flying squirrels, the only nocturnal members of the group, have furred folds of skin along their sides that spread like wings, allowing them to glide through the air. The prairie dogs are large, gregarious, chunky-bodied burrowers that live in vast, often highly structured colonies called towns.

148 Yellow-pine Chipmunk, 7⅛–9⅝", *p. 414*

149 Gray-collared Chipmunk, 8¼–9⅞", *p. 417*

50   Cliff Chipmunk, 7⅝–10⅞", *p. 418*

152 Least Chipmunk, 6⅝–8⅞", p. 420

154 Panamint Chipmunk, 7½–8⅝″, *p. 424*

156 Lodgepole Chipmunk, 7¾–9½", p. 433

158 Eastern Chipmunk, 8½–11¾", *p. 434*

159 Eastern Chipmunk filling cheek pouches, *p. 434*

160   Harris' Antelope Squirrel, 8⅝–9⅞", *p. 446*

161   White-tailed Antelope Squirrel, 7⅝–9⅜", *p. 449*

162 Nelson's Antelope Squirrel, 9–10½", *p. 451*

164 Uinta Ground Squirrel, 11–11⅞″, p. 453

166 Belding's Ground Squirrel, 9–11¾", p. 455

168 Wyoming Ground Squirrel, 10–12⅛″, *p. 459*

169 Franklin's Ground Squirrel, 15–15⅝″, *p. 461*

70 Golden-mantled Ground Squirrel, 9–12″, *p. 462*

71 Golden-mantled Ground Squirrel hibernating

172 Mexican Ground Squirrel, 11–15″, *p. 463*

173 Mohave Ground Squirrel, 8⅝–9″, *p. 464*

174 Arctic Ground Squirrel, 11¾–15½", *p. 464*

175 Richardson's Ground Squirrel, 9¾–14", *p. 466*

176 Townsend's Ground Squirrel, 6⅝–10¾", *p. 471*

177 Thirteen-lined Ground Squirrel, to 11⅝", *p. 472*

178   Round-tailed Ground Squirrel, 8–11″, *p. 469*

179   Round-tailed Ground Squirrels, *p. 469*

180 Rock Squirrel, 17–21″, *p. 474*

181 Abert's Squirrel, 18¼–23″, *p. 484*

182 Eastern Fox Squirrel, 17⅞–28", *p. 492*

184 Eastern Gray Squirrel, 17–19¾", *p. 487*

186 Western Gray Squirrel, 17½–23", p. 490

188 Douglas' Squirrel, 10⅝–14″, p. 494

190 Red Squirrel, 10⅝–15¼", p. 496

91 Red Squirrel cone cache, p. 496

192   Northern Flying Squirrel, 10⅜–14½″, *p. 498*

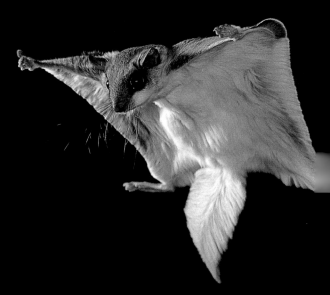

193   Northern Flying Squirrel in flight, *p. 498*

194 Southern Flying Squirrel ♀ with young, *p. 500*

195 Southern Flying Squirrel in flight, to 10⅛″, *p. 500*

196   **Utah Prairie Dog,** 12–14¼″, *p. 483*

197   **White-tailed Prairie Dogs,** 13⅜–14½″, *p. 478*

198 Black-tailed Prairie Dog, 14–16⅜″, *p. 480*

99 Black-tailed Prairie Dog, ♀ and young, *p. 480*

Rabbits and Hares

Immediately recognizable by their long ears, hares and rabbits have large eyes, small, often cottony tails, and long hindlegs with large feet. Hares and jackrabbits generally have longer ears and longer hindlegs than rabbits and are better adapted for jumping. Relying on their ability to make a speedy getaway, hares and jackrabbits occupy open country, resting on the ground or in shallow depressions. Rabbits seek concealing shelter in dense vegetation and, especially in winter, may hide in burrows abandoned by other animals.

200 Swamp Rabbit, 17¾–22″, *p. 380*

201 Brush Rabbit, 11–14¾″, *p. 382*

202 Marsh Rabbit, 13⅞–17¾″, *p. 387*

203 Mountain Cottontail, 13¾–15⅜″, *p. 385*

204 Eastern Cottontail, 14¾–18¼", *p. 383*

205 Eastern Cottontail young in nest, *p. 383*

206 Desert Cottontail, 13¾–16½", *p. 381*

208 Snowshoe Hare, 15–20″, *p. 391*

210 Arctic Hare, 18⅞–27", p. 394

212 Antelope Jackrabbit, 22–26″, *p. 390*

214 White-tailed Jackrabbit, 22–26″, p. 400

Large Rodents

The mammals in this group are the largest of the rodents: the Common Muskrat, the Nutria, the American Beaver, the Mountain Beaver, and the marmots. The Common Muskrat, with a long, vertically flattened, scaly tail, the Nutria, with a long, round, scantily haired tail, and the American Beaver, with a broad, paddle-shaped tail, all favor lakes, ponds, streams, or marshes. The burrowing marmots, including the Woodchuck, are members of the squirrel family and have moderately long, bushy tails. Another burrower, the virtually tail-less Mountain Beaver, is found near streams in its Pacific Northwest range.

216 Common Muskrat, 16⅛–24″, *p. 649*

218 Nutria, 26–55″, *p. 677*

219 Nutria family feeding, *p. 677*

220 American Beaver, 3–4', *p. 568*

221 American Beaver lodge, *p. 568*

222   American Beaver, *p. 568*

223   Tree gnawed by American Beaver, *p. 568*

224   Mountain Beaver, 9⅜–18½″, *p. 405*

225   Woodchuck, 16½–32″, *p. 441*

226 Hoary Marmot, 17¾–32", *p. 439*

228 Olympic Marmot, 17¾–31", *p. 443*

Armored Mammals

Although the Nine-banded Armadillo and the Common Porcupine belong to separate orders, each sports a unique armor that protects it against predators. The Nine-banded Armadillo is clad in hinged bony plates, which serve as a suit of armor, shielding it from harm. The Common Porcupine has its rump and tail covered with sharp quills that can become painfully embedded in the flesh of a predator.

230 Common Porcupine, 26–37", *p.* 672

232 Nine-banded Armadillo, 24–32″, *p. 370*

233 Nine-banded Armadillo group, *p. 370*

Opossum

The Virginia Opossum is the only North American representative of the pouched mammals, or marsupials, whose young are born in an embryonic state and then complete their development in a pouch on the mother's belly, each attaching itself to one of her nipples. The Virginia Opossum's small black ears and dark eyes contrast with its white face. It is the only mammal in our range with a long, naked prehensile (grasping) tail.

234 Virginia Opossum, 25–40″, *p. 274*

235 Virginia Opossum ♀ and young, *p. 274*

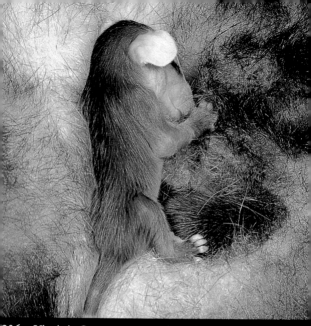

236 Virginia Opossum, 8 weeks old, *p. 274*

Raccoons and Their Kin

The raccoon family is represented by three species in our range, all with fairly long, ringed tails. They are good climbers, tend to be social, and have blunt cheek teeth, an adaptation to their varied diets. The Common Raccoon is distinguished by its black-and-white striped tail and its black face mask. The smaller Ringtail has a longer banded tail and a light-colored eye ring. The White-nosed Coati is the least distinctively marked of the three, with faint rings on its longer, thinner tail.

239 Common Raccoon, 24–37", *p. 747*

240 Common Raccoon with snake, *p. 747*

41 Common Raccoon family, *p. 747*

243 Ringtail ♀ and young, 24–32″, *p. 745*

Weasels, Skunks, and Their Kin

These mammals all belong to the
mustelid family, relatively small
carnivores with long, low-slung bodies,
short legs, short ears, and a thick, silky
coat. The slow-moving skunks, easily
identifiable by their bushy black-and-
white coats and tails, use their strongly
scented anal secretions for defense.
The sleek, agile weasels and martens,
including the Mink, Fisher, and Black-
footed Ferret, have less well developed
anal scent glands, which are used
mainly for social and sexual marking.

245 Short-tailed Weasel, 7½–13½", *p. 759*

246 Least Weasel, 6¾–8⅛", *p. 766*

247 Long-tailed Weasel, 11–22″, p. 761

249 Black-footed Ferret, 19¾–23", *p. 764*

51 American Marten, 19¼–27″, *p. 755*

253 Western Spotted Skunk, 13⅝–18¾", *p. 775*

55 Hooded Skunk, 22–31″, *p. 778*

257   Wolverine, 31–44", *p.* 770

Otters

Otters are fairly large, slender, playful mammals belonging to the same family as the skunks and weasels. Northern River Otters spend much time in freshwater rivers and streams, but are also quite agile on land. The larger Sea Otter eats, sleeps, mates, and gives birth at sea, usually taking to land only during storms.

261 Sea Otter, 30–71″, *p.* 785

262 Sea Otter ♀ and young, *p.* 785

Cats

Although varying in size and color, all
North American wildcats have lithe,
powerful bodies, short heads with
prominent ears, vertically slit pupils,
sensitive whiskers, and fleshy paws with
retractile claws. The fur may be plain,
mottled, or boldly spotted, and the tail
long or short. With the exception of the
Mountain Lion, Lynx, and Bobcat, most
wildcats are extremely rare in our range.

263   Mountain Lion, 6–9′, *p. 790*

264   Mountain Lion ♀ and cub, *p. 790*

65 Mountain Lion ♀ and cub, *p. 790*

267   Jaguar, 5′2″–7′11″, *p.* 798

69 Margays, 31–51″, *p. 796*

70 Jaguarundi, 3′–4′6″, *p. 797*

Cats

271 Lynx, 30–42″, *p. 799*

272 Lynx, *p. 799*

73   Lynx, *p.* 799

276 Bobcat, 28–49″, *p. 801*

277 Bobcat ♀ and kitten, *p. 801*

78 Bobcat, *p. 801*

'9 Bobcat swimming ↑ *801*

Wolves, Foxes, and Coyote

Resembling the domestic dog, these wild canids all have large eyes, erect ears, a pointed snout, slender legs, and a long, bushy tail. Fur varies in coloration and in thickness, with more northerly populations having the most luxuriant coats. Some species that inhabit Arctic regions acquire white coloring in winter.

280 Coyote, 3'5"–4'4", ht. 23–26", p. 684

281 Coyotes playing, p. 684

285 Gray Wolf, 4′3″–6′9″, *ht.* 26–38″, *p.* 687

286 Gray Wolf ♀ and young, *p.* 687

287 Gray Wolf, white phase, *p. 687*

290   Arctic Fox, blue phase, 30–36″, *ht.* 11¾″, *p. 693*

291   Arctic Fox, winter coat, *p. 693*

292 Kit Fox, 24–31″, *ht.* 11¾″, *p. 695*

293 Kit Fox ♀ and pup, *p. 695*

294 Red Fox, 35–41", ht. 15–16", p. 696

896 Common Gray Fox, 31–44″, *ht.* 14⅛–15″, *p. 699*

97 Common Gray Fox ♀ nursing pups, *p. 699*

Hoofed Mammals

These mammals all have hooves; most
have either two or four toes on each foo[t]
The Feral Pig and the Peccary, related t[o]
the domestic pig, have stocky bodies
covered with grizzled dark hair, long
heads with a cartilaginous disk-shaped
snout, and short legs. The Feral Pig has
a moderately long, uncoiled tail and
upper tusks that curl up, while the
much smaller Peccary is virtually tail-
less, with upper tusks pointing down.
The deer, including the Elk, Moose,
and Caribou; the Pronghorn; and the
bovids, which include goats, sheep, the
American Bison, and the Muskox, bear
head ornamentation in the form of
antlers or horns in all males and some
females. The feral horse is also included
in this group.

298   Feral Pig, 4' 4"–6', *ht.* to 3', *p. 816*

299   Collared Peccary, 35–40", *ht.* 20–24", *p. 820*

00   Feral Horses, Assateague race, *ht.* to 4′4″, *p.* 810

302 Feral Horses, western race, *p. 810*

303   Fallow Deer, 4′7″–5′11″, *ht.* 3′3″, *p. 824*

05 White-tailed Deer ♀ and fawns, *p. 834*

307 Sika Deer ♀ and young, to 5′, *ht.* to 6′3″, *p. 830*

309 Mule Deer ♂, 3′10″–7′6″, *ht.* 3′–3′5″, *p. 832*

311   Caribou ♂, 4′6″–8′4″, ht. 27–55″, p. 842

3 Moose ♀ and calf, 6′9″–9′2″, *ht.* 6′5″–7′5″, *p. 839*

316   Elk ♂, 6′8″–9′9″, *ht.* 4′6″–5′, *p. 826*

8 Pronghorns, 4′1″–4′9″, *ht.* 35–41″, *p. 846*

320 Mountain Goat ♀ and young, *p. 854*

Mountain Goat, 6', 5'10", ht 35–47", p. 854

22 Mountain Goat climbing, *p. 854*

23 Mountain Goats, *p. 854*

324   Dall's Sheep ♂, 3′5″–5′, *ht.* 30–41″, *p. 862*

325   Barbary Sheep ♂, 4′3″–6′3″, *ht.* 3′–3′5″, *p. 864*

326 Bighorn Sheep ♀ and ♂, *p. 859*

329 American Bison ♂, 7′–12′6″, *ht.* 5–6′, *p.* 850

331 American Bison crossing river, *p. 850*

332 American Bison, *p. 850*

333 American Bison, *p. 850*

334 Muskox ♂, 6′4″–8′1″, *ht.* 3–5′, *p. 856*

Bears

The largest terrestrial carnivores, bears have massive, densely furred bodies; five claws on each foot; small, rounded ears; small, close-set eyes; and tiny tails. Their coloration varies with the species: Polar Bears are white; Grizzlies may be tawny to blackish brown; and Black Bears range from black through brown to nearly white.

336   Black Bear, 4′6″–6′2″, *ht.* 3′–3′5″, *p. 703*

337   Black Bear, white phase, *p. 703*

38 **Black Bear** ♀ and cub, *p. 703*

340 Grizzly Bear ♀ and cubs, *p.* 706

345 Polar Bear ♀ and cubs, *p. 709*

47 Polar Bears swimming, *p. 709*

Marine Mammals

Sea lions, seals, and the Walrus are aquatic carnivores well adapted to their habitat, with large, streamlined bodies, flippers for limbs, and vestigial ears and tails. They often have whiskers, and the Walrus has long ivory tusks. Although all spend some time out of the water, they vary in degree of dexterity on land. Also included here is the Manatee, a large, warm-water creature with no hindlegs and a large, paddle-shaped tail.

350　California Sea Lions, 4′11″–8′2″, *p. 719*

351　California Sea Lion ♀ and pup, *p. 719*

52 Northern Sea Lion ♂ and harem, to 10′6″, *p. 716*

53 Guadalupe Fur Seal, 4′7″–6′3″, *p. 715*

354   Northern Fur Seal ♂, 3′7″–7′3″, *p. 713*

355   Northern Fur Seals with pups, *p. 713*

358 Harbor Seals, 3′11″–6′2″, *p. 728*

359 Ringed Seal, 4′–5′5″, *p. 730*

61 Ribbon Seal, to 6′, *p.* 736

62 Bearded Seal, 6′11″–7′11″, *p.* 738

63 Hooded Seal, 5′11″–9′10″, *p.* 739

364 Gray Seal, 7′7″–9′10″, *p. 732*

365 Northern Elephant Seal ♀ ♂ and pup, *p. 743*

66   Manatee ♀ and calves, 9′10″–11′6″, *p. 807*

The number(s) preceding the species description in the following pages corresponds to the number of the photograph in the color plates section. If the description has no number, there is no photograph.

OPOSSUMS
Order Didelphimorphia
Family Didelphidae

All pouched mammals, or marsupials,
were once placed in one order, the
Marsupialia, which contained several
families. However, the original order
has now been divided into seven orders:
four in the Australian region, with 203
species, and three in the New World,
with 69 species. The Australian species
of marsupials have adapted to many
niches filled by higher mammals in
other parts of the world. The major
New World order of marsupials is the
Didelphimorphia, the opossums, which
includes one family (Didelphidae),
15 genera, and 63 species. Only one
species in the order, the Virginia
Opossum, is found in North America
north of Mexico.

Opossums and other marsupials are
mammals that usually have a pouch, or
marsupium, on the belly in which the
female carries her young, although in
some species the pouch is reduced or
absent. After a gestation period of only
one to two weeks, the young are born
when still in a very undeveloped, or
embryonic, state. The tiny newborn,
about the size of a dime in most species,
must make its way from the base of the
mother's tail to the pouch, where it
attaches itself to a nipple. In higher
mammals, the specialized placenta
provides continued nourishment to the
embryo, which allows for a longer
gestation period and greater development
of the young.

234–238 Virginia Opossum
Didelphis virginiana

Description: House cat size. Grizzled white above; long white hairs cover black-tipped fur below. In some areas, individuals may appear grayish or blackish. *Long, naked prehensile tail.* Head and throat whitish; *ears large, naked, black with pinkish tips.* Legs short; *first toe of hindfoot opposable (thumb-like) and lacks claw.* Female has fur-lined abdominal pouch. L 25–40" (645–1,017 mm); T 10⅛–21" (255–53? mm); HF 1⅞–3⅛" (48–80 mm); Wt 4–14 lb (1.8–6.3 kg).

Breeding: After 12–13 day gestation, 1–14 young attach themselves to mother's nipples for 2 months; 2 or 3 litters per year.

Sign: *Tracks:* In mud, hindprint approximately 2" (50 mm) wide, with 5 toes printing: large, thumb-like toe slanted inward or backward, 3 middle toes close together, and remaining toe separate. Foreprint slightly smaller, with 5 toes printing in star-like fashion. Hindprints and foreprints parallel and close together; straddle 4" (100 mm); walking stride 7" (180 mm).

Habitat: Deciduous forests, open woods, brushy wastelands, and farmlands.

Range: Most of e U.S., except n Minnesota, n Michigan, and n New England; extends southwest to e Wyoming, Colorado, and c New Mexico. Also s British Columbia south to Baja California and east into c Idaho; and se Arizona.

A solitary nocturnal animal, the Virginia Opossum is terrestrial and arboreal, and climbs well. Although it does not hibernate, during very cold weather it may hole up for several days at a time, risking frostbite on its naked ears and tail to seek food when hunger strikes. Carrion forms much of its diet, and many individuals are killed on highways while attempting to feed on roadkill. The diet also includes insects, frogs, birds, snakes, small mammals,

earthworms, and berries and other fruits; persimmons, apples, and corn are favorite foods. Opossums scent-mark, especially during the breeding season, licking themselves and rubbing the sides of their heads against tree trunks or other objects. Because the penis is forked, there is a myth that this species mates through the female's nose. The Virginia Opossum makes a leaf nest in a hollow log, fallen tree, abandoned burrow, or other sheltered place. After a gestation of less than two weeks, the "living embryos," each the size of a navy bean, climb up through the hair of the female and enter the vertical opening of her pouch. Each takes one of her 13 nipples in its mouth and remains thus attached to the mother for two months. Those who do not obtain a nipple perish. Several defensive behaviors have been described in opossums. When threatened, an individual may roll over, shut its eyes, and allow its tongue to loll, feigning death, or "playing possum," for some time. More often, it tries to bluff its attacker by hissing, screeching, salivating, opening its mouth wide to show all of its 50 teeth, and sometimes excreting a greenish substance. Many of these behaviors occur in encounters between males. Clicks, used in aggressive displays by males during mating season, are also employed in communications between mother and young. The Virginia Opossum is hunted for its fur, which is not considered particularly valuable but is used for trimming, and for its meat, which many consider a delicacy.

INSECTIVORES
Order Insectivora

The order Insectivora contains seven families and 428 species. Five of the families are small and geographically restricted, but the other two, the shrews (Soricidae), with 312 species, and the moles (Talpidae), with 42 species, are common and widespread. There are 33 species of shrews and 7 species of moles in North America north of Mexico.

North American insectivores are small mammals with short, dense fur, five clawed toes on their forefeet and hindfeet, and small eyes and ears. The ears usually are hidden beneath the fur, yet in many species hearing is quite acute. As the name of the order implies, shrews and moles eat insects and their larvae, but they also consume many other invertebrates. In fact, they are not even the most exclusively insectivorous of North American mammals; many bats, for example, feed almost entirely on insects. The order Insectivora includes land-dwellers, burrowers, and species that spend much of their life in water.

SHREWS
Family Soricidae

Shrews look somewhat like mice but have a long, slender, pointed snout, a continuous row of needle-sharp teeth, tiny eyes, and five clawed toes on both forefeet and hindfeet (mice have four toes on their forefeet). The velvety fur of shrews, as in moles, can lie forward or backward, facilitating movement in burrows. Shrews range in size from less than 1⅜ inches (35 mm) long to 7⅛ inches (180 mm). Among the world's smallest mammals, weighing no more than 1/16 ounce (2 g), are Savi's Pygmy Shrew *(Suncus etruscus),* found in Africa and the Mediterranean region, and a population of the Pygmy Shrew found in southern Indiana. Many shrews have high-pitched, squeaky voices, and some employ ultrasonic sounds for echolocation, like bats. All 33 species of North American shrews have teeth with chestnut-colored tips.

Exceedingly active and nervous, shrews dart about constantly. Their life span is one to two years; most live less than a year. If a shrew is excited or frightened, its heart may beat 1,200 times per minute. These small animals may be literally frightened to death by capture or a loud noise. A shrew's energy output surpasses that of any other mammal. Because of its high metabolism, it feeds frequently, both day and night, resting during intervals between meals. Some species may consume more than their own weight in food per day, eating mainly invertebrates but also plants, fungi, and small mammals, usually carrion or nest young. Shrews in the wild often undergo daily torpor, reducing their metabolic rate and resulting food requirements. Often several species of shrews occur in size-ranked communities, in which large, medium, and small species generally

feed on different-size foods, thus helping to avoid competition among species.

Generally solitary, shrews fight to defend their nest areas and viciously battle any animal they perceive as an attacker, including their own kind. They must engage in complex courtship behavior to progress from aversion to copulation. Musk glands on their flanks exude an odor thought to be repellent to some carnivores, such as cats, which often kill shrews but won't eat them. However, the glands are probably more important for territorial marking and sexual recognition than for discouraging predators, as shrews are consumed by owls, snakes, and some mammals.

Females may have one or more litters per year, each containing 2 to 10 young, which are born blind, pink, and hairless. An unusual habit, probably unique to some shrews, has been observed in several species of Old World Shrew, including the Asian Musk Shrew *(Suncus murinus)* and the common shrew of Europe *(Sorex araneus),* and is suspected in other species: When a litter of unweaned young is disturbed, sometimes even at the slightest upset such as the sound of rain or a change in temperature, each juvenile uses its mouth to grasp the base of the tail of its closest neighbor, forming a caravan with the mother in the lead. From a distance, the moving chain might be mistaken for a small snake.

In the species accounts that follow, information about breeding is included for the species for which specific information is available. Tracks and other signs are described for only a few species; of limited usefulness in identifying shrews, both elements are tiny and not distinctive. Burrows are often so small they might be mistaken for tunnels of large earthworms, and prints are often

blurred dots. Those interested in
identifying the whereabouts of shrews
should look for tiny burrow openings in
the ground, in a bank, or in moss, or for
runways of freshly disturbed soil under a
log or under the lip of a bank.

Arctic Shrew
"Saddle-backed Shrew"
Sorex arcticus

Description: Tricolored: *back dark brown, sides light
brown,* and *belly grayish* tinged with buff;
pattern more distinct in winter. Long
tail. Juvenile dull brown. L 4–5″ (101–
126 mm); T 1⅛–1¾″ (30–46 mm);
HF ⅜–½″ (12–15 mm); Wt ¼–⅜ oz
(6–11 g).

Breeding: Litters of 4–8 born late April–
September; female seldom reproduces in
the first year.

Habitat: Swamps, bogs, marshes, and grass-sedge
meadows.

Range: From se Yukon southeastward across
much of Canada to North Dakota,
extreme ne South Dakota, Minnesota,
Wisconsin, and Michigan's Upper
Peninsula; northeast through sc Quebec
to Newfoundland, New Brunswick, and
Nova Scotia.

Arctic Shrews make a low, rapid chatter.
Caterpillars, centipedes, and beetles and
their larvae are their chief foods. Unlike
other shrews, this species is quite docile,
can be handled, and seldom attempts to
bite. The distinctive tricolored pattern
of the Arctic Shrew's fur has given rise
to its other common name, "Saddle-
backed Shrew."

Arizona Shrew
Sorex arizonae

Description: A small shrew. Brown above; slightly
paler below. *Long tail indistinctly
bicolored:* brownish above, paler below.

Third upper incisor equal to or larger than fourth. First incisors have tines on inner side. L 3⅝–4½" (93–114 mm); T 1¼–1¾" (37–46 mm); HF ½" (11–13 mm).

Similar Species: No other shrew in range has tines on inner sides of first incisors.

Habitat: Dry mountainous regions.

Range: Huachuca, Santa Rita, and Chiricahua mountains of se Arizona; Animas Mountains, Hidalgo County, in sw New Mexico.

This species was first described in 1977, and its habits are still unknown. They are probably similar to those of other long-tailed shrews.

Baird's Shrew
Sorex bairdii

Description: *A small shrew. Brown above and below. Tail indistinctly bicolored.* Tine on inside edge of first upper incisors. L 3⅞–5⅝" (100–143 mm); T 1¼–2½" (32–64 mm); HF ½–¾" (12–18 mm).

Similar Species: Larger Fog Shrew lacks tine on inner edge of first upper incisors. Vagrant Shrew can be distinguished only by pattern of pigmentation on teeth.

Habitat: Marshes, streamsides, and moist forests.

Range: Northwestern Oregon.

This shrew, once classified as the same species as the Vagrant Shrew, from which it is nearly indistinguishable, was recently given species status.

1 **Pacific Water Shrew**
"Marsh Shrew"
Sorex bendirii

Description: *A large shrew. Dark brown above and below. Long tail. Hindfeet lightly fringed with hair.* L 5¾–6⅞" (147–174 mm); T 2⅜–3⅛" (61–80 mm); HF ¾–⅞" (18.5–21 mm); Wt average ½ oz (16.1 g).

Similar Species: Water Shrew is silvery white below; fringe of hair on hindfeet is better developed.

Habitat: Marshes and streamsides; sometimes moist forests.

Range: Extreme sw British Columbia south to coastal n California.

The largest member of the *Sorex* genus in North America, the Pacific Water Shrew eats earthworms, sow bugs, spiders, centipedes, and other invertebrates on land, and readily enters water to take aquatic arthropods. It can run on top of the water for several seconds, gaining buoyancy from the air trapped in its partially fringed toes, and can also dive to the bottom. Air trapped in its fur gives this shrew a silvery appearance when submerged. This species has a penetrating scent, which suggests that it engages in some form of chemical communication.

2 Masked Shrew
"Cinereus Shrew"
Sorex cinereus

Description: *A very small shrew. Brownish above; belly silvery or grayish. Long tail* brown above, buff below; tail tufted, with terminal hairs of underside dark. Third and fourth unicuspids usually about same size. L 2¾–4⅜" (71–111 mm); T 1–2" (25–50 mm); HF ⅜–½" (10–14 mm); Wt 1/16–¼ oz (2.4–7.8 g).

Similar Species: In East, smaller Southeastern Shrew has shorter, less-hairy tail and shorter, less-pointed snout. In West, Vagrant and Dusky shrews usually have smaller third unicuspid than fourth.

Breeding: Mates from spring to fall. Gestation about 18 days. Each female probably has only 1 litter of 2–10 (usually 5–7) young.

Habitat: Numerous habitats; most common in moist fields, bogs, marshes, and moist woods. Where Hayden's and Masked shrews occur together, Masked inhabit

woody or high-grass areas, Hayden's short grass.

Range: Throughout most of Canada and n U.S. south to Washington, Idaho, sc Utah, nc New Mexico, and Nebraska in West, and to Illinois, n Kentucky, and throughout Appalachians in East.

One of the most widely distributed mammals in North America, this small, primarily nocturnal shrew is secretive and rarely seen. Although all shrews are noted for their large appetites, the Masked Shrew is particularly voracious. Its daily consumption of moth and beetle larvae, slugs, snails, and spiders often equals or exceeds its own weight. Like other shrews, the Masked Shrew does not hibernate; it seeks out dormant insects for sustenance even during the coldest part of winter. It makes a nest of leaves or grass under a log or in a stump or clump of vegetation. The young are born blind. At 18 days, their eyes open; at 19 days they are weaned and become independent. Few young shrews are noted in the field because they reach nearly full size before leaving the nest.

---

3 Long-tailed Shrew
"Rock Shrew"
Sorex dispar

Description: *A long, slender shrew. Entire animal, including long tail, uniformly dark gray.* L 4⅛–5½″ (104–139 mm); T 1⅞–2⅝″ (48–66 mm); HF ½–⅝″ (13–16 mm); Wt ⅛–¼ oz (5–6 g).

Similar Species: Smoky Shrew is paler, with shorter tail that is paler below. Water Shrew is larger, with fringe of stiff hairs along sides of feet; occurs along mountain streams. Very similar Gaspé Shrew occurs farther north.

Breeding: Breeds April–August. Probably 2–5 young per litter, born May–August.

Habitat: Usually deep, dark, cool, moist recesses among boulders or talus slopes.

Range: Mountainous areas from Maine southward through Smoky Mountains of Tennessee and North Carolina.

The Long-tailed Shrew is rare; only a few living specimens have been observed. It eats small invertebrates, especially spiders, centipedes, and beetles.

Maryland Shrew
Sorex fontinalis

Description: *Body and tail brown above, paler below.* Third and fourth unicuspids same size. L 3⅛–3¾″ (78–97 mm); T 1–1¼″ (27–39 mm); HF ⅜–½″ (10–13 mm).

Similar Species: Masked Shrew is larger, with longer tail. Southeastern Shrew is very similar, but third unicuspid is often smaller than fourth.

Habitat: A wide variety of habitats, including fields, hedgerows, and forests.

Range: Southeastern Pennsylvania, Maryland, Delaware, and n West Virginia.

For many years the Maryland Shrew was considered a subspecies of the Masked Shrew. However, the two occur together in Pennsylvania apparently without interbreeding, and thus are now considered separate species. The Maryland Shrew's habits are little known, but presumably are similar to those of the Masked Shrew.

4　**Smoky Shrew**
Sorex fumeus

Description: *Body brownish in summer, grayish in winter.* Long tail dark brown above and paler, sometimes yellowish, below. L 4¼–5″ (110–127 mm); T 1¼–2″ (37–52 mm); HF ½–⅝″ (12–15 mm); Wt ¼ oz (6–9 g).

Similar Species: Masked Shrew is usually smaller, browner, and paler below.

Breeding: Begins mating in March. Gestation about 20 days. Litters in early or late May, late August, and occasionally in autumn, of 2–8 (usually 4–7) young, in the female's second year.

Sign: Dime-sized burrow openings along stream and road banks, under logs, and at bases of large trees.

Habitat: Various types of moist wooded areas, deep woods, and swamps, and along streams where woods and fields meet.

Range: Southern Ontario, Quebec, New Brunswick, Nova Scotia, and ne U.S. south through mountains to e Tennessee, ne Georgia, and nw South Carolina.

Active in even the coldest weather, the Smoky Shrew commonly lives in extensive burrows in the leaf mold of the forest floor. Small invertebrates, such as insects and their larvae, earthworms, sow bugs, and centipedes, form its diet. It makes a leaf nest in a hollow log or stump, or under rocks. Smoky Shrews do not breed in their birth year; they overwinter as immature animals, becoming breeding adults in the spring. The young are weaned and become adult size by 30 days of age. They can be distinguished from adults by the hairs extending from the tip of the tail, which wear away as they age. Few Smoky Shrews live long enough to produce litters in more than one year. The life span of this species is estimated at a maximum of 14 to 17 months.

Gaspé Shrew
Sorex gaspensis

Description: *Body dark gray,* slightly paler below. Tail long, bicolored. L 3¾–4½″ (95–115 mm); T 1⅞–2⅛″ (47–55 mm); HF ½″ (12–13 mm); Wt ⅛ oz (3.5–5 g).

Similar Species: Closely related Long-tailed Shrew is slightly larger and occurs farther south.

Habitat: Near streams in coniferous forests.

Range: Gaspé Peninsula, Quebec; Mount
          Carleton, New Brunswick; and Cape
          Breton Island, Nova Scotia.

          Little is known about the habits of the
          Gaspé Shrew. They are thought to be
          similar to those of the Long-tailed
          Shrew, which resembles the Gaspé
          morphologically and occupies the same
          type of habitat.

Hayden's Shrew
Sorex haydeni

Description: Very similar to Masked Shrew, but
               smaller, and paler brownish in color. Tail
               shorter, with less of a dark tuft at the
               end. 4 medium-size unicuspids, then
               1 small one. L 3–4″ (77–101 mm);
               T 1–1¼″ (25–38 mm); HF ⅜–½″
               (9–12 mm); Wt ¹⁄₁₆–⅛ oz (2–5 g).

Similar Species: Masked Shrew is darker, and has longer
                    more tufted tail, with terminal hairs of
                    underside dark. Preble's Shrew is
                    smaller. Vagrant and Dusky shrews are
                    larger, with third unicuspid smaller
                    than fourth. Pygmy Shrew's third and
                    fifth unicuspids are tiny.

Habitat: Grassy habitats and prairies. Where
            Hayden's and Masked shrews occur
            together, Masked inhabits woody or
            high-grass areas, Hayden's short grass.

Range: Southwestern Alberta, s Saskatchewan,
          and s Manitoba southeast to Nebraska
          and Iowa.

          Until recently, Hayden's was thought to
          be a subspecies of the Masked Shrew.
          However, the two apparently remain
          distinct when they occur together, and
          thus are now considered separate species

5 Pygmy Shrew
Sorex hoyi

Description: *A tiny shrew,* often weighing no more
               than a dime. Brownish to grayish above

pale or silvery below. *Third and fifth
unicuspids very small.* L 3⅛–3⅞" (78–98
mm); T 1–1⅜" (27–35 mm); HF ⅜–½"
(9–12 mm); Wt ¹⁄₁₆–⅛ oz (1.8–3.8 g).

Similar Species: Other long-tailed shrews are larger,
with only the fifth unicuspid reduced.

Habitat: Deep woods; open, brushy fields; moist
sphagnum moss bogs.

Range: Alaska and most of Canada south to
e Idaho and nw Wyoming, Great Lakes
states, and n New England; also in
c Colorado and s Appalachians.

The Pygmy Shrew was thought to be
one of the rarest North American
mammals, although many individuals
have been captured in recent years in
pitfall traps. Little is known of its
behavior in the field. It forages among
the surface litter, feeding on tiny insect
larvae, spiders, and other small
invertebrates. The smallest individuals
of this species, and the tiniest mammals
in North America, are found in southern
Indiana, where they average only ¹⁄₁₆
ounce (2 g). The Pygmy Shrew is often
placed in a separate genus, *Microsorex*.
An isolated group of very small Pygmy
Shrews in the southern Appalachians
is classified as a separate subspecies,
S. h. winnemana.

Pribilof Island Shrew
Sorex hydrodromus

Description: Brownish above, paler below. *Tricolored
coat;* bicolored in winter. L 3⅝–3¾"
(93–97 mm); T 1¼–1⅝" (33–43 mm);
HF ½" (12–13 mm).

Similar Species: No other *Sorex* species occurs in same
range.

Habitat: Unknown.

Range: Saint Paul Islands, in the Pribilof
Islands, west of Alaska.

Very little is known about the biology
of the Pribilof Island Shrew. Records
of it in Unalaska, in the Aleutians, are

thought to be incorrect. Because of its tricolored coat, it was once believed to be related to the Arctic Shrew, but is now thought to be closer to the Masked Shrew on geographical and morphological grounds. It is sometimes called *S. pribilofensis,* but the name *S. hydrodromus* was used earlier and has priority.

St. Lawrence Island Shrew
Sorex jacksoni

Description: *Tricolored coat.* Slightly smaller than Arctic Shrew. Averages: L 2⅜″ (60 mm); T 1⅜″ (34 mm); HF ½″ (12 mm).
Habitat: Unknown.
Range: Saint Lawrence Island, nw Alaska.

Like the Pribilof Island Shrew, the St. Lawrence Island Shrew is thought to be related to the Masked rather than the Arctic Shrew. Because this tiny animal inhabits such a remote area, very little is known about its habits and behaviors.

6 Southeastern Shrew
"Bachman's Shrew"
Sorex longirostris

Description: *Brownish above;* buff below. *Long tail.* Third unicuspid often smaller than fourth. L 2⅞–4¼″ (72–108 mm); T 1–1⅝″ (26–40 mm); HF ⅜–½″ (9–13 mm); Wt 1⁄16–⅛ oz (2–4 g).
Similar Species: Masked Shrew has longer snout, longer and hairier tail.
Breeding: Several litters of 4–6 young from spring to fall. Subspecies in Dismal Swamp of Virginia and North Carolina reported to bear 6–10 young.
Sign: Pencil-size burrows, especially in mossy areas.
Habitat: Moist areas; also fields, brushy areas, and woods.
Range: Southeastern U.S. from Mississippi River eastward (except Appalachians); south

from s Indiana, Illinois, and s Maryland to n Florida; west of Mississippi only in Arkansas and Louisiana.

The Southeastern Shrew eats spiders, moth larvae, slugs and snails, centipedes, and vegetation, but little else is known of its habits, although they are thought to be similar to those of the Masked Shrew. It is sometimes called "Bachman's Shrew," after the naturalist John Bachman, who discovered it in 1837. One of the first two specimens was retrieved from the gullet of a hooded merganser, a duck that preys on a wide variety of small aquatic wildlife.

Mt. Lyell Shrew
Sorex lyelli

Description: *Brown above, gray below.* First and second unicuspids distinctly larger than third and fourth; third unicuspid larger than fourth (in contrast to other similar shrews in region). L 3⅞–4" (100–103 mm); T 1⅜–1⅝" (36–43 mm); HF ⅜" (11–12 mm).

Similar Species: Similar Masked Shrew occurs farther north and east. Preble's Shrew is smaller.

Habitat: At high altitudes (over 6,600'/2,000 m) in grass or around willows.

Range: Tuolumne and Mono counties, California.

Apparently very little is known about the habits of the Mt. Lyell Shrew. It is closely related to the Masked Shrew, which it strongly resembles.

Merriam's Shrew
Sorex merriami

Description: *Pale grayish above,* with paler flanks; *whitish below. Long, distinctly bicolored tail.* L 3½–4¼" (88–107 mm); T 1¼–1⅝" (33–42 mm); HF ½" (11–13 mm); Wt ⅛–¼ oz (4.4–6.5 g).

Breeding: Little is known; probably breeds in spring and summer, bearing perhaps 5–7 young per litter.

Habitat: Prefers a much drier habitat than most shrews: sagebrush, grasslands, and woodlands.

Range: Isolated colonies throughout w U.S.

Little is known about the habits of Merriam's Shrew, which often is found in deserts with the Sagebrush Vole, whose runways it uses. The diet of this shrew includes caterpillars, adult and larval beetles, cave crickets, and ichneumonid wasps. Glands on the flanks of males, prominent from April through June (the reproductive season), give the animal a strong odor that probably serves to attract females but is also believed to repel predators. These glands meaure ¹⁄₁₆–¼" (3–7 mm) in Merriam's, larger than those of all other shrews except the Desert Shrew.

Dusky Shrew
"Montane Shrew"
Sorex monticolus

Description: In summer: fur short and *rust brown abov* slightly paler and less reddish below. In winter: fur longer, and darker above and below. *Tail indistinctly bicolored.* Third unicuspid smaller than fourth. L 4⅛– 4⅞" (105–125 mm); T 1⅝–2⅛" (43–5 mm); HF ⅜–½" (11–14 mm).

Similar Species: Pacific Shrew is larger, with longer tail. Vagrant Shrew is usually shorter, with shorter tail.

Breeding: Little is known; probably has 2 breeding periods in southern parts of range (April and August), and only 1, spring, in northern areas.

Habitat: Forests near streams; wet meadows with sedges or willows.

Range: Western North America from Alaska south to n and c Oregon; most of Idaho Utah, and New Mexico; isolated populations in ne and s California.

The Dusky Shrew feeds on insects, insect larvae, spiders, snails, and other invertebrates. In some Montana alpine and subalpine populations, births begin in March and April (later than for the Vagrant Shrew), probably due to the higher elevation.

Dwarf Shrew
Sorex nanus

Description: *A tiny shrew. Grayish brown above; gray below.* Long tail slightly darker above than below. Third unicuspid smaller than fourth. L 3¼–4⅛″ (82–105 mm); T 1¼–1⅝″ (38–42 mm); HF ⅜″ (10–11 mm); Wt about ¹⁄₁₆ oz (2.5 g).

Similar Species: Masked and Merriam's shrews have distinctly bicolored tails.

Habitat: Rocky habitats in alpine tundra and subalpine coniferous forests; also sedge marsh and dry brushy slopes. Primarily montane spruce forests in Grand Canyon area; Douglas fir and white fir forests at 7,000–9,000′ (2,100–2,700 m) elevation in New Mexico; rocky gutters alternating with patches of alpine tundra vegetation on Beartooth Plateau, in Montana.

Range: Montana, Wyoming, and sw South Dakota south to Arizona and New Mexico.

The habits of the Dwarf Shrew are little known, but they are presumed to be similar to those of other long-tailed shrews. It feeds on soft-bodied insects and spiders.

7 Ornate Shrew
Sorex ornatus

Description: *Grayish brown above;* slightly paler below. *Tail dark brown above and below,* darkest at tip. Fourth unicuspid larger than third. L 3⅜–4¼″ (86–110 mm); T 1⅛–1¾″ (30–44 mm); HF ⅜–½″ (10–14 mm).

Breeding: Generally shrews born previous year
             breed in spring (in San Francisco Bay
             Area, breeding occurs from late
             February to early October). Litter of
             4–6 young born after 21-day gestation.

Habitat: Brackish water marshes; along streams;
             brushy areas of valleys and foothills;
             yellow pine forests. Especially in low,
             dense vegetation that forms cover for
             invertebrates.

Range: Southern two-thirds of coastal
             California; also c California.

Little is known about the habits of the
Ornate Shrew, but they are probably
similar to those of other small, long-
tailed shrews. As with many shrews, the
adult animals in spring are mostly the
young born the previous summer;
summer populations consist of these
adults, now old (Ornate Shrews seldom
live more than 12 to 16 months), and the
young of the year. By autumn, the old
adults have died and populations consist
almost entirely of the year's young.
Populations of Ornate Shrews at Tolay
Creek, on the north side of San Pablo
Bay, California, were once thought to
interbreed with the Vagrant Shrew, but
they are now believed to be slightly
differentiated populations of *S. ornatus.*

Pacific Shrew
Sorex pacificus

Description: In summer: dark reddish brown to dark
                  brown. In winter: cinnamon above,
                  slightly paler below. *Long tail usually tan*
                  or brown above and below, but sometimes
                  indistinctly bicolored. Fourth unicuspid
                  larger than third. L 5⅛–6⅜" (129–160
                  mm); T 2⅛–2⅝" (54–68 mm); HF ⅝–
                  ¾" (15–19 mm).

Similar Species: Dusky and Vagrant shrews are smaller,
                  with shorter tails.

Breeding: Breeds February–August, occasionally
                  to November. 2–6 (usually 4–5) young
                  per litter.

Habitat: Spruce and redwood forests; stands of alder-skunk cabbage along stream edges.

Range: Pacific Coast from s Oregon to n California.

Unlike most North American shrews, the Pacific Shrew is nocturnal. It feeds on slugs and snails, centipedes, amphibians, insect larvae, and other invertebrates.

8 Water Shrew
Sorex palustris

Description: Very dark or black above; belly silvery white. *Long tail. Hindfeet have fringe of stiff hairs.* L 5⅝–6¼″ (144–158 mm); T 2½–3⅛″ (63–78 mm); HF ¾–⅞″ (18–21 mm); Wt ⅜–⅝ oz (12–17 g).

Similar Species: Most other long-tailed shrews are much smaller. Pacific Water Shrew is dark brown above and below; fringe of hair on hindfeet is poorly developed.

Breeding: May begin breeding as early as January and continue through August. 2–3 litters per year of usually 6 young, born after 21-day gestation. Females born early in year may reproduce later the same year.

Habitat: Among boulders along mountain streams, or in sphagnum moss along lakes.

Range: Southeastern Alaska and most of Canada south throughout ne California, through Utah and central states to ne South Dakota, n Minnesota, Wisconsin, and Michigan; New England south through Appalachians to Nantahala Mountains of North Carolina. Isolated population in White Mountains of e Arizona.

At home in water, the Water Shrew can dive to the bottom; when it stops swimming, air trapped in the fur pops the animal back to the surface like a cork. Owing to the fringe of hairs on the hindfoot, which increases the foot's

surface area and traps air bubbles, this shrew can actually run on the water's surface. Its velvety fur is water-resistant if it does become wet, the shrew takes to shore and dries itself thoroughly with rapid strokes of its hindleg, with the hair fringe functioning as a comb. It eats aquatic food, such as mayfly and stonefly nymphs, as well as terrestrial invertebrates. Enemies include weasels, minks, and fish, such as pickerel, pike, bass, and large trout. Now included in this species is the Glacier Bay Water Shrew (formerly *S. alaskanus*), known only from Point Gustavus, Glacier Bay, Alaska.

Preble's Shrew
Sorex preblei

Description: A tiny shrew. *Brownish above, paler below.* Tail bicolored. Third unicuspid larger than fourth. L 3⅜–3¾″ (85–95 mm); T 1⅜″ (35–36 mm); HF ⅜″ (11 mm).

Similar Species: Mt. Lyell and Hayden's shrews are very similar but larger. Vagrant and Dwarf shrews have fourth unicuspid larger than third. Masked Shrew is larger and paler, and has shorter, blunter snout.

Habitat: Usually, arid and semi-arid habitats.

Range: Southeastern Washington, c and e Oregon, c Idaho, and much of Montana.

Although Preble's Shrew prefers arid and semi-arid habitats, individuals have been recorded in most habitats. However, such information is sparse, and little is known about the species.

Fog Shrew
Sorex sonomae

Description: *Medium brown above and dark brown below.* Tail same color as body. Medial tine lacking on first upper incisor. L 4⅛–7⅛″ (105–180 mm); T 1⅜–3⅜″ (36–85 mm); HF ⅜–⅞″ (11–22 mm).

Habitat: Moist meadows and woods.
Range: Southwestern Oregon through nw
California.

The Fog Shrew is the largest of the long-tailed brown shrews of the Pacific Northwest. Little is known of its biology or habits.

Inyo Shrew
Sorex tenellus

Description: In summer: fur drab gray above, smoke gray tinged with buff below. In winter: coat noticeably paler. *Tail indistinctly bicolored,* dark above, buff below. Third unicuspid smaller than fourth. L 3⅜–4″ (85–103 mm); T 1⅜–1⅝″ (36–42 mm); HF ⅜–½″ (9–12 mm).
Habitat: Probably alpine zones.
Range: Southwestern Nevada and adjacent Mono and Inyo counties, in California.

Little is known of the biology or the habits of the Inyo Shrew. It is closely related to the Ornate Shrew.

9 Trowbridge's Shrew
Sorex trowbridgii

Description: Grayish to brownish above; slightly paler below. *Long, distinctly bicolored tail,* dark above; nearly white below. *White feet.* Third unicuspid smaller than fourth. L 4¼–5¼″ (110–132 mm); T 1⅞–2⅜″ (48–62 mm); HF ½–⅝″ (12–15 mm); Wt ¼ oz (6–9 g).
Breeding: Little is known; in British Columbia, first litter born late April–early May.
Sign: Tiny burrows in leaf mold.
Habitat: Mature forests of West Coast.
Range: Extreme sw British Columbia south through w Washington, w Oregon, and much of California.

This small shrew feeds upon various insects and other invertebrates as well as

seeds, especially those of the Douglas fir.
Individuals reach maturation in February,
just before the breeding season, which
extends through May. Most individuals
probably live no longer than 18 months.

Tundra Shrew
Sorex tundrensis

Description: A medium-size shrew. *In summer: fur
tricolored; back brown, sharply contrasting
with paler brownish sides;* underparts pale.
Tricolor pattern indistinct or lacking in
winter and in young individuals. L 3¼–
4¾" (83–120 mm); T ⅞–1⅜" (22–36
mm); HF ⅜–⅝" (11–15 mm); Wt ⅛–⅜
oz (5–10 g).

Similar Species: Arctic Shrew is slightly larger. Most
other shrews are smaller and lack
tricolor pattern.

Breeding: 8–12 young born June (or earlier)
through September; probably several
litters per female.

Habitat: Varies: tundra, grassy areas, shrubby
areas, and swampy forest with willows,
alders, and birches.

Range: Alaska, w Yukon, and nw Northwest
Territory.

Active day or night, the Tundra Shrew
eats insects and earthworms. Nothing is
known of its winter behavior. It has a
high reproductive rate, which may be a
adaptation to the rigors of the Arctic. It
is very closely related to the Arctic
Shrew, but is presently considered
separate from that species.

Barren Ground Shrew
Sorex ugyunak

Description: A small, short-tailed shrew for this
genus. Distinctively colored: *pale color
of belly extends far up sides; brown on
back forms well-defined dorsal stripe.* Tail
bicolored: pale brown above, paler
below. Juvenile darker than adult, with

less distinct demarcation between pale sides and brown color of back. L 2⅞–4″ (74–103 mm); T ⅞–1¼″ (22–31 mm); HF ⅜–½″ (10–13.5 mm); Wt 1/16–⅛ oz (2.9–5.3 g).

Similar Species: Most similar species is Hayden's Shrew, which lives in grasslands and prairies and lacks pale color extending up sides. On Masked Shrew, pale color on sides does not extend as far up and lacks abrupt demarcation; tail dark buff below. Larger Dusky Shrew lacks distinct color pattern. Tundra Shrew much larger, lacks tricolor pattern.

Habitat: North of tree line in Arctic tundra; low, wet sedge-grass meadows and thickets of dwarf willow and birch.

Range: Northern Alaska from Point Barrow east across Arctic Canada to Hudson Bay.

Little is known about the habits of the Barren Ground Shrew. It has been observed using food caches, a behavior not usually seen in shrews.

10 **Vagrant Shrew**
"Wandering Shrew"
Sorex vagrans

Description: In summer: *brownish to grayish above;* grayish tinged with brown or red below. In winter: entirely grayish or blackish. Long tail uniform in color or grading to paler below. Fourth unicuspid larger than third. L 3¾–4⅝″ (95–119 mm); T 1⅜–2″ (34–51 mm); HF ⅜–½″ (11–14 mm); Wt 1/16–¼ oz (3–8.5 g).

Similar Species: Pacific and Dusky shrews have longer tails. Baird's Shrew is best distinguished by different pigmentation pattern on teeth.

Habitat: Mixed forests.

Range: Southern British Columbia and nw U.S., south through much of Arizona and New Mexico.

This tiny creature spends much of its time in the runways of voles, where it

finds such foods as insect larvae, slugs, snails, spiders, and other invertebrates, as well as the subterranean fungus *Endogone.* Its common and scientific names allude to its extraordinary activity in pursuit of food rather than to its wanderings, which are no greater than those of other shrews.

11, 12 Northern Short-tailed Shrew
Blarina brevicauda

Description: *The largest shrew in North America.* Solid gray above and below. *Short tail.* L 3¾–5" (96–127 mm); T ¾–1" (20–25 mm) HF ½–¾" (12–20 mm); Wt ½–1 oz (14–29 g).

Similar Species: Southern Short-tailed Shrew is smaller. Least Shrew is grayish brown or brownish.

Breeding: 4–8 young born from spring through fall, sometimes thoughout the year. Gestation 17–21 days.

Sign: Burrows less than 1" (25 mm) across, wider than high. Piles of snail shells under logs. Bulky nests, 6–8" (150–20 mm) wide, of shredded grass or leaves beneath a log or stump.

Habitat: In the north, a variety of habitats; in warmer, drier parts of range, more confined to woods and wet areas.

Range: Southeastern Canada and ne U.S. south to Nebraska, Missouri, Kentucky, and mountains to Alabama. Isolated populations in ne North Carolina and wc Florida.

The shrews of the genus *Blarina* are unique among mammals in producing poison in their salivary glands. This poison is apparently used to paralyze prey, such as snails and earthworms, which can then be stored for future use. The saliva is not dangerous to humans but a bite may be painful for several days. A voracious eater, consuming from half to more than its own weight per day (in captivity), the Northern Short-

tailed Shrew also feeds on centipedes, beetles, and other invertebrates, and quantities of the tiny subterranean fungus *Endogone.* It sometimes even feeds on mice, particularly nest young when mouse populations are very high, and it will occasionally eat smaller shrews. Using its sturdy snout as well as its powerful forefeet, this species excavates underground runways, which it patrols for prey mainly in early morning and late afternoon. Males mark their burrows with secretions from well-developed glands on the hips and belly; other males looking for mates will not enter burrows so marked. This system of territorial marking helps prevent meetings between individuals of this species, which often result in fierce combat. However, fights usually end when one shrew assumes the submissive posture of lying on its back, allowing the other to flee. Mates, however, may form unions that are more or less permanent. The Northern Short-tailed and Southern Short-tailed shrews were previously considered the same species, but the two are now recognized as separate since they remain distinct where they occur together in Nebraska. The Northern is one of the most common North American mammals.

13 Southern Short-tailed Shrew
Blarina carolinensis

Description: *Gray above and below. Short tail.* Total length seldom over 4″ (100 mm). L 3½–6″ (88–151 mm); T ¾–1⅛″ (19–28 mm); HF ⅜–¾″ (11–20 mm); Wt ⅛–¾ oz (5.5–22 g).

Sign: Small burrows that are wider than high. Snail shells piled under logs.

Habitat: Primarily woodlands.

Range: Southeastern U.S.: s Nebraska and s Illinois south to Gulf of Mexico; east of Appalachians from Maryland south to Florida.

Little is known of the habits of this species; they are probably similar to those of the Northern Short-tailed Shrew.

Elliot's Short-tailed Shrew
Blarina hylophaga

Description: Very similar to Northern Short-tailed Shrew, but *smaller. Slate gray to brownish gray above;* paler below. Ears nearly concealed. L 4–4¾″ (103–120 mm); T ¾–1″ (19–25 mm); HF ½–⅝″ (13–16 mm); Wt ⅜–½ oz (13–16 g).

Similar Species: Northern Short-tailed Shrew is larger. Southern Short-tailed Shrew is smaller. Least Shrew is brownish.

Breeding: Several litters per year of 5–8 young from spring to fall. Gestation 21 days.

Habitat: Oak-elm floodplain forest, wooded ravines, and grassy or weedy fields; sometimes in marshy areas or wet woods.

Range: Southern Nebraska and sw Iowa south to Texas; east to Missouri, nw Arkansas and Oklahoma, extending into Louisiana

Elliott's Short-tailed Shrew feeds on insects, worms, millipedes, arachnids, and other invertebrates. Its saliva is poisonous, like that of other species of the genus. Owls are a major predator, but Coyotes, foxes, Bobcats, hawks, and snakes are other enemies. Domestic cats often capture but do not eat shrews, often depositing them on their owner's doorstep instead.

14 Least Shrew
Cryptotis parva

Description: *Grayish brown or brownish above; paler below. Short tail.* Only 3 unicuspids visible (fifth absent; fourth is hidden behind third). L 2¾–3½″ (69–89 mm); T ½–⅞″ (12–22 mm); HF ⅜–½″ (9–1 mm); Wt ⅛–¼ oz (4–6.5 g).

Similar Species: The only other North American shrews with tails less than half the head and body length are species of the genus *Blarina,* most of which are gray and have 5 unicuspids.

Habitat: Grassy or weedy fields, sometimes in marshy areas or wet woods.

Range: Southern Minnesota and s South Dakota east to c New York and s New England, south to e Texas, and east throughout se U.S.

Most active at night, but sometimes moving about by day, the Least Shrew feeds on moth and beetle larvae, earthworms, spiders, and internal organs of large grasshoppers and crickets. It probes loose soil and leaf litter for prey, which it detects chiefly with the stiff hairs around its mouth. It is sometimes called a "bee mole" for its habit of entering beehives to feed on the brood. While most shrews tend to be solitary, this species may be fairly social: 25 Least Shrews were found in a leaf nest under a log in Virginia, and in Texas one nest contained 12 individuals and another 31. In a laboratory, two individuals were observed cooperating in excavating a burrow, with one digging and the other removing dirt from the burrow and packing the tunnel walls.

15 Desert Shrew
Notiosorex crawfordi

Description: *Grayish, washed with brown above;* pale gray below. Long grayish tail, paler below. *Ears more noticeable than in most shrews.* Prominent flank glands; larger than in any other North American shrew. Only 3 unicuspids; teeth pigmentation orange. L 3–3⅝" (77–93 mm); T ⅞–1¼" (22–32 mm); HF ⅜" (9–11.5 mm); Wt 1/16–⅛ oz (2.9–5 g).

Similar Species: No other North American shrew has 3 unicuspids or such pale-colored teeth; others have chestnut-colored teeth.

Breeding: Little is known; apparently breeds throughout warmer months, probably bearing litters of 3–5 young.

Habitat: Arid regions, especially in areas dominated by sagebrush and prickly pear. Sometimes found in woodrat nests or in large masses of vegetation at the base of agave, cactus, or other plants in desert areas.

Range: Southern California east through Arizona, New Mexico, and s Colorado to w Texas and w Arkansas.

Like many desert animals, the Desert Shrew can exist solely on the water obtained from its food, usually the soft inner parts of larger insects. Young Desert Shrews are able to fend for themselves by the time they are 40 day old. The most common predators of thi species are owls.

MOLES
Family Talpidae

Larger than shrews, usually with proportionally shorter tails, moles are among the most subterranean of mammals. Their bodies are streamlined, head and body merging together with almost no indentation at the neck. The pelvis is narrow, enabling moles to change direction easily in burrows, often by somersaulting. The fur is velvety and grainless, allowing the animals to move as easily backward as forward in tight burrows. Hearing is well developed in moles. Their ear openings are concealed within the fur and thus kept from becoming clogged with dirt. Their eyes are light-sensitive, pinhead-size dots, and their vision is poor. The most important sensory organ is the flexible and often naked snout. The teeth are white, differing from those of shrews, which are chestnut-tipped.

The enormously enlarged, long-clawed forefeet, turned outward, execute a kind of breaststroke, enabling a mole to virtually swim through porous soil at about a foot per minute. Mole tunnels are of two types: subsurface, which appear as ridges; and deep, which are generally marked by cone-shaped molehills. Usually 6 to 8 inches (150–200 mm) high, the molehills are most common in wet weather. Some burrows are used just once for foraging, whereas others are main travelways and are used many times. The tunneling activity, often considered a nuisance in lawns, is beneficial to the environment because it aerates soil, allowing rain to penetrate, thereby reducing erosion.

The seven species of moles in North America north of Mexico are active day and night and throughout the year. They eat several kinds of invertebrates and small amounts of vegetable matter,

but earthworms are their major food. They locate food chiefly through their keen senses of scent and touch and the vibrations picked up by their whiskers Most moles are solitary, except during the mating season, when a male will seek out a female in her burrow. Owing to their underground habitat, moles have a low predation rate, which allow them to maintain populations by producing only one annual litter of two to six offspring. The young are born in early spring in a nest chamber supplied with a maternity cradle of dry vegetation.

In the species accounts that follow, information about tracks and other sig and about breeding is included for the species for which specific information known and available.

16 Shrew-mole
Neurotrichus gibbsii

Description: *The smallest mole in North America,* and the only one with *no major development digging forelegs.* Gray fur. Long, hairy t is about half total head and body leng White-tipped teeth, 9 on each side of upper jaw. L 4–5″ (103–126 mm); T 1¼–1⅝″ (32–43 mm); HF ½–¾″ (14–18 mm); Wt ¼–⅜ oz (9–11 g).

Similar Species: All other moles are much larger, with much more developed forelimbs. Shre have chestnut-tipped teeth.

Habitat: Deep, soft soil in coniferous and deciduous rain forests; sometimes brushy areas or moist, weedy places.

Range: West Coast, from sw British Columbi to nw California.

This creature is aptly named, for it is characterized by the size and forefeet a shrew and the large head and dental structure of a mole. It is unique amon moles in being able to climb low bushes, which it explores for insects.

Earthworms, sow bugs, beetles, insect larvae, and some vegetation make up its diet.

Broad-footed Mole
Scapanus latimanus

Description: Usually *brownish or grayish,* seldom black. *Short, hairy tail.* Short hairs at base of snout. Unicuspids unevenly spaced and usually crowded. L 5¼–7½″ (132–190 mm); T ⅞–1¾″ (21–45 mm); HF ¾–1″ (18–25 mm).

Similar Species: Townsend's and Coast moles have nearly naked tails.

Breeding: Little is known; presumably 1 litter per year in early spring.

Sign: Typical ridges pushed up by burrowing, with molehills (mounds of dirt) where the moles have pushed dirt up through the surface.

Habitat: Moist soil in various habitats from valleys to mountains.

Range: South-central Oregon south through most of California.

Although the forefeet of the Broad-footed Mole are broad, they are no more so than those of most other moles. This species eats mainly earthworms, but also consumes snails, slugs, insects, and some vegetative matter.

17 Coast Mole
Scapanus orarius

Description: *Velvety black fur. Nearly naked tail.* Naked pink snout. Eyes tiny but visible. Unicuspid teeth evenly spaced and uncrowded. L 5¾–6⅞″ (147–175 mm); T 1–1⅝″ (26–43 mm); HF ¾″ (19–23 mm).

Similar Species: Townsend's Mole is much larger, with unevenly spaced, crowded unicuspids. Broad-footed Mole is brownish or grayish.

Breeding: Mates January–March. 1 litter per year around mid-April, averaging 4 young (younger females may have only 2 or 3)

Sign: Ridges along ground surface raised by burrowing moles; molehills where excess dirt has been pushed up and out of the burrow. Ridges and hills are both much more abundant during wet weather.

Habitat: Moist deciduous woods.

Range: Southwestern British Columbia south through w Washington and coastal Oregon and n California. Also n Oregon, se Washington, and wc Idaho.

This mole's inky fur is a good example of the frequent correlation in mammals between moist habitat and dark coloration. Earthworms are its main food, but it also eats insects, centipedes, snails, slugs, and some vegetation. The Coast Mole's nests are built in an underground cavity about 8 inches (20 cm) in diameter that is lined with grass. Nests have several entrances.

Townsend's Mole
Scapanus townsendii

Description: *A large mole. Black fur.* Short, thick, *nearly naked tail.* Nearly naked snout. Eyes tiny but visible. 11 upper teeth on each side, with unicuspids crowded and uneven. L 7¾–9¼″ (195–237 mm); T 1¼–2″ (34–51 mm); HF ⅞–1⅛″ (24–28 mm).

Similar Species: Other western moles smaller: Coast Mole has evenly spaced, uncrowded unicuspids. Broad-footed Mole is brownish or grayish.

Sign: Same as for Coast and Broad-footed moles.

Habitat: Meadows, fields, and lawns.

Range: West Coast, from extreme s British Columbia to n California.

Named for John Townsend, who first described it, this is one of eight North

American mammals whose scientific name honors the 19th-century naturalist. It eats earthworms but also snails, slugs, centipedes, insects, and vegetation.

18 Hairy-tailed Mole
Parascalops breweri

Description: Dark gray or nearly black above; slightly paler below. *Only eastern mole with short, hairy tail.* Older individuals may have white snout, tail, and feet. L 5½–6¾″ (139–170 mm); T ⅞–1⅜″ (23–36 mm); HF ¾″ (17–21 mm); Wt 1⅜–2¼ oz (40–64 g).

Similar Species: Eastern Mole is often larger, with nearly naked tail.

Sign: Typical mole runs of rounded ridges in dry woodland soils are likely to have been made by this species. They may be hard to see, as these moles sometimes live in hard-packed soil.

Habitat: Woods with well-drained, light soil; also brushy areas; occasionally lawns or golf courses adjacent to woods.

Range: Extreme se Canada and New England southwest through mountains of North Carolina and Tennessee, and west through e Ohio.

Tunneling beneath the surface by day, the Hairy-tailed Mole often emerges at night to feed. If abroad by day, it may be so oblivious to human presence that it can be caught by hand; occasionally it is captured by cats and other predators. Like most other moles, it eats earthworms, but also grubs, beetles, ants, and other invertebrates. In the laboratory it has been known to eat food in quantities up to three times its own weight per day. It is active all winter but frequents deeper, hence better-insulated, tunnels at that time. In the wild it may live for four to five years, and the extensive tunnels it builds may be used by succeeding generations.

21, 22 Eastern Mole
"Common Mole"
Scalopus aquaticus

Description: Short, velvety fur; gray in northern parts of range, brownish or tan in southern and western parts. *Forefeet broader than long, with palms turned out; toes slightly webbed.* Tail very short and nearly naked. Snout long, flexible, and naked. No visible eyes. L 3¼–8¾" (82–223 mm); T ¾–1¼" (18–38 mm); HF ⅝–1⅛" (15–29 mm); Wt 3–5 oz (82–140 g).

Similar Species: Hairy-tailed Mole has short, hairy tail.

Breeding: Mates February–March. 1 litter of 2–5 young born after gestation of 30–42 days.

Sign: Ridges and molehills created by burrowing moles; usually found in relatively dry areas, but much more evident in wet weather.

Habitat: Open fields, waste areas, lawns, gardens, and sometimes woods, in well-drained loose soil.

Range: Most of e U.S. from s Minnesota, extreme se Wyoming, Kansas, and c Texas east to Atlantic and Gulf coasts.

The Eastern Mole spends nearly all of its time underground, becoming most active at dawn or dusk. It feeds mainly on earthworms, but also eats larvae and adults of many kinds of insects and other invertebrates. The Latin species name *aquaticus* was given because the first North American individual described in records was found drowned in a well and presumed—in error—to be aquatic. Ironically, this is the least likely of North American moles to live in moist conditions. The forefeet spade alternately, shoving earth under the body, which the hindfeet then push behind into the tunnel. Nests are underground, usually beneath a log, stump, or boulder; Florida individuals do not build nests. In winter, this mole inhabits deeper burrows.

19, 20 Star-nosed Mole
Condylura cristata

Description:	Black fur. *Long, hairy tail.* Large digging forelegs. *22 pink fleshy projections on nose.* L 6–8¼" (152–211 mm); T 2⅛–3¼" (53–84 mm); HF 1–1⅛" (26–30 mm); Wt 1–2⅝ oz (30–75 g).
Breeding:	1 litter per year of 4–7 young born in an underground nest, usually April–May but occasionally as late as August. Gestation about 45 days.
Sign:	Mole burrows in muck, with 2–2½" (50–65 mm) openings into streams, lakes, or ponds. Openings surrounded by mounds of excess dirt.
Habitat:	Wet woods, fields, or swamps; sometimes relatively dry areas and lawns.
Range:	Southeastern Canada and ne U.S. from much of Minnesota east through Appalachians and along coastal Virginia. Isolated populations along Georgia coast.

This mole's tentacle-like nose projections are mobile and very sensitive, apparently helping the animal to find its way about the burrow and locate food. It has been hypothesized that the tentacles act as an electrical sensing device for detecting prey. When the Star-nosed Mole forages in the muck near its main burrows, presumably after earthworms, its favorite food, its tentacles are constantly in motion. When it eats, however, they are clumped together out of the way. An adept diver and swimmer, this mole also eats many aquatic animals, including fish. It propels itself in water, even under ice, by moving its feet and tail in unison. The Star-nosed Mole is more dependent on water during winter, when the frozen ground makes obtaining its usual foods difficult. Its nests are constructed of leaves, grass, or other vegetation, usually in a hummock or other raised area above the moist habitat. The young develop

rapidly and by three weeks leave the nest to hunt for themselves. Most shrews and moles are not often seen out of the nest until they are nearly full size, but smaller young of this species have occasionally been seen aboveground.

BATS
Order Chiroptera

Members of this order inhabit most of the temperate and tropical regions of the world, except for a few remote islands. Bats are exceeded only by rodents as the most numerous mammals, both in number of species and number of individuals. Chiroptera is the second-largest order, with 17 families and 925 species worldwide. Four families with 46 known species occur in North America north of Mexico.

The only mammals that truly fly, bats probably evolved from primitive shrew-like creatures that lived in trees and whose forelimbs eventually evolved to become wings (the order name means "flying hand"). The bones and muscles of the arms are enlarged in bats. Wing membranes are attached to four greatly elongated fingers, which spread when in flight and draw together when at rest, folding the wing along the *forearm;* the thumb projects from the end of the forearm as a small but sharp claw that is used as the animal crawls about. A small bone called the *calcar* juts back from the anklebone to help support a membrane between the legs that is known as the *interfemoral membrane,* or tail membrane. The calcar is said to be "keeled" when there is a flat projection of cartilage protruding from its side that looks like the keel of a boat.

Bats' bodies are generally well furred, but their wing membranes are often naked and translucent. Unlike birds, which flap their wings up and down, bats "swim" through the air, rotating their wings to catch the air with the membrane. Some species fly with the mobility of swifts or swallows and occasionally hunt with such birds at dusk. Wingbeats may be as rapid as 20 per second. With wings folded, bats can

walk or climb, and some can swing from branch to branch. At rest, they hang head downward, sometimes by one hindfoot, using the other to groom their fur and clean their teeth.

Although most North American bats have very small eyes (giving rise to the ill-founded expression "blind as a bat"), they have excellent vision. However, it is not known how much they rely on their sight at night. When flying in darkness, most bats locate food and avoid objects by means of *echolocation,* which is similar to radar or sonar. While flying, the bat emits through its mouth or nose a continuous series of supersonic sounds (some 30 to 60 squeaks per second) ranging in frequency or pitch from below 20 kilohertz—about the maximum frequency that the human ear can hear—up to about 100 kilohertz. Most North American bats emit the sounds through the mouth. Bats that emit sounds through the nose usually have a vertical projecting flap (a *noseleaf*) on the face; one such group of bats is the New World fruit bats (Phyllostomidae). These sounds bounce off objects and are picked up by the bat's complex ears. This means of navigation allows bats to avoid bumping into things and to zero in on small insects. In many bats, a projection from the ear's inner base, called the *tragus* (a useful feature for species identification), is thought to enhance directional input during echolocation. Muscles in the ears contract and relax in synchronization with the vocalization, blocking the emitted sounds (avoiding self-deafening) and receiving the echoes. By interpreting the sound waves, the bat determines the size, location, density, and movement, if any, of the object it approaches. Insectivorous bats emit a series of clicks in their echolocation when hunting; these clicks become much closer together, forming a "feeding buzz," as

the bat closes in on its quarry. In addition to ultrasonic vibrations, bats produce many harsh, shrill chirpings and screechings within the range of human hearing.

Nocturnal mammals, bats leave daytime roosts around dusk and usually fly to a stream, pond, or lake, where they dip their lower jaws into the water to drink. Nearly all North American bats are insectivores, although a few are plant feeders, consuming nectar, pollen, or fruit. Insectivorous bats may "glean" or pick insects from a leaf or other surfaces, or they may locate swarms of midges or other small insects by echolocation, then fly back and forth through the swarm to obtain them. They catch the insects in their mouths, or scoop them into their wings or tail membrane. Bats may feed as they fly, turning a somersault as they extract food from the wing or tail membrane. They capture large insects by individual pursuit; if their meal is too large to eat on the wing, they alight to dine. The legend that bats fly into people's hair is based on the fact that they often fly very close to animals, including humans, seeking the insects that sometimes fly about their heads.

Solitary bats tend to live among the leaves of trees; social bats cluster in *colonies* in caves, in hollow trees, in buildings, or in other protected places, such as under slabs of rock or loose bark. Many bats live in northern parts of the continent, where in winter few insects are active and little pollen and nectar is available. Most of these bats hibernate for the winter, although some species migrate great distances (as much as 1,000 miles/1,600 km) to the south, where they remain active and feed all winter. Hibernating bats may die if the temperature drops much below freezing, and too-warm winter temperatures can speed up their metabolism to the point

that all stored body fat is burned up and they starve to death. Unlike other hibernators, bats awaken very rapidly from hibernation.

Bats spend much of their time *roosting,* resting or sleeping in a particular spot, and exhibit many different roosting behavior patterns, depending on the species. Males and females often roost separately, and one or both sexes may be solitary roosters for part or all of the year. Since bats are active at night, their main roost in summer is the day roost. During the time that the young are born and developing, the day roost may also serve as a *maternity or nursery roost,* and is often occupied by thousands of females and young. Bats generally leave the day or maternity roost at dusk to feed. Some species then return to their day roost, while others occupy a night roost near the feeding area for part of the night, feed again before dawn, and then return to their day roost. For some species, the day roost becomes a hibernating roost (or *hibernaculum*) in winter, although most bats migrate to a separate hibernation roost. Roost sites are often included in the habitat sections of the species accounts. For most bats, more is known about the roosting habitat than about the feeding habitat, which may change as insect distribution changes.

Most North American bats mate in fall. In many hibernating species, sperm is stored in the female over the winter and fertilization is delayed until spring. The actual gestation in spring varies with the species, but a typical period might be about 40 days. Delayed fertilization allows mating to proceed in fall, when energy supplies are ample, and prevents birth during the unfavorable winter months. Ovulation and births occur nearly simultaneously in all members of the colony. Most species give birth to one (rarely two) offspring annually,

although a few have three or four. Young are born feet first, a presentation probably unique in mammals.

The life spans of bats vary, but some species may live more than 30 years. Bats are susceptible to rabies, although antibodies found in healthy bats indicate that they sometimes recover from the disease. Rabid bats rarely attack humans or other animals, but bats may be found lying on the ground when rabid or otherwise sick. Most bats found on the ground are not rabid, but they should be treated as if they were, and never touched or picked up. Due to the loss of habitat, the use of pesticides and insecticides on crops, and direct persecution—such as bombing and other disturbances of caves—most species of bats are declining in number. However, bats are the most important predators of night-flying insects, feeding on many serious pest species and consuming hundreds of thousands of tons of destructive insects annually. They are essential in maintaining the balance of nature and thus deserve legislative protection. Bat Conservation International, in Austin, Texas, has done much to protect bats and has worked tirelessly to improve the image of these fascinating creatures.

In the species accounts that follow, the forearm length (abbreviated as "FA"), which is the best means for comparing the sizes of bats, is given for each bat in the description section. For those species in which the calcar is important as a means of identification, it is mentioned in the description. Information about the feeding, roosting, and mating habits is given for those species for which it is available.

LEAF-CHINNED BATS
Family Mormoopidae

Of the eight species of leaf-chinned bats,
which are primarily restricted to
tropical habitats of the New World,
only one is found in our range. The
bizarre faces of mormoopid bats have
flaps and grooves on the chin that are
believed to enhance echolocation. In
these bats, the tragus, a projection from
the base of the ear, is a horizontal,
pocket-like fold of skin.

23 Ghost-faced Bat
Mormoops megalophylla

Description: Reddish-brown to dark brown bat;
immediately recognizable by its bizarre
face, with *folds of skin across chin* from ear
to ear, a feature unique among North
American bats. Short tail projects from
upper side of interfemoral membrane.
L 2⅜–2⅝" (59–66 mm); T average 1"
(27 mm); HF average ½" (12 mm);
FA 1¾–2¼" (46–56 mm).

Breeding: 1 young, probably born late May–early
June.

Habitat: Desert or scrub. Usually roosts in very
hot and humid caves or mines; seldom
in buildings.

Range: Southeastern Arizona and s Texas south
into Mexico.

Strong, swift fliers, Ghost-faced Bats
emerge late in the evening to hunt just
above the ground. They roost in loose
colonies. While colony size varies, up to
500,000 bats have been found in one
roost in Mexico. These bats occur in
much lower numbers in the United
States. Large areas are needed to house
great numbers of Ghost-faced Bats, as
they tend to space themselves about 6
inches (150 mm) apart over the available
space. These bats apparently do not
hibernate in winter.

NEW WORLD FRUIT BATS
Family Phyllostomidae

There are 140 species of bats in this family, most of which are found in tropical regions of the New World, where many feed on fruit. Six species occur in North America north of Mexico, five in the Southwest and only one, the Jamaican Fruit-eating Bat, in the eastern U.S. Phyllostomid bats generally have a *noseleaf* (a vertical projecting flap). The tail is variable; it may be absent or reduced, or may extend beyond the interfemoral membrane. Fruit bats usually have relatively large eyes and poorly developed echolocation.

24 California Leaf-nosed Bat
Macrotus californicus

Description: Grayish to dark brownish above, with fur nearly white at base; paler below. *Large ears. Erect triangular flap on nose.* L 3¼–3⅝" (84–93 mm); T 1¼–1⅝" (33–41 mm); HF ½" (12–14 mm); FA 1⅞–2⅛" (47–55 mm).

Breeding: Mating, fertilization, and implantation occur in fall; fetal development slows until March; young born in June.

Habitat: Desert scrub. By day, roosts in abandoned mine tunnels.

Range: Southern California, extreme s Nevada, and w Arizona.

The California Leaf-nosed Bat is the only bat in North America north of Mexico with large ears and leaf-like projections on the nose. It roosts by day, usually fairly close to the entrance of a mine tunnel, in small groups of up to 100 bats, which do not touch each other. This species cannot crawl on thumbs and toes like most bats, but instead often dangles by one leg from a mine tunnel's ceiling, which it can cross in a swinging stride, using its hindlegs

alternately. After dark, this species drop
from its perch into flight. Leaf-nosed
Bats eat various insects, including some
flightless forms, such as crickets and
some beetles, which they probably detec
as they hover, swooping down to seize
them from the ground. After feeding fo
about an hour, they retreat to their
night roosts in a sheltered area. They do
not hibernate. Male California Leaf-
nosed Bats occupy bachelors' quarters ir
July and August, soon thereafter joining
the females for the mating season.

25   Mexican Long-tongued Bat
Choeronycteris mexicana

Description:	Gray or brownish above; paler below. *Very long nose and tongue; small noseleaf.* Large eyes. *Tiny tail* extends less than half length of interfemoral membrane. L 2⅛–3⅛″ (55–78 mm); T ¼–⅜″ (6–10 mm); HF ⅜–½″ (10–13 mm); FA 1⅝–1¾″ (43–44 mm); Wt ⅜–⅝ oz (10–20 g).
Similar Species:	Southern Long-nosed and Mexican Long-nosed bats lack tail.
Breeding:	Pregnant females separate from males into small groups in early spring, giving birth in June or July.
Habitat:	Arid scrub, mixed forest, and canyons i mountain ranges rising from desert. By day, usually roosts in caves and mines, sometimes in buildings; tends to hang near entrances.
Range:	Extreme s California, s Arizona, and extreme sw New Mexico.

In roosts, these bats do not cluster but
remain an inch or two (25–50 mm)
apart. Extremely wary and easily
disturbed, they are able to hang by one
foot in such a way that they can rotate
and watch any intruder. The long
tongue, tipped by a brush of tiny
nipple-like projections, and the lack of
lower incisors make it easy for this bat
to lap up flower nectar and fruit juices.

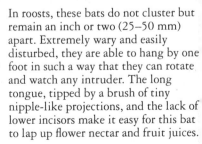

It also eats insects and pollen, and may pollinate certain plants. Mexican Long-tongued Bats in Arizona probably migrate to Mexico for the winter.

31 Southern Long-nosed Bat
Leptonycteris curasoae

Description: Reddish brown on back; brownish on belly. *Long nose has erect leaf-shaped projection on tip.* Large eyes. *No visible tail.* L 2¾–3¼″ (69–84 mm); HF ½–¾″ (13–17 mm); FA 2–2¼″ (51–56 mm).

Similar Species: Mexican Long-nosed Bat is less reddish, and usually larger, with larger forearm. Mexican Long-tongued Bat has small tail.

Habitat: Areas where mountains rise from desert. Roosts in caves and mines.

Range: Southern Arizona and extreme sw New Mexico.

The Southern Long-nosed Bat, formerly known as *L. sanborni,* emerges late in the evening to feed on flower nectar and pollen, as well as insects and the nectar and fruit of various agaves and cacti. It either alights on vegetation while feeding or hovers somewhat like a hummingbird, inserting its long snout into flowers. Hair-like projections on the long tongue help sweep food into its mouth. During late pregnancy, females congregate in large maternity colonies. In the fall, this bat apparently migrates to Mexico, where it spends the winter.

27 Mexican Long-nosed Bat
Leptonycteris nivalis

Description: Grayish brown above; paler on shoulders and underparts. *Long nose has leaf-like projection.* Tongue long and protrusible. Medium-size ears. *No tail.* L 3–3½″ (76–88 mm); HF average ⅝″ (15–17 mm); FA 2⅛–2⅜″ (55–60 mm); E average ⅝″ (15 mm); Wt ¾ oz (21 g).

Similar Species: Southern Long-nosed Bat is usually smaller, with shorter forearm. Mexican Long-tongued Bat has small tail.

Breeding: 1 young born in early summer.

Habitat: Forests of pine and oak. Usually roosts in caves.

Range: Emory Peak, in Big Bend National Park, Texas, and south into Mexico.

This colonial bat is a cave dweller, but can also be found in mines, vacant buildings, and hollow trees in clusters of as many as 150 individuals per square foot (1,600 per square meter). Inhabited caves often have a peculiar musty smell. The wall of the cave in Emory Peak is covered with their yellowish, sweet-smelling, semi-liquid droppings. These bats emerge late in the evening from their roosting caves, apparently moving to lower elevations to feed on nectar and pollen, and probably some insects. Nectar and pollen feeding is facilitated by the elongated nose and long, flexible tongue, which can be extended as far as 3″ (75 mm). The beginning of the rainy season prompts a southward migration. The sexes apparently segregate geographically, with the males occupying the southern part of the range in summer. This bat is listed as endangered by the U.S. government.

26 Jamaican Fruit-eating Bat
Artibeus jamaicensis

Description: Only bat in e U.S. with *noseleaf and no tail*. Upperparts various shades of brown; underparts often grayish and usually somewhat paler than back. Often has pale facial stripes above and below eye. Averages: L 3⅛″ (80 mm); HF ⅝″ (17 mm); FA 2¼″ (54–60 mm); E ⅞″ (22 mm); Wt 1½ oz (42 g).

Breeding: Studies in Panama show that births peak in March and April, and again in July and August. Mating and implantation follow, but the pregnancy remains

dormant from September to November. Development then proceeds normally until birth in March or April. 1 (rarely 2) young born to each female.

Habitat: Wide range of forest types. In Florida Keys roosts mostly in buildings.

Range: In North America found only in Florida, at Key West, Cudjoe Key, and Ramrod Key; may occur in other western Keys.

Even though no specimen has yet been taken, it seems clear that the Jamaican Fruit-eating Bat occurs in the Florida Keys. As it is widespread and often common in Cuba and on oceanic islands in the Caribbean, its occurrence in the Keys is not improbable, especially since it feeds primarily on fig trees, which fruit year-round in the western Keys. The Monroe County Extension offices on Stock Island in the western Keys report that big, tailless, leaf-nosed bats are regularly brought in. However, the records frequently are overlooked or ignored, and no specimens have ever been saved. The Jamaican Fruit-eating Bat was first reported in the Keys in 1872, but the record was discounted. On February 3, 1983, L. Page Brown photographed one of several flushed from East Martelo Tower, Key West. The bat in the photograph was then identified by two different biologists, James D. Lazell and Karl Koopman, as a Jamaican Fruit-eating Bat. The bat was not collected, and this record, too, was discounted. In 1984, a small colony of fruit bats (about six) moved into a house on Ramrod Key, but they were evicted before a specimen could be examined. On January 28, 1986, one landed on a screen in a Cudjoe Key house, where James Lazell was investigating a bat report. Although it soon departed, it was there long enough to be identified as a Jamaican Fruit-eating Bat. It was also seen to fly into a large fig tree *(Ficus citrifolia)* near the house, from which ripe figs were falling at the time.

Jamaican Fruit-eating Bats feed primarily on the fruit of fig trees (reproduction is tied closely to maximum abundance of the fig), but will also eat other kinds of fruit, including mango, avocado, and banana, as well as pollen, nectar, flower parts, and perhaps insects. Their fecal material often has the odor of the fruit consumed as food passes through the digestive tract in 15 to 20 minutes. These bats may forage in small groups; captured individuals may produce stress calls, inducing "mobbing" behavior by other members of the group. They tend to be much less active on bright moonlit nights. These bats roost in buildings, caves, and hollow trees; they sometimes break the midribs of large leaves by biting them so the leaves fold up, forming tents. Females have a daytime roost and leave their young in crèches near the feeding tree during nightly feeding forays. The species is polygynous; males may defend harems of up to 25 females. Besides harems, this species also forms groups of bachelor males and of nonreproductive females. The harems often reside in tree hollows; the males and nonreproductive females often roost in solitary fashion or form small groups among foliage. It is suspected that owls may be nocturnal predators, and snakes and the bat falcon may be diurnal predators. The life span of one individual of this species was 7 years; that of another was 10 years.

Hairy-legged Vampire Bat
Diphylla ecaudata

Description: Usually gray-brown (sometimes dark brown) above; paler below. Surface of interfemoral membrane well furred. *No tail.* Short, rounded ears; small tragus. Large eyes. *Thickened, M-shaped fold of skin over nostrils.* Outside lower incisor has 7 fan-shaped lobes (unique among

bats); cheek teeth reduced. L 2½–3⅜"
(65–87 mm); HF average ⅝" (18 mm);
FA 2–2¼" (50–56 mm); Wt ⅞–1½ oz
(24–43 g).

Breeding: Female bears 1 young, apparently at any
time of year.

Habitat: Mainly humid forest. Usually roosts in
caves, but also in mines and hollow
trees.

Range: Central Mexico south into South
America. In U.S., possibly in s Texas.

Discovered in Mexico in the 16th
century by the Spanish conquistadores,
who named them after the blood-
sucking creatures of eastern European
legend, vampire bats feed entirely upon
blood and occur only in the New World.
The only record north of Mexico is one
Hairy-legged Vampire Bat taken from
an abandoned railroad tunnel near
Comstock, Val Verde County, Texas, in
1967. The vampires are among the few
bats that use their wings for walking on
the ground. Sleeping by day in caves or
other dark, protected places, Hairy-
legged Vampire Bats feed at night,
chiefly on the blood of birds, such as
chickens, turkeys, guinea fowl, ducks,
and geese, but also of mammals,
including pigs, cattle, and horses. After
locating prey, this bat often lands
nearby, then scuttles over to it, and may
climb up its leg (if a large animal) or
jump on it. Perching lightly on or near
the sleeping or motionless prey, it uses
sharp upper incisors to quickly inflict a
wound so shallow that the victim rarely
notices. It bites chickens on the lower
legs or near the cloaca (anal opening),
where there are few feathers. Curling its
tongue into a tube that fits a V-shaped
notch in its lower lip, the bat sips the
blood, which flows freely due to an
anticoagulant in its saliva. Unless the
victim is very small, the amount of
blood taken by the bat is not harmful;
more serious, after the bat has flown
away, are the potential for blood loss

owing to slowed clotting, the vulnerability of the open wound to infection, and the possibility of rabies transmission. These large-eyed bats have good vision, but a poorly developed system of echolocation. They usually congregate in small groups of only 1 to 3 per cave, rarely more than 12, though in one cave 35 were found.

VESPERTILIONID BATS
Family Vespertilionidae

Most North American bats are members
of this family, which is found in
temperate and tropical regions
throughout the world. There are 10
genera and 32 species of these bats in
North America north of Mexico.
Vespertilionid bats have plain noses,
their earlobes form a tragus, and their
tails extend only slightly beyond the
edge of the interfemoral membrane.
Most have a well-developed sense of
echolocation, and send out ultrasonic
vibrations through their mouths or
noses. Seen chiefly in the evening, these
"vesper bats" are insectivorous. On
evenings in late summer they often
swarm at cave entrances. The significance
of this swarming is not completely
understood, but it is thought that its
major purpose is to bring members of
the opposite sex together for mating.
It is also speculated that the bats may
be exploring possible hibernation sites.
Although a few migrate to southern
regions, most members of this family
hibernate in winter.

Many vespertilionid bats exhibit
delayed fertilization. They mate in fall,
but the sperm remain free in the female
reproductive tract, and fertilization then
occurs in spring. Most species bear only
one or two young, except the red bats,
which have three or four. Most of the
vespertilinoid bats are colonial, forming
colonies in summer habitats and
undergoing relatively short migrations
to winter quarters, often in caves. A
number of the *Lasiurus* species, the
solitary or tree bats, undergo very
long migrations between summer and
winter homes. Within the family
Vespertilionidae is the largest genus of
bats of North America, *Myotis,* the
mouse-eared bats, with 15 species. One
feature that distinguishes the myotises

from other bats is the configuration of
its teeth. It has two tiny teeth between
the canine and the larger cheek teeth,
forming a space, while other genera
have only one reduced tooth or no tooth
there and thus have a smaller space.

Southwestern Myotis
Myotis auriculus

Description: Dull brownish fur. *Large brown ears;
long, thin tragus. Calcar not keeled.*
L 3⅛–3½″ (78–88 mm); T 1⅜–1¾″
(36–45 mm); HF ¼–⅜″ (7–9 mm);
FA 1¼–1⅝″ (38–40 mm); E ¾″ (18–
20 mm).

Similar Species: Long-eared Myotis has longer ears.
Fringed Myotis has fringe of hair along
edge of interfemoral membrane.

Breeding: 1 young born in June or July.

Habitat: Primarily ponderosa pine forests, but
also deserts, mesquite, and chaparral,
especially areas of rocky outcroppings.
Roosts at night in buildings and caves.

Range: Southeastern Arizona and sw New
Mexico.

Little is known of the habits of the
Southwestern Myotis. It roosts in
buildings and caves at night, but its
daytime roosts have not been found. It
feeds on insects, especially moths,
primarily by gleaning, but apparently
produces no feeding buzz during this
activity.

28 Southeastern Myotis
Myotis austroriparius

Description: Dull brown, with short, thick, woolly
fur; hairs show little or no contrast in
color between base and tip. *Toes have l[o]
hairs extending beyond tips of claws. Skul[l]
has sagittal crest* (bony ridge extending
along top from front to back) that can
be felt through the skin. Calcar not
keeled. Typical short, rounded tragus.

In Indiana and Illinois, underparts buff in summer, white in winter. L 3¼–3⅞" (84–99 mm); T 1⅜–1¾" (36–45 mm); HF ⅜–½" (10–12 mm); FA 1⅜–1⅝" (35–42 mm); Wt ¼–⅜ oz (7–12 g).

Similar Species: Little Brown Myotis has longer silky fur with burnished tips. Keen's Myotis has longer ears and longer, thinner tragus. Indiana Myotis has keeled calcar. Eastern Small-footed and Western Small-footed myotises have smaller foot and keeled calcar.

Breeding: Nursery colonies begin forming in March in caves where water is plentiful. 2 young are born late April to mid-May.

Habitat: Wooded and open areas. In North, roosts in caves; in South, in buildings or hollow trees, also in caves in Florida.

Range: Southeastern U.S.: w Kentucky; Arkansas and Louisiana east to Georgia and n Florida; isolated populations in s Illinois and s Indiana.

In recent years, Florida caves have had some of the largest colonies of the Southeastern Myotis, with up to 90,000 bats packed 150 per square foot (1,600 per square meter). This is the only North American myotis that produces more than one young. In winter, most of these bats leave the maternity caves to seek shelter in other protected places, such as culverts, bridge beams, and buildings, where they gather in small numbers. Contrary to popular opinion, North American bats seldom eat mosquitoes, although this species often does so in Florida, where mosquitoes are abundant. Predators include corn snakes and rat snakes.

29 California Myotis
Myotis californicus

Description: Dull fur, light to dark brown with yellowish or orangish cast above, paler below; base of fur blackish. Ears, wings, and interfemoral membrane dark. *Foot*

tiny. Calcar keeled. L 2⅞–3⅜″ (74–85 mm); T 1⅜–1⅝″ (36–42 mm); HF ¼″ (5–7 mm); FA 1⅛–1⅜″ (29–36 mm).

Similar Species: Western Small-footed Myotis has glossy fur and blackish face and ears.

Breeding: Maternity colonies are usually small. 1 young born late May–early June.

Habitat: Desert to semi-desert areas; in Southwest, rocky canyons. Roosts by day in buildings, under bridges, under bark, and in hollow trees; by night, in buildings.

Range: Western North America southward from sw British Columbia, and east to Idaho, Colorado, and Texas.

The California Myotis feeds on small flies, moths, and a few other insects. A ability to veer suddenly sideways, up, or down makes its flight conspicuously erratic. Some of these bats hibernate in mines in winter, while others remain active. Maternity colonies form in buildings or under bridges, and also in hollow trees or under bark.

Western Small-footed Myotis
Myotis ciliolabrum

Description: Small, yellowish-brown myotis, with black face and ears; buff to nearly white below. Fur glossy. Wings black; interfemoral membrane dark brown. Foot small. *Calcar keeled.* L 3–3½″ (76–90 mm); T 1¼–1⅜″ (31–34 mm); HF ¼″ (6–7 mm); FA 1⅛–1⅜″ (30–3 mm); Wt ¹⁄₁₆–¼ oz (2.8–7 g).

Similar Species: Western Pipistrelle has yellowish fur and a short, blunt tragus. California Myotis lacks black face and ears. Little Brown Myotis usually lacks or has only weak keel on calcar.

Breeding: Maternity colonies are small, often in man-made structures. 1 young born late May–early June.

Habitat: Arid and shortgrass prairie regions; cliffs, talus, or clay buttes or riverbank in prairie areas.

Range: Southern British Columbia, Alberta, and Saskatchewan south thoughout most of w U.S.

The Western Small-footed Myotis forages along cliffs and rocky slopes, where it feeds on a variety of small insects, including moths, true bugs, and flies. Little else is known about its habits.

32 Long-eared Myotis
Myotis evotis

Description: Long, glossy fur, light brown to brown. *Ears dark, usually black; longer than in any other myotis;* when laid forward *extend ¼" (7 mm) beyond nose.* Tragus long and thin. Calcar keeled. L 3–3¾" (75–97 mm); T 1⅜–1¾" (36–46 mm); HF ¼–⅜" (7–10 mm); FA 1⅜–1⅝" (35–41 mm); E ¾–⅞" (18–22 mm); Wt ⅛–¼ oz (5–8 g).

Similar Species: All other myotises have shorter ears.

Breeding: Females form small maternity colonies in early summer; 1 young born late June–July.

Habitat: Variety of habitats, from sage to high-altitude coniferous forests; mostly found in forested regions. Sometimes roosts in buildings.

Range: Western North America, from s British Columbia, Alberta, and Saskatchewan south through much of California, ne Arizona, and nw New Mexico.

The Long-eared Myotis feeds heavily on small moths and also eats flies, beetles, and other insects. By day, this bat roosts singly or in small clusters in buildings and perhaps under tree bark; night roosts are sometimes in caves. Males are solitary, but groups of 12 to 20 females form small nursery colonies, often in buildings but also in other protected places, such as in hollow trees and behind slabs of bark. Nothing is known of the hibernating sites of this species.

33 Gray Myotis
Myotis grisescens

Description: A medium-size bat. Grayish or brownish; *hairs uniform in color from base to tip.* Calcar not keeled. Ears relatively short; tragus relatively short and rounded. *Wing membrane attaches at ankle rather than base of toes.* L 3⅛–3¾" (80–9 mm); T 1¼–1¾" (32–44 mm); HF ¼–⅜" (8–11 mm); FA 1⅝–1¾" (41–46 mm); Wt ¼ oz (6–9 g).

Similar Species: While many other myotises have bicolored hairs, with contrasting base and tip, no other bat has wing membrane attached at ankle.

Breeding: Females form very large maternity colonies in wet caves. 1 young born in June.

Sign: In caves, rocks and water covered with stinking feces may indicate active summer colonies.

Habitat: Areas near rivers and streams. Roosts i caves, often containing much water.

Range: Northwestern Oklahoma, n Arkansas, and Missouri, east through Kentucky and Tennessee, and south through Alabama and w Georgia.

A true cave species, the Gray Myotis feeds on various insects and often eats mayflies. Females and young congrega in huge numbers apart from the males in large caves containing much water. The maternity colonies disband in late July or August after the young are weaned. These bats move in large floc between summer and winter caves, with hibernating colonies forming in October. Most members of this species hibernate in only a very few specific locations. Eastern individuals have three main wintering caves, one each in northeastern Alabama, central Tennessee, and eastern Tennessee. The one maternity colony in Indiana, with 1,100 individuals (and on the increase is in a water-filled quarry with several

entrances. Such factors as disturbances from cave exploration, the opening of caves to the public, and the flooding of caves through dam building have caused a serious decline in populations of this bat. The Gray Myotis is now classified as endangered by the U.S. government.

Keen's Myotis
Myotis keenii

Description: Variable shades of brown; paler and somewhat buffy below. Dark shoulder spots. Posterior border of tail membrane has scattered hairs. *Long ears* protrude about ⅛″ (4 mm) beyond nose when laid forward. *Tragus thin, long* (⅜″/9–11 mm), and pointed. Calcar not keeled. L 2½–3⅝″ (63–93 mm); T 1⅜–1¾″ (35–44 mm); HF ¼–⅜″ (8–10 mm); FA 1⅜–1½″ (35–38 mm); E ⅝–¾″ (16–20 mm); Wt ⅛–¼ oz (4.3–9 g).

Similar Species: Long-eared Myotis ears are blackish, extending ¼″ (7 mm) beyond nose when laid forward; fur lighter-colored. Little Brown Myotis has shorter ears and shorter, more rounded tragus. Fringed Myotis has fringe along edge of interfemoral membrane.

Breeding: 1 young born June–early July in British Columbia.

Habitat: Dense forests. Summer roosts probably in rock cavities and crevices.

Range: West Coast of British Columbia south to nw Washington.

Little is known of the biology and habits of Keen's Myotis. Thought to be solitary and to roost in tree cavities and rock crevices, it feeds along forest edges and over ponds and clearings. The only known maternity colony is under rocks heated by a natural hot spring in the Queen Charlotte Islands, off British Columbia.

34 Eastern Small-footed Myotis
Myotis leibii

Description: *Glossy fur, light tan to golden brown above;* buff to nearly white below. Black ears; black mask. Wings and interfemoral membrane dark brown. *Calcar keeled.* L 2¾–3¼″ (71–82 mm); T 1⅛–1¼″ (30–38 mm); HF ¼″ (6–8 mm); FA 1⅛–1⅜″ (30–36 mm); Wt ¼ oz (6–9 g)

Similar Species: Eastern Pipistrelle has tricolored reddish fur. Little Brown Myotis lacks or has weak keel on calcar.

Breeding: 1 young born June–early July in small nursery colonies (12–20 individuals) in buildings.

Habitat: Wooded areas. Winters in caves and mines; has been found beneath rock slabs, a sliding door, and wallpaper, and in crevices.

Range: Southern Ontario and ne U.S. from Maine south through Appalachians. Isolated populations in Oklahoma, Arkansas, Missouri, and Kentucky.

The hindfoot of the Eastern Small-footed Myotis is slightly smaller than that of most other members of the genus. It hibernates in small numbers in caves, often wedged into crevices, sometimes under rocks on the cave floor. The feeding and other habits of this species are little known.

37, 38 Little Brown Myotis
Myotis lucifugus

Description: Variable shades of glossy brown above, with tips of hairs burnished brown; buff below. *Tragus rounded and short* (about ¼″/7–8 mm). *Calcar lacks keel* or sometimes has weak keel. *Hairs on toes project beyond ends of toes.* L 3⅛–3⅝″ (79–93 mm); T 1¼–1⅝″ (31–40 mm); HF ¼–⅜″ (6–10 mm); FA 1⅜–1⅝″ (34–42 mm); E ⅝″ (14–16 mm); Wt 1/16–½ oz (3.1–14.4 g).

Similar Species: In East, Southeastern Myotis often has white belly, as well as sagittal crest on skull; Indiana Myotis has prominent keel on calcar, and hairs do not project beyond ends of toes; Keen's Myotis has longer ears and longer, thinner tragus. In West, Long-legged and California myotises have keeled calcar; Long-eared and Southwestern myotises have longer ears and longer, thinner tragus; Yuma Myotis has slightly shorter forearms and duller fur.

Breeding: Mates in fall and sometimes again in winter or spring. Sperm remains in female's reproductive tract until spring, when eggs are fertilized. 1 young born late May–early July, usually in a building, occasionally in a hollow tree.

Habitat: Areas along streams and lakes. In summer, forms nursery colonies, usually in buildings or other structures. In winter, hibernates in caves and mines in the East.

Range: Much of North America from middle Alaska south throughout most of Canada and U.S. except s California and much of se and sc U.S.

The Little Brown Myotis, whose nitrate-rich guano was sold as fertilizer in the first half of this century, is one of the most common bats in the U.S. and Canada. Nursery colonies begin forming in April or May and disperse from late July through October. They may number in the thousands (one observed maternity colony had 6,700 individuals, others have had 4,000). The first two to three days after the young are born, their mothers suckle them constantly, except while foraging. Until they are ready to fly on their own, at about four weeks, the young remain in the roost while the mother hunts for small insects, especially flies and moths. Bats usually do not carry their young in flight. However, if disturbed, the mother may take flight with the young,

carrying it crosswise, with the infant's mouth grasping one teat and its hindleg tucked under the opposite armpit. Besides echolocation clicks, this species produces warning "honks" when on a collision course with other bats during feeding or near roosts. In the fall, these bats may fly several hundred miles to a hibernating site; they often can be seen swarming at cave entrances. From September, October, or early November through March or April, they hibernate in irregular clusters, some tight, some loose. They wake an average of once every two weeks during hibernation and may fly about outdoors on warm winter nights, but without feeding. They store about $\frac{1}{16}$ ounce (2 g) of fat as winter sustenance, using nearly three-quarters of it during winter awakenings and emergence. The remainder must sustain them through the winter.

35 Northern Myotis
Myotis septentrionalis

Description: Dull brown; paler below, belly somewhat yellowish. Hairs bicolored, with base and tips contrasting. *Ears long;* when laid forward they protrude $\frac{1}{8}$" (4 mm) beyond nose. *Tragus thin and long* ($\frac{3}{8}$"/9–10 mm). Calcar not keeled. L $3\frac{1}{8}$–$3\frac{1}{2}$" (79–88 mm); T $1\frac{3}{8}$–$1\frac{3}{4}$" (36–45 mm); HF $\frac{1}{4}$–$\frac{3}{8}$" (7–9 mm); FA $1\frac{3}{8}$–$1\frac{1}{2}$" (34–39 mm); E $\frac{3}{4}$" (17–19 mm); Wt $\frac{1}{8}$–$\frac{3}{8}$ oz (5–10 g).

Similar Species: Little Brown, Southeastern, and Indiana myotises have shorter ears and shorter, more rounded tragus. Very similar Keen Myotis occurs only on nw Pacific Coast.

Breeding: Maternity colonies form April–May. 1 young born in June; flies on its own at about 4 weeks.

Habitat: Woods and wooded streams. Hibernate in caves and mines in winter; usually roosts under loose bark, shutters, and shingles, but sometimes in buildings in summer.

Range: Southern Canada from Saskatchewan east to Newfoundland, and e U.S. south to Nebraska, Arkansas, w Georgia, and Virginia.

Sometimes Northern Myotises swarm in the hundreds at cave entrances in the fall. They feed on small insects, particularly flies and small moths. This species hibernates singly or in clusters of four to six individuals, often wedged into tiny crevices in caves or mines. It may fly about outside on warm winter nights, but without feeding. A few small maternity colonies have been found in summer, including 30 bats under the bark of an elm tree and about 100 in a very large building. Males of this species are often found in caves and mines in summer.

36 Indiana Myotis
"Social Myotis"
Myotis sodalis

Description: Uniformly dark pinkish brown. Lips, nose, and forearms pinkish. Ears moderate in size; tragus relatively short and rounded. *Feet small; hairs on toes short and inconspicuous. Calcar keeled.* L 2¾–3⅝" (71–91 mm); T 1–1¾" (27–44 mm); HF ¼–⅜" (7–9 mm); FA 1⅜–1⅝" (35–41 mm); E ⅜–⅝" (10–15 mm); Wt ⅛–¼ oz (5–8 g).

Similar Species: Little Brown Myotis has unkeeled or weakly keeled calcar; hairs on toes project beyond ends of toes.

Breeding: Apparently mates in fall at cave entrances when swarming or in winter when hibernating. 1 young born in June; able to fly at 4 weeks.

Habitat: Wooded or semi-wooded areas along streams in summer. Hibernates in cold caves in winter.

Range: Midwestern U.S. from extreme ne Oklahoma, n Arkansas, and Missouri north to s Michigan, east to New England, and south to Alabama.

In October, Indiana Myotises congrega in huge numbers (as many as 125,000 per cave) in a few large caves with low temperatures and high humidity. The hibernate in tightly packed clusters of hundreds of bats, only one row deep ar so neatly aligned that the noses, lips, wrists, and ears of each bat can be seer A cluster observed in Indiana in 1991 contained an estimated 38,000 bats. Maternity colonies of up to 125 female congregate under loose tree bark. In spring, they leave their caves and spre into wooded areas over their range, where they feed on a variety of small insects, particularly moths. The U.S. government lists this species as endangered throughout its range.

30 Fringed Myotis
Myotis thysanodes

Description: Reddish brown or brown above; sligh paler below. Unique among myotises having *fringe of stiff hairs along edge of interfemoral membrane.* L 3⅛–3¾" (80–95 mm); T 1½–1⅝" (37–42 mm HF ¼–⅜" (8–11 mm); FA 1½–1¾" (39–46 mm); E ⅝–¾" (16–20 mm); Wt ⅛–¼ oz (5–8 g).

Similar Species: No other myotis has fringe of stiff hai on interfemoral membrane.

Breeding: Up to 300 females and young congreg in nursery colonies in caves and buildir 1 young born in June or early July.

Habitat: Oak, piñon, and juniper forests; dese scrub. Roosts in caves, mines, buildir and other protected locations.

Range: Throughout much of w U.S., south fr British Columbia to California and e; to Montana, Colorado, and Big Bend area of Texas; also in e Wyoming, sw South Dakota, and w Nebraska.

This colonial bat is active from April through September. It roosts by day protected spots, and may rest betwee foraging bouts in night roosts in tigl

packed clusters. Its diet includes moths, crickets, and daddy longlegs. This bat is known to migrate to a winter roost, but its winter habits are unknown. Mating is in fall, with ovulation occurring between late April and mid-May. As with many bat species, the birth of the young is synchronized within the colony. Nursery colonies, from which males are usually absent, sometimes number in the low hundreds.

39 Cave Myotis
Myotis velifer

Description: A large myotis. Pale brown in eastern part of range to black in western part. Fur dull; often has *bare patch on midback and rather pale facial skin. Calcar not keeled.* Ears reach tip of nose when extended forward. L 3½–4½" (90–115 mm); T 1⅝–1⅞" (41–49 mm); HF ⅜–½" (10–12 mm); FA 1½–1⅞" (37–47 mm); E ⅝" (15–17 mm); Wt ¼–⅜ oz (7–12 g).

Similar Species: Yuma, Eastern Small-footed, and Western Small-footed myotises are smaller. Southwestern, Long-eared, and Fringed myotises have longer ears. Long-legged and California myotises have keeled calcar.

Breeding: Mates in fall; fertilization is delayed until spring. Maternity colonies form in caves or mines or under bridges. Gestation 60–70 days. Female is upright giving birth, and the young is caught in the flight membrane as it emerges.

Habitat: Arid Southwest. In summer, roosts in caves, mines, and sometimes buildings; in winter, roosts in caves.

Range: Extreme se California east through much of Arizona, New Mexico, and w Texas, and north to sc Kansas.

The Cave Myotis, which feeds mostly on beetles and moths, sometimes forms very large colonies, commonly of 2,000 to 5,000 individuals, although Kansas nurseries are estimated to contain

15,000 to 20,000. In winter, these bats
hibernate in tight clusters in caves,
sometimes on walls or ceilings or in
crevices. Nursery colonies may be
located in hibernating or other caves or
under bridges. Except for those in
Kansas, these bats migrate between
summer and winter quarters.

40 Long-legged Myotis
Myotis volans

Description: A large myotis. Tawny or reddish to
nearly black above; grayish to pale buff
below. *Calcar has well-developed keel.*
Underarm and interfemoral membrane
furred to elbow and knee. Ears short.
L 3⅜–4″ (87–103 mm); T 1½–1⅞″
(37–49 mm); HF ¼″–⅜″ (8–10 mm);
FA 1⅜–1⅝″ (35–42 mm); E ⅜–⅝″
(10–15 mm).

Similar Species: Other large western myotises lack well-
developed keel on calcar.

Breeding: Mates in fall; fertilization delayed until
spring. Ovulates March–May; 1 young
born May–August.

Habitat: Mainly coniferous forests. In summer,
roosts in trees, crevices, or buildings.

Range: Western North America from w and s
British Columbia and Alberta south
to Mexico and east to w Dakotas,
w Nebraska, Colorado, and w Texas.

The Long-legged Myotis feeds primarily
on small moths, but also eats other small
insects. These bats form nursery colonies
of up to several hundred members,
which disperse in fall. Their winter
behavior is unknown. Bats are very
long-lived; the maximum age known for
an individual of this species is 21 years.

41, 42 Yuma Myotis
Myotis yumanensis

Description: *Short, dull fur;* variable shades of brown
above, paler below; throat sometimes

whitish. *Calcar not keeled.* L 3¼–3⅞"
(84–99 mm); T 1¼–1¾" (32–45 mm);
HF ⅜" (9–11 mm); FA 1¼–1½" (33–37
mm); E ¼–⅝" (8–16 mm); Wt ⅛–¼ oz
(5–7 g).

Similar Species: Cave Myotis is larger. Little Brown
Myotis, very difficult to distinguish,
usually has longer forearm. California,
Eastern Small-footed, Western Small-
footed, and Long-legged myotises have
keeled calcar.

Breeding: Females congregate in maternity roosts,
probably in April. 1 young born in late
May or early June in Texas.

Habitat: Areas near ponds, streams, or lakes.
Roosts under sidings or shingles by day,
often in buildings at night.

Range: Western North America: British
Columbia to California, east to Colorado
and New Mexico.

Closely associated with water, the Yuma
Myotis feeds by flying very low over the
surface. Its principal foods are midges,
moths, termites, and other small insects.
During the breeding season, the males
usually remain alone. Nursery colonies
form in places that have high, stable
temperatures in the range of 86 to
131°F (30–55°C). Usually assembled in
caves, mines, buildings, tree cavities,
rock crevices, or under bridges or the
bark of trees, these colonies may contain
thousands of individuals.

Western Red Bat
Lasiurus blossevillii

Description: Very similar to Eastern Red Bat, but
slightly larger, with *rusty-red coloration*
on back, fewer frosted hairs, and much
less hair on posterior margin of tail
membrane. Males are brighter red than
females. White patches on shoulders.
L 3⅜–4¾" (87–120 mm); T 1⅜–2⅜"
(35–60 mm); HF ¼–½" (6–13 mm);
FA 1⅜–1⅝" (34–42 mm); E ⅜–½"
(9–14 mm); Wt ¼–⅝ oz (7–19 g).

Similar Species: Eastern Red Bat is slightly smaller and occurs farther east; its tail membrane is furrier. Hoary Bat is larger, browner, more frosted. Yellow bats are yellowish, with fur only on basal half of interfemoral membrane, none on back half.

Breeding: Females bear 2 or 3 young in branches of deciduous trees.

Habitat: Trees, hedgerows, and forest edges. In summer, roosts among foliage.

Range: From se British Columbia south to Baja California, and east to Utah and Arizona

The Eastern Red and Western Red bats were long considered to be a single species, until recent findings revealed genetic and morphological differences that prompted mammalogists to consider them as separate species. The Western Red Bat is presumed to have habits similar to those of its eastern counterpart.

43, 44 Eastern Red Bat
Lasiurus borealis

Description: *Males bright red* or orange-red; *females dull red,* brick, or chestnut; *both sexes frosted white on back and breast,* with whitish patch on each shoulder. Ears small and rounded; tragus small. Interfemoral membrane furred above. L 3¾–5″ (95–126 mm); T 1¾–2⅜″ (45–62 mm); HF ⅜″ (8.5–10 mm); FA 1½–1⅝″ (37–42 mm); Wt ⅜–½ o (9.5–15 g).

Similar Species: Hoary Bat is larger, browner, more frosted. Seminole Bat is deep mahogan brown. Yellow bats are yellowish, with fur only on basal half of interfemoral membrane, none on outer half.

Breeding: Mates in fall, apparently often on the wing. 3 or 4 young born among foliag about mid-June.

Habitat: Forests, forest edges, and hedgerows. Roosts among foliage, usually in trees.

Range: Across s Canada and most of U.S. east Rockies.

By day, this solitary bat hangs 4 to 10 feet (1–3 m) above the ground among dense foliage that provides shade from above and at the sides but is open below, allowing a downward fall into flight. Emerging early in the evening, these fast fliers often use the same route each night in foraging for many kinds of insects, especially moths, beetles, plant hoppers, leafhoppers, ants, and flies. On two different occasions, radio-tracking of females for several nights in Indiana revealed that the bats would remain in the same general vicinity of 2 acres (1 ha) or so, but would roost in different trees on different nights. It is possible to identify the Eastern Red Bat, one of the few mammals in which coloration differs between male and female, on the wing before dusk. Occasionally it alights on vegetation to pluck off insects. Red bats commonly have three or four offspring at a time, and are the only bats with four nipples. Females with young usually roost 10 to 20 feet (3–6 m) off the ground. While evidence indicates that the young are not carried about during flight, females with young attached are often found on the ground; most likely they have been blown down or scared out of a tree by a predator and were unable to take off due to the weight of the young. This species migrates, often in flocks, to the southern parts of its range for the winter, where it hibernates. It sometimes emerges from hibernation on warm days, but whether it feeds at these times in the northern parts of its range has yet to be determined. It does feed in winter in the Dismal Swamp of Virginia and North Carolina.

45, 46 Hoary Bat
Lasiurus cinereus

Description: *The largest bat in its range in the East. Pale brown above, with tips of fur heavily frosted white; throat buffy yellow. Ears short and*

rounded, with black, naked rims.
Interfemoral membrane well furred above.
L 4–6″ (102–152 mm); T 1¾–2½″
(44–65 mm); HF ¼–½″ (6–14 mm);
FA 1⅝–2⅜″ (42–59 mm); Wt ⅝–1¼ c
(20–35 g).

Similar Species: Silver-haired Bat is blackish with pale
frosting. Red bats, much smaller and
more reddish, do not have yellowish
throat or black-rimmed ears.

Breeding: Mates in fall and perhaps in winter. 2
young born late May–early June amon;
foliage.

Habitat: Both deciduous and coniferous forests,
as well as desert canyons. Roosts in
foliage.

Range: Across southern half of Canada
southward throughout continental U.S
except peninsular Florida.

The most widely distributed bat in the
U.S. and Canada, the Hoary Bat emerg
late in the evening to feed, mostly on
moths. In summer, male Hoary Bats ar
found in the southwestern U.S., while
the females are spread over the rest of
the range. In fall, the females join the
males in the Southwest and both sexes
migrate to Mexico, lower California, ar
South America, where they spend the
winter. A few individuals winter in
coastal South Carolina; some may
remain in northern areas and hibernate
Presumably mating takes place in
winter or during migration. In spring,
the males move back to the Southwest
and females return to the North and
East. The only bat in the East that
surpasses the Hoary Bat in size is the
rare Jamaican Fruit-eating Bat, found
only in the Florida Keys and southwar
The endangered subspecies *L. c. semottu*
the Hawaiian Hoary Bat, is the only b
in the Hawaiian Islands, where it feed
on many types of insects rather than
primarily on moths.

Southern Yellow Bat
Lasiurus ega

Description: A large bat. *Yellowish buff. Inner half of interfemoral membrane furred above. Rather large pointed ears;* higher than wide, partially furred outside; reach end of nose when laid forward. L 4¼–5" (109–126 mm); T 1⅞–2⅛" (47–53 mm); HF ⅜" (10 mm); FA 1¾–1⅞" (45–48 mm); E ⅝" (17 mm) long; Wt ¼–¾ oz (9.2–22.5 g).

Similar Species: Nearly identical Western Yellow Bat occurs farther west in U.S. Northern Yellow Bat is larger. Red bats are more reddish. Western Pipistrelle is much smaller.

Habitat: Forests. By day, roosts among vegetation, commonly in palm trees.

Range: Extreme s Texas.

Little is known about this bat except that it emerges early from its day roost in leafy vegetation and feeds on insects. It appears to be a solitary rooster. Biologists have recently deemed it a species separate from the Western Yellow Bat, as genetic differences have been found between the two. It may be migratory in parts of its range, but it appears to be a permanent resident of southern Texas.

---

47 Northern Yellow Bat
Lasiurus intermedius

Description: A large bat. *Yellowish brown, washed with black above.* Long, silky fur. *Inner half of interfemoral membrane well furred above.* Rather large, pointed ears. Tragus broad below, tapering near tip. L average 4½" (115 mm); T 2" (51 mm); HF ⅜" (10 mm); FA 1¾–2¼" (45–56 mm); E ⅝" (15–17 mm).

Similar Species: Southern Yellow Bat is smaller.

Breeding: Mates in fall; fertilization delayed until spring. Litters of 2–4 young born late May–June.

Habitat: Semi-forested or forested areas. By day, often roosts in clumps of Spanish moss in areas of longleaf pine and turkey oak; in Rio Grande Valley, roosts in palm trees.

Range: Southeastern U.S., primarily from s and e Texas to coastal North Carolina.

This tree-loving species congregates to feed about 15 to 20 feet (4.5–6 m) above the ground. Females form loose colonies during the breeding season; otherwise, this species is a solitary rooster.

48 Seminole Bat
Lasiurus seminolus

Description: *Mahogany brown, with slight silver frosting. Interfemoral membrane furred above.* Ears short and rounded; tragus very short. L 4¼–4½" (108–114 mm); T 1¾–2" (44–52 mm); HF ¼–⅜" (8–9 mm); FA 1⅜–1¾" (35–45 mm); E average ⅜" (9 mm); Wt ¼–½ oz (7–14 g).

Similar Species: Red bats are much more reddish. Yellow bats are yellowish. Hoary Bat is much larger and heavily frosted.

Breeding: Usually 3 or 4 young, born late May–early June.

Habitat: Forested or semi-forested areas. Roosts in trees in clumps of Spanish moss.

Range: Southeastern U.S. from e Texas to e North Carolina.

Little is known of the habits of this solitary bat, which is found in the homelands of the Seminole Indians and is locally abundant wherever Spanish moss occurs. By day, it generally hangs 3½ to 5 feet (1–1.5 m) above the ground in clumps of Spanish moss that have clear areas beneath, permitting a downward fall into flight. Its foods consist of true bugs, beetles, flies, and dragonflies.

Western Yellow Bat
Lasiurus xanthinus

Description: Externally nearly *identical to Southern Yellow Bat.* L 4¼–5⅛" (109–131 mm), T 1⅝–2¼" (42–58 mm); HF ¼–⅜" (7–11 mm); FA 1⅝–1⅞" (42–47 mm); E ¼–⅝" (7–17 mm).

Similar Species: Red bats are more reddish. Western Pipistrelle is much smaller. Southern Yellow Bat is found only in extreme s Texas and southward.

Breeding: Presumably, solitary female bears 2–4 young in early summer among foliage.

Habitat: Forested and semi-forested areas. By day, roosts among vegetation.

Range: Southern California through s Arizona to extreme sw New Mexico.

As with the Eastern Red and Western Red bats, mammalogists have found the Southern Yellow and Western Yellow bats to be sufficiently different genetically and morphologically to designate them as separate species. Some mammalogists question these new designations, however, as it is unclear whether or not these species might be similar enough to interbreed and produce viable offspring if they were to occur together (in which case they should be designated as subspecies rather than separate species). Little is known of the habits of the Western Yellow Bat, but they are presumably similar to those of the Southern Yellow Bat.

49 Silver-haired Bat
Lasionycteris noctivagans

Description: A medium-size bat. *Nearly black, with silvery-tipped hairs on back,* giving frosted appearance. Interfemoral membrane lightly furred above. Short, rounded, naked ears. L 3⅝–4¼" (92–108 mm); T 1½–1¾" (37–45 mm); HF ⅜" (9–10 mm); FA 1½–1¾" (37–44 mm); Wt ¼–½ oz (9–15 g).

Similar Species: Much paler-colored overall, the solitary tree bats (genus *Lasiurus*) have whitish-tipped fur and much more fur on upper part of interfemoral membrane.

Breeding: Mates in fall; implantation delayed until spring. Usually 2 young born in early summer, probably in hollows or crevices in trees. Although this is thought to be a solitary bat, females apparently may form maternity colonies.

Habitat: Both deciduous and coniferous forests. In summer, roosts in protected spots such as under bark or in dead trees, woodpecker holes, or bird nests. In winter, hibernates in trees, crevices, buildings, and other protected places.

Range: Across southern half of Canada and southward through most of U.S. In summer, found in much of w U.S., but in East only in n U.S. and Canada.

Generally regarded as a solitary, tree-roosting species, the Silver-haired Bat is usually found roosting under a slab of bark or in some other protected spot, although there are a few records of it roosting among foliage. It emerges in early evening, flying very slowly to feed on a variety of insects, especially moths, caddis flies, and flies. Generally migrating south for the winter, it possesses a well-developed homing instinct; one bat traveled 107 miles (172 km) to its home roost. Over major parts of its range it is present only during the spring and fall migrations. In winter, it hibernates from the latitude of middle Indiana and southward. In southern Illinois, it occasionally hibernates in silica mines but rarely enters caves. There is no reliable evidence as to whether females giving birth are solitary or colonial.

50 Western Pipistrelle
Pipistrellus hesperus

Description: *The smallest bat in the U.S.* Pale yellow or grayish to reddish brown above.

Belly whitish. Wings, interfemoral membrane, ears, nose, and feet blackish. *Calcar keeled.* Tragus short and blunt. 1 tiny premolar behind canine. L 2⅜–3⅜" (60–86 mm); T 1–1¼" (27–32 mm); HF ¼" (6–7 mm); FA 1–1¼" (27–33 mm); Wt ⅛–¼ oz (5–6 g).

Similar Species: Eastern Pipistrelle has tricolored fur. California, Eastern Small-footed, and Western Small-footed myotises usually are larger, and have much longer tragus, keeled calcar, and 2 tiny premolars behind canine.

Breeding: 2 young born in June in small maternity colony (up to about 12 individuals).

Habitat: Deserts, rocky areas, and scrub. Roosts in caves, mines, and buildings.

Range: Southwestern U.S., from se Washington and e Oregon south to California, Utah, Arizona, w New Mexico, and nc and Big Bend areas of Texas.

Usually the first bat to appear in the evening, the Western Pipistrelle often flies before dark and is even seen in broad daylight. Its flight is erratic and slow, less than 6 mph (10 km/h). It skims ponds or streams, dipping its lower jaw into the water to sip while flying. In colder areas, Western Pipistrelles generally hibernate in caves, mines, and crevices, although some migrate southward. Maternity colonies and individual females have been seen on a cliff, and one colony was discovered under shutters in California. The jerky flight of the Common Pipistrelle (*P. pipistrellus*) of Great Britain and Europe inspired the name "flittermouse" (*Fledermaus* in German) for bats.

51, 52 **Eastern Pipistrelle**
Pipistrellus subflavus

Description: *The smallest bat in the East.* Reddish to light brown. *Hairs tricolored:* dark at base, pale in middle, dark at tip. Calcar not keeled. Tragus blunt. 1 tiny

premolar behind canine. L 3⅛–3½″ (81–89 mm); T 1⅜–1¾″ (36–45 mm); HF ¼–⅜″ (7–10 mm); FA 1¼–1⅜″ (31–35 mm); Wt ⅛–¼ oz (3.5–6 g).

Similar Species: Myotises lack distinctly tricolored hairs; most have longer, thinner tragus and have 2 tiny premolars behind canine. Western Pipistrelle also lacks tricolored fur.

Breeding: Maternity colonies start forming late April–early May. Gestation about 44 days; 2 young born June–early July.

Habitat: Primarily woodlands. Hibernates in caves, mines, or crevices. Summer colonies may be in buildings or hollow trees; there is at least one record in a cave.

Range: Eastern U.S. from se Minnesota, Iowa, Kansas, Oklahoma, and Texas eastward

Emerging early from its daytime hiding place in a building or hollow tree, the Eastern Pipistrelle feeds on tiny insects especially leafhoppers, plant hoppers, beetles, and flies. Maternity colonies are often found in buildings, but the great majority are probably in tree hollows. The number of bats in these colonies is very small, usually not more than 30 or 35 individuals. Each colony appears to use several alternate roosts, moving from one to the other during the season even when young are present. In fall, pipistrelles migrate to a small mine or cave to hibernate. During hibernation they are often covered with water droplets, which sparkle and give the bats a whitish appearance.

53 Big Brown Bat
Eptesicus fuscus

Description: *A large bat. Brown above,* varying from light (in deserts) to dark (in forests), usually glossy; belly paler, with hairs dark at base; wings and interfemoral membrane black. Calcar keeled. No reduced premolar behind canine. *No fu*

on wings or interfemoral membrane. Tragus short, broad, and rounded. L 4⅛–5″ (106–127 mm); T 1⅝–2″ (42–52 mm); HF ⅜″ (10–11.5 mm); FA 1⅝–2″ (42–51 mm); Wt ⅜–⅝ oz (13–18 g).

Similar Species: Evening Bat is much smaller. Myotises usually are smaller, with 2 tiny premolars behind canine, and longer, thinner tragus.

Breeding: Mates in fall, winter, or spring. 2 young born in maternity colony, usually in a building.

Sign: Dirty markings on building exteriors at entry points; droppings on floor.

Habitat: Variety of habitats, such as farmland, cities, parks, and forests. In summer, roosts in buildings and sometimes hollow trees; in winter, mainly in buildings, less frequently in caves, mines, storm sewers, and other protected places.

Range: Throughout s Canada and U.S., except s Florida and sc Texas.

In much of the northern U.S. and Canada, this is the bat most often seen in winter, and the one usually found hibernating in buildings in urban areas. Very common throughout much of their range, Big Brown Bats spread out over the available area individually or in small groups. Those that reside in buildings, usually one to five per structure, hang from rafters, hibernate in cracks or under objects, or burrow in insulation. Some individuals retreat to caves or mines, where they hang near the entrance. This species often emerges from the hibernaculum and flies outside, but—at least in northern parts of its range—it does not feed in winter; it depends for energy on fat reserves, which constitute up to one-third of its body weight. A relatively fast flier, at speeds of 40 mph (65 km/h), the Big Brown Bat eats mostly beetles but also takes wasps, ants, plant hoppers, and leafhoppers. It rarely eats flies or moths. Particularly important foods in the

Midwest are June bugs, green
stinkbugs, and cucumber beetles; all
are agricultural pests, especially the
latter, whose larva, the corn rootworm,
probably inflicts more crop damage
than any other agricultural pest in the
country. All farms benefit from the
presence of Big Brown Bats. Maternity
colonies, numbering up to about 600
individuals, are usually located in
buildings, but sometimes under a
bridge or in a hollow tree. Pups often
fall to the floor or the ground, but those
that are able to climb partially up a wall
or other structure are often retrieved by
their mothers, who find them by their
continual sharp squeaking notes.

54 Evening Bat
Nycticeius humeralis

Description: *Reddish brown* to dark brown above;
tawny below. No fur on wings or
interfemoral membrane. Calcar not
keeled. *Tragus short and curved.* 1 upper
incisor; no reduced premolar behind
canine. L 3⅛–3⅝″ (78–93 mm);
T 1⅜–1½″ (35–37 mm); HF ¼–⅜″
(7–10 mm); FA 1¼–1½″ (33–39 mm);
Wt ⅛–¼ oz (5–9 g).

Similar Species: Big Brown Bat is much larger. Myotises
usually have long narrow, straight
tragus, and 2 tiny premolars behind
canine.

Breeding: Mates in fall; fertilization delayed until
spring. 2 young born June–early July.
Young can fly in about 4 weeks.

Habitat: Woodland or mixed woodland and open
areas. In summer, roosts in buildings
and hollow trees; winter residences not
known.

Range: Southeastern U.S., from se Nebraska,
s Iowa, and s Michigan east to
s Pennsylvania (but not Allegheny
Mountains) and Atlantic states, and
south to e Texas and Gulf Coast.

The Evening Bat almost never enters caves, roosting mainly in buildings and tree hollows. Its food preferences are similar to those of the Big Brown Bat, primarily beetles and true bugs, although it eats more moths. It spends much time foraging over cornfields for cucumber beetles, and also eats many green stinkbugs, June bugs, and leafhoppers. Maternity colonies often form in buildings, but sometimes in hollow trees, and may include hundreds of individuals; they disperse by fall, but their winter whereabouts are not known. This species is becoming very scarce in the northern parts of its range and should be protected there.

55 Spotted Bat
"Death's Head Bat," "Jackass Bat"
Euderma maculatum

Description: Black above, with *3 large white spots on back* (1 on each shoulder and 1 at base of tail); white below. *Huge pink ears.* L 4¼–4½″ (107–115 mm); T 1¾–2″ (46–50 mm); HF average ½″ (12 mm); FA 1⅞–2″ (48–51 mm); E almost 2″ (50 mm).

Similar Species: No other bat has 3 large white spots on back, nor ears as long.

Breeding: Apparently 1 young born in June.

Habitat: Mountainous regions with ponderosa pines. Apparently roosts primarily in crevices in rocky cliffs and canyons.

Range: Southern California and s Nevada through Arizona, w New Mexico, and sw Colorado; scattered records as far north as Idaho and Montana.

One of the rarest North American bats, the Spotted Bat emerges late in the evening, carrying its huge ears forward during flight and giving a loud, high-pitched call. It feeds almost entirely on moths. The ears are held erect when the bat is alert or just before it takes flight,

and are curled backward when it rests.
This species appears to be relatively
solitary, but sometimes hibernates in
small clusters. It is sometimes called the
"Death's Head Bat" because of its
striking color and pattern.

56 Rafinesque's Big-eared Bat
Plecotus rafinesquii

Description: Brown, with *white-tipped fur on belly.*
Wings and interfemoral membrane
naked. *Large ears* extend to middle
of back when laid down. 2 large
glandular lumps on nose. L 3⅝–4⅛″
(92–106 mm); T 1⅝–2⅛″ (41–54 mm);
HF ¼–½″ (8–12 mm); FA 1⅝–1¾″
(40–46 mm); E 1¼–1⅜″ (32–36 mm);
Wt ¼–⅜ oz (9–12 g).

Similar Species: Townsend's Big-eared Bat has buff fur
on belly.

Breeding: Mates in fall; fertilization is delayed.
1 young born in late May–early June
in small maternity colony, usually in a
building or other man-made structure.

Habitat: Forested regions. Often roosts in
buildings, sometimes in caves and
mines.

Range: Southeastern U.S., from Kentucky and e
Texas east to North Carolina and
Florida.

When Rafinesque's Big-eared Bat is
resting, its large ears are coiled against
the side of the head, rather like a ram's
horns, which reduces the ear's surface
area, minimizing water loss. Upon
being disturbed, the bat unfolds its ears
The ability to hover like a butterfly
enables this moth specialist to pluck
insects from foliage. This species roosts
singly and hibernates in winter. It is
becoming very scarce, especially in the
northern parts of its range, and is
considered extirpated in Indiana.

57, 58 Townsend's Big-eared Bat
Plecotus townsendii

Description: Pale gray or brown above; *buff below.*
Wings and interfemoral membrane
naked. *Enormous ears* extend to middle of
body when laid back; 2 large glandular
lumps on nose. L 3½–4¼" (89–110
mm); T 1⅜–2⅛" (35–54 mm); HF ⅜–
½" (10–13 mm); FA 1½–1⅞" (39–47
mm); E 1¼–1½" (31–37 mm); Wt ¼–⅜
oz (9–12 g).

Similar Species: In East, Rafinesque's Big-eared Bat has
white underparts. In West, Allen's Big-
eared Bat has pair of leaf-like flaps
projecting from base of ears.

Breeding: Mates in fall and winter. Fertilization is
delayed; ovulation occurs before or after
female leaves hibernation. Gestation
60–100 days. 1 young born late
spring–early summer in maternity
colony in a cave, mine, or building.

Habitat: In West, scrub deserts and pine and
piñon-juniper forests. In East, generally
oak-hickory forests. Usually roosts in
caves, sometimes in buildings, especially
on West Coast.

Range: Western U.S., from Washington
southeast through Idaho, Wyoming,
Colorado, New Mexico, Oklahoma, and
Texas. Scattered populations in
Missouri, Arkansas, Kentucky, West
Virginia, and Virginia.

This bat emerges late in the evening to
feed almost entirely on moths. In
summer, females form nursery colonies
of up to about 200 bats in the West and
1,000 in the East; males are solitary.
Young are large, weighing about 25
percent of the body weight of the
mother. They can fly at two and a half
to three weeks of age. During winter,
when Townsend's Big-eared Bat
hibernates in a cave, its great ears are
folded back; if the bat is disturbed, the
ears unfold and move in circles like
antennae. Even in the coldest weather,

these bats often move from one cave to another. Two subspecies of Townsend's Big-eared Bat are classified as endangered by the U.S. government.

59 Allen's Big-eared Bat
Idionycteris phyllotis

Description: *Tawny above,* with hairs dark brown at base; underparts slightly paler. No fur on wings or membranes. Tragus ⅝" (16 mm) long. White patches behind *enormous ears; flap projects forward from base of ear.* L 4–4⅝" (103–118 mm); T 1¾–2⅛" (46–55 mm); HF ⅜" (10–11 mm); FA 1⅝–1⅞" (42–48 mm); E 1⅝" (40 mm); Wt ⅜ oz (10.4–13.2 g).

Similar Species: No other big-eared bat has similar ear flaps.

Breeding: 1 young born in maternity colony in early summer.

Habitat: Forested mountains. Roosts in caves and mines.

Range: Southeastern California west through Arizona and w New Mexico.

Like other big-eared bats, this species emerges long after dark to feed. Its ears lie back when it is resting, becoming erect if it is disturbed. Females separate from males and form maternity colonies in protected places in rocks or mines. Three colonies observed had 25, 30, and 97 individuals.

60 Pallid Bat
Antrozous pallidus

Description: A large bat. *Creamy to beige above; nearly white below. Big ears,* separated at base. Wings and interfemoral membrane essentially naked. L 4¼–5⅛" (107–130 mm); T 1⅜–1⅞" (35–49 mm); HF ⅜–⅝" (11–16 mm); FA 1⅞–2⅜" (48–60 mm); E 1–1⅜" (25–35 mm); Wt 1–1¼ oz (28–37 g).

Similar Species: All other big-eared bats are darker.

Breeding: Mates on horizontal surfaces or while
hanging upside down in fall and
perhaps in winter. Maternity colonies
form in rock crevices, in buildings, and
in other man-made structures. 1 or 2
young born usually May–June. Female
is upright while giving birth, and young
are caught in the flight membrane as
they emerge.

Sign: Droppings (guano) on hay in barns or
on floors of buildings; they somewhat
resemble mouse droppings, but are
segmented and consist chiefly of
insect fragments.

Habitat: Deserts and canyons. Daytime roosts in
buildings and crevices; less often in
caves, mines, hollow trees, and other
shelters. Nighttime roosts in buildings,
caves, mines, and cliff overhangs.

Range: South-central British Columbia and ec
Washington east to s Colorado, and
south to s California and sw Texas.

Emerging long after dusk and beating its
wings more slowly than many bats (only
10 or 11 beats per second), this stately
bat is unusual in that it often feeds on
the ground; several have been caught in
mousetraps. Its food consists of many
large insects, including flightless beetles,
crickets, scorpions, and grasshoppers; in
captivity, it has captured and consumed
lizards. Its several calls include an insect-
like buzz; high-pitched dry, rasping,
thin double notes; single, clear, resonant
high-pitched notes; and clicks. The first
two generally are given when the colony
is disturbed, the latter two at night in
flight. The summer colonies, consisting
of from 30 to 100 bats, are unusual in
that they include members of both sexes
and young. Night roosts, often near day
roosts but distinct from them, are
commonly used by this species. A
skunk-like odor given off by glands on
the muzzle is most pronounced when
the bat is disturbed. Pallid Bats are
presumed to hibernate, but few have
been found in winter.

FREE-TAILED BATS
Family Molossidae

Some 80 species of free-tailed bats are distributed throughout the warmer parts of the world. There are four genera and seven species in North America north of Mexico, including the Little Mastiff Bat recently found in the Florida Keys. Free-tailed bats have naked tails extending well beyond the edge of the interfemoral membrane; usually at least half the tail is free. The feet have long, stiff hairs. Their wings are narrow, and their flight swift and straight. As their faces somewhat resemble those of mastiff-type dogs, these snub-nosed, thickset species are also known as mastiff bats. They groom themselves with spoon-shaped bristles on the hindtoe of each broad foot.

62 Brazilian Free-tailed Bat
Tadarida brasiliensis

Description: *The smallest free-tailed bat.* Dark brown or dark gray above, with hairs whitish at base. *Ears separated at base.* Calcar pointed backward. Tail extends beyond interfemoral membrane for more than half its length. L 3½–4¼″ (90–110 mm); T 1¼–1¾″ (33–44 mm); HF ¼–½″ (7–14 mm); FA 1⅜–1¾″ (36–46 mm); Wt ⅜–½ oz (11–14 g).

Similar Species: Except for Little Free-tailed, which occurs only in Florida Keys and has ears joined at the base, all other free-tailed bats are considerably larger.

Breeding: Mates February–March; ovulates in March. Females form very large maternity colonies usually in caves or man-made structures. 1–3 (usually 1) young born in June. Female hangs head downward during birth, but flight membrane is not used to receive young.

Habitat: Deserts, canyons, farmlands, and other habitats. Roosts in buildings on West Coast and in Southeast, and in caves from Texas to Arizona.

Range: Throughout s U.S.; in West, south from
s Oregon and s Nebraska; in East, south
from n Louisiana, Alabama, and South
Carolina. A few are scattered farther
north.

The Brazilian Free-tailed Bat is by far
the most common bat in the Southwest;
with a U.S. population previously
estimated at a minimum of 100 million,
it is also one of the most numerous
mammals in the country. In the East and
on the West Coast, it hibernates in
winter rather than migrate. From Texas
through the Southwest, it lives in huge
colonies in caves, packed 250 per square
foot (2,700 per sq m); a few of the
Southwestern bats hibernate, but most
migrate to Mexico for the winter,
usually toward the end of October,
returning northward in March to mate.
The young hang, sometimes among
millions of others, in a nursery, yet pups
and mothers are capable of finding one
another by their calls and probably odor.
Mothers make no attempt to save young
that lose their grip on the ceiling,
however; such pups perish on the cave
floor, where they are consumed by
tenebrionid beetles. The Carlsbad
Caverns in New Mexico were discovered
when these bats were seen emerging
from them. Although the cavern
population has declined from an
estimated 8 to 9 million in the 1930s to
several hundred thousand, the bats'
daily emergence is still a major tourist
attraction. At sunset, bats begin flitting
about inside the cave, causing a slight
rise in temperature and humidity. After
circling for several minutes, they begin
to emerge from the depths of the caverns
in a counterclockwise spiral, ascending
150 to 180 feet (45–55 m) into the
night air. They emerge in various ways:
as one continuous wave; split into two
groups with an interval of half an hour
in between; or in bursts of several
hundred to several thousand bats that

give way after 15 to 20 minutes to a continuous stream. They make a great roar and form a dark cloud visible miles away; when the egression is at its heaviest, 5,000 to 10,000 bats emerge each minute. While they may roam up to 150 miles (240 km), most Carlsbad bats feed within a 50-mile (80-km) radius. Generally they fly throughout the night, at 10 to 15 mph (15–25 km/h), feeding on a variety of small insects, especially moths, ants, beetles, and leafhoppers captured in the tail membrane. Each night, a bat eats up to one-third its own weight; 250,000 bats can consume half a ton of insects. The return to the caves, at sunrise, is even more spectacular than the emergence, as the bats plummet straight down from heights of 600 to 1,000 feet (180–300 m) at speeds of more than 25 mph (40 km/h) to a reported maximum speed of 60 mph (96 km/h). Bat droppings in Carlsbad Caverns over the past 17,000 years have formed guano deposits covering several thousand square feet to a depth of almost 50 feet (15 m). Guano was used during the Civil War as a source of sodium nitrate for gunpowder and mined as fertilizer from the turn of the century through the 1940s. A few small-scale guano mines are still in operation. These bats may have a life span of up to 18 years. Hawks and owls sometimes sit at cave entrances and prey on them as they emerge. Black snakes, Common Raccoons, house cats, and other predators sometimes manage to gain access to their roosts. A hazard of entering free-tailed bat caves in the Southwest is the possibility of contracting rabies, which can be transmitted by a bat bite or by the airborne virus. People have also contracted histoplasmosis, a fungal infection of the respiratory tract, from bat caves of the Southwest.

Pocketed Free-tailed Bat
Nyctinomops femorosaccus

Description: Dark gray or brown above and below; *lower half of hairs nearly white.* Wings long and narrow. Tail free about half its length. *Ears joined at base.* L 3⅞–4⅝" (98–118 mm); T 1⅛–1⅝" (30–42 mm); HF ⅜–½" (10–12 mm); FA 1¾–2" (44–51 mm); Wt ⅛–⅝ oz (5–18 g).

Similar Species: Brazilian Free-tailed Bat's ears are not joined. Little Mastiff Bat has shorter forearm. Other free-tailed bats have larger forearm.

Breeding: Mates just before ovulation in spring. 1 young born late June–July. Lactation may continue until September.

Habitat: Deserts. Roosts in rock outcrops.

Range: Southern California, s Arizona, se New Mexico, and Big Bend area of Texas.

By day, the Pocketed Free-tailed Bat roosts in rock crevices or other shelters in rocky areas, in small colonies usually composed of fewer than 100 members. At night, as it drops from its perch into the air it gives a loud, high-pitched call; it also calls frequently while in its day roosts. This bat feeds primarily on moths, but eats ants, wasps, leafhoppers, and other insects as well. The femoral "pocket" that gives it its common and species names is inconspicuous.

---

61 Big Free-tailed Bat
Nyctinomops macrotis

Description: A large bat. Fur reddish brown, dark brown, or black; *hairs white at base. Tail extends 1" (25 mm) or more* beyond interfemoral membrane. Ears joined at base; extend beyond tip of nose when laid forward. L 5⅛–5⅝" (129–144 mm); T 1⅝–2" (43–50 mm); HF ¼–⅜" (7–11 mm); FA 2¼–2½" (58–64 mm); Wt average ¾ oz (20.6 g).

Similar Species:	Underwood's Mastiff and Western Mastiff bats are larger. Other free-tailed bats are smaller.
Breeding:	1 young born in late spring or early summer.
Habitat:	Rocky areas. Roosts by day in rocky cliffs (in crevices or under rock slabs); sometimes caves, buildings, or tree holes.
Range:	Locally abundant in parts of California and Nevada east to Kansas and Texas.

This colonial bat emerges late in the evening to feed primarily on moths, though it also will eat crickets, grasshoppers, ants, and other insects, sometimes chattering loudly when feeding. It is widespread but usually uncommon; there have been scattered autumn records in British Columbia and Iowa, far from its normal range.

Wagner's Mastiff Bat
Eumops glaucinus

Description:	A large bat. *Gray or nearly black;* slightly paler below. Free tail extends far beyond interfemoral membrane. Ears joined at base. Wings long and narrow. L 3⅛" (80 mm); FA 2¼–2⅝" (57–66 mm).
Similar Species:	Brazilian Free-tailed Bat and Little Mastiff Bat, only other free-tailed bats in East, are much smaller (forearm less than 2"/50 mm long); ears not joined at base in Brazilian Free-tailed Bat.
Breeding:	Presumably 1 young born June–July.
Habitat:	Little known; in Florida, found under roofing tiles.
Range:	Extreme se Florida in the Miami, Coral Gables, and Fort Lauderdale region.

Although considered rare, Wagner's Mastiff Bat flies about over the city of Miami, making a high, piercing noise that can be heard even over downtown traffic. Of specimens taken in Florida, many were found under "Cuban tile," often used as roofing material in Coral Gables, Florida.

63 Western Mastiff Bat
"Bonnet Bat"
Eumops perotis

Description: *The largest bat in North America.* Body sparsely furred; fur dark brown; hairs white at base. *Enormous ears,* joined at base, protrude over forehead. L 5½–7¼" (140–185 mm); T 1⅜–3⅛" (35–80 mm); HF average ⅝" (17 mm); FA 2⅞–3¼" (72–82 mm); E 1–1⅝" (25–40 mm); Wt average 2¼ oz (65 g).

Similar Species: Underwood's Mastiff Bat is smaller, with smaller ears.

Breeding: Mates in early spring. Gestation is estimated at 80–90 days. 1 young born usually June–July.

Sign: Crevices used for day roosts often marked by yellowish urine stains on rocks below and by large-size droppings.

Habitat: Rocky cliff and canyon areas. Roosts in crevices, also in buildings.

Range: Southern California, extreme s Nevada, s Arizona, extreme sw New Mexico, and Big Bend area of Texas.

By day, these bats form small colonies, usually with fewer than 100 members. Because of their large size and long wings, they require considerable space to launch themselves into flight, so roosting sites are usually situated to permit a free downward fall for at least 10 feet (3 m). They roost in crevices, choosing long vertical slits at least 2 inches (50 mm) wide, from which they climb rapidly to a narrower spot and wedge themselves in. Roosts are sometimes alternated throughout the year. As night approaches, loud squeaks may be heard near roost entrances. During flight, cries are frequent and can be heard more than 1,000 feet (300 m) away. Foraging high and at great distances from roosting sites, the bats feed primarily on moths, which constitute 80 percent of their diet, but also eat ground-living crickets and long-horned grasshoppers. The Western

Mastiff often crawls on all fours with its tail sticking straight up and perhaps serving as a tactile organ. The way its large ears cover its head has earned this creature the nickname "Bonnet Bat."

64   Underwood's Mastiff Bat
Eumops underwoodi

Description: A large, free-tailed bat. Dark brown above; grayish below. Body hairs paler at base. *Large ears* are joined at base and project over forehead. L 6⅜–6⅝″ (160–167 mm); T 2–2⅜″ (52–60 mm); HF ⅝–¾″ (15–20 mm); FA 2⅝–2¾″ (67–70 mm); E 1⅛–1¼″ (29–33 mm); Wt 1⅞–2⅛ oz (53–61 g).

Similar Species: Brazilian Free-tailed and Big Free-tailed bats are much smaller. Western Mastiff Bat is larger, with bigger ears.

Breeding: Females produce 1 young in July.

Habitat: Deserts.

Range: Pima County, Arizona.

This little-known bat makes frequent, ear-piercing calls during flight. Its range extends southward to northern Central America. Most of the specimens from North America north of Mexico have been taken from a watering tank near the Mexican border. The population there apparently lives in a deep cave.

Little Mastiff Bat
Molossus molossus

Description: *A small, free-tailed bat with bases of ears joined on forehead.* Color varies from reddish brown, dark chestnut brown, o dark brown to rusty blackish or black. 2 coat phases, one with long, bicolored fur, the other with short, velvety, unicolored fur. 1 upper premolar per side. Averages: L 4″ (101 mm); T 1⅜″ (36 mm); HF ⅜″ (10 mm); FA 1½″ (39 mm); Wt ½ oz (14 g).

Similar Species:	Brazilian Free-tailed Bat has large, broad ears separated at base and 2 upper premolars per side.
Sign:	Odor and squeaking in colony; guano deposits.
Habitat:	Semi-wooded areas. Individuals have been known to roost in palm fronds; nursery colonies found in buildings, hollow trees, and caves. Florida colonies are found in building roofs.
Range:	Throughout Central America, Caribbean, and much of South America. Isolated colonies in Boca Chica, Stock Island, and probably Marathon in Florida Keys.

As of this writing there has been no published record of the Little Mastiff Bat in North America north of Mexico, but we have been given permission by mammalogists Phillip Frank and James D. Lazell to include here information they have gathered about Florida colonies. At least two, and probably three, colonies of these bats have been found in the Florida Keys, all in the roof spaces of flat-topped buildings. One individual from each of the Boca Chica and Stock Island colonies has been identified by Karl Koopman of the American Museum of Natural History. An additional bat colony in Marathon is apparently also of this species. In the Boca Chica colony, located in the roof of a dormitory, 268 bats were counted on September 7, 1994. Bats of the Stock Island colony, in the roof of an apartment building in a densely populated area, were plainly visible leaving the roost at dusk on February 14, 1994, but an accurate count was not possible. Seventy bats were counted in the Marathon colony, found in the roof of a condominium in Key Colony Beach, adjacent to the ocean, on September 8, 1994.

PRIMATES
Order Primates

Primates comprise a variety of forms, such as lemurs, lorises, and galagos (bush babies), through monkeys, apes, gorillas, and humans. Some of the major evolutionary trends in this order are toward increased brainpower, awareness of the environment, use of the eyes, use of the hands (including an opposable thumb on the front feet and sometimes on the hindfeet), and upright position of the body. These characteristics have allowed primates to move about in trees and exploit the arboreal herbage, and to develop manual dexterity, stereoscopic vision, and complex social behavior and communication. Most primates are arboreal, but some—such as baboons and chimpanzees—have become secondarily terrestrial; humans alone are fully terrestrial. Although some primates do move about on two legs, only humans are fully upright and bipedal. Primates are basically herbivorous, but some are omnivorous. The teeth are usually relatively simple and flat-topped. Other than *Homo sapiens,* the Rhesus Monkey is the only primate established in North America.

OLD WORLD MONKEYS
Family Cercopithecidae

Members of this family, which includes more than 90 species worldwide, have nongrasping tails (the New World monkeys, all found south of our range, have grasping, or prehensile, tails). Old World monkeys generally are well covered with fur but have bare faces. They have good vision, hearing, and sense of smell, and use facial expressions and many calls. Mostly diurnal, these monkeys have a high awareness of their environment.

Rhesus Monkey
Macaca mulatta

Description: A large monkey; *heavyset, with strong limbs.* Fur somewhat long, grayish brown to rufous yellow above; paler below. Tail about half length of head and body. L 16–22″ (40–55 cm); T 7–10″ (17.8–25.4 cm).

Breeding: 1 young born every other year in Asia; as often as every 8 months in Florida. Birth weight about 1 lb (450 g).

Habitat: In their native Asia, mountain forest; in Florida, floodplain dominated by bald cypress, pumpkin ash, red maple, Florida elm, cabbage palm, and black gum.

Range: Native from India and Afghanistan to China and Vietnam. 2 populations in North America, both in Florida: at the headwaters of the Silver River, near Silver Springs, and in the Ocala National Forest.

Rhesus Monkeys were introduced on an island in the Silver River in Florida in the late 1930s, probably as a tourist attraction. The animals lived along the route of a "jungle cruise" boat in the woods along the river, and were fed by tourists. One of the populations residing in Florida today lives at the endpoint of the old cruise. Most primates live in tropical areas, but the Rhesus Monkey

may live in areas with freezing temperatures in winter. It sleeps in trees, but spends much of the day foraging for herbaceous plants on the ground. In hot weather, it may begin searching for food well before dawn, and rest at intervals throughout the day. Although primarily vegetarians, Rhesus Monkeys will eat some invertebrates and may scavenge vertebrate prey. They feed on many plant foods in sequence as they appear; the tip of cabbage palm (*Sabal palmetto*) frond is one of the most important foods in Florida. In winter, they seek out roots and tubers as available. Rhesus Monkeys live in troops of 11 to 70 animals, half of them adults, with two to four times as many adult females as males. The troop's hierarchy is one of matriarchal dominance: The dominant female's daughters dominate all other adult and juvenile females. The youngest daughter of the dominant female is the number-two individual, and ultimately becomes the dominant female. Sometimes troops divide. Males may disperse and form their own hierarchies, which are much less stable. In India, reproduction may be seasonal, with females bearing young at most every other year. In Florida, however, the animals mature more quickly and may give birth as often as every eight months; 82 percent of the females had young over a three-year period. Females first produce young at age four or five. The baby rides on the female's belly, and later sometimes rides on her back as well. The Rhesus Monkey has a number of calls, including shrill alarm barks and screeches in response to aggressive behavior, a scream when attacked, an aggressive growl, and a squawk when surprised. The life span of a Rhesus Monkey in Florida is about 28 years; in 1978, there were about 76 monkeys present. The Rhesus Monkey has been used extensively for medical research and in space flight.

XENARTHRANS
Order Xenarthra
Family Dasypodidae

Members of this order are distinguished
from other living mammals by their
unique vertebral joints, with extra
articulations between some vertebrae—
hence the name Xenarthra, meaning
"extra articulation" or "strange joint."
Sometimes also referred to as the
Edentata (Latin for "without teeth"),
this order includes armadillos, sloths,
and anteaters. The misleading name
Edentata originated in the 18th
century, when members of the order
were classified with such truly toothless
creatures as the pangolins and the
Duck-billed Platypus. In the order
Xenarthrans, only the anteaters are
toothless, and the Giant Armadillo
(Megatherium), with up to 100 teeth,
is among the toothiest of mammals.

Members of this diverse group are found
only in the New World; they developed
in South America when that continent
was separated from the rest of the New
World. While today there are four
families, with a total of 29 species,
throughout the Americas, the only
family that occurs north of Mexico,
Dasypodidae, the armadillo family, is
represented by a single species.

232, 233 Nine-banded Armadillo
Dasypus novemcinctus

Description: The only North American mammal
armored with *heavy, bony plates.* Scaly-
looking plates cover head, body, and tail.
Body has wide front and back plates;
midsection has 9 (sometimes fewer)
narrow, jointed armor bands that permit
body to curl. Head small. Underparts
and upright ears soft. Sparsely haired
body is brown, tan, or sometimes
yellowish; depending on where it
burrows, may be stained dark, even
black, by earth or mud. Teeth are
simple pegs. L 24–32″ (61.5–80 cm);
T 9⅝–14½″ (24.5–37 cm); HF 3–3⅞″
(7.5–10 cm); Wt 8–17 lb (3.6–7.7 kg).

Breeding: Mates in summer; after delay of 14
weeks, embryo is implanted in uterine
wall in November. Single egg divides
into 4 identical quadruplets, born in
March, each weighing 3 oz (85 g).
1 litter per year.

Sign: Burrows, with entrance holes about
6–8″ (150–200 mm) across, often along
creek banks.
Scat: Looks like clay marbles and
consists chiefly of clay, for an armadillo
consumes much soil as it feeds on
insects.

Tracks: Often bird-like in appearance,
with prominent marks made by each
hindfoot. Foreprint about 1¾″ (45 mm)
long, 1⅝″ (40 mm) wide; hindprint
more than 2″ (50 mm) long, 1⅝″ (40
mm) wide. In sand or dust, tracks are
blurred and appear almost hoof-like. In
soft earth or mud, occasionally all toes
show more or less clearly. Forefoot has
toes: long middle ones closely spaced,
much shorter outer 2 spread wide.
Hindfoot has 5 toes more evenly spread;
middle 3 long, outer 2 short, with no
separation from heelpad. Trail
sometimes shows only occasional
footprints, as some are obliterated by
the drag marks of the armor shell.

Habitat: Often determined by quality of soil for burrowing: favors areas with soft soil and rotting wood, and abundant in sandy soils; less common in clay, where digging is more difficult. Locally abundant in areas with shallow soils and rocky substrates (limestone) in the Edwards Plateau in c Texas.

Range: Texas, Oklahoma, and se Kansas southeastward to s Georgia and most of Florida.

The Spanish conquistadores first encountered this strange creature and named it "little man in armor." Around the turn of the century, the Nine-banded Armadillo occurred in the U.S. only in semi-arid areas of southern Texas. It was introduced into Arkansas and Florida, and has expanded its range throughout much of the Southeast. It spends most of its active hours digging for food and building burrows, snuffling almost constantly. For such a clumsy-looking animal, the armadillo is surprisingly swift. It can swim short distances, gulping air to inflate its intestines for increased buoyancy, and can cross small streams or ponds by walking underwater on the bottom. The armadillo is primarily nocturnal during hot weather and diurnal during cooler weather. It does not hibernate and cannot survive prolonged below-freezing weather. The Nine-banded Armadillo goes about its business with a steady, stiff-legged jog. When approached, it escapes by running away or curling its body to protect its vulnerable belly; it can also burrow underground with amazing speed. When startled it may jump vertically and erratically, then run. This animal produces several grunts, and other low-volume sounds that appear to derive from breathing or sniffing. It will sometimes stand nearly upright, supported mainly by its tail, to sniff the air for danger or food. It roots

and searches rotting logs for insects and snuffs about in vegetation for ants and invertebrates. It also eats crayfish, amphibians, reptile and bird eggs, and carrion. Armadillos sleep in nests place underground, in a crevice, or on the ground. Underground burrows may be up to 3 feet (1 m) in depth, with one or more entrance tunnels. Dens of several adults may be clumped together, and one adult may have more than one den Armadillos sometimes construct very short burrows without nests, probably escape routes. Breeding burrows contai a nest of leaves or grass. Armadillo embryos are often used in experiments requiring identical animals. Well formed at birth and with eyes open, newborns can walk about within hours their skin is soft during infancy and slowly hardens. The young resemble miniature piglets as they trail after the mother. The Nine-banded Armadillo i hunted for its meat and its decorative shell. Armadillos can carry a form of Hansen's disease (leprosy) and have been used extensively in research on the disease.

PIKAS, RABBITS, AND HARES
Order Lagomorpha

This order contains two families and 80 species distributed over most of the world, including the Australian region, where they have been introduced. In the U.S. and Canada, the order is represented by two families: the pikas (Ochotonidae), which have small rounded ears and nonjumping hindlegs that are only slightly longer than the forelegs, and the rabbits and hares (Leporidae), which have long ears and long jumping hindlegs.

Lagomorphs, or "hare-shaped" animals (from the Greek *lagos,* "hare," and *morpha,* "form"), possess two pairs of upper incisors. The first pair is enlarged and chisel-like; the second is small and directly behind the first, and lacks cutting edges. As in some other animals, the incisors continue to grow throughout life but are constantly worn down by use. Lagomorphs are also characterized by the location of the scrotum in front of the penis rather than behind it; except for marsupials, no other mammals have this genital structure.

Strict vegetarians, lagomorphs make the most of their meals by eating their food twice, in a process called reingestion or coprophagy. Vulnerable to predators, they quickly fill the stomach and then hurry to a hiding place, where they defecate pellets of soft, green, partially digested material, which they then eat and digest normally to obtain maximum nutrition. Coprophagy may have evolved as a protective mechanism or it may be an adaptation to compensate for their relatively inefficient, nonruminant stomachs.

Lagomorphs and rodents, with their similar gnawing teeth, were once lumped in one order, called Glires. This

relationship was later disputed, as the groups' similarities were thought to result from convergent evolution rather than inheritance from some distant common ancestor. Lately, however, studies based on other data (skeletal structure, formation of the placenta, etc.) suggest to some mammalogists that these two orders (Rodentia and Lagomorpha) are indeed each other's nearest relatives; these scientists group the two orders together in the superorder Glires.

PIKAS
Family Ochotonidae

The family Ochotonidae contains 26 species worldwide, most of which occur in Eurasia. Two species are found in western North America. Pikas, similar in shape to guinea pigs, have prominent, evenly rounded ears and no visible tail. The soles of their feet are furred. They are active by day and, unlike most members of the order Lagomorpha, are highly vocal and social. They live in large colonies and constantly communicate with their fellows in shrill nasal bleats and barking chatters. The name "pika" comes from the language of the Tunga people of northeastern Siberia; originally pronounced "peeka," it has been Americanized to "pie-ka."

138 Collared Pika
Ochotona collaris

Description: Similar to American Pika. Brown above, gray on sides, with *pale gray collar* on neck and shoulders; belly pale. Ears dark, edged with buff. No visible tail. L 7–7¾" (178–198 mm); HF 1⅛–1¼" (29–31 mm).

Similar Species: American Pika is brownish, lacks collar.

Breeding: Mates in spring; 1 litter per year; 2–6 young.

Sign: Grass piles among rocks.

Habitat: Talus slopes.

Range: Southeastern Alaska to sw Northwest Territories (Mackenzie district).

Like the American Pika, which occurs farther south, this animal piles summer plant cuttings into haystacks to serve as its winter food. Presumably its other habits are similar to those of the American Pika.

139, 140 American Pika
"Cony," "Whistling Hare,"
"Piping Hare"
Ochotona princeps

Description:
Brownish. *Small, rounded ears. No visible tail.* L 6⅜–8½" (162–216 mm); HF 1⅛–1⅜" (30–36 mm); Wt 3¾–4½ oz (108–128 g).

Similar Species:
Collared Pika has pale gray collar.

Breeding:
Mates in early spring. 2–6 young born May–June; a second litter may be produced in late summer.

Sign:
Grass piles among rocks.
Scat: Small, sticky, round, black pellets.
Tracks: Seldom visible except in patches of snow or mud. Complete cluster, showing all 4 feet, is no more than 3" (75 mm) wide, and less than 4" (100 mm) from hindprints to foreprints. Frontfoot has 5 toes, but reduced fifth toe often does not print. Hindfoot has 4 toes; is slightly larger than forefoot, but not greatly elongated. Track likely to be confused with that of a rodent rather than that of a rabbit.

Habitat:
Talus slides; rocky banks; steep, boulder-covered hillsides; usually at elevations of 8,000–13,500' (2,500–4,100 m).

Range:
Western North America from c British Columbia and w Alberta south to c California and n New Mexico.

The American Pika feeds on many species of green plants, eating some on the spot and, in late summer, when foraging may continue into evening, scurrying away with cuttings to boulders near its home. It spreads them to dry in the sun, curing its "hay" as a farmer does; haystacks are not high but may contain as much as a bushel of vegetation, primarily grasses and sedges and also including fireweed, stonecrop, sweetgrass, and thistles. Even when large, piles are moved frequently for better drying or to shelter them from rain. Later the dried vegetation is stored

in the pika's den deep among the rocks. In winter, the American Pika does not hibernate; kept warm by its long, thick fur, it remains active, feeding on stored hay and lichens. Like all pikas, this species is highly vocal. The naturalist Thomas Nuttall, who described the call as "a slender, but very distinct bleat, so like that of a young kid or goat," was astonished when "the mountains brought forth nothing much larger than a mouse." The animal characteristically jerks its body upward and forward with each call, which perhaps explains why calls tend to be ventriloquial, sometimes seeming to come from far off when, in fact, they echo from sources almost underfoot. An observer of pikas once noted a fascinating sequence of events, when a weasel, attempting to capture an American Pika, was chasing it among the rocks. When the pika began to tire, another pika emerged and ran between the weasel and the first pika. The weasel then pursued the newcomer until the larger animal tired and withdrew to find easier prey. While it is not known if such behavior is widespread, it is easy to see how it could have evolved as a defensive response beneficial to the entire community.

RABBITS AND HARES
Family Leporidae

The family Leporidae, with 54 species worldwide, is represented in the U.S. and Canada by 18 species. Rabbits and hares are medium-size grazing mammals with long ears, long hindlegs and bulging eyes on the sides of the head, which enable them to watch for danger over a wide arc. Does (females) are larger than bucks (males), which is unusual in mammals. Hares (*Lepus* species) are generally larger than rabbits and have longer ears and longer, more powerful hindlegs; they usually live in a more open habitat than rabbits (the Snowshoe Hare is an exception) and attempt to outrun predators. Rabbits, less proficient runners, often elude enemies by hiding in dense cover.

Primarily nocturnal, leporids can detect enemies by scent. They thump the ground with a hindleg when alarmed and, in turn, sense the ground vibration caused by the thumps of other rabbits or hares nearby. They freeze when threatened and can instantly switch direction when running. Although rabbits and hares are generally silent, they give a piercing distress call (the hare's is louder and deeper than the rabbit's). The young squeal almost inaudibly, and does may grunt or purr while nursing. Most rest in "forms," or shallow depressions, in the ground. Their astonishing reproductive powers help compensate for their vulnerability to predation.

Hares make no maternity nests—their young are born well developed, fully furred, and with eyes open. The young leave the birthing area usually within a day after birth, and the female remains far from them. The young congregate, and the female returns to feed them once per day, allowing them to suckle

for about five minutes apiece. They are weaned at three or four weeks. Cottontail rabbits (*Sylvilagus* species) do make nests, because their young, born naked and with eyes closed, need an extended period of maternal care. About a week before giving birth, the pregnant cottontail finds a suitable spot, usually where tangled brush or high grass provides a screen, and makes a saucer-like depression in the ground, about 3 to 4 inches (75–100 mm) deep and 8 inches (200 mm) across. She then pulls off downy fur from her breast and belly, mixes it with soft, dead grasses and leaves, and uses this to line the nest. She makes a second layer to cover the young while she is away foraging. Not only does this soft coverlet help keep the young dry and conceal the nest, but by removing her fur, the mother exposes her nipples, facilitating nursing.

The signs, including tracks, of all members of this family are similar, although they differ in size, and are presented in detail in the account for the most common species, the Eastern Cottontail. Tracks of hares are generally much larger than those of cottontails.

Pygmy Rabbit
Brachylagus idahoensis

Description: *The smallest rabbit in North America.* Buffy grayish or blackish above. *Tail gray above and below. Whitish spots at sides of nostrils.* L 9⅞–11⅜″ (250–290 mm); T ¾–1⅛″ (20–30 mm); HF 2¼–2⅞″ (58–72 mm); E 1⅜–1⅞″ (36–48 mm); Wt 8¾–16 oz (246–458 g).

Similar Species: Cottontails are larger, with tail white below.

Breeding: Mates in spring and summer. Gestation about 27–30 days. 4–8 young per litter.

Sign: Burrows have 3 or more entrances with 3″ (75 mm) openings. Scattered quantities

of tiny, round fecal pellets slightly more than ¼″ (8 mm) in diameter.

Habitat: Associated with sagebrush throughout its range, especially tall stands of *Artemisia tridentata.*

Range: Southeastern Oregon, s Idaho, and extreme sw Montana south to nw Utah, n Nevada, and small adjacent sections of California. Isolated population in se Washington.

The Pygmy Rabbit differs from all other native North American rabbits in digging its own burrow system. (Others either use the burrows of other animals or don't use burrows at all.) Its primary food is sagebrush, which constitutes up to 99 percent of its diet in winter. Grasses become important in summer, forming as much as 30 to 40 percent of the diet. Unlike other rabbits, it scampers rather than leaps.

200 Swamp Rabbit
"Cane-cutter Rabbit"
Sylvilagus aquaticus

Description: *The largest cottontail.* Short, coarse fur; *brownish gray* mottled with black above, whitish below. Thin tail white below. Feet rust-colored. L 17¾–22″ (452–552 mm); T 2⅝–2¾″ (67–71 mm); HF 4¼″ (110 mm); E 2⅜–3⅛″ (60–80 mm); Wt 3½–6 lb (1.6–2.7 kg).

Similar Species: Marsh Rabbits are much smaller, with tail brown below. Cottontails have whitish feet. Eastern Cottontail is smaller and has rufous nape.

Breeding: 2 litters per year of 1–6 young (most often 3); timing varies with latitude. Gestation about 38 days.

Sign: Fecal pellets on logs and stumps, especially in winter.

Habitat: Bottomlands, swamps, and canebrakes; needs high ground during flooding.

Range: Eastern Texas and e Oklahoma northeast to s Indiana and east to n Georgia and w South Carolina.

All rabbits will take to water when pursued, but this excellent swimmer will swim simply to get about. To elude predators, it may remain submerged except for its nose. It feeds on green and young woody plants, both terrestrial and aquatic, including cane (*Arundinaria*; a favorite), sedges, grasses, tree seedlings, and greenbrier, as well as corn and other field crops. It rests in a form under thick brush and will hide in hollow logs or in the burrows of other animals. The young are born in a shallow, fur-lined form. Unlike other cottontails, which are born after only a month's gestation naked and with eyes closed, Swamp Rabbits are born after a gestation of almost six weeks. Their bodies are furred, and their eyes open almost immediately. The most important predators of this species are humans, dogs, and alligators.

206, 207 Desert Cottontail
Sylvilagus audubonii

Description: Buff-brown above; white below. *Nape bright rust. Moderately long ears,* sparsely furred inside. L 13¾–16½" (350–420 mm); T 1¾–2⅞" (46–74 mm); HF 3–3⅞" (75–100 mm); E 2⅛–2¾" (55–70 mm); Wt 1¾–2¾ lb (835–1,191 g).

Similar Species: Mountain Cottontail usually has slightly shorter ears. Brush Rabbit is smaller, with shorter tail and hindlegs. Pygmy Rabbit is much smaller, with tail gray above and below. Eastern Cottontail is often larger, and has proportionally shorter ears.

Breeding: At least 2 litters per year, born year-round; 1–6 young (average 3) per litter.

Sign: Fecal pellets deposited on logs and stumps, especially in winter.

Habitat: Grasslands to creosote brush and deserts.

Range: California to Texas, north to e Montana and sw North Dakota.

Unlike most cottontails, this species seldom uses forms. In areas of sparse vegetation, it occasionally rests in the burrows of other animals, such as prairie dogs or skunks. It readily climbs sloping trees, and is thought to use logs and stumps as lookout posts after dark. The home range of the Desert Cottontail extends up to 15 acres (37 ha) for males and 9 acres (22 ha) for females. Its running speed has been timed at 15 mph (25 km/h). Grasses, mesquite, other green plants, cacti, bark, and twigs are its chief foods. The young of this species are born in a fur-lined depression.

201 Brush Rabbit
Sylvilagus bachmani

Description: *A small rabbit.* Reddish brown mottled with black in summer; paler, but still mottled, in winter. *Short legs. Small tail. Short, dark ears.* L 11–14¾″ (280–375 mm); T ¾–1⅝″ (20–43 mm); HF 2⅝–3⅜″ (67–85 mm); E 2–2½″ (50–64 mm). Wt 1–2 lb (450–965 g).

Similar Species: Desert Cottontail, the only cottontail in the same range (in California), is larger and has longer tail and hindlegs.

Breeding: Mates February–August. Gestation about 22 days. 5 litters per year, each of 1–7 young (average 3).

Sign: Close-cropped feeding sites; a maze of runways connecting forms.

Habitat: Thick, brushy areas, especially where some brush has been cut.

Range: West Coast from Oregon to Baja California, Mexico.

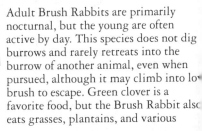

Adult Brush Rabbits are primarily nocturnal, but the young are often active by day. This species does not dig burrows and rarely retreats into the burrow of another animal, even when pursued, although it may climb into low brush to escape. Green clover is a favorite food, but the Brush Rabbit also eats grasses, plantains, and various

berries, and, in winter, woody vegetation, primarily salal and Douglas fir. The mother covers the nest with a blanket of grass before leaving it. The young mature in four to five months. Lynx, Coyotes, hawks, and snakes are among the predators of this species.

204, 205 Eastern Cottontail
Sylvilagus floridanus

Description: The familiar cottontail of eastern U.S. *Grayish brown above,* grizzled with black; forehead often has white spot. Distinct *rust-colored nape.* Tail cottony white below. Feet whitish above. Long ears. L 14¾–18¼" (375–463 mm); T 1½– 2½" (39–65 mm); HF 3⅜–4⅛" (87–104 mm); E 1⅞–2⅝" (49–68 mm); Wt 2–4 lb (900–1,800 g).

Similar Species: New England Cottontail usually lacks rust-colored nape and has black patch between ears. Desert Cottontail is usually smaller, with slightly longer ears. Marsh and Swamp rabbits have rust-colored feet. Snowshoe Hare is usually larger and uniform dark brown in summer.

Breeding: Mates February–September. Gestation 28–32 days. 3 or 4 litters per year, each of 1–9 young (usually 4 or 5).

Sign: Small woody sprigs cut off cleanly and at an angle (sprigs browsed by deer, which lack upper incisors, are raggedly torn). Young trees stripped of bark to height of 3–4" (75–100 mm) when snow is deep.
Scat: Dark brown, slightly flattened, pea-size pellets, usually in piles.
Tracks: In clusters of 4. Foreprints almost round, about 1" (25 mm) wide; hindprints ahead of foreprints, oblong, about 3–4" (75–100 mm) long, depending on size and speed of rabbit. When sitting or standing: 2 hindprints side by side, just behind 2 more closely spaced foreprints. When moving: 1 foreprint slightly ahead of the other;

hindprints ahead of foreprints, as forefeet act as fulcrums for hops. Hindprints relatively short when moving fast, as less of leg touches down. Straddle 4–5″ (100–125 mm); stride varies with speed.

Habitat: Brushy areas, old fields, woods, and cultivated areas; especially around thickets and brush piles.

Range: Eastern U.S. except for New England; west through North Dakota, Kansas, Texas; also in n New Mexico and Arizona.

The most common rabbit in much of the U.S., the Eastern Cottontail is primarily nocturnal, but is abroad near dawn and dusk and often on dark days. On midwinter nights, groups of cottontails have been seen frolicking on crusted snow. Cottontails usually hop, but they can leap 10 to 15 feet (3–4.5 m); sometimes they stand on their hindfeet to view their surroundings. When pursued, they usually circle their territory and often jump sideways to break their scent trail. They dislike getting wet but will swim if pressed. In winter, where brush is strong enough to hold a covering blanket of snow, they may make a network of runways beneath it. In cold weather, they often take shelter in Woodchuck burrows. This species feeds on many different plants, mainly herbaceous varieties in summer and woody varieties in winter. As is the case with many lagomorphs, in addition to producing typical fecal pellets, the Eastern Cottontail will feed rapidly, then retreat to the safety of a brush pile or other shelter and defecate soft green pellets, to be eaten at leisure. During the breeding season, males fight one another and perform dance-like courtship displays before the territorial females. These displays involve face-offs and much jumping, including females jumping over males. Individuals often jump straight up into the air, and pairs

or small groups often engage in active chases. The young are born in a nest lined with plant material and fur from the mother's breast. The nest cavity, in a hollow in the ground, is about 7 inches (180 mm) deep and 5 inches (125 mm) wide. The top of the nest is capped over with vegetation; nests in lawns are often exposed when lawn mowers take the tops off. The young are nursed at dawn and dusk. Within hours after giving birth the female mates again. If no young were lost, a single pair, together with their offspring, could produce 350,000 rabbits in five years. However, this rabbit's death rate vies with its birth rate; few individuals live longer than one year.

203 Mountain Cottontail
"Nuttall's Cottontail"
Sylvilagus nuttallii

Description: Grayish brown above; white below. *Black-tipped ears,* densely furred inside. L 13¾–15⅜" (350–390 mm); T 1¾–2" (44–50 mm); HF 3½–3⅞" (88–100 mm); E 2⅛–2½" (54–65 mm); Wt 1½–2¼ lb (700–1,000 g).

Similar Species: Desert Cottontail's ears, sometimes longer, lack black tips and are sparsely furred inside.

Breeding: Mates February–July; 2–5 litters per year, each of 3–8 young.

Sign: Tracks and scat similar to those of Eastern Cottontail.

Habitat: Rocky wooded or brushy areas, often with sagebrush.

Range: Extreme sc British Columbia and e Washington south to e California, and east through s Saskatchewan, Montana, and n New Mexico.

The Mountain Cottontail rests in forms when dense vegetation is available; otherwise it uses burrows and rocky crevices for shelter. It prefers grass but lives most of the year on sagebrush and

will eat juniper berries. The young are born in a hair-lined cup nest and weaned at about one month. Predators include the Bobcat, the Coyote, owls, and hawks.

Allegheny Cottontail
Sylvilagus obscurus

Description: *Closely resembles the New England Cottontail;* differs genetically, and in such characteristics as cranial shape. Usually has dark spot between ears. Measurements similar to those of New England Cottontail.

Similar Species: Nearly identical New England Cottontail occurs farther north.

Breeding: Presumed to be similar to New England Cottontail.

Sign: Typical rabbit pellets and cuttings.

Habitat: Woods and brushy areas at higher elevations in Appalachian Mountains.

Range: Scattered distribution in Appalachian Mountains from Hudson River southward through Pennsylvania, Maryland, West Virginia, Virginia, Tennessee, North Carolina, South Carolina, Georgia, and Alabama.

Until recently the New England Cottontail was thought to occur over a large range from southern Maine through the Appalachians to northern Georgia and Alabama. However, individuals in the southern part of the range were discovered to have a different number of chromosomes from the northern ones. Currently the two are treated as different species, the Allegheny Cottontail and the New England Cottontail, as there is no evidence that they interbreed. The habits of the Allegheny Cottontail are assumed to be generally similar to those of the New England Cottontail. It feeds heavily on green vegetation in summer and on woody vegetation in winter.

202 Marsh Rabbit
Sylvilagus palustris

Description: *Dark brown above; nape dark cinnamon;* belly white. *Very small tail;* gray mixed with brown below. *Short, broad ears.* Feet small, reddish brown above. L 13⅞– 17¾″ (352–450 mm); T 1¼–1½″ (33–39 mm); HF 3½–3⅝″ (88–91 mm); E 1¾–2″ (45–52 mm); Wt average 3½ lb (1.6 kg).

Similar Species: All other cottontails in region of overlap have tails that are white below. Very similar Swamp Rabbit is larger.

Breeding: Begins breeding in February; several litters per year, each of 2–5 young.

Sign: Trails in marshy vegetation.
Tracks: In addition to the typical rabbit tracks, alternating tracks are left when this species walks on its hindlegs. Marsh Rabbit tracks (HF about 3½″/90 mm) are smaller than those of Swamp Rabbit (HF about 4¼″/110 mm).

Habitat: Bottomlands, swamps, lake borders, and coastal waterways.

Range: Southeastern Virginia south through Florida to s Alabama.

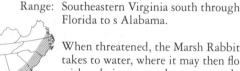

When threatened, the Marsh Rabbit takes to water, where it may then float with only its eyes and nose exposed. To elude a pursuer if cut off from water, it will run a zigzag trail, but its shorter legs make it less agile on land than other rabbits. The Marsh Rabbit sometimes walks on its hindlegs, an unusual habit for a leporid. It eats many types of green vegetation, including cane, greenbrier, grasses, subterranean bulbs, and leaves and twigs of deciduous trees, and uses rushes, grasses, and leaves to build a large covered nest. The great horned owl and northern harrier are important predators. A subspecies, the Lower Keys Rabbit *(S. p. hefneri),* is classified as endangered in Florida by the U.S. government.

New England Cottontail
Sylvilagus transitionalis

Description: *Brownish, heavily sprinkled with black.*
Black patch between ears. No white spot
on forehead. Tail whitish below; feet
whitish above. Ears edged with black
stripe. L 14¼–19″ (363–483 mm);
T 1¼–1⅞″ (31–49 mm); HF 3½–4″
(90–102 mm); E 2¼–2¾″ (56–71 mm);
Wt 1½–3 lb (750–1,350 g).

Similar Species: Nearly identical Allegheny Cottontail
occurs farther south. Eastern Cottontail
lacks black patch between ears, often
has white spot on forehead.

Breeding: Several litters per year in spring and fall,
each of 3–8 young. Gestation 28 days.

Sign: Tracks and scat similar to those of
Eastern Cottontail.

Habitat: Woods or brushlands.

Range: New England (except Maine) southwest
through Allegheny Mountains.

The habits of this rabbit are similar to
those of the Eastern Cottontail, but this
species is much more secretive and rarely
ventures from cover. Its home range is
about half to three-quarters of an acre
(1–2 ha). Nests are in depressions in the
ground, about 4 inches (100 mm) deep
and 5 inches (125 mm) wide. They are
lined and capped with fur and grass,
then covered with twigs and grass.
Reingestion of fecal pellets (coprophagy)
has been observed in the New England
Cottontail. In recent years, its numbers
have been declining and its range
shrinking, probably because of loss of
habitat and the widespread introduction
of other rabbit species.

European Rabbit
"Domestic Rabbit," "San Juan Rabbit"
Oryctolagus cuniculus

Description: Usually *brownish to grayish,* with tail
dark above and white below, but
coloration almost as variable as in

domestic cat. L 17¾–24″ (450–600
mm); T 2⅝–3½″ (66–88 mm); HF 3⅝–
4⅜″ (92–112 mm); E 2⅜–3⅞″ (60–100
mm); Wt 3–5 lb (1.4–2.3 kg).

Similar Species: Cottontails generally smaller.
Breeding: 6 or more litters per year, each of 4–12
young.
Sign: Warrens (extensive burrow systems with
many entrances), located in areas
denuded of vegetation, are littered with
dark round scat.
Habitat: Brushy areas and open fields.
Range: Present range very scattered, with
mainland releases in Pennsylvania,
Indiana, Illinois, New Jersey,
Wisconsin, and Maryland. Escaped
individuals may be encountered in
rural and suburban areas throughout
North America. Introduced on
Farallon Islands, California; San Juan
Islands, Washington; and Middelm
Island, Alaska.

European Rabbits were introduced on
San Juan Island and in several other
North American areas in an effort to
establish a new game species larger than
the cottontail and with a reproductive
capacity at least as great. These rabbits
have often thrived in the wild. A high
population of European Rabbits can be
destructive in agricultural areas, not
only because they overbrowse but also
because, unlike cottontails, they dig
extensive burrows, forming colonial
networks called warrens. The lighthouse
keeper who introduced them on San
Juan Island was almost literally
overcome with success. There were soon
so many rabbits and so many rabbit
holes that the lighthouse nearly
collapsed, and thousands of the animals
had to be poisoned. Some individuals
escaped to a nearby island, where their
tunneling activities caused chunks of
land along the shore to crumble into the
sea. These incidents earned notoriety for
the "San Juan Rabbit," as this species
came to be called. It prefers short grass

and leafy plants, but when hungry will eat any available vegetation. This is the rabbit that "breeds like a rabbit," often producing more than six litters per year. Each litter may contain as many as a dozen young, which are born in a grass nest in a special chamber in an underground den.

---

212 Antelope Jackrabbit
Lepus alleni

Description: Grayish brown above; *lower sides largely white.* Face, throat, and ears brownish. Tail black above. *Very long ears with pale tips.* L 22–26" (553–670 mm); T 1⅞–3" (48–76 mm); HF 5–5⅞" (127–150 mm); E 6–8" (152–203 mm); Wt of 3 animals ranged from 6½ to 9½ lb (3–4.3 kg).

Similar Species: White-sided Jackrabbit has shorter ears.

Breeding: 3 or 4 litters per year, each of 1–5 young, born after gestation of 6 weeks.

Sign: Scat and tracks similar to those of Eastern Cottontail, but larger. Greatest width of straddle about 8" (200 mm).

Habitat: Variety of habitats: mostly grassy slopes at moderate elevations; also deserts with little grass.

Range: South-central Arizona.

The Antelope Jackrabbit, named for its ability to make enormous leaps, can "flash" its white sides as it runs, moving the white fur higher or lower at will. Flashing white at one spot, it comes to rest elsewhere with brown showing. This presumably confuses a predator, which heads for the last glimpse of white. The long ears play an important role in regulating body temperature: In hot weather, they stand erect, and their dilated blood vessels give off heat, thus cooling the jack; in cold weather, the ears lie back close to the body, and blood vessels constrict to maintain body warmth. Despite this adaptation, in the hottest part of the day the Antelope

Jackrabbit avoids direct exposure to the sun and rests under the shade of plants in a form, which it may make simply by backing into vegetation. In the form, the hare is in full view on three sides but concealed from above and protected from the sun. The ears are laid back when the animal is at rest, but erect when it is alarmed. This species feeds on various coarse grasses, mesquite, prickly pear, cat's-claw, and other desert vegetation. When feeding on mesquite *(Prosopis)* it will stand on its hindfeet, or even its toes, to reach the desired leaves, bark, or buds, especially the green leaves growing in the plant's axils. The animal may hold its front feet limp or may support itself on the mesquite branch. There is no evidence that this species needs free water (water other than from its food). In this species, as with all hares, the young are born fully haired, with eyes open, and able to hop. The young apparently scatter at birth, but the mother returns to the area at night, and the young gather to feed. Predators of the Antelope Jackrabbit include the Bobcat, Coyote, and golden eagle.

208, 209 **Snowshoe Hare**
"Varying Hare"
Lepus americanus

Description: *In summer, dark brown,* with small tail dark above and dusky to white below. *In winter, white,* sometimes mottled with brown; white hairs remain dark at base. *Black-tipped, moderately long ears.* Large hindfeet, with soles well furred, especially in winter. This species stays brown all year in w Washington and Oregon. Some black (melanistic) hares in Adirondack Mountains remain black all year. L 15–20" (382–520 mm); T 1–2¼" (25–56 mm); HF 3⅞–5⅞" (100–150 mm); E 2⅝–3⅛" (66–79 mm); Wt 2–3 lb (896–1,400 g).

Similar Species: Cottontails are smaller and do not turn white in winter. Most other hares are larger and occur in open habitats. Arctic Hare and White-tailed Jackrabbit have all-white tails. Antelope Jackrabbit has white lower sides and lacks black on ears. Black-tailed Jackrabbit has much longer ears. European Hare is larger, with larger ears.

Breeding: 2 or 3 litters per year, beginning in March, each of 1–6 young (average 3).

Sign: In snow, trails packed down as much as 1′ (300 mm) deep.
Scat: Brown, slightly flattened spherical pellets, similar to those of Eastern Cottontail but larger.

Tracks: Hindprints 4–5″ long, wider than those of other hares in snow, due to snowshoe effect of widely spread toes. Straddle more that 6″ (150 mm).

Habitat: Northern forests.

Range: Alaska and most of Canada south to n California, n New Mexico, n Minnesota, n Michigan, n New Jersey, and southward through Allegheny Mountains.

One of the smallest and shiest of the hares, the Snowshoe Hare rests by day in a form, hollow log, or Woodchuck or Mountain Beaver burrow, although it may venture out in overcast weather. If disturbed, it may run in a circle covering several acres, traveling at speeds up to 30 mph (50 km/h), with bounds to 12 feet (3.5 m), and usually passing close to its point of departure. It often tries to hide in brush like a cottontail instead of running into the open like most hares. When alarmed, it may thump its hindfeet. Although a good swimmer, it avoids water. It frequently bathes in dust and often uses the dusting wallows of grouse. In summer, it feeds on grasses, green vegetation, willow, and berries when available; in winter, it resorts to conifer buds and the bark of aspen, alder, and willow. It will also eat carrion, and som

individuals become nuisances to trappers by stealing bait. The seasonal molt, when the coat of the Snowshoe Hare (and other species displaying seasonal coloration) changes, is a photoperiodic phenomenon governed by lengthening or shortening periods of daylight. As daylight diminishes in autumn, the hare begins to grow a white-tipped winter coat that at first is patchy—excellent camouflage against patchy snow; by the time daylight is at the minimum and large expanses of ground are blanketedwith snow, the hare has turned white to match. As the days lengthen in spring, the winter coat is gradually shed and replaced with brown. But when snow comes unusually late in fall or lasts unusually long in spring, the hare will have molted nonetheless; at such times it is conspicuous to predators and becomes less active than usual, seeking cover. Young Snowshoe Hares can run within hours of birth, but may nurse for almost a month. Northern populations of Snowshoe Hares seem to be very cyclic, undergoing major highs and lows at fairly regular intervals. Snowshoe Hares become exceedingly plentiful every nine or ten years, then swiftly plummet in number. The causes of these population fluctuations are not fully understood, though many explanations have been proposed. The fluctuations appear to be related to the crowding that exists at higher population densities. One likely explanation is that the crowding initiates a "stress syndrome" believed to retard the reproduction process and cause a dramatic decline in population. The Snowshoe Hare's predators include weasels, foxes, Minks, owls, hawks, Wolverines, Bobcats, and especially Lynx, which in Canada depend on these hares so heavily as a food supply that their population levels parallel the hare's, following them by one year.

210 Arctic Hare
Lepus arcticus

Description: *The largest hare in North America. In summer, grayish brown above;* underparts, tail, and back edges of ears white; ear tips black. *In winter, white,* with hairs white to base; ear tips remain black. Remains white all year in northernmost areas. Ears relatively short. L 18⅞–27″ (480–678 mm); T 1⅜–3⅛″ (34–80 mm); HF 5¼–6⅞″ (132–174 mm); E 2¾–3¼″ (70–84 mm); Wt 6–15 lb (2.7–6.8 kg).

Similar Species: Snowshoe Hare is smaller, with tail dark above in summer; white hairs are dark at base in winter.

Breeding: Mates early April; 2–8 young born late May–July; 1 litter per year.

Sign: Cut willow shoots.
Tracks: Similar to those of other hares; occasionally makes kangaroo-like print when hopping erect on hindfeet only.

Habitat: Tundra and rocky slopes.

Range: Northern Canada, from Mackenzie River to Newfoundland, primarily above the tree line.

Arctic Hares live in the coldest parts of Canada, including Ellesmere Island, where temperatures in winter average −22°F (−30°C). This hare is equipped for the bitterly cold Arctic winters with its relatively short ears (which help conserve body heat), densely furred feet and strong claws, which are used to dig through crusted snow for twigs and roots of dwarf willow (its chief food) and other woody species, as well as sedges and saxifrages. It uses its incisors, straighter than those of other leporids, like tweezers to extract tiny plants from rocky crevices. More gregarious than other hares, this species occurs in groups of 10 to 60 individuals, particularly in windswept areas; on Canada's Arctic islands, it can be found in herds of several thousand. It takes shelter in depressions especially on the leeward side of rocks,

and faces into the wind so the cold air
flows over its fur. In storms, it may
tunnel in the snow. On several occasions,
it has been observed walking upright on
its hindfeet. During the breeding season,
the first two weeks in April, males box
with their forefeet (jackrabbits mainly
use the hindfeet), slashing out with
sharp claws. Females bear their single
litter in a simple depression in the
tundra vegetation or in a nest made of
mosses or grasses among the rocks. The
mother remains with the young full-
time for the first two or three days and
will defend them against adversaries,
even those far stronger than she. After
the first few days, the young are able to
protect themselves by hiding motionless
among stones or vegetation. Young
Arctic Hares are cared for much longer
than other hares, remaining near the
mother and nursing for about two
weeks, by which time they turn either
gray or white, depending on latitude.
They then disperse and form nursery
bands of up to 20 individuals, which
congregate at a nursing site and nurse
when the mother arrives. They are
weaned as late as eight or nine weeks of
age. When abundant, Arctic Hares may
trample the snow in large areas and
compete for food with Muskoxen and
Caribou. High population levels of this
species occur in a nine- or ten-year cycle
and often coincide with a similar rise in
population of its principal predator, the
Lynx. Other important predators include
wolves and the Arctic Fox. South of the
tundra, Arctic Hares are replaced by
Snowshoe Hares.

213 Black-tailed Jackrabbit
Lepus californicus

Description: Buffy gray or sandy above, peppered
with black; white below. *Tail has black
stripe above, extending onto rump,* with
white border. *Very long ears* brownish

with black tips. Very large hindfoot.
L 18¼–25″ (465–630 mm); T 2–4⅜″
(50–112 mm); HF 4⅜–5¾″ (112–145
mm); E 3⅞–5⅛″ (99–131 mm); Wt 4–
8 lb (1.8–3.6 kg).

Similar Species: White-tailed Jackrabbit lacks black on
top of tail. Antelope Jackrabbit has
white lower sides; lacks black on ears.

Breeding: 1–4 litters per year, each of 1–8 young
(usually 2–4); births are year-round, bu
more frequent in milder weather.
Gestation 41–47 days.

Sign: Trails worn between feeding and restin
sites; scarred or freshly nipped prickly
pear cactus or pale tufts of fur on thorn
Nesting forms are shallow scrapes, ofte
beneath sagebrush or rabbit bush.
Scat: Dark brown, slightly flattened
spherical pellets about ½″ (12 mm) in
diameter.

Habitat: Barren areas and prairies, meadows,
cultivated fields; also areas where
vegetation exceeds 2′ (60 cm) in heigh

Range: Western U.S., from sc Washington
south to California, east to Nebraska,
w Missouri, and Texas. Introduced in
New Jersey and Kentucky.

This most abundant and widespread
jack was originally called a "jackass
rabbit," after its very large ears. Like
other jacks, it is not really a rabbit but
hare, as its young are born well furred
and with their eyes open. By day, it
generally rests in dense vegetation or i
a form, becoming active in late afternoo
Somewhat social, it often feeds in loose
groups. In summer, it eats many kinds
of plants, favoring alfalfa when availab
in winter, it depends on woody and dri
vegetation. When alarmed, it remains
very still, but may move its ears to cat
sounds. It rarely walks, but hops 5 to]
feet (1.5–3 m) at a time, up to 20 feet
m) when panicked, and reaches speeds
30 to 35 mph (50–55 km/h) over shor
distances. When it runs at moderate
speeds, every fourth or fifth leap is
exceptionally high, allowing it a bette

view of surroundings or a pursuing predator; at top speed there are no such special jumps. When escaping from a predator, it flashes the white underside of its tail, perhaps alerting other jacks to danger and confusing its enemy. After fleeing a short distance, it stops and looks back, evidently to see if it is still being pursued, and may then give a danger signal by thumping its hindfeet. It can swim, dog-paddling with all four feet. The young are born in a relatively deep form lined with hair from the mother's breast. The mother then places them in separate forms, thus decreasing a predator's chance of taking her entire litter. To avoid attracting attention, she keeps her distance by day but comes several times a night to nurse. The young can fend for themselves in less than a month. Generally silent, these hares can squeal and give distinctive calls when fighting or distressed and when assembling their young. Coyotes, Bobcats, foxes, hawks, owls, and snakes are predators.

White-sided Jackrabbit
Lepus callotis

Description: Grayish brown above; *white on sides* and belly. *Ears long,* with little or *no black on tips.* Large hindfoot. L 17–24″ (432–598 mm); T 2–3⅝″ (51–92 mm); HF 4¾–5½″ (121–139 mm); E 4⅝″ (117 mm); Wt 5½–6 lb (2.5–2.7 kg).

Similar Species: Antelope Jackrabbit is larger, with longer ears.

Breeding: 2 or 3 young per litter; born from mid-April to mid-August.

Sign: Depressions in grass.

Habitat: Flat grassy plains.

Range: Extreme s Hidalgo County, New Mexico, into Mexico.

This hare hides in grass by day and is active after dark, especially on clear, moonlit nights. It constructs shelter

forms, which are usually surrounded by
grass. Its food consists almost entirely o
grasses. This species usually occurs in
male-female pairs, behavior that is mos
evident during the breeding season;
partners nearly always stay within 15
feet (5 m) of each other. This species
produces grunts and screams.

---

211   European Hare
"Brown Hare"
Lepus europaeus

Description: *A large hare.* Thick, kinky fur. *In summe
brownish above* interspersed with black
hairs, *white below;* in winter, grayish
above. *Tail black on top,* pale below. Ear
moderately long. L 24–30″ (600–750
mm); T 2¾–3⅞″ (70–100 mm);
HF 5⅛–6¾″ (130–170 mm); E 3⅛–
3⅞″ (79–100 mm); Wt 6–12 lb
(2.7–5.4 kg).

Similar Species: Black-tailed Jackrabbit has black on
rump above tail. Snowshoe Hare is
smaller, with smaller ears; tail is browr
in summer.

Breeding: Mates at any time of year, chiefly from
late winter through early summer. 3 or
litters per year, each of 2–4 young.
Gestation 30–40 days.

Sign: Tracks and scat similar to those of
Eastern Cottontail, but larger; hindfoo
track about 5″ (125 mm) long.

Habitat: Open fields in rolling country;
sometimes sparse woods.

Range: Great Lakes region east to New York
and New England.

Introduced into Dutchess County, Nev
York, in 1893, the European Hare is th
largest hare in its range. By day, it rest
in a form scratched out in an area with
an unobstructed view and a gentle dra
that carries sound so predators may be
heard before they are glimpsed. When
pursued, it seeks to elude a predator
with confusing movements involving
sharp turns, backtracking, and

recrossing of its path. Normally taking about 4 feet (1.2 m) per leap, when in a hurry it bounds to 12 feet (3.7 m) and can clear obstacles 5 feet (1.5 m) high. It has been timed at 30 mph (50 km/h) over a quarter-mile (.5 km) stretch and in bursts can make 45 mph (75 km/h). Like many lagomorphs, the European Hare is a good swimmer and will not hesitate to cross a river. It feeds on grass, clover, wheat, corn, apples, berries, and many green plants in summer, resorting to twigs and buds in winter. During the mating season, males battle fiercely. The young are born in a grass-lined nest in a form. The mother scatters the newborns in separate forms and visits each form nightly to nurse. To assemble the newborns, she gives a low, bugle-like call. European Hares scream when distressed and give a warning noise by grating their teeth. Foxes and Bobcats are their major predators. Many individuals are killed by farm machinery, however, because they often respond to danger by freezing in their tracks. This hare is the original Easter Bunny. According to a Germanic legend, Eostre, goddess of spring, created the first hare from a bird, and to show its gratitude, the hare has ever since laid eggs during the Easter festival in her honor.

Alaska Hare
"Mountain Hare," "Blue Hare,"
"European Varying Hare,"
"Northern Hare"
Lepus othus

Description: *In summer, reddish brown or brownish gray above;* underparts, legs, tail, edges of ears, and *eye ring white;* ear tips black. *In winter, white* except for black ear tips. L 22–27" (565–690 mm); T 2⅛–4⅛" (53–104 mm); HF 5¾–7⅜" (147–189 mm); E 3–3⅛" (75–78 mm); Wt 7–10 lb (3.2–4.5 kg).

Similar Species: Snowshoe Hare has smaller ears.

Breeding: Mates in April or May. 1 litter per year of 3–8 young, born in June. Gestation about 46 days.

Sign: Tracks and scat similar to those of Eastern Cottontail, but larger; hindfoot track about 7" (180 mm) long.

Habitat: Open tundra; brush when available.

Range: Northern and w Alaska.

The Alaska Hare is larger and has shorter ears than hares in more southerl areas; these are both energy-saving adaptations, for they reduce the ratio of the animal's surface area (through whicl heat is lost) to its body mass. This hare feeds primarily on green plants in summer, and on woody vegetation, such as bark, twigs, and shoots, in winter. Its cries include puffing and hissing sounds and a *hoo-hoo* given only during the mating season. The young are born in June in a hollow in an open area or near brush and are fully grown by mid-August.

214, 215 White-tailed Jackrabbit
Lepus townsendii

Description: *Buffy gray above;* white or pale gray below. In winter, white or very pale gra in most of range (except most southerl parts). *Tail white above and below;* sometimes with dusky stripe on top, bu not extending onto rump. Long ears; buff or gray on fronts; on backs, whitish with black stripe on tip. L 22–26" (565–655 mm); T 2⅝–4⅜" (66–112 mm); HF 5¾–6¾" (145–172 mm); E 3¾–4⅜" (96–113 mm); Wt 5¾–9½ lb (2.6–4.3 kg).

Similar Species: Black-tailed Jackrabbit has black on ta continuing up rump. Snowshoe Hare is smaller, dark brown in summer.

Breeding: Up to 4 litters per year, each of 1–6 young (average 4); born late April, earl June, July, and August–September, afte gestation of a month or more.

Sign: Tracks and scat similar to those of
Eastern Cottontail, but larger; hindfoot
track about 6″ (150 mm) long.

Habitat: Barren, grazed, or cultivated lands;
grasslands.

Range: Eastern Washington, e Oregon, and ne
California east through Minnesota,
Iowa, and Kansas.

One of the least social of hares, the
White-tailed Jackrabbit tends to be
solitary except during the mating
season, when three or four individuals
may group together. A nocturnal
animal, it hides in forms during the day.
In winter, it may hide by day in hollows
in the snow connected by burrows.
Traveling in 12- to 20-foot (3.7–6 m)
leaps, this jackrabbit can maintain a
speed of 35 mph (55 km/h), with spurts
up to 45 mph (75 km/h). When
cornered, it will swim, dog-paddling
with all four feet. In summer, it eats
grasses, clover, and other green
vegetation; in winter, it feeds on twigs,
buds, and dried vegetation. Bucks fight
furiously during the mating season,
mostly by kicking out with their
hindfeet, and biting when they can. The
young are born in a form or in a nest
lined with hair on the ground. The
young soon forage for themselves and
are independent in four to eight weeks,
depending on locality. This species may
produce up to four litters, but in the
northern part of its range it has only one
litter per year.

RODENTS
Order Rodentia

The nearly 3,000 species of rodents form the world's largest mammalian order. More than half of all mammal species are rodents; by sheer numbers of individuals they probably represent the majority of mammals on earth. North American species range from mice weighing a fraction of an ounce (less than 10 g) to the American Beaver, which may weigh up to 86 pounds (39 kg), but most members of this order are relatively small. They are distinguished by having only two pairs of incisors—one upper and one lower—and no canines, which leaves a wide gap, the diastema, between the incisors and the molars. (Lagomorphs have a similar diastema, but they have two pairs of upper incisors, the large front pair and a small pair directly behind them.) Because the incisors are enameled on the front only, the working of the upper teeth against the lower ones wears away the softer inner surfaces more rapidly, producing a short, chisel-like beveled edge ideal for gnawing. Incisors grow throughout the animal's life (if they didn't they shortly would be worn away), and rodents must gnaw enough to keep them from growing too long. If wear fails to keep pace with growth, a malocclusion may occur, which means the teeth fail to meet and wear off properly. Such teeth may grow completely out of the animal's mouth and prevent eating, or they may curve inward, growing back into the skull or jaw and eventually causing death. Malocclusions occur mostly in captivity, when adequate gnawing material is lacking. Rodents have bulbous eyes on the sides of the head, enabling them to see forward or behind, and they can detect danger over a wide arc. Most, but not all, have four toes on each forefoot, and five on each hindfoot.

Most rodents are nocturnal, emerging at night to feed. The majority are active throughout the year, but a number hibernate in winter, and a few that live in habitats with intense heat and sparse vegetation estivate (become dormant in summer). In both hibernation and estivation, body temperature adjusts to within a degree or two of environmental temperature, and all other bodily functions are also greatly reduced, thus conserving energy. Most hibernators survive by metabolizing fat stored in their bodies, but some, such as the chipmunks, wake to eat stored food. Because the surface area of a small animal is large in proportion to its bulk, rodents lose body heat rapidly through radiation. Most are extremely active, but while activity temporarily warms an animal, it also expends energy. Rodents make up for their high energy losses by feeding proportionally more than do larger or less active animals. A reproductive capacity unmatched by other mammals helps some rodents compensate for a high rate of predation.

While a few species of rodents are among humankind's worst pests, carrying disease, eating or spoiling stored grain and other foods, and destroying vast amounts of property, the majority feed heavily on weed seeds and help keep insects in check by eating them in great numbers.

MOUNTAIN BEAVER
Family Aplodontidae

The one living member of this family—
found only in the western U.S. and
British Columbia—is often considered
the most primitive living rodent. The
family name refers to its simple, rootless
molars. The American Beaver belongs to
a different family.

224 Mountain Beaver
"Aplodontia," "Sewellel"
Aplodontia rufa

Description: Woodchuck-like but smaller, with a
short heavy body. Dark brown above; paler
brown below. Blunt head; small ears and
eyes; distinctive white spot below ear.
Tiny tail. Short legs, with 5 toes on each
foot: first toe on front foot has flattened
nail; other front toes have very long,
strong claws. L 9⅜–18½" (238–470
mm); T ¾–2⅛" (20–55 mm);
HF 2⅛–2½" (55–63 mm); Wt 1–3 lb
(0.5–1.4 kg).

Similar Species: Woodchuck has longer, bushy tail.
Common Muskrat has much longer,
naked, scaly tail.

Breeding: Mates in late winter; 1 litter per year of
3–5 young (usually 4 or 5) born in early
spring, after gestation of 28–30 days.
Young are weaned by autumn and
disperse soon after. Females breed at 2
years of age.

Sign: Burrows up to 19" (480 mm) in diameter,
with entrances surrounded by fan-
shaped earth mounds and pathways, or
sometimes by cylindrical earth cores, up
to 6" (150 mm) in diameter, made
during winter tunneling and left on
surface after snow cover melts. In very
wet areas, a "tent" of sticks covered with
leaves and fern fronds is erected over
burrow entrances. In late summer, "hay
stacks" of ferns and other vegetation are
piled up to 2' (600 mm) high on logs or
on the ground.

Tracks: Narrow footprints, each less tha[n] 2″ (50 mm) long; hindprints are larger than foreprints and may overlap them. Small first toe on forefoot may not prin[t]

Habitat: Moist forests, especially near streams.

Range: Extreme sw British Columbia, w Washington, w Oregon, n California and extreme wc Nevada.

The common name "Mountain Beaver" is misleading, as this rodent is not a beaver nor does it prefer a mountainou[s] habitat. The name may have derived from its beaver-like habit of diverting streams into its tunnels or from its occasional gnawing of bark and cutting of tree limbs. Active throughout the year, this mostly nocturnal animal occasionally browses during the day, especially in autumn. Its home range i[s] small, averaging about ⅓ acre (⅛ ha). The Mountain Beaver is mainly fossori[al] (adapted to digging) and stays within [a] few yards of a burrow entrance. Althou[gh] it does not hibernate, it seldom appear[s] aboveground in winter. Its labyrinthin[e] burrow system is usually shallow and often near cover, and may be used by many other animals. Radiating from a nest chamber 1 to 5 feet (300–1,500 mm) deep and about 12 inches (300 mm) in diameter, the burrow system has numerous openings to the surface, only a few of which have earth mound[s] outside. It contains a nest of coarse vegetation lined with fine, soft vegetation, dead-end tunnels where excrement and rejected bits of food are deposited, and separate chambers for storage of food and so-called Mountai[n] Beaver baseballs, which are baseball-si[ze] chunks of stone or clay encountered in digging; the animal occasionally gnaw[s] upon these balls to sharpen its teeth a[nd] uses them to close off vacated nesting [and] feeding areas. The Mountain Beaver feeds solely on vegetation, particularly sword fern, bracken, grasses, and the bark of coniferous and hardwood trees.

The animal usually builds a burrow to a new food supply. It cuts green vegetation, allows it to wilt, then transports it to the burrow, where it is eaten or stored; in storage the food remains moist in the water-saturated air of the burrows. This species excretes both soft and hard fecal pellets and reingests the former. Not a particularly good climber, it does climb trees to cut off small limbs, leaving stubs that it uses as a ladder when it descends headfirst, carrying the wood; occasionally it lets the small limbs drop to the ground. Although a relatively solitary species, the Mountain Beaver sometimes becomes so abundant that its burrows honeycomb the ground. A booming sound was once attributed to this species, but recent research indicates that its only vocalizations are whining or sobbing sounds in response to pain and a squeal or shrill whistle when agitated. The animal will defend its own territory; if disturbed it grates its incisors and may bite. The Bobcat, Long-tailed Weasel, and Mink are its major predators. The Mountain Beaver sometimes causes severe damage to conifers by peeling off the bark, and it may raid nearby gardens or damage them with its burrowing activity.

SQUIRRELS
Family Sciuridae

The squirrel family, which contains 27
species worldwide, is well represented
in the U.S. and Canada, with 66 specie
in eight genera. This large, diverse
group of animals, showing great variety
in both appearance and habits, include
not only the common tree squirrels
and the flying, antelope, and ground
squirrels, but also chipmunks, marmots
(including the Woodchuck), and prairie
dogs. The family name means "shade-
tail" and alludes to the bushy tails, ofte
held over the back, of many species.
Many members of the family lack such
a tail, but all share certain cranial
characteristics. Most of the Sciuridae
are diurnal, many hibernate, and some
estivate (become dormant during
summer). Some live in trees and some
in burrows, usually in open areas.

The 22 North American species of
chipmunks, formerly split in two
genera—with the western chipmunks
in the genus *Eutamias* and the Eastern
Chipmunk the sole member of the gen
Tamias—are all now considered to be
the genus *Tamias*. The 21 species of
western chipmunks are very similar to
one another, with the shape of the pen
bone (or baculum) and the call often
serving as the basis for identification.
The Eastern Chipmunk is larger than
its western relatives and is distinguish
from them by its teeth. All the wester
chipmunks have five upper cheek teeth
on each side, including a reduced first
premolar. The Eastern Chipmunk,
lacking this reduced premolar, has fou
large upper cheek teeth.

Chipmunks, generally forest creatures
have large, fur-lined internal cheek
pouches for carrying nuts and seeds,
black and white facial stripes, and five
dark stripes (separated by four pale

ones) on the back and sides of their small bodies. The Latin *tamias* means "storer," and these hibernators cache a great deal of food—a variety of seeds, nuts, fruits, and sometimes green vegetation and insects. Instead of relying on stored body fat to sustain them during hibernation, they awaken periodically throughout the winter and early spring to eat from their stores. Chipmunks are diurnal and are usually most active in early morning and late afternoon. They live mostly on the ground, but their nests may be in either an underground burrow or the hollow limb of a tree. The young are born blind and naked after a gestation of about 30 days. Many species inhabit overlapping ranges from lowland to mountaintop, with the palest chipmunks in the driest areas and the darkest in the most humid. Most chipmunks are highly vocal and have several calls—primarily chattering, twittering chirps, trills, chuck notes, and growls, sometimes repeated very rapidly.

Field identification of chipmunks is often difficult, perhaps even impossible at times in the West. Range and habitat data can be very helpful, although often more than one species of chipmunk occur together. Many western species can be distinguished by their calls. The Least Chipmunk is the only western species whose range extends into that of the Eastern Chipmunk.

The largest members of the family are the marmots *(Marmota),* including the solitary Woodchuck. These burrowers generally live in open areas and feed on green vegetation. They vary in degree of sociability—ranging from asocial to highly social. Most hibernate in winter.

Most antelope squirrels and ground squirrels (the genera *Ammospermophilus* and *Spermophilus,* respectively) have no

stripes. Those that do, such as the Golden-mantled Ground Squirrel, have none on the head or around the eyes as chipmunks do. Members of the two genera are distinguished from one another by the way they hold their tails when running: Spermophiles hold them horizontally, while ammospermophiles hold theirs vertically, exposing their white undersides, just as the Pronghorn or American Antelope, flashes its white rump. All species in the two genera are burrowers, nearly all hibernate, and many are colonial. Most feed heavily on seeds, which is why John James Audubon and his contemporaries named them "spermophiles," or "seed-lovers"—although many or most also feed heavily on insects. (Later mammalogists regrouped both genera with their Old World relatives of the genus *Citellus,* a classification still seen in some books, but the use of *Spermophilus* and *Ammospermophilus* is now generally accepted.)

The antelope squirrels *(Ammospermophilus)* are true ground squirrels, living in burrows that they dig for themselves. While most remain active throughout the year, those in higher and colder areas may become inactive in winter, although they do not truly hibernate. These squirrels feed mainly on nuts, seeds, fruits, stems, and roots, but consume insects as well from time to time.

There are 19 species of ground squirrel *(Spermophilus)* in North America north of Mexico. Most are very common in the West; east of the Mississippi, the Thirteen-lined Ground Squirrel is common east to northern Indiana and Ohio. Small to medium-size squirrels with long bushy tails, and sometimes with striped or spotted fur, ground squirrels are often seen standing upright along the roadside or running across the road in areas with patches of good

habitat. When danger appears, they dart down their burrows, often churtling softly just inside the entrance. When several species live in the same general area, they often divide it up by habitat and soil moisture levels. Like most squirrels, ground squirrels are strictly diurnal. Most hibernate, and most store energy in the form of body fat, using it for maintenance during hibernation. Some species estivate (become dormant) during hot or dry periods. The timing and length of hibernation varies greatly with latitude and altitude, with species at high elevations or latitudes having short seasons. Males typically emerge before females, which maximizes their access to females, and adults emerge before juveniles, which maximizes the time available for breeding and growth.

Prairie dogs (*Cynomys* species), stocky animals found in open prairies or plains, are also colonial burrowers. Their social structure is quite complex, consisting of often huge towns that are subdivided into coteries. Each coterie includes one male, three or four adult females, and several young up to two years old. These sociable animals feed on plants, including roots. Prairie dogs, ground squirrels, the Woodchuck, and marmots are all sometimes called "picket pins" because of the way they sit bolt upright to survey their domain.

The tree squirrels include the relatively large gray squirrels (*Sciurus* species), the smaller red squirrels (*Tamiasciurus* species), and the most arboreal and the only nocturnal members of the family: the flying squirrels (*Glaucomys* species). Along the sides of their slender bodies, the flying squirrels have furred folds of skin that spread to enable them to glide through the air. Tree squirrels have larger ears and eyes than ground squirrels, and large bushy tails, useful for balance on tree limbs and stability

and maneuverability in the air. They feed on a great variety of foods, including nuts, seeds, fruits, and insects.

The members of the sciurid family have evolved differing degrees of social activity. The Olympic Marmot and the Black-tailed Prairie Dog, for example, are very social; at the other end of the scale is the strictly asocial and solitary Woodchuck. Some animals are asocial but aggregate in favorable habitats, as do Thirteen-lined Ground Squirrels. Even asocial animals may use calls as warnings to one another. Some Sciuridae species live in single-family female kin clusters. In this arrangement, males and females occupy separate ranges after breeding, and adult males do not live in proximity to offspring. Young males disperse from their natal area, but females often remain in their mother's range throughout their lives. Juveniles from different litters do not intermingle, and females and their offspring form a fairly cohesive group that does not socialize with other groups. Round-tailed, Belding's, and Richardson's ground squirrels, and the White-tailed Prairie Dog exhibit this type of living arrangement. In a slightly different variation, the female kin cluster coexists with male territoriality; adult males maintain territories outside of the breeding season that overlap the smaller ranges of several females (composing one or more kin clusters) and their offspring. The females of these species, including Arctic and Columbian ground squirrels, tolerate offspring of adjacent females. In species that exhibit male dominance and polygynous harems, such as the Yellow-bellied Marmot, the dominant male maintains a territory throughout the active season in which several females and their offspring live. Juvenile males disperse, but juvenile

females may remain and become part of the harem. Finally, in egalitarian polygynous harems, seen in the Olympic Marmot and the Black-tailed Prairie Dog, a male and several females form a unit that defends a common territory. The male is not dominant and maintains amicable relations with juveniles. Litter distinctions are not maintained after emergence from the natal burrow. Juvenile males disperse, and juvenile females remain.

In the following species accounts, breeding and sign information are given for those species for which it is known. There are illustrations and descriptions of the tracks of the Eastern Chipmunk, the Eastern Gray Squirrel, and the Red Squirrel. The tracks of these species are generally similar to those of other members of their genera.

Alpine Chipmunk
Tamias alpinus

Description: A small chipmunk. Generally yellowish gray, lightly contrasting stripes giving overall pale coloration. *Dark side stripes reddish or brownish,* not blackish; dark stripe down middle of back may be black. *Undersides bright orange. Tail grayish white to yellowish above, bright orange-yellow below. Underside of tail has broad black border and black tip.* L 6½–8″ (166–203 mm); T 2½–3⅜″ (63–85 mm); HF 1⅛–1¼″ (28–31 mm); E ½″ (12–14 mm); Wt 1–1¾ oz (28–50 g).

Similar Species: Least Chipmunk has longer tail that is grayish yellow below. Yellow-pine Chipmunk is larger and more brightly colored. Lodgepole Chipmunk is larger and browner.

Breeding: 1 litter per year of 4 or 5 young born in early summer, probably in a den among rocks. Breeds later at higher elevations.

Habitat: Talus slopes and subalpine forests, from timberline to 8,000' (2,400 m) elevation.

Range: Sierra Nevada mountains of ec California.

In late summer the Alpine Chipmunk puts on a great deal of fat in preparation for hibernation, which, depending on latitude and altitude, occurs from about October to June. Its food consists mainly of seeds of small sedges and other alpine plants, but may also include berries and fungi. This chipmunk pulls plant stalks down within reach to collect seeds, which it stuffs into its cheek pouches. It has also been seen in pine trees gathering seeds from pinecones. This species is thought to build nests among or under rocks, though none has been found. The Alpine's call is a weak, high-pitched, repeated *sweet, sweet, sweet,* but the animal has other calls and chitterings as well.

148 Yellow-pine Chipmunk
Tamias amoenus

Description: *Brightly colored,* from tawny to pinkish cinnamon, with 5 *distinct longitudinal dark stripes, usually black, that are evenly spaced and about equal in width.* Central 3 dark stripes extend to rump; lateral 2 only to mid-body. Pale stripes are white or grayish. Distinct black lower eye stripe. Sides of body and *underside of tail grayish yellow.* Top of head brown. Ears blackish in front, whitish behind. L 7⅛–9⅝" (181–245 mm); T 2⅞–4¼" (73–108 mm); HF 1⅛–1⅜" (29–35 mm); Wt 1–2½ oz (30–73 g).

Similar Species: Least Chipmunk is smaller and paler, with underside of tail grayish yellow. Alpine Chipmunk is smaller, with shorter tail and less-contrasting stripes. Uinta Chipmunk is larger and less reddish, with dark brown side stripes and grayer head and shoulders. Panamint Chipmunk has tawny ears;

top of head is brown. Lodgepole Chipmunk has larger ears and more-contrasting dark back stripes, with the outer wider and the inner narrower.

Breeding: Mates April or May; 1 litter per year of 4–7 young born May–early June in a nest of leaves, grass, or lichens.

Sign: Remnants of nuts; burrow openings with no loose soil at entrances.

Habitat: Brush-covered areas in coniferous forests, particularly yellow pine. Has a much broader range of habitats than do Allen's, Lodgepole, and Long-eared chipmunks, with which it is often found.

Range: Much of British Columbia and extreme w Alberta south to n California; east to w Montana and nw Wyoming.

In open forests where the sun casts sharp shadows, the well-defined stripes of the Yellow-pine Chipmunk afford protective coloration. Individuals of this species that were observed in Washington State remained active about seven months and hibernated about five, waking to eat about every two weeks and emerging in April and May. One study indicated that 97 percent of the individuals survived—a phenomenal rate of winter survival for a small mammal. Some individuals are active even on snow. This chipmunk lives in underground burrows, usually about 1½ to 3 feet (450–900 mm) long and 7 to 21 inches (180–540 mm) deep in an open area within the forest; there is generally one entrance, though there may also be short side openings. Seeds, its most important food, are eaten as they are available—early in the season when green and later when ripe. When pinecones open in the fall, this chipmunk climbs trees to get the seeds. It also eats some insects and fungi. In Washington it apparently finds the thorns of the thistle no deterrent: First it eats the seeds from the head; then it cuts the head, which falls to the ground, and consumes it with impunity. In the

fall the animal stuffs its cheek pouches
with food to be stored in its burrows;
one food cache contained an estimated
67,970 items, including 15 kinds of seeds,
corn, and part of a bumblebee. It has at
least 10 different calls; one sounds like a
robin's chirp and another, among the
most common, is a sharp, accented *kwis*.

Gray-footed Chipmunk
Tamias canipes

Description: A small, grayish-appearing chipmunk.
whitish and 5 brownish stripes on back
Nape and shoulders have grayish cast,
especially in winter. Sides are pale. *Tops
of hindfeet gray.* Male is smaller than
female. L 8¼–10⅜″ (210–264 mm);
T 3⅝–4½″ (91–115 mm); HF 1¼–1⅜″
(32–36 mm); E ½–⅝″ (14–17 mm).

Similar Species: No other chipmunk has gray surfaces
on hindfeet. Closely related Colorado
Chipmunk has tawny-colored wash on
upperparts and narrower, less-black
eye stripe. Gray-collared Chipmunk,
another close relative, is less gray
overall, with darker sides and blacker
back stripes.

Breeding: 1 litter per year of about 4 or 5 young,
presumably born between mid-May an
end of August.

Sign: Remnants of acorns; burrow openings
with no loose soil at entrances.

Habitat: Great variety of habitats, but especially
forested areas, both coniferous and
hardwood, and around logs at edges of
clearings, in dense stands of mixed
timber, and on brushy, rocky hillsides.

Range: South-central New Mexico and adjacer
w Texas.

The Gray-footed Chipmunk, primarily
forest species, is especially active soon
after daybreak. Like other chipmunks,
the Gray-footed does not store much
body fat but presumably hibernates an
wakes periodically to feed on stored
food. This species eats a wide variety o

seeds, nuts, and fruits, favoring acorns and pine seeds when available and carrying its food supplies in its cheek pouches. It often establishes nests in cavities of logs. Its calls include chipper notes and a low *chuck-chuck-chuck*. When alarmed the animal stops calling and disappears down a burrow. In New Mexico, lava field populations are darker than individuals in the nearby Sacramento Mountains.

149 Gray-collared Chipmunk
Tamias cinereicollis

Description: Dark, grayish, with paler gray neck and shoulders. *Prominently gray cheeks, upper back, and shoulders* are unique among chipmunks. 5 well-defined black or brown stripes and 4 pale stripes usually present on back; outer stripes may be hard to discern. L 8¼–9⅞″ (208–250 mm); T 3⅛–4½″ (80–115 mm); HF 1¼–1⅜″ (32–36 mm); Wt 2–3 oz (57–85 g).

Similar Species: No other chipmunk has gray cheeks. Gray-footed Chipmunk has paler sides and browner outer back stripes. Least Chipmunk is smaller and less gray; lacks gray "collar." Cliff and Merriam's chipmunks have indistinct striping on back. Colorado Chipmunk has gray only on shoulders.

Breeding: Apparently mates late April–early May; gestation is at least 30 days. 1 litter per year of 4–6 young is born June–July and is aboveground by late July. Young begin eating solid food at 36–40 days and are weaned at 41–45 days.

Habitat: Ponderosa pine forests extending into spruce forests, especially around logs near clearings.

Range: East-central Arizona and sw New Mexico.

The common and Latin species names of this small chipmunk refer to its pale gray "collar." Although the Gray-

collared stores food prior to winter, it may or may not hibernate. If it occurs, hibernation is short, extending from late November to mid-March. The diverse diet of this species includes acorns, Douglas-fir seeds, currants, gooseberries, green vegetation, insects, and mushrooms and other fungi. A good climber, the Gray-collared Chipmunk is often seen in dense foliage. Its nests are located under logs, stumps, or roots, or in tree hollows. One chipmunk's nest was found in a woodpecker hole.

150 Cliff Chipmunk
Tamias dorsalis

Description: *Grayish,* with *stripes on body indistinct* or absent; often more distinct on sides of head. Bushy *tail rust-red below.* L 7⅝–10⅞" (195–277 mm); T 3⅜–5½" (85–140 mm); HF 1¼–1½" (32–39 mm); Wt 2–3 oz (57–85 g).

Similar Species: Least, Uinta, and Gray-collared chipmunks have distinct striping on back. The two other chipmunks with indistinct striping, Merriam's and California chipmunks, occur farther west in California.

Breeding: Mates April–May, producing 4–8 young; some individuals have more than 1 litter.

Habitat: Rocky areas and cliffs, especially piñon-juniper zones, but also in ponderosa pine, oak, and maple.

Range: Eastern Nevada, Utah, and extreme nw Colorado south to Arizona and w New Mexico.

Like most of its kin, the Cliff Chipmunk hibernates in winter (except apparently in Arizona), but it may emerge during warm spells. This species forms feeding aggregations of up to 10 individuals, mostly females. These groups move slowly through a feeding area, each member separated from the next by 10 to 30 feet (3–10 m), collecting many

kinds of nuts, seeds, fruits, and berries. Storable items are taken directly to the main den or to a cache; many of the cached items are later taken to the main den for winter use. Dens are usually in cliffs or rock piles, but may be in underground burrows or in trees. This vociferous species gives a sharp bark about 160 times per minute, each accompanied by a tail twitch. Its other common calls are a chirping *whsst* or *psst* given when the animal is excited, and a high-pitched mixture of sounds employed when the animal is surprised or threatened. As with Merriam's Chipmunk and others of brushy areas where small twigs cast indistinct shadows, the Cliff Chipmunk's indistinct striping serves as protective coloration. This species may occur in close proximity with other chipmunk species, each occupying a slightly different ecological niche.

151 Merriam's Chipmunk
Tamias merriami

Description: Grayish brown above; belly white. *Stripes indistinct,* nearly equal in width: *dark stripes gray or brown,* usually not black; *pale stripes grayish.* Dull black spots in front of and behind eyes; brownish stripe below ear. Long tail edged with buff or white. L 8¼–11″ (208–280 mm); T 3½–5½″ (89–140 mm); HF 1¼–1½″ (32–39 mm); Wt 2½–4 oz (71–113 g).

Similar Species: California Chipmunk is paler and less yellowish, with more reddish in dark stripes. Long-eared Chipmunk has dark stripe below ear and white patch behind it. Both Lodgepole and Yellow-pine chipmunks have distinct back stripes.

Breeding: 1 litter of 3–6 young born April–July (earlier at lower elevations), in a nest in an underground tunnel or cavity or in a woodpecker hole. Gestation about 30 days. Nest young make soft mewing

notes; they become active outside the
nest at about 30 days.

Habitat: Most abundant at upper edge of chaparral
and in rocky scrub; also brushlands,
forested foothills, or coniferous or piñon-
juniper forests in lower mountains to
elevations of about 7,000′ (2,100 m).

Range: Central and s California.

Evidence indicates that over part of its
range Merriam's Chipmunk occurs in
two very similar forms that remain
distinct. They can be distinguished by
skull and penis bone characteristics, and
by their appearance. The form found in
the Eagle, San Bernardino, Little San
Bernardino, San Jacinto, and Santa Rosa
mountain areas is paler in summer,
with the dark stripes on its back more
reddish, than the typical form of
Merriam's Chipmunk.

152 Least Chipmunk
Tamias minimus

Description: A small chipmunk. Color varies: in drier
regions, muted yellowish gray above
with dark stripes tan; in moister areas,
brownish gray with black side stripes.
Pale stripes white; all of equal width.
*2 white areas on flanks continue to base of
tail. Sides orange brown;* belly grayish
white. Long tail light brown above,
grayish yellow below, with hairs black-
tipped. Ears tawny in front. L 6⅝–8⅞″
(167–225 mm); T 2¾–4½″ (70–114
mm); HF 1–1⅜″ (26–35 mm); Wt 1–3
oz (30–85 g).

Similar Species: Ears of Uinta, Yellow-pine, Lodgepole,
and Colorado chipmunks are dark in
front, whitish behind. Gray-collared
Chipmunk is larger and more grayish.
Townsend's Chipmunk, also larger, lacks
white area on flanks; backs of ears are
bicolored: dusky and gray. Eastern
Chipmunk is larger; its median pale
stripes are much paler and darker than
outer ones, which are white. Cliff

Chipmunk has indistinct or absent side stripes. Alpine Chipmunk's tail is bright orange-yellow below.

Breeding: Mates within 10–20 days of emerging from hibernation in spring. Gestation about 31 days. 1 litter of 2–7 (usually 5 or 6) young born in May in a nest, often grass-lined, in an underground cavity, stump, log, or tree. A second litter sometimes follows.

Sign: Remnants of acorns, nuts, or other foods on ground or log. Burrows usually concealed.

Habitat: Pastures, piney woods, rocky cliffs, and sagebrush deserts; often abundant in open coniferous forests.

Range: Most of s Canada from s Yukon through Ontario; w U.S. from se Washington and nw California east to w North Dakota and New Mexico; also n Minnesota, Wisconsin, and Upper Peninsula of Michigan.

The Least Chipmunk, palest in color of all the western chipmunks, often lives in the most desert-like habitats and generally enters hibernation much later than the Eastern Chipmunk. Acorns, seeds, fruits, berries, and grasses are its main foods, but it also eats fungi, invertebrates, and (rarely) small vertebrates. An excellent climber, it ascends trees to sun itself and may even nest in them. Its distinctive call, a series of high-pitched chipping notes, is similar to that of the Eastern Chipmunk.

---

153 California Chipmunk
Tamias obscurus

Description: Top of head pale gray shaded with pinkish cinnamon. *Facial stripes brown;* eye stripe black. Somewhat indistinct *dark dorsal stripes are brown or russet;* median stripe is black behind. *Median light stripes are smoke gray, the outer pair paler.* Hindfeet cinnamon-buff. L 8¼–9⅜" (208–240 mm); T 3⅝–3⅞" (91–99

mm); HF 1¼–1⅜″ (33–35 mm); Wt 2–3⅛ oz (56–90 g).

Similar Species: Merriam's Chipmunk is darker; dark stripes are less reddish.

Breeding: Sometimes mates as early as January. Females call from perches prior to estrus; males may come from distances of 200 yards (meters). 1 or 2 litters per year are produced; gestation period and number of young unknown.

Sign: Gnawed acorns or other food litter.

Habitat: Piñon-juniper or pine-oak forests, often with manzanita or sage and near rock outcrops.

Range: South-central California.

The California Chipmunk is active throughout the year. It feeds on a variety of seeds, fruits, and flowers, including piñon nuts and acorns. The burrows, which have short side branches that serve as turn-around spots or fecal pellet depositories, are usually protected by large boulders. One burrow was a straight tunnel, 3 feet (.9 m) long and 3 inches (75 mm) in diameter. Another burrow's entrance into the ground was through a log. The California Chipmunk often calls from bushes, but not from trees. Its repertoire of calls includes the chip, chuck, and chipper, and a trill. An individual may chip, usually from the top of a boulder, for up to five minutes at a rate of 100 to 165 chips per minute, each one accompanied by a tail flip. The chip stimulates other chipmunks to call, while the chuck, which can also occur at a rate of 160 per minute, appears to quiet other chipmunks. The chipper call is given as the animal runs for cover.

Yellow-cheeked Chipmunk
"Redwood Chipmunk"
Tamias ochrogenys

Description: *Very large, dark-colored chipmunk, with prominent alternating dark and pale longitudinal stripes on face and back.*

2 lateral pale stripes most prominent; middle stripe is longest, widest, and most conspicuous of dark stripes. Pale-colored patch behind ear. L 9⅛–11⅝″ (233–297 mm); T 3¾–5⅛″ (97–130 mm); HF 1¼–1⅝″ (33–42 mm); E ¾–1″ (20–26 mm); Wt 2⅛–4⅛ oz (60–117 g).

Similar Species: Merriam's and Sonoma chipmunks are smaller and paler, with more contrasting fur. Although very difficult to distinguish in the field, Townsend's, Allen's, and Siskiyou chipmunks are paler, with bushier tails, less distinct back stripes, and smaller head and body length.

Habitat: Humid coastal redwood forest.

Range: Only in nw coastal California.

The Yellow-cheeked Chipmunk has the largest head and body length of any of the western chipmunks, and is even darker in color than Townsend's Chipmunk, of which it was formerly considered a subspecies. The Yellow-cheeked is difficult to observe as it is found in the deep forest. It feeds on a wide variety of nuts, seeds, fruits, and fungi. It is conspicuous only by its shrill, whistling chipper, and its chuck calls.

Palmer's Chipmunk
Tamias palmeri

Description: Closely related and similarly colored to Uinta Chipmunk, but has *browner or more reddish dark back stripes;* is more *tawny or orangish on underside of tail;* has more gray on shoulders. L 8¼–8¾″ (210–223 mm); T 3⅜–4″ (86–101 mm); HF 1¼–1⅜″ (32–34 mm); Wt 1¾–2⅜ oz (50–69 g).

Similar Species: Panamint Chipmunk (the only other chipmunk in its range) is somewhat smaller, with shoulders less gray and back stripes less distinctly dark and light. Uinta Chipmunk has darker stripes that are less brown or reddish in color, and is white below.

Breeding: Mates April–May; 3–7 young born June–early July, after a gestation of at least 33 days. Weight at birth is about ⅛ oz (4 g).

Habitat: Found up to the timberline, most often along rock cliffs or downed logs in various conifer and mountain mahogany, manzanita areas.

Range: Spring and Potosoi mountains of Clark County in s Nevada.

Palmer's Chipmunk generally occurs at higher elevations in Clark Canyon than the Panamint Chipmunk. Whether this species hibernates is not known, but it utilizes stored food and ceases most activity during winter, though it occasionally appears aboveground. Male are active earlier in spring than females, but overall the species becomes active later than Panamint Chipmunks at Clark Canyon, probably because of the higher elevations it occupies. Palmer's feeds on a variety of seeds, fruits, nuts, fungi, green vegetation, and insects; fruits of conifers are very important. Its calls include chips, chucks, and growls.

154 Panamint Chipmunk
Tamias panamintinus

Description: Brightly colored, *with reddish or tawny back; gray head and rump.* Outer dark stripes on back indistinct; inner ones reddish or grayish. Head gray on top; upper eye stripe black, lower one brown. Ears tawny in front. L 7½–8⅝″ (192–220 mm); T 3⅛–4″ (80–102 mm); HF 1⅛–1¼″ (28–33 mm); Wt 1½–2¼ oz (42.5–65.2 g).

Similar Species: Yellow-pine Chipmunk is more brightly colored, with top of head brown, and lower eye stripe black and distinct. Least Chipmunk is grayer and generally smaller, with rump similar in color to back. Long-eared Chipmunk is larger and more reddish, with less gray and less-distinct stripes. Cliff Chipmunk averages

much larger and has indistinct back stripes. Lodgepole and Uinta chipmunks' ears are blackish in front, whitish behind. Palmer's Chipmunk is somewhat larger, with more gray on shoulders and more distinct striping on back.

Breeding: Mates primarily April–early May. 1 litter per year of 3–7 young born May–June. Gestation period is about 36 days. The young first appear aboveground in July, and weaning is completed by August.

Sign: Parts of piñon nuts and other food items.

Habitat: Piñon-juniper forest in rocky areas.

Range: Southwestern Nevada and sc California.

Although this chipmunk's range is virtually identical to that of the region's piñon pine, their coincidence probably does not depend so much on piñon nut crops, which can vary in abundance, as on other factors, such as temperature, the extremes of which sometimes limit species distribution. Although it is considered an only occasional hibernator, the Panamint Chipmunk does become inactive at lower temperatures. In addition to piñon nuts, the Panamint Chipmunk feeds on juniper berries, forb seeds, fruits, green vegetation, and insects. This chipmunk will sometimes climb piñon trees and bushes, and even establish nests there, although it generally remains on the ground and appears to depend on rocks for shelter. It will call from the tops of rocks, from which it can see in all directions, and may shell seeds there. Its calls consist of sharp chattering notes, chucks, and whistles.

Long-eared Chipmunk
Tamias quadrimaculatus

Description: A fairly large chipmunk. *Brightly colored;* reddish overall. Body stripes indistinct. Tail reddish brown below, edged with white. *Ears have large white patch behind*

and dark, almost black, stripe below.
L 7⅞–10⅛" (200–255 mm); T 3⅜–4⅝" (85–118 mm); HF 1⅜–1½" (34–37 mm); E ¾–1" (18–26 mm); Wt 2½–3½ oz (71–100 g).

Similar Species: Townsend's and Merriam's chipmunks have brown stripe below ear. Other chipmunks in range are smaller, with less-distinct stripes.

Breeding: Mates late April–June; 1 litter per year of 2–6 young born May–July after 31-day gestation.

Sign: Parts of conifer cones and small pits dug for fungi.

Habitat: Openings, brushy areas, and edges in pine and fir forests of Sierra Nevada mountains at elevations from 3,600 to 7,300' (1,100–2,200 m).

Range: Northern Sierra Nevada of ne California and extreme w Nevada.

Numbers of Long-eared Chipmunks slowly increase beginning in late April and May as individuals emerge from hibernation; reduced numbers in November indicate the beginning of hibernation. The primary foods of the Long-eared Chipmunk are conifer seeds, fungi, cedar and gooseberry seeds, manzanita flowers and fruits, various nutlets, and arthropods—mainly caterpillars, but also termites. It gleans leftover pine or cedar seeds from cone middens left by Douglas Squirrels and seeds that Douglas and Western Gray squirrels have dropped from trees. It sometimes eats exclusively underground fungi similar to truffles. Although this species usually remains near the ground, it may climb cedar trees to a height of 30 feet (9 m), cut cones directly from the twigs, gnaw out the seeds, and let the seeds and cones drop to the ground. It then caches this food for consumption in winter and spring. The call note of this species is sharper than that of the Lodgepole Chipmunk, which occurs within the same range.

155 Colorado Chipmunk
Tamias quadrivittatus

Description: A medium-size chipmunk, basically
*orange overall. Head cinnamon, shading to
gray.* Creamy white below; tawny wash
on sides. *3 median stripes on back black,*
with yellowish-orange margins; *outer
stripes brownish.* Shoulders gray. Rump
and thighs cinnamon-buff. Tail black-
tipped, white-bordered, tawny below.
Ears blackish in front, whitish behind.
L 8⅛–9⅜″ (207–240 mm); T 3⅜–4⅛″
(85–105 mm); HF 1¼–1⅜″ (31–34
mm); Wt about 2 oz (60 g).

Similar Species: Very similar Uinta Chipmunk is
basically brown. Hopi Chipmunk is
usually paler and redder, with back
stripes less black. Gray-collared
Chipmunk has gray shoulders. Gray-
footed Chipmunk has gray on upper
part of hindfoot.

Breeding: Mates in spring; 1 litter of 2–6 young
born in late spring. Young are nearly
full grown by July–August. Presence of
some small young in October may
indicate an occasional second litter.

Sign: Remnants of piñon nuts, acorns, fruits,
and other food items.

Habitat: In Colorado, a variety of habitats
including desert scrub, grassland-
chaparral, tundra, and spruce-fir and
piñon-juniper forests. In New Mexico,
primarily ponderosa pine areas, but not
restricted to forested habitats.

Range: Southeastern Utah, Colorado, ne
Arizona, and n New Mexico.

Active in early morning and late
afternoon, the Colorado Chipmunk
gives a short warning bark when
alarmed. More arboreal than most
chipmunks, it tends to occupy
coniferous areas and has been seen
eating seeds at the tops of spruce trees.
It feeds on a great variety of seeds,
fruits, fungi, and insects, and its caches
often contain grain seeds. In Colorado
this chipmunk is often associated with

the more abundant Least Chipmunk,
which occupies meadows. In New
Mexico, it occurs at higher elevations—
in ponderosa pine, Douglas fir, and
aspen—than the Cliff Chipmunk, which
occurs in lower-elevation piñon-pine
and juniper woods. This species occurs
in male-female pairs during the
breeding season.

Red-tailed Chipmunk
Tamias ruficaudus

Description: A large, brightly colored chipmunk.
Deep tawny above and on sides; *gray
rump* contrasts with front part of body.
Tail rufous above, dark reddish below.
3 median stripes on back black; *outer
stripes brownish.* 2 white and 3 brown
stripes on cheeks. L 8¾–9¾″ (223–
248 mm); T 4–4¾″ (101–122 mm);
HF 1¼–1⅜″ (32–36 mm); Wt about
2 oz (60 g).

Similar Species: Least Chipmunk is smaller, with shorter
tail and ears, and without contrasting
rump. Yellow-pine Chipmunk is grayish
yellow on underside of tail; ears and tail
are shorter.

Breeding: Mates April–May (later farther north).
Bears 4–6 young per litter. Gestation
probably about 30 days.

Habitat: Spruce-fir, pine-larch-fir, or yellow pine
coniferous forests; boulder-covered
slopes and mountains below timberline,
particularly in dense forest areas.

Range: Southeastern British Columbia, ne
Washington, n Idaho, and w Montana.

Little is known about the habits of the
Red-tailed Chipmunk, which is more
arboreal than most chipmunks. During
cold weather the Red-tailed Chipmunk
stays in its burrow but apparently does
not hibernate. Its known foods are the
seeds of fir trees, honeysuckle, black
locust, cranberries, and knotweed. In
northeastern Washington three dried-
grass nests were discovered in spruce

trees, resting on limbs close to the trunk about 20 to 60 feet (6–18 m) above the ground. Two ground nests, built of dried grass and lichens, were found in tunnels in the same area. This species sandbathes, rolling its body from side to side in the sand while simultaneously moving forward. It gives a short warning bark when alarmed.

Hopi Chipmunk
Tamias rufus

Description: Marked reduction of black in back stripes gives an *overall buffy to pale orangish-red appearance.* Pale gray crown, *large white spot behind ear. Center stripe chestnut or rufous, black in middle part of its posterior portion. Nearly invisible outer pair of stripes rufous, without black.* Tail mixed black and chestnut above, bright chestnut with an indistinct border below. 2 white and 3 rufous stripes on cheeks. L 3⅝–5⅞″ (93–148 mm); HF 1⅛–1⅜″ (30–35 mm); Wt about 2 oz (60 g).

Similar Species: Least Chipmunk is smaller, lacking the bright tawny to orangish-red coloring; its back stripes are usually black rather than reddish. Colorado Chipmunk is usually brighter, though less reddish in color, with blacker back stripes.

Breeding: Mates late February–March; 1 litter per year of 4–7 young, usually born in April after gestation of 30–33 days. Lactation usually continues through May.

Sign: Leavings of food items such as piñon pine nuts.

Habitat: Bare or vegetated rocky areas where juniper, piñon pine, blackbrush, and Mormon tea are found.

Range: Eastern Utah, w Colorado, and nw Arizona.

Active from late February or early March until late November, the Hopi Chipmunk apparently has a short hibernation period. It feeds heavily on berries of one-seeded juniper, Indian

ricegrass, mountain mahogany, and piñon pine nuts. In late spring and summer it also consumes green vegetation. This species apparently needs free drinking water, which it probably finds in rock pools. An accomplished climber, the Hopi Chipmunk often can be seen in trees or running about on cliffs. Nests have been found in broken rock or crevices in solid rock in Arches National Park, Utah. The Hopi Chipmunk has recently been separated from the Colorado Chipmunk, because the two show no sign of interbreeding where they occur together. Its range overlaps with that of the Least Chipmunk, but the Hopi occurs at lower elevations; one way to distinguish the two is that the Hopi carries its tail more horizontally than the Least.

Allen's Chipmunk
Tamias senex

Description: *A large chipmunk,* with grayish fur, especially on head, rump, and thighs. Varies from overall dark coloration with obscure back stripes along coast to olive grayish coloration with conspicuous stripes in Nevada mountains. Usually *markings on body relatively indistinct, except dark middle stripe on back.* In winter, dark back stripes are black mixed with brown, with middle stripe darkest. In summer pair of light stripes often tinged with pinkish buff in middle of back. *Conspicuous white spot behind ears.* Tail pale tawny underneath with indistinct buffy edging. L 9–10¼" (229–261 mm); T 3¾–4⅜" (95–112 mm); HF 1⅜–1½" (35–38 mm); Wt 2⅜–3⅞ oz (67–109 g).

Similar Species: Closely related Townsend's Chipmunk differs primarily in bacular (penis bone) morphology and vocalizations. Yellow-cheeked and Siskiyou chipmunks are difficult to distinguish, but Yellow-cheeked is darker, with more distinct back stripes.

Breeding: Mating season begins about 1 month after emergence from hibernation and lasts about 4 weeks. 1 litter per year of 3–5 young.

Sign: Remnants of food items such as cedar cones. Numerous pits in ground where animal has foraged for underground fungi.

Habitat: Dense, moist brushy areas and coniferous forest with red and white fir trees (*Abies magnifica* and *concolor*, respectively), from sea level to 9,500′ (2,900 m) elevation in Yosemite National Park.

Range: Central Oregon south through n and e California and extreme nw and wc Nevada.

Allen's Chipmunk hibernates from November to mid-March. Before hibernation it gains an average of 20 percent of its body weight, depositing layers of fat under the skin and around internal organs. In contrast to other chipmunks, Allen's is a heavy fungus feeder, but it also eats, in much lower amounts, insects and some vegetative material. This chipmunk was previously considered a subspecies of Townsend's Chipmunk, but is now recognized as a separate species based on penis-bone morphology and its calls. The calls include an excited bark consisting of three to five notes in a series and a single-syllable chip. These calls also help distinguish Allen's from the Yellow-cheeked and Siskiyou chipmunks.

Siskiyou Chipmunk
Tamias siskiyou

Description: Closely resembles Townsend's Chipmunk. *In winter, stripes on middle of back black.* In summer, outer pair of light stripes on back are clear grayish white; inner pair tinged with pinkish cinnamon. L 8⅝–10⅞″ (220–277 mm);

T 3⅝–5″ (93–126 mm); HF 1¼–1½″ (32–39 mm).

Similar Species: Very similar Townsend's Chipmunk differs primarily in vocalization and the shape of the penis bone. California Chipmunk has paler fur in winter. Yellow-cheeked Chipmunk is less grayish and more brownish. Allen's Chipmunk is paler. Yellow-cheeked Chipmunk is darker.

Habitat: Unknown.

Range: Coastal nw California and adjacent sw Oregon.

The species formerly known as Townsend's Chipmunk has been broken down into separate species, based on differences discovered in the shape of the penis bone (bacular morphology) and the differences in their calls. These species—Townsend's, Yellow-cheeked, Allen's, and Siskiyou—remain distinct without interbreeding, even when they occur together geographically. They are virtually identical in appearance, with only subtle external differences.

Sonoma Chipmunk
Tamias sonomae

Description: A large chipmunk. *Dark brown, with outer dark stripes on back indistinct, pale stripes yellowish. Tail reddish below, becoming paler toward base and edged with buff.* On head, dark stripes reddish brown, or brownish black; black spots behind eyes and below ears. Backs of ears nearly uniform brownish. L 8⅝–10⅞″ (220–277 mm); T 3⅝–5″ (93–126 mm); HF 1¼–1½″ (32–39 mm).

Similar Species: Townsend's, Allen's, and Siskiyou chipmunks are darker, with shorter ears, legs, and tail, and cheeks gray in winter. Siskiyou has paler hindfeet; its middle pair of light stripes is heavily mixed with pinkish cinnamon. Yellow-cheeked Chipmunk is dull-colored rather than tawny, with dark stripes

paler and less distinct, and pale stripes
more buffy.

Breeding: 1 litter per year of 3–5 young born in
spring. Female raises the litter alone,
suckling the young for at least 3 weeks
after they emerge, mostly at night.

Habitat: Brushy open ground in redwood,
ponderosa pine, yellow pine, and other
coniferous forests, from sea level to
6,000′ (1,800 m).

Range: Northwestern California.

Chipmunks that occur together may
have distinctive calls that allow them to
distinguish members of their own
species and help to keep related
individuals in contact. The Sonoma
Chipmunk's call, a chirp that drops then
rises in pitch, is lower and slower than
that of many other chipmunks. The
Sonoma Chipmunk forages in small
branches of bushes as well as on the
ground. It spends much time perched on
a rock or log. Males travel about during
the breeding season in their search for
mates. Females stay with the young at
least three weeks after they emerge from
the nest; the young remain together
after the mother leaves.

156 Lodgepole Chipmunk
Tamias speciosus

Description: *Brown above, with distinct stripes:* median
dark back stripes black, but *outer ones
brown or often missing; outer pale stripe
bright white and broader than inner pale
stripes. Top of head brown; stripes on front of
head often lacking.* Black spots in front of
and behind eyes. Ears blackish in front,
white behind. Tail has black band about
1″ (25 mm) wide on underside near tip.
L 7¾–9½″ (197–241 mm); T 2⅝–4½″
(67–114 mm); HF 1⅛–1⅜″ (30–36
mm); Wt 1¾–2⅛ oz (51–62 g).

Similar Species: Uinta Chipmunk has narrower black
band beneath tail; top of head is gray.
Yellow-pine Chipmunk has black side

stripes. Least, Alpine, and Panamint chipmunks are smaller. Merriam's, Long-eared, and Townsend's chipmunk are larger.

Breeding: Mates about 1 month after emergence from hibernation, in May or June. 1 litter per year of 3–6 young born in spring or early summer.

Habitat: Lodgepole pine and red fir stands; often associated with manzanita.

Range: Much of e and sc California.

The Lodgepole Chipmunk apparently hibernates for five or six days at a time between October and mid-April. Manzanita flowers and berries, nutlets, subterranean fungi, and caterpillars are important foods. In the San Gabriel Mountains, the Lodgepole Chipmunk often sits on the upper stems of a snowbush to scan its environment. It occurs in the mountains with Merriam Chipmunk, which it replaces at higher elevations. The call of the Lodgepole is not as sharp as that of the Long-eared Chipmunk.

158, 159 Eastern Chipmunk
Tamias striatus

Description: Reddish brown above; belly white. *1 white stripe bordered by 2 black stripes on sides; stripes end at rump.* 2 white stripes on back much thinner than side stripes. Dark center stripe down back; pale facial stripes above and below eyes. Tail brown on tip, edged with black. Prominent ears. L 8½–11¾" (215–299 mm); T 3⅛–4⅜" (78–113 mm); HF 1¼–1½" (32–38 mm); Wt 2¼–5 (66–139 g).

Similar Species: Least Chipmunk has 4 white stripes of equal width on back.

Breeding: Mates in early spring; 1 litter per year 3–5 young born in May. First-year females not breeding in early spring may produce a litter late July–August.

Sign: Burrow entrances 2″ (50 mm) wide, without piles of dirt, often on a woody slope or bank. Occasional sprinklings of nutshells opened on one side. Bits of chaff on logs, stumps, and rocks. *Tracks:* In mud, hindprint 1⅞″ (48 mm) long, foreprint considerably smaller; straddle 1¾–3½″ (45–90 mm); stride 7–15″ (180–380 mm), with hindprints closer together and printing ahead of foreprints.

Habitat: Open deciduous woodlands, forest edges, brushy areas, bushes and stone walls in cemeteries and around houses.

Range: Southeastern Canada and ne U.S. east from North Dakota and e Oklahoma, and south to Mississippi, nw South Carolina, and Virginia.

The Eastern Chipmunk hibernates from late fall to early spring, waking to eat every two weeks or so. Individuals may occasionally appear on the surface in the snow, especially in mild weather. Essentially a ground species, this pert chipmunk, like the gray and fox squirrels, often feeds on acorns and hickory nuts. It does not hesitate to climb large oak trees when acorns are ripe, and will also scale *Corylus* bushes to harvest hazelnuts. The cutting sounds it makes as it eats nuts can be heard for some distance. In addition to nuts, its diet includes seeds and other types of vegetation, some invertebrates such as slugs and snails, and small vertebrates, probably found as carrion. This species is single-minded in its food gathering, making trips from tree to storage burrow almost continuously. It was estimated that over three days one chipmunk stored a bushel of chestnuts, hickory nuts, and corn kernels. Burrows, consisting of single tunnels or more complex systems, are up to 10 feet (3 m) long and less than 3 feet (1 m) deep. They may include enlarged cavities for nests (made of pieces of leaves) and food

caches, which are often large enough to last into the following spring and summer. Eastern Chipmunks of both sexes vocalize prominently, using one of two chattering calls: a trilling *chip-chip-chip* repeated very rapidly (about 130 trills per minute) and a lower-pitched, slower *chuck . . . chuck . . . chuck.* The Long-tailed Weasel is the Eastern Chipmunk's major predator, but hawks, foxes, the Bobcat, and house cats also take their share.

Townsend's Chipmunk
Tamias townsendii

Description: A large chipmunk. *Dark brown,* often with rather *wide diffuse or indistinct blackish and pale stripes on head and continuing down body*. Paler overall in summer than winter. *Backs of ears bicolored: dusky on forepart; gray on hindpart. Brownish stripe below ears.* Tail long and bushy; blackish above, with many white-tipped hairs; bright reddish brown below, bordered with black and finely edged with white-tipped hairs. L 8¾–14¼″ (221–363 mm); T 3½–6″ (90–152 mm); HF 1¼–1⅝″ (32–40 mm); Wt 1¾–3⅞ oz (51–109 g).

Similar Species: Long-eared Chipmunk has prominent white patch behind ear. Sonoma Chipmunk has unicolored backs of ears. Other chipmunks in range are smaller, with more distinct stripes.

Breeding: Mates in spring; 1 litter per year of 2–6 young born May–June.

Sign: Remnants of food items such as acorns, maple seeds, and conifer cones.

Habitat: Rank vegetation, usually among dense hardwood or humid coniferous forests.

Range: Extreme sw British Columbia south through most of w Oregon.

One of the largest western chipmunks and the darkest in color (as is common among species in moist climates), Townsend's Chipmunk has a looser, less

sleek coat than other species. It is active all day, but rather shy. In northern parts of its range it may put on a layer of fat and remain in a nest burrow all winter, but in milder climates it puts on little or no fat and may be abroad most of the winter. It forages within a home range of 1½ acres (½ ha), eating many types of berries in summer, switching in late fall to acorns, maple seeds, and seeds of various conifers, and in winter to numerous types of subterranean fungi. It also eats some insects. Townsend's Chipmunk lives in a burrow about 2 inches (50 mm) across and only 5 feet (1.5 m) long. A very good climber, it often suns itself in trees and may run up a tree to flee a predator. Its major predators are the Long-tailed Weasel and the Mink. One tagged specimen lived in the wild for at least seven years, unusual longevity for a small rodent. Chipmunks previously recognized as belonging to this species have been split into several species: Townsend's, Allen's, Yellow-cheeked, and Siskiyou. These divisions were based primarily on differences in the penis bone and the call. The four species do not interbreed where they occur together.

157 Uinta Chipmunk
Tamias umbrinus

Description: Forehead light brown, washed with grayish; shoulders and sides brownish, with no gray wash. *Back stripes very distinct; 3 middle stripes very dark; wide, dark brown side stripes.* White below; tawny wash on sides. Tail black-tipped, white-bordered; tawny below, with narrow black band. *Ears blackish in front, whitish behind.* L 7¾–9½" (196–243 mm); T 2⅞–4½" (73–115 mm); HF 1⅛–1⅜" (30–35 mm); Wt 2–3 oz (57–85 g).

Similar Species: Least Chipmunk is smaller, with its underparts more grayish yellow and ears tawny in front. Townsend's, Long-eared,

and Cliff chipmunks have indistinct side stripes. Panamint Chipmunk has tawny shoulders, sides, and fronts of ears. Yellow-pine Chipmunk has distinct black side stripes. Lodgepole Chipmunk is brown above, with top of head brown. Gray-collared Chipmunk has gray cheeks and more gray on shoulders.

Breeding: 1 litter per year of about 5 young born late June–early July; time of births varies with latitude and elevation.

Sign: Remnants of nuts, acorns, and cones.

Habitat: Coniferous forests, mixed woods, open areas; dwells in yellow pine, white pine, juniper, and scrub oak.

Range: West-central California, Nevada, Utah, nc Arizona, Wyoming, and nc Colorado.

This tree-dwelling species often occurs with the Golden-mantled Ground Squirrel. Nuts, seeds, fruits, and berries are its chief foods. It accumulates much fat, which is used in hibernation during the long winters of its range.

Alaska Marmot
Marmota broweri

Description: Body appears tricolored: front gray, middle black, rump reddish. *Black "cap" on head and face,* extending from nose to neck. Fur on back softer than in other species of marmots. Guard hairs (long, often coarse hairs projecting beyond shorter and denser underfur) black-tipped with pale band near end. Feet grizzled or black and white. Averages: L 23–24″ (580–605 mm); T 6–6⅜″ (154–163 mm).

Similar species: All other marmots (except for Vancouver Marmot, which is found only on Vancouver Island) have guard hairs with pale tips.

Breeding: Probably mates before leaving hibernation quarters; young born 2 weeks after emergence from hibernation.

Sign: Large burrows with 8–12″ (200–300 mm) openings.

Habitat: Rocky outcrops and talus slopes near vegetation above timberline.

Range: Northern Alaska.

This species is very similar to the Hoary Marmot, and was once thought to belong to the same species. Recent studies, however, have shown it to be more closely related to a Eurasian species, *M. camtschatica.* Little is known about the habits of this far northern species. Individuals hibernate together in a burrow.

---

226 Hoary Marmot
"Rockchuck," "Mountain Marmot," "Whistler"
Marmota caligata

Description: A large marmot. Silver-gray above, with brownish rump; whitish belly. Distinctive *black and whitish markings on head and shoulders:* nose and large patch between eyes whitish; patches on forehead around eyes and ears black; often has black band on snout above nose. Tail large, reddish brown, bushy. Ears small. Feet black or very dark brown; forefeet may have white spots. L 17¾–32″ (450–820 mm); T 6¾–9⅞″ (170–250 mm); HF 3⅝–4″ (91–102 mm); Wt 8–20 lb (3.6–9 kg).

Similar Species: Olympic Marmot, found only in Olympic Mountains of Washington, is brown on head and chest, and has dark brown feet. Yellow-bellied Marmot is yellowish. Woodchuck has grizzled brown head and body.

Breeding: Mating occurs soon after emergence in spring; 4 or 5 young are born about a month later. 1 litter per year.

Sign: Large burrows 9–15″ (230–380 mm) wide with fans or mounds of dirt at openings.

Habitat: Talus slopes in mountains, alpine meadows, and cliffs.

Range: Alaska and Yukon Territory south to Washington, n Idaho, and w Montana.

The Hoary Marmot's silvery fur, which offers good camouflage in its rocky habitat, gave rise to its common name, and its dark feet to its Latin species name (*caligata* means "booted"). Its shrill alarm whistle, louder than that of other marmots and similar to a human whistle accounts for the nickname "Whistler." In late summer, the Hoary Marmot puts on a great deal of fat, which sustains the animal through hibernation. In the more southerly parts of its range it hibernates from October to February, in British Columbia from September to April. This marmot feeds almost entirely on grasses and many other kinds of green plants. An individual may chase others from feeding grounds it considers its own. These marmots often engage in wrestling matches in which two animal stand erect on their hindlegs, place thei forefeet together, then push at each other. Many carnivores prey on the Hoary Marmot; while its rocky habitat provides good escape cover, bears often dig up individuals that are still hibernating in early spring. Native peoples of the far north use its pelt, which has soft, dense underfur, to make parkas, and they eat its meat.

227 Yellow-bellied Marmot
"Yellow-footed Marmot," "Rockchuck,"
"Mountain Marmot"
Marmota flaviventris

Description: A heavy-bodied marmot. *Yellowish brown, with yellowish belly.* Feet buff to light brown. *Whitish spots between eyes.* Buff or yellowish patches below ear to shoulders. Tail bushy. L 18½–28″ (470–700 mm); T 5⅛–8⅝″ (130–220 mm); HF 3–3⅛″ (76–80 mm); Wt 5–10 lb (2.2–4.5 kg).

Similar Species: Woodchuck, usually somewhat larger and darker with darker feet, has no white spots between eyes and occurs farther north and east. Hoary Marmot

mainly black and white on head and forepart of body.

Breeding: 1 litter per year, of about 5 young, born March–April.

Sign: In open areas, burrow entrances 8–9″ (200–230 mm) wide, with mounds or fans of packed earth.

Habitat: Rocky areas, talus slopes, valleys, and foothills to elevations of 11,000′ (3,300 m).

Range: British Columbia, s Alberta, and Montana south through e California and east to Colorado and n New Mexico.

The Yellow-bellied Marmot feeds entirely on green vegetation of many kinds. In the fall, it puts on a layer of fat, which sustains it through hibernation from August (October in the mountains of New Mexico) through February or March. It lives in a den in a hillside, under a rock pile, or in a crevice or rock shelter. If alarmed, the animal returns to its den and often chirps or whistles from its position of safety. This marmot is a host for the tick that carries Rocky Mountain spotted fever.

---

225 **Woodchuck**
"Groundhog," "Marmot"
Marmota monax

Description: A large marmot. *Grizzled brown* (with variations from reddish to blackish); *uniformly colored.* Prominent *bushy tail.* Small ears. Short legs. *Feet dark brown or black.* Incisors white. L 16½–32″ (418–820 mm); T 3⅞–6″ (100–152 mm); HF 2⅞–3⅞″ (75–100 mm); Wt 4½–14 lb (2–6.4 kg).

Similar Species: Yellow-bellied Marmot, usually paler and yellower, has whitish spots between eyes and occurs farther south and west.

Breeding: 1 litter per year of 4 or 5 young born April–early May, after 28-day gestation. Blind and naked at birth, they open their eyes and crawl at about 1 month and disperse at 2 months.

Sign: Large burrow openings, 8–12″ (200–300 mm) across, with mounds of dirt just outside main entrance; often additional escape openings with no mounds. When hay is high, woodchucks tramp down trails, which radiate from burrows.

Habitat: Pastures, meadows, old fields, and wooded areas.

Range: East-central Alaska and British Columbia south to n Idaho, east through most of s Canada, and south to e Kansas, n Alabama, and Virginia.

The name Woodchuck comes from a Cree Indian word, *wuchak,* used to identify several different animals of similar size and color, including other marmots; it denotes nothing about the Woodchuck's habits or habitat. This sun-loving creature is active by day, especially in early morning and late afternoon. In late summer or early fall, the Woodchuck puts on a heavy layer of fat, which sustains it through hibernation. It digs a winter burrow with a hibernation chamber, where it curls up in a ball on a mat of grasses. The animal's body temperature falls from almost 97°F (36°C) to less than 40°F (4°C), its breathing slows to once every six minutes, and its heartbeat drops from more than 100 beats per minute to four. The Woodchuck emerges in early spring (according to legend, on February 2, Groundhog Day, but much later in northern parts of its range). A male at once seeks a mate; its brief stay in the burrow of a receptive female is almost the only time that two adults share a den. A good swimmer and climber, the Woodchuck will ascend a tree to escape an enemy or obtain a vantage point, but never travels far from its den. Its burrow, up to 5 feet (1.5 m) deep and 30 feet (9 m) long, has one or more tunnels terminating in a chamber containing a large grass nest. Other mammals, including cottontail rabbits,

Virginia Opossums, Common Raccoons,
skunks, and foxes, may use a vacant
Woodchuck burrow, sometimes
enlarging it to create a nursery den.
Green vegetation, such as grasses, clover,
alfalfa, and plantain, forms its diet; at
times it will feed heavily on corn and
can cause extensive damage in a garden.
If alarmed, the Woodchuck often gives a
loud, sharp whistle, followed by softer
ones as it runs for its burrow, from which
it then peeks out. When agitated, it
chatters its teeth, and it can hiss, squeal,
and growl. The human hunter is the
Woodchuck's major enemy, but the
automobile and large predators, especially
the Red Fox, also take their toll. While
an overpopulation can damage crop fields,
gardens, and pastures, Woodchucks are
beneficial in moderate numbers. Their
defecation inside the burrow, in a special
excrement chamber separate from the
nesting chamber, fertilizes the earth.
Their digging loosens and aerates the
soil, letting in moisture and organic
matter while bringing up subsoil for
transformation into topsoil (in New
York State they turn over 1.6 million
tons of soil each year).

228 Olympic Marmot
Marmota olympus

Description: Head and forepart of body mainly brown
(fading to yellow by August), mixed
with white guard hairs. Small ears.
Prominent *bushy tail.* Short legs; *dark
brown feet.* White incisors. L 17¾–31″
(450–785 mm); T 7⅝–9⅞″ (195–252
mm); HF 2⅝–4¼″ (68–110 mm).

Similar Species: Hoary Marmot has prominent black and
white coloration on head and forepart of
body. Woodchuck is uniformly grizzled
blackish or dark brown on head and
upper body, and has dark brown to
black feet. Yellow-bellied Marmot has
conspicuous yellowish patches on sides
of head and neck, and buffy or pale feet.

Breeding:	Generally 1 litter every other year, averaging 4 offspring. Gestation is 30–32 days.
Sign:	Large burrow openings, 8–12″ (200–300 mm) across.
Habitat:	Subalpine meadows and rocky slopes.
Range:	Olympic Mountains, Clallam County, Washington.

The Olympic Marmot is highly social and tolerant of close proximity to other individuals. It lives in colonies that usually consist of one male, two female a litter of yearlings, and the new litter. Mating occurs shortly after emergence from hibernation, which lasts generally from September to March. The young disperse in their second year. This species eats green vegetation, including sedges, lupines, lilies, heather blossom and mosses. The Olympic Marmot's overall color changes during the active season from brown in spring to yellow by August, apparently because of bleaching by sunlight.

---

229 Vancouver Marmot
Marmota vancouverensis

Description:	*Top of head, shoulders, and upper back uniformly dark brown,* without pale tips. Some whitish guard hairs on back. *White patch on nose; white spotting on abdomen.* Prominent, *bushy tail.* Small ears. Short legs. *Feet pale or buffy.* Incisc white. Males generally larger and heavier than females. Averages: L male 27″ (695 mm), female 26″ (661 mm); T male 8⅝″ (220 mm), female 7⅝″ (19 mm); HF 3¾″ (96–97 mm). Weights range from about 6½ lb (3 kg) in May to 14¼ lb (6.5 kg) in September, with males much heavier than females.
Similar Species:	No other marmots occur on Vancouver Island.
Breeding:	1 litter of about 3 young born after gestation of 28–33 days. Young first emerge from burrows late June–July.

Females reproduce in alternate years, beginning in their third year.

Sign: Large burrow openings, 8–12″ (200–300 mm) across.

Habitat: Most abundant in subalpine herbaceous communities with scattered hemlock and spruce, at altitudes of 3,600–4,750′ (1,100–1,450 m); also occurs in woodlands, cutover land, and along road banks and ski runs.

Range: Vancouver Island, British Columbia.

The Vancouver Marmot is very dark, with white guard hairs, and lacks the multicolored or banded hairs that often occur in other marmots. It has been suggested that the dark coloration, which may occur as an individual genetic variant in other marmots, has risen in frequency by chance in this species to become the exclusive coat type. Especially active in morning and evening, the Vancouver Marmot spends most of its time outside the burrow resting or feeding on green vegetation and fruits, including bearberry, blueberry, bracken fern, bluebells, sedges, and grasses. Hibernation occupies about seven to eight months of the year, from early October to early May. In September, the animals are nearly twice their springtime weight. Mating presumably occurs aboveground during the first three weeks after emergence from hibernation. This marmot is colonial, forming groups that average seven or eight individuals before the emergence of the year's young in June or July. A colony typically consists of an adult male and an adult female, plus two-year-olds and yearlings. The dominance hierarchy is as follows: adult males, adult females, two-year-old females, and yearling females. (Young males probably leave at an early age.) Adult Vancouver Marmots are territorial and use scent produced by their cheek glands to mark territories. This behavior is especially common in adult males, but

two-year-old females also scent-mark.
Social interactions observed in this
species are quite varied, with greeting
and play-fighting the most common
behaviors exhibited when two
individuals meet. In play-fighting,
both animals stand on their hindlegs
and push against each other with their
forefeet. The marmots greet one
another by touching noses or sniffing
the other's cheek, ear, or flank. Anal
sniffing, mounting, tail raising, social
grooming, and play-chasing are other
forms of social interaction; lunging
and chasing are indicators of
dominance. Vocalizations include
whistles, hisses, chirps, screams, and
growls. The Vancouver Marmot is
listed as endangered, with just 11
colonies counted in 1979–1980, and
only 231 animals found in 1984. One
problem has been habitat loss due to
the proliferation of ski resorts on
Vancouver Island. However, much of
the island has not been surveyed
properly, and additional work may show
this animal to be more abundant than
current records indicate. Captive
breeding programs with this species
have had some success.

160 Harris' Antelope Squirrel
Ammospermophilus harrisii

Description: Upperparts pinkish buff in summer,
gray in winter; underparts white. Single
white stripe on each side. *Tail* grayish
above, *mixed black and white below*. Ears
small. L 8⅝–9⅞″ (220–250 mm);
T 2⅞–3¾″ (74–94 mm); HF 1½–1⅝″
(38–42 mm); Wt 4–5¼ oz (113–150 g)

Similar Species: In other antelope squirrels, underside of
tail is all white.

Breeding: Mates December–February; apparently
produces 1 litter per year. 4–9 young
born January–March after 30-day
gestation. At birth, young weigh about
⅛ oz (3.6 g).

Sign: Burrows with diverging pathways under desert shrubs; also, shallow holes resulting from digging.

Habitat: Variety of desert habitats, especially low deserts with little vegetation.

Range: Western and s Arizona and extreme sw New Mexico.

The pale coloration of Harris' Antelope Squirrel helps it blend in with its arid environment. Individuals of this solitary species are found scattered about and are sometimes seen up in cactus plants, although it is not known how they avoid the spines. Active at any time of day, even the hottest, this omnivorous animal feeds heavily on the fruit and seeds of cactus and yucca, but eats many other plants and insects as well. It metabolizes its water from food. Harris' Antelope Squirrel can carry considerable amounts of food in its cheek pouches; one animal was found carrying a cache of 44 mesquite beans. It shells the beans before storing them in the burrow, which is usually under a desert shrub, especially mesquite *(Prosopis),* creosote bush *(Larrea),* or paloverde *(Cercidium).* When disturbed, Harris' Antelope Squirrel runs with its tail straight up in the air, often chippering as it runs. Before making its final escape, it often stops, calls, and stomps its feet.

Texas Antelope Squirrel
Ammospermophilus interpres

Description: Fur uniformly dark, with no apparent reddish tint. Single white stripe on each side. *Underside of tail white; upperside has 2 distinct blackish bands. Bold white eye ring.* L 8⅝–9¼" (220–235 mm); T 2⅝–3¾" (68–94 mm); HF 1⅜–1⅝" (36–42 mm); Wt 3¼–4¼ oz (94–121 g).

Similar Species: White-tailed Antelope Squirrel has just 1 black band on tail. Harris' Antelope Squirrel has mixed black and white on underside of tail.

Breeding: Mating begins in February. 1 litter of 5–14 young born late February–April; some evidence of a second litter.

Sign: Burrows with diverging pathways under desert shrubs.

Habitat: Rough, hard-surfaced terrain, such as gravelly washes, canyons, and rocky foothills near juniper woodlands. In Chihuahuan Desert, associated with a variety of vegetation, including creosote bush *(Larrea)* and sotol *(Dasylirion)*. In New Mexico, mostly restricted to canyons and rocky foothills, rarely venturing very far into the desert floor. In Texas, occupies middle-elevation rocky areas, as well as creosote bush flats and oak areas; uncommonly found in higher canyons in Big Bend area. Also associated with tarbush *(Flourensia)* and lechuguilla *(Agave).*

Range: Central New Mexico south into w Texas.

Generally uncommon and spotty in distribution, the Texas Antelope Squirrel is most active during the hottest part of the day. It probably does not hibernate, although it puts on a winter layer of fat; it will emerge and collect food on warm or even cold days during the winter. Its diet includes insects, parts of cacti, berries, mesquite beans, and the seeds of creosote bush, sotol, juniper, and yucca. Green vegetation is added to the diet in spring and early summer. This species is often seen sitting atop boulders or shrubs, and running from bush to bush or across rocks with its tail curled over its back. Its alarm call, like that of all antelope ground squirrels, is a trill. This species usually occupies a burrow (its own or that of another animal) but sometimes nests in a crevice. One Texas Antelope Squirrel dug a burrow with three entrances under hardpan soil in a roadbed. The main tunnel was 10 feet (3 m) long and 3⅜ inches (87 mm) in diameter. The nest chamber, 6⅞ by 4⅞ inches (175 by 125 mm) and 3⅞ inches

(100 mm) high, featured an accessory loop behind the nest and two blind pockets. The nest was constructed of rabbit fur, shredded bark, feathers, dry grass, and bits of cotton.

161 White-tailed Antelope Squirrel
Ammospermophilus leucurus

Description: Upperparts buff in summer, gray in winter; 1 narrow white stripe on each side. Underparts white. *Underside of tail pure white,* with black-tipped hairs forming narrow black border. Upperside of tail has 1 black band. Ears small. L 7⅝–9⅜" (194–239 mm); T 2⅛–3⅜" (54–87 mm); HF 1⅜–1⅝" (35–43 mm); Wt 3–5½ oz (85–156 g).

Similar Species: Very similar Nelson's Antelope Squirrel is larger and more yellowish. Harris' Antelope Squirrel has tail that is black and white underneath. Mohave Ground Squirrel lacks stripes.

Breeding: 1 litter of 5–14 young born in early spring; sometimes a second litter later in the year.

Sign: Burrows with pathways radiating from them; no mounds at entrances.

Habitat: Deserts and foothills, from valley floor to juniper belt; hard, gravelly surfaces. Associated with sage, greasewood, shad scale, and creosote bush *(Larrea)*.

Range: Southeastern Oregon and sw Idaho south to s California; east through s Nevada, Utah, and n Arizona to w Colorado and nw New Mexico.

The most widespread member of its genus in North America, the White-tailed Antelope Squirrel is most active in midmorning and late afternoon, with a lull during the heat of the day. During winter, it is active throughout midday, spending much time basking in the sun. It hibernates in northern parts of its range, but may or may not hibernate in more southerly areas. This species, like other antelope squirrels, runs fast,

with its tail held over its back, and exposing its white underparts. Usually foraging on the ground, but sometimes in yuccas or cacti, it feeds on green vegetation, seeds, insects, and vertebrate flesh, in descending order of importance. Green vegetation, especially evening primrose and storksbill, makes up the largest part of this animal's diet from December to May; it eats seeds, particularly those of ephedra, yucca, and opuntia, all year, but in smallest quantities in March and April. Grasshoppers contribute to the diet in spring and summer, beetles in autumn, and vertebrates all year, probably as carrion. The White-tailed Antelope Squirrel usually lives in burrows, but may also take up residence in rock crevices or abandoned burrows of other animals, often kangaroo rats. Burrows are about 18 inches (450 mm) deep, their entrances under shrubs or in the open. Inside are food caches and, near the center of the system, a nest measuring about 5 to 8 inches (130 to 200 mm) in diameter. It uses available material to build a nest, which may incorporate grasses, fur, and bark. Many burrows throughout the home range allow refuge from danger and from heat. Escape burrows are shorter than home burrows, with no nest or food caches. The White-tailed Antelope Squirrel forms dominance hierarchies, which are maintained by visual and tactical cues. Young animals may lie on the ground and face off with one another, and then proceed to box or wrestle until dominance is established. This sparring is accompanied by growls, chattering, and chirps. Greetings are by oral or nasal touching. This species mates between February and June. The mating season peaks in February and March in southern Nevada, though in California it apparently takes place entirely within a two-week period. This desert creature has several adaptations to survive in extreme heat. When exposed to critical

temperatures, it will enter a burrow and crawl on the floor with its sparsely furred belly in contact with the ground, which quickly draws out excess heat. While out of the burrow, this species uses available shade or shades itself with its tail, and climbs into bushes where the airflow is greater.

162, 163 Nelson's Antelope Squirrel
Ammospermophilus nelsoni

Description: *Yellowish brown* or buffy above, with *1 white stripe on each side; tail white below.* L 9–10½″ (230–267 mm); T 2⅝–3⅛″ (66–78 mm); HF 1½–1¾″ (37–44 mm); Wt 5½ oz (155 g).

Similar Species: White-tailed Antelope Squirrel is smaller and less buffy. Tail of Harris' Antelope Squirrel is black and white underneath.

Breeding: Mates in late winter or early spring; 1 litter per year of 6–11 young born after gestation of 26 days. Average newborn weighs less than ¼ oz (4.9 g).

Sign: Burrows under desert shrubs or in the sides of small gullies.

Habitat: Open, rolling, or hilly desert country and sandy washes; with shrubs in San Joaquin Valley, without shrubs in Kern County, California. Associated plants are orach, Mormon tea, ephedra, and juniper.

Range: Kern, Kings, and w Fresno counties in southern California.

Nelson's Antelope Squirrel is most active in early morning and late evening, showing little movement at midday. It apparently does not hibernate, and there is no evidence that it estivates (becomes dormant during periods of high temperatures), although it often disappears from view in the hottest weather. Its pale coloration is an adaptation to the severe desert conditions in which it lives. Using its cheek pouches to transport food to the

burrow, this species feeds on many different items as available. Green vegetation is a prominent food source i spring, while insects (June bugs, tenebrionid beetles, Jerusalem crickets camel crickets, grasshoppers, and ants) constitute up to 90 percent of the diet from April to December. Seeds, particularly of grasses and Mormon tea comprise 10 to 20 percent of the diet. This species also consumes vertebrates, mostly lizards and rodents. It climbs into shrubs to gather seeds or fruits, ar may leap with grasshoppers until both come down at the same time and the insects can be captured. Most young ar born in March and first appear aboveground in April, the only time of year when green vegetation is present. Weaning may begin before emergence from the nest, and is complete by mid late April. Forming colonies of six to eight individuals, Nelson's Antelope Squirrels dig their burrows in loamy or alluvial soils (which often become brick hard), as well as in sandy or gravelly soils. Very similar to those of kangaroo rats, these burrows have openings unde shrubs and often honeycomb the sides gullies. This squirrel is cautious in exiting from the burrow, relying on sounds more than visual cues. Like all antelope squirrels, it scurries about wit its tail curled over its body, giving the impression of thistledown blowing ove the ground. When the animal is alarmed, its tail twitches back and fort not side to side. Its alarm call is a trill. Even though it is a desert species, Nelson's Antelope Squirrel cannot survive extremely high temperatures. the animal is exposed to sun at 90°F (32°C), its actions become frenzied, an it froths at the mouth and then dies. T annual mortality rate of established adults is 60 to 80 percent, with summe the most critical period, although cold also a problem. California Ground Squirrels may compete with Nelson's

Antelope Squirrels, squeezing them out of their habitat, and kangaroo rats also will take over their burrows. Predators are American Badgers, Coyotes, and Kit Foxes, though land cultivation by humans is the biggest threat to this species. The state of California lists Nelson's Antelope Squirrel as a threatened species.

164 Uinta Ground Squirrel
Spermophilus armatus

Description: Brownish to buff above; sides paler; belly buff. *Tail buff mixed with black above and below,* with pinkish-buff edge. Head, front of face, and ears cinnamon, with *grayish dappling on top of head;* sides of face and neck gray. Forelegs and forefeet buff; *hindlegs cinnamon;* hindfeet buff. L 11–11⅞″ (280–303 mm); T 2½–3⅛″ (63–81 mm); HF 1⅝–1¾″ (42–46 mm); Wt 10–15 oz (284–425 g).

Similar Species: Richardson's Ground Squirrel's tail is clay-colored, buff, or light brownish below. Townsend's Ground Squirrel usually has shorter tail that is reddish or tawny below. Belding's Ground Squirrel has brown streak down back.

Breeding: 1 litter per year, born in May after 28-day gestation. First-year females bear an average of 4 or 5 young, older females 6 or 7. Young emerge from burrow in 24 days.

Sign: Burrow entrances.

Habitat: Dry sage and sage grass; also lawns.

Range: Southwestern Montana, e Idaho, w Wyoming, and nc Utah.

The Uinta Ground Squirrel both estivates (becomes dormant in summer) and hibernates. Adults begin estivation in July, juveniles later; by September, all individuals have disappeared. From estivation, they directly enter the long hibernation. In Utah, adult males emerge first, in late March to mid-April, and remain active only about 3½

months before reentering their sleep cycles. These ground squirrels eat seeds, green vegetation, invertebrates, and some vertebrates. The American Badger appears to be the major predator of this and several other ground squirrels.

---

165 California Ground Squirrel
Spermophilus beecheyi

Description: Brownish, with prominent buff flecks; *whitish wash from sides of neck across shoulders and forelegs to haunches,* enclosing a *dark brown or black, forward-pointing V pattern* on upper back. Rather bushy tail, brownish gray above and below, edged with white. L 14–19¾" (357–500 mm); T 5¾–8⅞" (145–227 mm); HF 1⅞–2½" (49–64 mm); Wt 9⅞–26 oz (280–738 g).

Similar Species: Other ground squirrels in range are smaller and lack V pattern and whitish wash.

Breeding: Mates in early spring. 1 litter per year of 5–8 young, born in May. Time of birth varies with locality.

Sign: Burrows with entrance mound and radiating pathways.

Habitat: Open areas, including rocky outcrops, fields, pastures, and sparsely wooded hillsides.

Range: South-central Washington, w Oregon, most of California, and wc Nevada.

Active from dawn to dusk, the California Ground Squirrel hibernates from November to February; first-year animals often remain aboveground through the winter. While this ground squirrel may climb into brush or a tree to bask in early-morning sunlight, it otherwise remains on the ground. It sometimes consumes insects and small vertebrates, but primarily feeds on plant material, including leaves, stems, flowers, bulbs, roots, seeds, fruits, and berries; it often damages grain, fruit, and nut crops. These squirrels form

loose colonies, but individuals tend to be antisocial. Young first begin to burrow at about eight weeks of age. Several animals may occupy one burrow, which typically is 3 to 6 inches (75–150 mm) wide and 5 to 200 feet (1.5–60 m) long. Each animal uses its own entrance, and it is usually to this hole, rather than the nearest, that it races when alarmed. Burrows are generally under a log, tree, or rock when one is available, but are otherwise in the open, with a mound at the main entrance. Some are used for many years by successive occupants. The fleas of this ground squirrel sometimes carry bubonic plague.

166 Belding's Ground Squirrel
Spermophilus beldingi

Description: Gray washed with reddish or pinkish above, with broad *brown streak down middle of back.* Top of head pinkish. Pale *gray underneath,* with pinkish wash (especially toward front). *Relatively long,* black-tipped *tail,* pinkish gray above, *reddish to hazel below,* edged with pinkish buff. L 9–11¾" (230–300 mm); T 1¾–3" (44–76 mm); HF 1½–1⅞" (39–47 mm); Wt 8–12 oz (227–340 g).

Similar Species: Most other similar ground squirrels have spotting or mottling on back, and lack reddish coloration on underside of tail. Richardson's Ground Squirrel's tail is clay-colored, light brownish, or buff below. Townsend's Ground Squirrel is smaller, with buff or whitish underparts.

Breeding: 1 litter per year of 3–8 young, born late June–early July after gestation of 23–28 days. Size of litter varies with maternal age, with yearlings and 6- to 8-year-olds bearing 3 or 4 young, and 2- to 5-year-olds bearing 6–8 young. Newborns weigh about ¼ oz (6 g).

Habitat: Subalpine meadows, old fields, roadsides, and other grassy areas with short vegetation; in Utah, hay and alfalfa fields.

Range: Eastern Oregon, sw Idaho, ne Californi
n Nevada, and extreme nw Utah.

The semi-colonial Belding's Ground
Squirrel has one of the longest
hibernation periods (seven to eight
months) of any North American
mammal. By early August, some adult
males have entered hibernation. Femal
follow in late September, and the youn
which need more time to accumulate
fat, at snowfall. In populations at the
summit of Tioga Pass (9,941 feet/3,03
m elevation) in the High Sierras of
California, and perhaps in other group
as well, the males apparently hibernat
alone, the females often in groups,
generally of close relatives. Since food
not stored in the burrow, Belding's
Antelope Squirrel forages voraciously
summer on weed and grain seeds, leav
and stems of green plants, and on
grasshoppers, crickets, caterpillars, an
other insects. By hibernation it has
nearly doubled its body weight. Three
fourths of the stored fat provides energ
during hibernation, and the remainde
is used after the animals emerge in
spring. Males emerge first, tunneling
through snow; females ascend one to
two weeks later, after snow has meltec
from the tops of their burrows. Withi
four to six days females are sexually
receptive, but they remain so for only
three to six hours. Males, ready for
reproduction since their emergence,
compete so fiercely for mates—even
interrupting copulating rivals—that
are injured, and some are killed.
Courtship and mating occur mostly
aboveground, which is unusual for
ground-living squirrels. On average,
females mate four times with three
different males, with most offspring
sired by more than one male, but mos
by the first, or first and second, partne
The female digs a nesting burrow tha
1 to 2 feet (about half a meter) below

ground and 10 to 15 feet (3–4.5 m) long, with at least two openings, and builds a grass-lined nest inside. Young are born less than a month later, nursed for 26 to 31 days, and appear aboveground in late July or early August. Males disperse after weaning, but females are sedentary, with several generations sharing an ancestral site. Close relatives are highly cooperative, seldom fighting over nest sites. They often share parts of the territory, give alarm calls, and will even chase intruders from one another's unguarded burrows. Like many of its kind, this species has a single note, a chirp, and a more extended trill. Females give alarm calls more often than males. This squirrel often stands up on its hindlegs to view its surroundings, and exhibits various forms of play behavior. Male Belding's Ground Squirrels have a life span of three to four years, while females usually live four to six years. The chief cause of death is severe weather: 54 to 93 percent of juveniles and 23 to 68 percent of adults perish during hibernation, and more may freeze or starve during snowstorms after their spring emergence. Among their predators are Coyotes, American Badgers, bears, weasels, and hawks, as well as their own kind. Females that have lost their young to predation sometimes migrate to other sites and kill (but do not eat) the offspring of unrelated females, probably as a means of acquiring safer breeding places; they then settle in that area. Infanticide is also practiced by the relatively carnivorous yearling males, which kill and eat offspring of unrelated females. This ground squirrel is sometimes a garden pest, especially in eastern Oregon and northeastern California, and large colonies may damage pastures and grain fields.

Idaho Ground Squirrel
Spermophilus brunneus

Description: *Dappled grayish brown above,* with lower
back mainly brown; belly grayish
yellow. Chin white; nose rust-brown.
Shoulders and forelegs golden buff;
outer hindlegs and underside of tail
rust-brown. *Tail has 5–8 alternating light
and dark bands.* L 8¼–8⅝" (211–220
mm); T 1¾–2" (46–50 mm); HF 1¼–
1⅜" (33–35 mm).

Similar Species: Much larger Columbian Ground
Squirrel lacks white on chin.

Breeding: Mates first or second day after female
emerges from hibernation; litter of
2–7 young born after gestation of
50–52 days.

Sign: Entrances to burrows under rocks and
logs.

Habitat: Dry open areas with low, sparse green
vegetation.

Range: Payette and Weiser valleys, in wc Idaho

The rarest and least-known North
American *Spermophilus,* this colonial
ground squirrel feeds on green plants
including seeds and onion bulbs. Active
only about four months of the year, it
begins hibernation in July or August,
emerging late the following March.
Juveniles are abroad by late May and
reach adult size by mid-July.

---

167 Columbian Ground Squirrel
Spermophilus columbianus

Description: Grayish mixed with black above,
with indistinct buff spotting. Front
of *face, forelegs, and belly reddish brown.*
Forefeet buff. *Bushy tail* mostly reddish
but edged with white and with some
black hairs above, especially at base
and tip. L 12⅞–16⅛" (327–410 mm);
T 3–4¾" (77–120 mm); HF 1¾–2¼"
(45–58 mm); Wt 12–28¾ oz (340–
812 g).

Similar Species: Idaho Ground Squirrel is much smaller, with white chin. Richardson's, Townsend's, and Washington ground squirrels tend to be smaller; none has reddish-brown forelegs and feet.

Breeding: 1 litter per year of 2–7 young (average 3 or 4) born May–June, after 24-day gestation.

Habitat: Variable: alpine meadows, brushy areas, and arid grasslands.

Range: Eastern British Columbia and w Alberta south to ne Oregon, n Idaho, and nw Montana.

This colonial estivator and hibernator sleeps seven to eight months of the year, starting estivation as early as July in a chamber it seals off from its main tunnels with a 2-foot-long (600-mm) plug of earth. While it puts on fat in summer, it also stores some seeds or bulbs in its hibernation chamber to eat after it awakens in spring. As with many hibernators, males emerge several days before females. The Columbian Ground Squirrel eats many kinds of food, including grasses, plant stems and leaves, seeds, bulbs and tubers, insects, and birds and other small vertebrates. Especially when in large colonies, these squirrels sometimes damage grainfields.

168 Wyoming Ground Squirrel
Spermophilus elegans

Description: Pale drab above with grayish or buffy flecks. Nose cinnamon; *ears large. Tail relatively long,* edged with buff or white; *underside buffy or light brown.* L 10–12⅛" (253–307 mm); T 2⅝–3⅞" (66–100 mm); HF 1½–1¾" (38–45 mm); Wt 7⅜–11⅛ oz (210–315 g).

Similar Species: Uinta Ground Squirrel's tail is buff mixed with black below. Belding's Ground Squirrel has less cinnamon on nose; shorter tail is reddish below. Townsend's Ground Squirrel is smaller,

whitish below, with white border on tail. Richardson's and Columbian ground squirrels are larger.

Breeding: 1 litter per year, averaging 6 or 7 young, born April–May. Newborn weighs about ¼ oz (6 g).

Sign: Burrows concealed in tall grass, with some mounds.

Habitat: Grasslands, sage areas, montane meadows, and talus slopes.

Range: Isolated populations in ne Idaho and sw Montana; se Oregon, sw Idaho, and ne Nevada; and s Wyoming, nw Colorado, and small parts of adjoining states.

Most active in midmorning and evening, the Wyoming Ground Squirrel retires to its burrow in hot weather. It begins hibernation between July and September. Males emerge first in early March and become sexually active in a few days. Mating occurs within five days of the females' emergence in late March. Young appear aboveground when one month old. They stay at the burrow entrance the first week, but in two to three weeks follow the mother up to 50 yards (meters) from the burrow. Males defend territories during the breeding season, but not during the gestation and lactation phases. The females protect only the burrow entrance area. Adult males disperse in April; juvenile males leave the area in August. Although relatively asocial, the Wyoming Ground Squirrel species forms aggregations in favorable habitats. It spends about 40 percent of its time aboveground feeding primarily on forbs and grasses, though will also eat carrion found along the road, including that of its own kind. Its calls include cricket-like chirps and trilled churrs. It is difficult to locate the squirrel by its call because of its weak projection and because the animal often calls from the burrow entrance. This species' predators include dogs, Coyote, American Badgers, weasels, Bobcats, foxes, and hawks.

169 Franklin's Ground Squirrel
"Whistling Ground Squirrel,"
"Gray Gopher"
Spermophilus franklinii

Description: *Brownish gray peppered with black above; almost as dark below.* Tail blackish mixed with buff above and below, bordered with white. Tail long. L 15–15⅝" (381–397 mm); T 5⅜–6" (136–153 mm); HF 2⅛–2¼" (53–58 mm); Wt 17¾–24¾ oz (500–700 g).

Similar Species: Richardson's Ground Squirrel is smaller, with paler underparts and shorter tail.

Breeding: 1 litter of 5–8 young born in May.

Sign: Burrows concealed in tall grass, with some mounds.

Habitat: Dense grassy areas, hedges, and brush borders.

Range: East-central Alberta, s Saskatchewan, and s Manitoba south to n Kansas, n Illinois, and nw Indiana.

The largest and darkest ground squirrel in its range, Franklin's is active on sunny days, but generally retires when the sky is overcast. In summer, it puts on a layer of fat, and hibernates from about October to April, the exact dates depending on latitude. Small birds and mammals form perhaps a third of its diet, with green vegetation, corn, clover, seeds, berries, and caterpillars and other insects making up the rest. Although a good climber, Franklin's Ground Squirrel is believed to spend more than 90 percent of its life underground. It sometimes gathers in small colonies but usually is solitary. The young appear aboveground at one month of age and disperse several weeks later. This squirrel gives a bird-like whistling call, usually heard in the mating season, with males doing most of the vocalizing. Males also fight frequently, biting one another on the rump; in spring, almost every male has a cut or bare spot near the base of its tail. The red-tailed hawk and northern harrier are among its predators.

170, 171   Golden-mantled Ground Squirrel
"Copperhead"
Spermophilus lateralis

Description: Back gray, brownish, or buff. *Head and shoulders coppery red,* forming golden mantle. Belly whitish. *1 white stripe bordered by black stripes on each side;* no stripes on face. L 9–12⅛″ (230–308 mm); T 2½–4⅝″ (63–118 mm); HF 1⅜–1¾″(35–46 mm); Wt 6–9¾ oz (170–276 g).

Similar Species: Chipmunks are smaller and have facial stripes.

Breeding: 1 litter per year of 4–6 young born in early summer, after gestation of 26–33 days.

Sign: Burrow entrances.

Habitat: Moist coniferous or mixed forest; in mountains to above timberline; sometimes in sagebrush country or rocky meadows.

Range: Southeastern British Columbia and sw Alberta south through much of w U.S., east to se Wyoming, w Colorado, and n and w New Mexico.

In the fall, the Golden-mantled Groun Squirrel puts on a layer of fat, which helps maintain it through winter hibernation; it also carries food in its well-developed cheek pouches to its de to be stored and presumably eaten in spring when it awakens. During hibernation, which usually lasts from about October to May (the exact dates depending on latitude), a few individua awaken periodically to feed, as chipmun do. The mainstays of this squirrel's varied diet are seeds, nuts (especially those of the piñon), and fruits; these ar supplemented by green vegetation and insects, as well as by large quantities o: subterranean fungi, which it locates by smell and digs out. Coniferous seeds constitute a third of the diet in fall. Th Golden-mantled Ground Squirrel digs shallow burrows up to 100 feet (30 m) long to nest in; openings are usually

under or near a log, tree roots, or a boulder. It cleans its brilliantly colored coat by rolling in dust, then combing itself with its teeth and claws. The animal has a variety of calls—it can chirp and squeal with fright and growls when fighting—but seldom uses them. It often occurs with the Uinta Chipmunk.

172 Mexican Ground Squirrel
Spermophilus mexicanus

Description: Brown, with about *9 rows of squarish white spots on back;* belly whitish or buff. Long, moderately bushy tail. Small rounded ears. Males larger than females. L 11–15" (280–380 mm); T 4¼–6½" (110–166 mm); HF 1½–2" (38–51 mm); Wt 4⅞–11⅝ oz (137–330 g).

Similar Species: Spotted Ground Squirrel has spots evenly scattered, not in rows. Thirteen-lined Ground Squirrel has stripes as well as spots.

Breeding: 1 litter per year of 1–10 young (usually about 5) born in May; gestation less than 30 days.

Habitat: Brushy or grassy areas; mesquite or cactus deserts, usually on sand or gravelly soil.

Range: Southeastern New Mexico and sw Texas.

A somewhat colonial species, the Mexican Ground Squirrel is thought to hibernate in winter in the cooler parts of its range. It feeds primarily on mesquite leaves and beans and other plants, along with insects in summer, and often on dead animals along the highway. One Mexican Ground Squirrel was observed killing a young cottontail rabbit almost one-fourth its own size. Each animal has several burrows, including a home burrow, usually with two entrances, and two or more secondary refuges (usually old home burrows). The burrows are about 2½ to 3 inches (60–80 mm) in diameter, with a 7- to 8-inch (180–200 mm) nesting chamber in the deepest

part of a side tunnel, in which the young are born. The nest is lined with grasses and leaves, often those of mesquite. Timid and usually silent, this species has an alarm call consisting of a short trilling whistle. Around small farms, these ground squirrels sometimes inflict crop damage by digging up germinating seeds and eating ripening grain.

173 Mohave Ground Squirrel
Spermophilus mohavensis

Description: *Pinkish gray above; belly white.* No mottling or stripes. Short, thin tail *cinnamon above, white below.* Cheeks brownish. L 8⅝–9″ (219–230 mm); T 2¼–2⅞″ (57–73 mm); HF 1¼–1½″ (32–38 mm); Wt 3–4⅝ oz (85–130 g).

Similar Species: Round-tailed Ground Squirrel has longer, cinnamon or drab-colored tail. White-tailed Antelope Squirrel has 1 narrow white stripe on each side.

Breeding: Mates in spring; 1 litter recorded consisted of 6 young.

Habitat: Creosote bush scrub in sandy desert.

Range: Mojave Desert of s California.

The mainly solitary Mohave Ground Squirrel estivates and hibernates from August to March, when food is scarce. (The White-tailed Antelope Squirrel, which occurs with it, remains active during this period.) This ground squirrel carries its tail over its back when running; the white underside helps reflect the sun's rays. It is preyed upon by American Badgers, foxes, Coyotes, hawks, and eagles.

174 Arctic Ground Squirrel
"Parka Squirrel"
Spermophilus parryii

Description: One of the largest North American ground squirrels. *Back reddish to grayish brown, with numerous whitish flecks;* head

and shoulders tawny to reddish. In winter, upperparts are grayer. Underparts and legs are yellowish or tawny. Male slightly larger than female. L 11¾–15½″ (300–395 mm); T 3–5⅜″ (77–135 mm); HF 2–2⅝″ (50–68 mm); Wt average: male 28 oz (791 g), female 25 oz (698 g).

Similar Species: No other ground squirrel occurs in the same range.

Breeding: Mates in May; 5–10 young born blind and hairless in late June.

Habitat: Subalpine brushy meadows, riverbanks, lakeshores, and sandbanks, but not in permafrost areas.

Range: Alaska, Yukon Territory, n British Columbia, and mainland Northwest Territories.

In the fall, after putting on a layer of fat, the Arctic Ground Squirrel enters hibernation—adults first, then juveniles, which need more time to accumulate fat. Hibernation lasts more than half the year, from September through April or May. Males emerge first, through the snow, and have been seen abroad at temperatures of 22°F (−6°C). Food stored in the burrow, such as seeds, willow leaves, and bog rush fruit, is eaten at this time. Despite the continuous light during most of its active summer season, this highly vocal, colonial ground squirrel keeps to a "daily" routine of activity, from about 4:00 A.M. to 9:30 P.M. The reduced intensity of light at night may serve as a sleep stimulus. This squirrel wanders far from its home range to forage, crawling through vegetation to feed on stems, leaves, seeds, fruits, and roots of grasses, sedges, and other green plants, as well as woody plants and mushrooms. It may occasionally pause to sunbathe, sandbathe, or swim. On rainy or cloudy days, it keeps to its burrow. Often used for many years, the burrow is an extensive series of tunnels, many just under the surface and most not more than 3 feet (1 m) deep, with

several entrances. Shorter temporary
burrows are often dug to provide refuge
from summer predators. These squirrels
mate soon after they emerge from
hibernation. The young are weaned at
20 days, and dig their own burrows by
late summer. This species is preyed upon
by weasels, wolves, Arctic Foxes, and
Grizzly Bears—which tear up the ground
to find the animals in their burrows—as
well as by native peoples, who eat them
and use the skins to line parkas.

175 Richardson's Ground Squirrel
"Picket Pin," "Wyoming Ground
Squirrel," "Flickertail"
Spermophilus richardsonii

Description: *Gray* or yellowish gray *above,* tinged
with brown or buff and indistinctly
mottled. Underparts whitish or pale
buff. *Tail* bordered with white or buff;
light brownish or *buff below.* L 9¾–14"
(248–355 mm); T 2½–3⅞" (65–100
mm); HF 1⅝–1⅞" (40–49 mm);
Wt 13–16½ oz (369–469 g).

Similar Species: Belding's Ground Squirrel has brown
streak down back; tail is reddish below.
Townsend's Ground Squirrel is smaller
with shorter tail that is reddish or tawny
below. Uinta Ground Squirrel's tail is
buff mixed with black above and below.
Columbian Ground Squirrel is larger,
with reddish-brown face, forelegs,
and belly. Franklin's Ground Squirrel is
larger, with longer tail and dark
underparts.

Breeding: 1 litter per year of 6–11 young (usually
7 or 8) born in May, after 22- to 23-day
gestation.

Habitat: Open prairies.

Range: Southern and e Alberta, s Saskatchewan
and s Manitoba south to ne Montana,
ne South Dakota, and extreme w
Minnesota.

Because it often stands erect on its
hindlegs to survey its surroundings,

Richardson's Ground Squirrel has acquired the nickname "Picket Pin." It is also called "Flickertail," for the way it accompanies its shrill whistle with a flick of its tail. This ground squirrel hibernates from September through January, February, or March. It also goes through a prehibernation period after the juveniles emerge, and sometimes estivates in the hottest months, often beginning in July. Its diet includes a variety of insects—especially crickets, grasshoppers, and caterpillars—as well as the seeds, leaves, and stems of many kinds of plants. It stuffs its cheek pouches with seeds (one animal was found with 162 oat, 140 wheat, and almost 1,000 wild buckwheat seeds), which it stores in the burrow and probably eats in spring, upon awakening from hibernation. Mating occurs about three to five days after the females emerge from hibernation; 90 to 100 percent of the females are bred. The young may be seen foraging with the mother in June, as they undergo a four-week emergence phase from the time they first appear at the burrow entrance to when they forage alone. Although a rather solitary species, Richardson's Ground Squirrel sometimes lives colonially in favorable habitats, and is especially abundant where vegetation is short. In addition to its shrill whistle, this species also produces chirps, churrs, squeals, and tooth chatters. Its major predators are American Badgers, weasels, gopher snakes, and hawks.

Cascade Golden-mantled Ground Squirrel
Spermophilus saturatus

Description: Back dark gray-brown; head and shoulders russet. *1 white stripe on each side, bordered above and below by faint black stripes.* L 11¼–12⅜" (286–315 mm);

T 3⅝–4⅝″ (92–118 mm); HF 1⅝–1⅞″ (43–49 mm).

Similar Species: Closely related Golden-mantled Groun‹ Squirrel is usually smaller; its "mantle" and side stripes are better defined. Chipmunks have eye stripes.

Breeding: 1 litter per year of 1–5 young (average 4) born in late May, after gestation of 28 days.

Habitat: Coniferous forests, meadows, and cleared areas, from alpine zone with abundant talus to yellow pine belt.

Range: Cascade Mountains of s British Columbia and c Washington.

Male Cascade Golden-mantled Ground Squirrels emerge from hibernation in spring a few days before the females, and the yearlings follow the adults one to three weeks later. The squirrel is the active aboveground for four to five months, entering hibernation from mi‹ August to late September. In spring, it feeds especially heavily on plant material, favoring vetch leaves, bark, huckleberries, seeds and leaves of grasses, berries of salal and mountain ash, and seeds of lupine and pine; in fall, it consumes large amounts of subterranean fungi. Underground burrows, with their entrances usually beside logs, stumps, rocks, or under bushes, are used for hibernation, sleeping, refuge, and raising young. Several tunnels lead to the nest, a cup ‹ dry grass with loose, fresh vegetation o top, which is at least 3 feet (1 m) underground. The Cascade Golden-mantled Ground Squirrel mates in mi‹ to late April, a few days after the females emerge from hibernation. One hundred percent of adult females and 5 to 100 percent of yearling females are bred. Both males and females disperse from the natal burrow, males traveling farther than females. The normal life span of these animals is at least four years. Their predators include owls, hawks, Coyotes, foxes, and Bobcats.

Spotted Ground Squirrel
Spermophilus spilosoma

Description: Grayish or brownish above, with small, squarish, *indistinct pale spots scattered on back.* Belly whitish. Rather scantily haired tail similar to back, with black tip; buff below. Ears small. L 7¼–10″ (185–253 mm); T 2⅛–3⅝″ (55–92 mm); HF 1⅛–1⅜″ (28–36 mm); Wt 3½–4⅜ oz (100–125 g).

Similar Species: Mexican Ground Squirrel has spots in definite rows. Thirteen-lined Ground Squirrel has stripes as well as spots.

Breeding: Litter of 5–7 young born March–April and abroad by late April. Second litter born July–August or earlier.

Sign: Burrows about 2″ (50 mm) wide, usually opening under bushes or overhanging rocks.

Habitat: Predominately dry, sandy areas, grassy areas, and pinewoods.

Range: Southwestern South Dakota south to Arizona, New Mexico, and w Texas.

Active in the morning and the late afternoon, the Spotted Ground Squirrel often retires to its burrow in the heat of the day. In southern parts of its range it is active all year, but may hibernate in northern areas. Green vegetation and seeds are its primary foods, but it also eats grasshoppers and beetles.

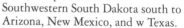

178, 179 **Round-tailed Ground Squirrel**
Spermophilus tereticaudus

Description: Various shades of *cinnamon,* with drab grayish cast above, slightly paler below. No stripes or mottling. *Tail round, long, slender, not bushy; cinnamon or drab below.* L 8–11″ (204–278 mm); T 2⅜–4⅜″ (60–112 mm); HF 1¼–1⅝″ (32–40 mm); Wt 5–6½ oz (142–184 g).

Similar Species: Mohave Ground Squirrel is pinkish gray above; shorter tail is white below.

Breeding: Mates late March–April; gestation is about 27 days. Litter size, ranging from

1 to 12 young, varies with abundance o
rainfall and vegetation. One dry year
brought average litters of 3.3 young,
while a wet year saw an average of 9.
Newborn weighs about ⅛ oz (3.7 g).

Sign: Burrow entrance under creosote or
mesquite bush.

Habitat: Flat, sandy desert areas; creosote scrub.

Range: Southeastern California, s Nevada, and
sw Arizona.

The Round-tailed Ground Squirrel is
most active during mornings and
evenings, avoiding the most intense hea
by retiring to its burrow at midday or
seeking shade under a plant. It will
climb into bushes not only to obtain
leaves, but also to get out of the sun an
off the hot sand. This species hibernate
from late September or early October t
early January, though in some areas it
remains active all year. Seeds, other
plant parts, and insects are this animal
chief foods, with green vegetation
constituting about 80 percent, seeds 1
percent, and insects 5 percent of its die
in spring. In summer, its food is almos
100 percent green vegetation, while in
fall greens drop to 75 percent of the
total, with seeds making up the other
25 percent. It will sometimes eat
cultivated crops, if easily available. The
Round-tailed Ground Squirrel digs its
own burrows or uses old burrows of
other species; entrances, usually at the
base of a bush, are not revealed by
mounds, as the dirt is scattered. Young
appear aboveground the first week in
May and are weaned mid-May through
June, at about five weeks. The males
disperse, but the females tend to remai
in their natal areas. Hawks, eagles,
Coyotes, foxes, American Badgers, and
Bobcats prey on this squirrel.

176 Townsend's Ground Squirrel
Spermophilus townsendii

Description: A small ground squirrel. Plain gray above, with pinkish tinge; belly whitish or buff. *Short tail reddish or tawny below, with white edge.* Face and hindlegs reddish. L 6⅝–10¾" (167–271 mm); T 1¼–2⅞" (32–72 mm); HF 1⅛–1½" (29–38 mm); Wt 4½–11½ oz (128–325 g).

Similar Species: Belding's Ground Squirrel has brown streak down back. Uinta Ground Squirrel is more brownish; tail buff mixed with black above and below. Richardson's Ground Squirrel is gray tinged with brown or buff above. Columbian Ground Squirrel is more blackish above. All four are larger. Washington Ground Squirrel has indistinct flecking on back and has black-tipped tail.

Breeding: 1 litter per year of 4–10 young, born by mid-March. Newborn is about 2" (50 mm) long and weighs about ⅛ oz (3.7 g).

Sign: Burrow openings rimmed with dirt piled 4–6" (100–150 mm) high.

Habitat: Primarily arid desert with open sagebrush, shad scale, or greasewood communities; one population in wc Idaho and ec Oregon occurs in a fertile river bottom.

Range: South-central Washington, e Oregon, s Idaho, extreme ne California, most of Nevada, and w Utah.

In June or July, Townsend's Ground Squirrel begins an estivation that continues right into winter hibernation, which ends in early spring, depending on locality. The squirrels emerge in late January or early February, breeding soon after. The young are abroad by mid-April, when half-grown. Seeds of grasses and other plants and green plant parts are the preferred foods of this squirrel, which often climbs into bushes. These animals sometimes form large colonies,

yet they are not very social. Each adult digs two burrows—a small one in the feeding area, evidently used as an escape hatch if predators approach, and a much bigger home burrow that is at least 50 feet (15 m) long and up to 6 feet (2 m) deep. Diggings often extend from the sagebrush flats where the animals forage onto nearby rock ridges. Townsend's Ground Squirrels have single and multi-note aboveground calls and also emit faint, high-pitched calls while underground or partially so. They often fall prey to American Badgers.

177 Thirteen-lined Ground Squirrel
"Striped Gopher"
Spermophilus tridecemlineatus

Description: Brownish, with *13 alternating brown and whitish longitudinal lines* (sometimes partially broken into spots) on back and sides. *Rows of whitish spots within dark lines.* L 6¾–11⅝" (170–297 mm); T 2⅜–5¼" (60–132 mm); HF 1–1⅝" (27–41 mm); Wt 3⅞–9½ oz (110–270 g).

Similar Species: Spotted Ground Squirrel has scattered spots, and Mexican Ground Squirrel has 9 rows of squarish spots; neither has lines. However, Mexican Ground Squirrel may be difficult to distinguish the two species occasionally hybridize when they have contact.

Breeding: Mates in April; 1 litter per year of 8–10 young born in May, after gestation of 27–28 days.

Sign: Burrow openings with radiating runways on surface; openings are often hidden under a clump, and no mound marks them, for excess dirt is spread evenly over the ground.

Habitat: Originally shortgrass prairies; now along roadsides, in yards, cemeteries, golf courses, and wherever grass is kept mowed.

Range: Much of c North America, from se Alberta and s Manitoba south to n New

Mexico, n and se Texas, and east through Minnesota and Missouri to Michigan and Ohio.

This handsome little spotted animal was known as the "Leopard-Spermophile" in Audubon's day. Also called the "Striped Gopher," it is the namesake of the University of Minnesota Gophers football team. Strictly diurnal, this ground squirrel is especially active on warm days. A solitary or only somewhat colonial hibernator, it often occurs in aggregations in suitable habitats. In late summer, it puts on a heavy layer of fat and stores some food in its burrow. It enters its nest in October (some adults retire much earlier), rolls into a stiff ball, and decreases its respiration from 100 to 200 breaths per minute to one breath about every five minutes. It emerges in March or early April. The burrow may be 15 to 20 feet (4.5–6 m) long, with several side passages. Most of the burrow is within 1 to 2 feet (about half a meter) of the surface, with only the hibernation nest in a special deeper section. Shorter burrows are dug as hiding places. The Thirteen-lined Ground Squirrel's home range is 2 to 3 acres (¾–1¼ ha). Grass and weed seeds, caterpillars, and grasshoppers, are its dietary staples, but it sometimes also eats bird flesh and even mice and shrews. This squirrel sometimes damages gardens by digging burrows and eating vegetables, but also devours weed seeds and harmful insects. It often stands upright to survey its domain, diving down into its burrow when it senses danger, then sometimes poking out its nose and giving a bird-like trill. It has a maximum running speed of 8 mph (13 km/h), and reverses direction if chased. Great numbers of these squirrels are killed by automobiles.

180 Rock Squirrel
Spermophilus variegatus

Description: The largest ground squirrel in its range. *Mottled above, grayish brown in front, brownish black behind;* buff-white or pinkish buff below. Long *bushy tail variegated buff and brown* with white edges. L 17–21″ (430–540 mm); T 6¾–9⅞″ (172–252 mm); HF 2⅛–2½″ (53–65 mm); Wt 21–28 oz (600–800 g).

Similar Species: Its large size distinguishes this species from most other ground squirrels.

Breeding: 2 litters per year of 3–9 young; first litter born April–June, second born August–September. Newborn weighs about ¼ oz (8 g).

Habitat: Nearly always associated with rocky locales such as cliffs, canyon walls, talus slopes, boulder piles, and hills along highways; oak-juniper growth in canyons.

Range: Southern Nevada, Utah, Colorado, and panhandle of Oklahoma, south through Arizona, New Mexico, and w Texas.

The well-named Rock Squirrel is often seen sitting on or running among rocks and makes its den in a burrow beneath them. Active in early morning and late afternoon, it sometimes estivates (becomes dormant) in hot weather. In the north it hibernates, but only for short periods, and is often abroad during winter warm spells. In the southern parts of its range it is active all year. This species climbs bushes and trees nearly as well as tree squirrels to feed on the fruit of juniper or mesquite. It gathers acorns, nuts, and currants, and the seeds of mesquite, cacti, agaves, and many other plants, eating them on the spot or carting them back to the den to be stored. It sometimes damages crops. Rock Squirrels are colonial, organized into maternal aggregations at main den sites, with one dominant male and some subordinate males at peripheral locations.

The dominant male defends the colony from other males, but allows females and juveniles to move about freely. Unlike most ground squirrels, males are not in full breeding condition upon emergence from hibernation, but are ready soon thereafter. The young nurse for about two months. They begin foraging for food about three days after emerging from the natal burrow. This species has a sharp, clear, sometimes quavering whistle; its alarm call is short, followed by a lower-pitched trill. Predators include golden eagles, Bobcats, Ringtails, Common Gray Foxes, and rattlesnakes.

Washington Ground Squirrel
Spermophilus washingtoni

Description: Back gray, indistinctly *flecked with whitish spots.* Underparts buffy. Reddish hindlegs and patch on snout. *Tail short* with blackish tip. L 7¼–9⅝″ (185–245 mm); T 1¼–2½″ (32–65 mm); HF 1⅛–1½″ (30–38 mm); Wt 5⅜–10 oz (152–284 g).

Similar Species: Townsend's and Belding's ground squirrels lack flecking; Belding's has brown streak down back. Idaho Ground Squirrel has brownish lower back and a banded tail, and occurs only in w Idaho. Columbian Ground Squirrel is larger, with longer, bushier reddish tail, and reddish-brown forelegs and face.

Breeding: Mates late January–early February. 1 litter per year of 5–11 young (average 8) born in mid-March; gestation probably about 28 days.

Sign: Burrow entrances in open ground with trails radiating from them.

Habitat: Open areas: grasslands, low sage areas, cultivated fields, and hillsides.

Range: East-central Washington and nc Oregon.

In summer, the Washington Ground Squirrel puts on a thick layer of fat that serves as its source of energy during

both estivation (summer dormancy), which begins in June or July, and hibernation, which follows immediately and lasts through February. Weed and grass seeds, and stems and leaves of such plants as mustards, mallows, plantain, and alfalfa, as well as some insects, form its diet. Mating occurs soon after emergence from hibernation. Abroad by mid-April, when half-grown, the young are difficult to distinguish from adults by late May. This squirrel often stands upright at its burrow entrance. Its call is faint and squeaky. American Badgers are its chief predator.

Gunnison's Prairie Dog
Cynomys gunnisoni

Description: Yellowish buff mixed with black above; slightly paler below. *Short, white-tipped tail. Terminal half of tail grayish white in center.* L 12⅛–14¾″ (309–373 mm); T 1½–2⅝″ (39–68 mm); HF 2–2½″ (52–63 mm); Wt 23–42 oz (650–1,200 g).

Similar Species: White-tailed and Utah prairie dogs have white in center of tail rather than grayish. Black-tailed Prairie Dog's tail has black tip.

Breeding: 1 litter per year of 1–8 young, born in early May; gestation 27–33 days.

Sign: Burrow systems, resembling those of ground squirrels, with mounds of earth at entrances; entrances are located on slopes or rises rather than in depressions, which tends to keep them from flooding.
Tracks: Similar to those of Black-tailed Prairie Dog.

Habitat: Shortgrass prairies in high mountain valleys and plateaus of southern Rocky Mountains at elevations of 6,000–12,000′ (1,800–3,600 m). Habitat is much more variable topographically and vegetationally than that of the Black-tailed Prairie Dog, which occurs at lower elevations.

Range: Southeastern Utah, s and c Colorado, ne and c Arizona, and nw New Mexico.

Gunnison's Prairie Dog, like the rest of its kin, is active only when the sun is up, and is most energetic near dawn and dusk. It is constantly vigilant while aboveground, often sitting upright on its hindfeet while it pursues its main activities: mainly feeding, but also grooming and playing. This animal generally is seen from April to October. It is inactive in winter, living on stored body fat, and probably hibernates. It usually emerges in April, though tracks were once found in the snow in late March. Gunnison's Prairie Dog feeds on green vegetation, particularly grasses, but also forbs, sedges, and shrubs, as well as a few insects. Its colonies are generally smaller and less closely knit than those of other prairie dogs, resembling ground squirrel aggregations, with fewer than 50 to 100 individuals. The animals in the colony cannot always see one another because their habitat is in such varied and patchy terrain, which is caused in part by human activities. On flat ground—and where this prairie dog is protected, such as at Blue Mesa Reservoir in Gunnison County, Colorado—where contact among colony members can be bettered managed, colonies are much larger and more extensive. This species' burrow systems are shallower than those of the Black-tailed Prairie Dog, and contain a single nest. Burrows have neither food storage nor excrement chambers, as these animals usually feed and defecate outside. Territoriality is not well developed in Gunnison's Prairie Dog, although old males may defend small areas outside their burrows. Mother-young relationships form the basic social unit. Newborns remain in the burrow about three weeks before emerging and are weaned about three weeks later. The female sits almost straight up on her haunches to nurse her

young, who suckle either pectoral or inguinal (hindleg) nipples. Gunnison's alarm call, distinctive among prairie dogs, is important to the survival and structure of the community. It is a series of high-pitched barks of one or two distinct syllables, with the second syllable lower and more guttural. The call may be repeated frequently and may continue for as long as half an hour. It increases in intensity as danger escalates and ends in chatter as the animal enters its burrow. Predators include American Badgers, Coyotes, weasels, and raptors. Plague *(Yersina pestis),* carried by fleas, can decimate populations of this species However, humans, through their extermination programs, are the chief enemy of Gunnison's Prairie Dog.

197   White-tailed Prairie Dog
Cynomys leucurus

Description: Stocky. Pinkish buff mixed with black above; slightly paler below. Short, *white-tipped tail;* terminal half entirely white. Dark patches above and below eyes; yellowish nose. Small ears. L 13⅜–14½″ (340–370 mm); T 1⅝–2½″ (40–65 mm); HF 2⅜–2½″ (60–65 mm); Wt 24–40 oz (675–1,125 g).

Similar Species: Black-tailed Prairie Dog has black-tipped tail and lacks dark patches over and under eyes. Gunnison's Prairie Dog has grayish hairs in center of terminal half of tail.

Breeding: Mates in March; 1 litter per year averaging 5 young. Gestation is about 30 days.

Sign: Burrow openings with mounds up to 3′ (1 m) high and 8–10′ (2.5–3 m) wide. Burrow mounds are less conspicuous than those of Black-tailed Prairie Dog, and are looser and not tamped down as thoroughly. An occasional burrow may have no discernible mound at all, probably because the loose earth has washed away.

Tracks: Similar to those of Black-tailed
Prairie Dog.

Habitat: Sagebrush plains at high elevations.
Range: Western Wyoming, ne Utah, and
nw Colorado.

Although the habits of the White-tailed
Prairie Dog are similar to those of the
Black-tailed, the White-tailed is less
colonial, with only a few of its burrows
interlinked with those of other
individuals. It also engages in fewer
social contacts such as "kissing," mutual
grooming, and cooperative burrow
building. Unlike the Black-tailed, it
lives at high, cool elevations (mountain
meadows and high pastures rather than
level plains) and hibernates throughout
the longer winters, entering its burrows
by late October and reemerging in March.
It is believed to awaken occasionally
during hibernation, at which times it
does not emerge but probably feeds on
stored roots and seeds. Its preferred
foods are grasses and forbs, and in some
areas it is forced to rely heavily on
saltbush. All prairie dogs gain weight
rapidly in summer; a White-tailed that
weighs 1½ pounds (680 g) when the
grasses begin to sprout may weigh twice
as much before hibernation. Because of
the cold winters in its range, the White-
tailed Prairie Dog breeds slightly later—
chiefly in March at higher elevations—
than the Black-tailed. Young do not
appear at burrow entrances until May or
June. This animal's vocalizations,
including alarm calls, resemble those of
the Black-tailed Prairie Dog. Circling
birds of prey elicit the most pronounced
alarm response: a few quick, terrified
yaps and every prairie dog in sight
disappears down a hole. The life span of
this species is four or five years. Its
predators, in addition to birds of prey,
include American Badgers, Bobcats, and
Coyotes; floods and fires also take a toll.
In some regions, burrowing owls and
rattlesnakes reside in vacated prairie

dog burrows (as do Black-footed Ferrets on the plains), and there is a widespread belief that rattlers kill many of the rodents. Without doubt, they kill a few pups in spring, but mature prairie dogs show surprisingly little fear of them. If rattlesnakes killed as many as some suppose, many "dog towns" would soon become ghost towns.

198, 199 Black-tailed Prairie Dog
Cynomys ludovicianus

Description:
A large prairie dog. *Pinkish brown above;* whitish or buffy white below. Slim, sparsely haired *tail with black tip* unique among prairie dogs. Short, rounded ears; large black eyes. L 14–16⅜" (355–415 mm); T 2⅞–4½" (72–115 mm); HF 2–2½" (50–65 mm); Wt 32–48 oz (900–1,360 g).

Similar Species:
No other prairie dog has black-tipped tail.

Breeding:
Mates February–March. 1 litter per year of usually 4 or 5 young born after gestation of about 30 days.

Sign:

Burrow has conical entrance mounds of hard-packed earth at least 1' (300 mm) high and often more than 2' (.6 m) wide, resembling a miniature volcano. *Tracks:* Hindprint 1¼" (31–33 mm) long, with 5 toes printing; foreprint slightly smaller, with 4 toes printing. White-tailed and other prairie dogs have similar tracks.

Habitat:
Shortgrass prairies.

Range:
Eastern Montana and sw North Dakota south to extreme se Arizona, New Mexico, and nw Texas.

During the hot summers in most of its range, the usually diurnal Black-tailed Prairie Dog is most active aboveground mornings and evenings, often sleeping in its burrow to escape midday heat. In cool, overcast weather it may be active all day, but it retreats to its burrow to wait out storms. While it does not

hibernate, during periods of severe cold or snowstorms this species undergoes a mild torpor and keeps to its burrow for a few days. About 98 percent of the Black-tailed Prairie Dog's diet consists of green plants, including various kinds of grasses, such as grama grass, bluegrass, bromegrass, and, in Texas, burro grass and purple needlegrass. It occasionally eats a few insects, especially grasshoppers, and may rarely eat meat. It habitually consumes all the green vegetation around its burrow, not only because it is convenient, but also to clear away protective cover that might shield predators. In the fall, prairie dogs put on a layer of fat that helps them through winter months when food is scarce. Among the most gregarious of mammals, the Black-tailed Prairie Dog lives in "towns," which may contain as many as several thousand individuals, covering 100 acres (40 ha) or more. The town is divided into territorial neighborhoods, or "wards," which in turn are composed of several "coteries," or family groups of one male, one to four females, and their young up to two years of age. Sociable animals, Black-tailed Prairie Dogs approach each other, touch noses, and turn their heads sideways to touch incisors; this "kissing" is not a part of courtship but a gesture of recognition and identification among ward members. The animals also groom one another and cooperate in the building of burrows. The burrows have conical entrance and exit mounds, which prevent flooding and serve as vantage points at which prairie dogs often sit on their haunches to survey their surroundings and scan for danger. The mounds are of different heights to facilitate airflow through the burrow. About 3 to 5 feet (1–1.5 m) below the entrance a short, lateral tunnel serves as a listening post and turn-around point, and at the bottom of the entrance shaft, which is up to 14 feet (4 m) deep, a

long, horizontal tunnel features several nesting chambers lined with dry grass, and an excrement chamber. A prairie dog covers its scat with dirt, and as one excrement chamber is filled up, a new one is excavated. The Black-tailed rarely wanders far from the safety of its burrow. Prairie dogs mate in late winter or early spring. The young, born deaf, blind, and hairless, don't emerge from the burrow until six weeks of age. They begin fending for themselves at about 10 weeks, and are fully grown at six months. Like many gregarious mammals, the Black-tailed Prairie Dog is highly vocal. Studies with a sound spectrogram indicate that it has nine distinctive calls, including chirps and chatters much like those of a tree squirrel; snarls, used when fighting; squeals of fright; and a shrill bark that gives this animal its common and genus names (*Cynomys* comes from Greek words meaning "dog mouse"). A staccato, double-noted call consisting of a chirp followed by a wheezing sound and accompanied by tail flicking is an alarm signal; it is chorused by other prairie dogs before all dive for safety. The "jump-yip" display, in which the prairie dog leaps into the air with head thrown back and forelegs raised as it gives a wheezing, whistling *yip,* seems to be an all-clear signal; this is also picked up by other prairie dogs, and soon the whole community is jumping and yipping. The average life span of the Black-tailed is seven to eight years. In the past, prairie dog towns covered vast expanses of the Great Plains. After great numbers of American Bison were killed off, the use of the prairie for agriculture and grazing allowed a prairie dog population explosion. However, as the rodents competed with cattle for grass—250 prairie dogs can consume as much grass each day as a 1,000-pound (450 kg) cow—they became the object of such fierce extermination campaigns

that their numbers declined by more than 90 percent (which, in turn, led to the decline in numbers of their chief predator, the Black-footed Ferret). Now foxes and the American Badger are this prairie dog's chief predators, but Coyotes, Bobcats, eagles, hawks, and snakes also take a share. In some areas, ranchers still attempt to eliminate prairie dogs with cyanide. However, a balanced population of the animals (controlled by predators, sport hunters, and modern methods of habitat modification) can actually improve rangeland, and today many ranchers take pains to maintain dog towns instead of eliminating them. Prairie dog meat once provided food for Native Americans and early settlers.

196 Utah Prairie Dog
Cynomys parvidens

Description: *In summer, reddish (ranging from tawny olive to clay) above,* mixed with black-tipped hairs; slightly paler below. Short, *white-tipped tail; terminal half has white center.* L 12–14¼" (305–360 mm); T 1⅛–2¼" (30–57 mm); HF 2⅛–2⅜" (55–61 mm).

Similar Species: White-tailed Prairie Dog is pinkish buff mixed with black above. Gunnison's Prairie Dog has grayish hairs in center of tail. Black-tailed Prairie Dog has black-tipped tail.

Sign: Burrow systems with dirt mounds at entrances.
Tracks: Similar to those of Black-tailed Prairie Dog.

Habitat: Shortgrass prairies.

Range: South-central Utah.

Some mammalogists think the Utah Prairie Dog should be recognized as a subspecies of its close relative the White-tailed Prairie Dog. However, it has not been demonstrated that the two types interbreed, so they are considered separate species. This prairie dog's habits

are presumably similar to those of the White-tailed. It is listed as a threatened species by the U.S. government.

181 Abert's Squirrel
"Tassel-eared Squirrel"
Sciurus aberti

Description: *A large tree squirrel.* Dark, grizzled gray above with darker sides and reddish back; *belly white.* Tail above similar to back, but bordered with white and with whitish cast. Tail white below. *Tasseled ears* reddish on back, about 1¾" (44 mm) long, with tufts or "tassels" extending about ¾" (18–20 mm) beyond ear tips. Tassels much reduced or even lacking in summer. In parts of Colorado, these squirrels are all black (melanistic). L 18¼–23" (463–584 mm); T 7¼–10⅛" (185–255 mm); HF 2⅝–3⅛" (66–78 mm); Wt 24–32 oz (681–908 g).

Similar Species: Arizona Gray Squirrel lacks ear tufts.

Breeding: Litter of about 4 young born April–May after gestation of 46 days; in some years more than 1 litter. Young are on their own by late June.

Sign: Nests, about 1' (300 mm) in diameter, placed high in trees, especially pine or juniper. Sign much the same as that of Eastern Gray Squirrel.

Habitat: Coniferous forests, including yellow or ponderosa pine, sometimes piñon or juniper.

Range: Isolated mountainous areas in Arizona, New Mexico, se Utah and Colorado.

Active throughout the winter, Abert's Squirrel may remain in its nest in very cold weather, venturing out only to retrieve buried seeds, especially at tree bases where there is no snow. Seeds and the inner bark of the ponderosa pine, as well as terminal buds, staminate flowers, and piñon nuts are its chief foods, but Abert's Squirrel also eats mistletoe and other vegetable

items. Nuts are buried in the ground,
but no food is stored in the nest, which
is built on a base of twigs placed in a
tree crotch, about 20 to 40 feet (6 to 13
m) high, typically in a ponderosa pine.
This squirrel uses its nest throughout
the year, as a refuge by day and for
sleeping at night. It either constructs a
ball-like mass of twigs from pine or
builds its nest within "witches'-brooms,"
growths of small pine twigs infected
by dwarf mistletoe. The outside
diameter of the nest measures from
1 to 3 feet (.3 to 1 m). The animal lines
the inside of the 6-inch by10-inch
(157 × 254 mm) chamber with dry
grass, leaves, shredded bark, or other
soft, dry material. The nest is repaired
as necessary. This species, like many
other tree squirrels, engages in mating
chases involving several males and one
female. This activity may be witnessed
in February or March, beginning in the
morning and lasting all day. "Chase"
is probably an inappropriate name for
this behavior, however, as it is not
really an active chase but rather the
males following the female as she
moves through the trees. The Kaibab
Squirrel *(S. a. kaibabensis),* formerly
considered a separate species, is now
classified as a subspecies of Abert's. It
has dark underparts, an all-white tail,
and is found only on the North Rim of
the Grand Canyon. Abert's squirrels
are killed by hawks, hunters, and
automobiles.

187 Arizona Gray Squirrel
Sciurus arizonensis

Description: A *plain gray, white-bellied* tree squirrel.
Relatively long tail fringed with white.
No ear tufts. L 20–22" (506–568 mm);
T 9⅜–12¼" (240–310 mm); HF 2½–
2⅞" (64–73 mm); Wt 22–25 oz
(605–706 g).

Similar Species: Abert's Squirrel has ear tufts.

Breeding: Mating activity may occur from January to June. 1 litter per year, usually born in early summer.

Sign: Leaf nests in trees. Other signs are same as those of Eastern Gray Squirrel.

Habitat: Deciduous lowland forests, including walnut, sycamore, and cottonwood trees.

Range: Isolated areas of e Arizona and wc New Mexico, mostly in canyons and valleys.

The Arizona Gray Squirrel's habits are apparently similar to those of the Eastern Gray Squirrel. It feeds on pinecones, nuts, acorns, berries, and seeds.

Mexican Gray Squirrel
"Red-bellied Squirrel"
Sciurus aureogaster

Description: Related to Eastern Gray Squirrel, but slightly larger. Gray above; *bright reddish-chestnut belly and flanks*. About half of the individuals in Florida are completely black (melanistic). L 18½–23" (470–573 mm); T 9¼–10⅞" (235–276 mm); HF 2½–2¾" (63–70 mm); Wt 15⅛–24 oz (430–680 g).

Similar Species: Eastern Fox Squirrel is larger and has yellow-tipped tail hairs. Eastern Gray Squirrel is slightly smaller, with pale underparts.

Breeding: Litter of 1 or 2 young, born any time of year. Both normal and black individuals may occur in same litter.

Sign: Large leaf nests in trees.

Habitat: Dense, jungle-type forests.

Range: Native of s Mexico and Guatemala; in U.S., only on Elliot Key, about 20 miles (32 km) south of Miami, Florida.

Two pairs of Mexican Gray Squirrels were released on Elliot Key, Florida, in 1938, and became well established there. Their leaf nests are found in a variety of trees, especially Florida poisonwood and West Indies mahogany; these animals rarely occur in hollow trees or in the abundant gumbo-limbo.

Important foods are the fruit of the mastic tree, mahogany seeds, coconut, sea grape, papaya, and thatch palm. The Mexican Gray Squirrel is threatened by few predators in Florida, but competition for tree holes may limit the population.

184, 185 Eastern Gray Squirrel
Sciurus carolinensis

Description: Gray above, with buff underfur showing especially on head, shoulders, back, and feet; underparts paler gray. Flattened *tail* bushy, *gray with silvery-tipped hairs.* In Canada, some have rufous bellies and tails. Black phase common in northern parts of range. L 17–19¾" (430–500 mm); T 8¼–9⅜" (210–240 mm); HF 2⅜–2¾" (60–70 mm); Wt 14⅛–25 oz (400–710 g).

Similar Species: Eastern Fox Squirrel is larger and has orange- or yellow-tipped tail hairs.

Breeding: 1 litter of 2 or 3 young born in spring; second litter born in late summer. Gestation about 44 days.

Sign: Gnawed acorn husks or other nutshells, especially hickory, walnut, beechnut, or pecan, littering the ground. Corncobs with only germ end of kernels eaten. In winter and spring, ragged little holes in snow or earth where squirrels have dug up nut caches. Gnawings on tree trunks and limbs, similar to Porcupine's but with smaller tooth marks and no droppings below. Leaf nests in high tree crotches or limbs (obvious in winter in bare branches); in summer, "cooling beds" or "loafing platforms" (flatter, smaller leaf nests without cavity). *Scat:* Small, dark, and oval; seldom conspicuous.

Tracks: Foreprints round, 1" (25–27 mm) long; hindprints more triangular, 2¼" (56–58 mm) long. When bounding, paired hindprints are slightly ahead of paired foreprints; sometimes foreprints are between rear parts of hindprints, often directly behind them,

leaving tracks like exclamation points
(!!); bounding stride ranges from a few
inches to over 3′ (900 mm). On snow,
foreprints 1½–1¾″ (37–46 mm) long,
hindprints nearly 3″ (75–77 mm) long,
with claws usually showing. On mud or
soft ground, hindprints shorter and
rounder because entire pad does not
always print, and long toes may print
more distinctly. (Rabbit tracks are
similar but longer, and foreprints are
not paired.)

Habitat: Hardwood or mixed forests with nut
trees, especially oak-hickory forests.

Range: Eastern U.S. east of s Manitoba, e North
Dakota, most of Iowa, e Kansas, e
Oklahoma, and e Texas.

Especially active in morning and
evening, the Eastern Gray Squirrel is
abroad all year, even digging through
snow in intense cold to retrieve buried
nuts. The only large squirrel in much of
the northeastern U.S., it feeds especially
heavily on hickory nuts, beechnuts,
acorns, and walnuts. It does not cache
nuts where it finds them, but carries
them to a new spot, burying each nut
individually in a hole dug with the
forefeet and then tamped down with the
forefeet, hindfeet, and nose. Most nuts
are buried at the surface, with few more
than ¼ inch (6–8 mm) below the
ground. In this fashion, many trees are
propagated, although the animal may
nip off the germinating end of the nut
before burying it, which prevents
germination. About 85 percent of the
nuts may be recovered. Nuts buried by
scientists conducting an experiment
were recovered by the squirrels at about
the same rate as nuts they buried
themselves, indicating that memory is
not involved in nut recovery. This
squirrel can smell buried nuts under a
foot of snow; when snow is deep, the
squirrel tunnels under it to get closer to
the scent. Besides nuts, the Eastern Gray
Squirrel feeds on a great number of

other items as available, including maple buds, bark, and samaras, tulip tree blossoms, apples, fungi, and a wide variety of seeds, as well as the occasional insect. These squirrels are ever on the move about their home ranges, so are always abreast of the many potential food items. They usually feed on just one food at a time, changing the item as additional sources come along. Buried nuts and other items are the mainstays in winter and in spring, but other foods are heavily consumed as they ripen. There is a great increase of activity in fall, when the squirrels spend most of their time cutting and burying nuts. Sometimes there is a rain of nuts on the forest floor, especially when the animals cut white oak acorns. The Eastern Gray Squirrel dens in trees year-round, using either natural cavities, old woodpecker holes, or leaf nests in stout mature trees or standing dead ones, especially white oaks, beeches, elms, and red maples. Tree cavities must be at least 12 inches (300 mm) deep and have an opening at least 3 inches (75 mm) in diameter. Both males and females build winter nests and more loosely constructed summer nests, which are likely to be near dens but are not always in the same trees. Rough population estimates have been made by assuming one and one-half leaf nests per squirrel. Leaf nests are difficult to spot in summer because they are made of green leaves, but nests are very obvious in winter. The more permanent nests are woven together well to weather the elements. Extremely ramshackle nests may have been damaged by the elements but are likely to have been built by juveniles or as temporary shelters near corn or other attractive crops. The Eastern Gray Squirrel mates in midwinter; a mating "chase" is often involved, with several males following a female as she moves about during the day. Frequently the spring litter of young is born in a tree

cavity, while the second, late-summer litter is born in a leaf nest. Females ofte move their litters back and forth betwee cavity dens and leaf nests, perhaps because of changes in the weather or to escape predation or parasite infestation. The young are weaned in about 50 days The second litter stays with the female over the winter. The characteristic aggressive bark of the Eastern Gray Squirrel—*que, que, que, que*—is usually accompanied by flicks of the tail. It makes other calls as well, including a loud, nasal cry. This animal's tail is use primarily for balance in trees, but serve as a sunshade, an umbrella, a blanket, and a rudder when swimming; it gives lift when the squirrel leaps from brancl to branch and slows descent should the squirrel fall. Overpopulation may trigger major migrations of this squirre species. In the early 19th century, when vast tracts of the East were covered by dense hardwood forest, observers reported migrations in which squirrels never touched ground but moved great distances from tree to tree. A major migration of thousands of squirrels too place in October 1968 in Tennessee, Georgia, and North Carolina. This movement was attributed to substantia nut production and a high reproductior rate in 1967, followed by a late frost an little nut production in 1968. Black an gray phases of this species often are found together, leading some to think they are two different species. There ar albino colonies in Olney, Illinois; Trenton, New Jersey; and Greenwood, South Carolina.

186   Western Gray Squirrel
"California Gray Squirrel"
Sciurus griseus

Description: *Gray with numerous white-tipped hairs above; belly white.* Backs of ears reddish brown. Long, *bushy tail with bands of*

gray, white, and black, especially below.
L 17½–23" (445–593 mm); T 9⅜–12¼"
(240–310 mm); HF 3–4" (76–101
mm); Wt 12–34 oz (340–964 g).

Similar Species: Eastern Fox Squirrel has bright rust
belly and yellow-tipped tail hairs; is
slightly larger.

Breeding: 1 litter per year of 3–5 young born
March–June.

Sign: Large leaf nests, obvious in winter, often
high in trees. Sign is same as that of
Eastern Gray Squirrel.

Habitat: Woodlands.

Range: Washington south to California.

The only large gray tree squirrel in its
range on the West Coast, this species is
active all year, although during bad
storms it may remain in its nest. In
summer it uses a nest of shredded bark
and sticks, usually placed at least 20 feet
(6 m) above the ground; in winter, it
probably lives in a tree hollow. Its chief
foods are pinecones, acorns, and other
nuts, and some fungi, berries, and
insects. This squirrel's hoarse barking
call is heard mostly in late summer.

Mexican Fox Squirrel
"Nayarit Squirrel"
Sciurus nayaritensis

Description: *Grayish above, washed with buff or yellow;
ocher below.* L 21–23" (530–575 mm);
T 9⅜–11¾" (237–298 mm); HF 2¾–
3¼" (70–84 mm).

Similar Species: No other large squirrel occurs in its range.

Sign: Leaf nests in oaks and pines or in hollow
trees. Gnawed cones of Douglas fir and
Arizona yellow pine.

Habitat: Wooded areas; Apache pine-oak forest
in Arizona.

Range: Chiricahua Mountains of extreme
se Arizona.

The Mexican Fox Squirrel is very similar
to the Eastern Fox Squirrel, and some
mammalogists consider them the same

species. The Mexican Fox Squirrel is seen frequently on the ground and may be seen in large groups in winter. It eats seeds of pines and Douglas fir, acorns, and walnuts. Its nests are similar to those of the Arizona Gray Squirrel, and may be on branches or in holes in trees.

182, 183 Eastern Fox Squirrel
Sciurus niger

Description: *The largest tree squirrel.* 3 color phases: in northeastern part of range, gray above, yellowish below; in western part, bright rust below; in South, black, often with white blaze on face and white tail tip (in South Carolina, typically black with white ears and nose). *Large, bushy tail with yellow-tipped hairs.* L 17⅞–28″ (454–698 mm); T 7⅞–13″ (200–330 mm); HF 2–3¼″ (51–82 mm); Wt 17¾–37½ oz (504–1,062 g).

Similar Species: Eastern Gray Squirrel is smaller and has silvery-tipped tail hairs.

Breeding: Litter of 2–4 young born late February–early March, sometimes June–July, occasionally August–early September. 2-year-old females may have 2 litters per year. Gestation about 45 days.

Sign: Large leaf nests in trees, fairly well hidden in summer but obvious in winter. Other signs are similar to those of Eastern Gray Squirrel, except that food debris is often much more evident because fox squirrels commonly carry nuts to a favorite feeding perch such as low branch, log, or stump, where the ground may be heavily strewn with shells. When squirrels raid cornfields, it may be possible to tell which species is the culprit: The Eastern Fox Squirrel usually cuts and hauls an entire cob to a feeding perch, which is strewn with husks and bits of cob, while the Eastern Gray Squirrel bites the kernels from the cob, then eats only the germ, dropping the remainder.

Habitat: Woods, particularly oak-hickory; in the South, live oak and mixed forests, cypress and mangrove swamps, piney areas. 1 or 2 tree holes per acre (.4 ha) are needed for good habitat.

Range: Eastern U.S. (except New England, most of New Jersey, extreme w New York, and n and e Pennsylvania); east to Dakotas, ne Colorado, and e Texas.

The Eastern Fox Squirrel is most active in morning and late afternoon, burying nuts that it will locate in winter with its keen sense of smell, even under snow. This animal eats mainly hickory nuts and acorns, but also tulip poplar fruit, winged maple seeds, ripening corn along wooded areas, open buds, and various berries in season. In the Southeast, its major foods are green pinecones, as well as surface and subterranean fungi. The latter, located by smell, have large numbers of spores, which the squirrel disperses through defecation over several days and much territory. These subterranean fungi are beneficial to the germination and growth of trees, and some contain nitrogen-fixing bacteria. The close relationship between these small mammals and their woodland environment—the forest, the fungi, bacteria—is destroyed by clear-cutting of forest; this constitutes a powerful, although little-understood, argument against the practice.

Although the Eastern Fox Squirrel will feed in common areas, and several individuals may den together in winter, this animal is not very social. It spends much time in trees feeding or cutting down nuts or sunbathing on a limb or in a tree crotch. In fall, it is often on the ground gathering and caching nuts, usually individually or in twos and threes, sometimes establishing larger caches in tree cavities. Where home ranges overlap, more than one squirrel may store nuts in the same general area, or even in the same tree hole. The

Eastern Fox Squirrel uses tree holes extensively, particularly in winter, often nesting in them with a family group of several other squirrels. Where tree holes are scarce, the Eastern Fox Squirrel builds leaf nests in tree crotches. These structures, up to 12 inches (300 mm) in diameter, have a side entrance and are lined with shredded material. Such nests are obvious in winter but more difficult to see in summer, when they are green, as they are usually made from the leaves of the tree in which they occur. However, they can be found if one looks for them carefully. Each squirrel usually maintains three to six active nests. Winter mating "chases" are begun by males, who are ready to copulate before females come into heat. A "chase" consists of several males following a female throughout the day (rather than actually chasing her). She mates with one or more of the males. The young, whose eyes remain closed for about a month, keep to the nest for the first seven or eight weeks and become independent of the mother when nearly three months old. These squirrels are relatively easy to locate: one can listen for falling nuts, the swishing of tree branches, chewing sounds, or their call, *que, que, que.* Eastern Fox Squirrels are becoming much less abundant throughout the Southeast. Fire is favorable to this species, as it helps to remove undergrowth and thus maintain spacing between trees, especially pines, whereas dense hardwood forests and thick underbrush are more suitable for the Eastern Gray Squirrel.

188, 189   **Douglas' Squirrel**
"Pine Squirrel," "Chickaree"
Tamiasciurus douglasii

Description: *Upperparts reddish gray or brownish gray,* grading into chestnut-brown on middle of back; *underparts grayish to orangish.* Upperparts grayer in winter. Blackish

line on sides in summer; indistinct or absent in winter. Upperside of tail similar to back, except terminal third blackish; underneath, tail is rusty in center, bordered by broad black band with whitish edge. Small ear tufts in winter. L 10⅝–14″ (270–355 mm); T 3⅞–6⅛″ (100–156 mm); HF 1¾–2¼″ (44–57 mm); Wt 5¼–10½ oz (150–300 g).

Similar Species: Red Squirrel has white to grayish-white underparts and is usually brighter red above.

Breeding: Mates in early spring; litter of 4–6 young born May–June; sometimes a second litter August–October.

Sign: Piles of cone remnants. Summer nests resembling large balls in trees. Other signs similar to those of Red Squirrel.

Habitat: Primarily coniferous forests.

Range: Southwestern British Columbia, w Washington, w and c Oregon, and n California.

Very active throughout the year, Douglas' Squirrel runs about in trees and on the ground, although during storms it usually remains in its nest. It eats new shoots of conifers, green vegetation, acorns, nuts, mushrooms, fruits, and berries. In late summer and fall, this squirrel cuts cones from tree limbs and feeds on the seeds at special feeding stations in trees, below which discarded scales pile into middens. It stores green cones in moist places to keep them tender and often caches mushrooms in forks of trees. Douglas' Squirrel builds a summer nest mainly of mosses and lichens, twigs, and shredded bark, but sometimes builds a top on a deserted bird's nest; in winter, it nests in tree holes. The young of this species first venture out in August, and families remain together for much of the first year. Noisier than most, Douglas' Squirrel has a large repertoire of calls, including a trill. American Martens, Bobcats, house cats, and owls are its

predators. People harvest the cone caches for seeds to plant nursery trees.

190, 191   **Red Squirrel**
"Pine Squirrel," "Chickaree"
Tamiasciurus hudsonicus

Description: The smallest tree squirrel in its range. *Rust-red* to grayish red above, brightest on sides; white or grayish white below. In winter, *black line separates reddish back from whitish belly.* Tail similar to back color, but outlined with broad black band edged with white. In summer, coat is duller. In winter, has prominent ear tufts. L 10⅝–15¼″ (270–385 mm); T 3⅝–6¼″ (92–158 mm); HF 1⅜–2¼″ (35–57 mm); Wt 5–8⅞ oz (140–252 g).

Similar Species: Douglas' Squirrel usually duller red, has grayish to orangish underparts.

Breeding: Litter of 3–7 young born March–April, sometimes a second litter August–September. Gestation 35 days. Newborns weigh about ¼ oz (7 g), and are 2¾″ (70 mm) long.

Sign: Piles of cones or remnants; acorns or hickory nuts with a ragged hole at one end (flying squirrels make a smooth opening; White-footed Mice make 2 or 3 openings). Small holes in earth, often near cone remnants. Tree nests, especially in conifer stands, often built of grass and shredded bark.
Tracks: In mud, hindprint about 1½″ (38 mm) long, with 5 toes printing; foreprints half as long, with 4 toes printing. In rapid bounds, front tracks appear between hind; in slow bounds, front tracks slightly behind hind. Straddle about 4″ (100 mm). Unlike ground squirrels *(Spermophilus),* which run with one forefoot in front of the other, the Red Squirrel tends to keep forefeet parallel when running.

Habitat: Often abundant in any kind of forest: natural coniferous forests, pine

plantations, mixed, or hardwood forests; often around buildings.

Range: Throughout much of Alaska and Canada; in U.S., south through Rocky Mountain states; in East, south to Iowa, n Illinois, n Indiana, n Ohio, n Virginia, and through Alleghenies.

The Red Squirrel is active all year, although it may remain inactive for a few days in inclement weather. In conifer forests, this squirrel feeds heavily on pine seeds, leaving piles of cone remnants everywhere. In the fall, it cuts green pinecones and buries them in damp earth. Like other North American tree squirrels, this species stores food in one or more large caches (sometimes up to a bushel's worth in each) in the ground, in a hollow tree, or at the base of a tree. The Red Squirrel is a prodigious and opportunistic feeder, moving through its home range and trying many different items; in this way it keeps abreast of where and when various foods become available. Among the additional foods it may eat or store are acorns, beechnuts, and other nuts; seeds of hickory, tulip, sycamore, maple, and elm; berries; birds' eggs and young birds; and fungi, including even the deadly (to humans) amanita mushrooms, which are often cached in trees. Red Squirrels also harvest maple sugar by biting into the trees' xylem, letting the sap ooze out, and returning when the water in the sap (which when fresh is only 2 percent sugar) has evaporated and the sugar content is about 55 percent. Red maples apparently suffice for this where sugar maples are absent. The Red Squirrel's nest, often constructed of shredded bark from a grapevine, is made in a hollow or fallen tree, a hole in the ground, a hummock, or a tree crotch (as are the leaf nests of gray squirrels). The female is in heat for only one day in late winter, at which time she will allow

males on her territory. Animated nuptial chases precede mating. The Red Squirrel's vocalizations include a slightly descending, drawn-out, rather nonmusical trill that can be heard for some distance, and a chatter of various notes and chucks.

192, 193   Northern Flying Squirrel
Glaucomys sabrinus

Description: *A small squirrel.* Very soft fur, rich brown above, white below; *abdominal hairs usually gray at base. Loose fold of skin between forelegs and hindlegs.* Tail has characteristic flattened appearance; dark brown above and white below. Large black eyes. L 10⅜–14½″ (263–368 mm); T 4½–7⅛″ (115–180 mm); HF 1½–1¾″ (38–45 mm); Wt 1½–2½ oz (45–70 g).

Similar Species: Southern Flying Squirrel is generally smaller and grayer, with abdominal fur white at base.

Breeding: Mates in late winter; 1 litter per year of 2–5 young born in spring after gestation of 40 days. Newborns weigh ⅛–¼ oz (4–6 g), and are weaned at 55 to 60 days.

Sign: Standing dead trees, especially stumps 6–20′ (2–6 m) high with woodpecker holes near top. Piles of cone remnants and nutshells.

Habitat: Coniferous and mixed forests; may occur in hardwoods where old or dead trees have numerous woodpecker-type nesting holes.

Range: Eastern Alaska, s Yukon, and s Northwest Territories, and southern tier of Canadian provinces east to Labrador; south in w U.S. through California, Idaho, Montana, Utah, and n Wyoming; in e U.S. to Minnesota, Wisconsin, Michigan, New England, and New York, and through Appalachians.

The Northern Flying Squirrel, unlike its southern counterpart, which sometimes enters torpor, is active throughout the

year. This animal is quite common, foraging on the ground a great deal, but because it is nocturnal it is seldom seen except when the old, hollow trees in which it lives are cut down. Flying squirrels glide from tree to tree by spreading their legs and stretching their flight skin, which acts as a sail. They pull upright at the last instant to land gently, using the tail as a rudder. The Northern Flying Squirrel feeds primarily on lichens and subterranean fungi, such as *Endogone* and its relatives. In feeding, it dispenses fungal spores and the nitrogen-fixing bacteria that help trees obtain nutrients and water. (This offers a compelling argument against clear-cutting, which breaks up this natural and necessary association for the germination and proper growth of the forest.) This species also eats various nuts, seeds, and insects, probably storing much food for winter use. The Northern Flying Squirrel makes a nest of shredded bark in tree hollows, sometimes capping an abandoned bird's nest to provide temporary shelter or, in summer, using a leaf nest. The young are often born in a hollow stump or limb, or sometimes in a bark nest in a conifer crotch. There is apparently only one litter per year, in contrast to the Southern Flying Squirrel, which sometimes has two. The Northern Flying Squirrel's chirping, bird-like notes are similar to those of night-flying warblers. The best way to locate this species is to look for and tap on dead vertical tree stubs with woodpecker holes. If these gregarious animals are present, they will peek out immediately and will scamper or glide away if tapping is continued. Although the Southern Flying Squirrel is more aggressive and sometimes displaces the Northern, when the two species occur together the Southern selects hardwoods and the Northern conifers. Owls are the Northern Flying Squirrel's main predators.

194, 195 Southern Flying Squirrel
Glaucomys volans

Description: *A very small squirrel. Very silky coat*
grayish brown above, white below, with ha
all white from tip to base. Loose fold of ski
between foreleg and hindleg. Flattened,
gray-brown tail. Large black eyes.
L 7¾–10⅛″ (198–255 mm); T 3⅛–4
(81–120 mm); HF ⅞–1¼″ (22–32 mn
Wt 1½–3½ oz (45–100 g).

Similar Species: Slightly larger Northern Flying Squirr
is a richer brown, with abdominal fur
usually gray at base.

Breeding: Mates in early spring; 2–7 young born
after gestation of 41 days. Often secon
litter August–September, usually by
females not breeding in spring.

Sign: Hickory nuts with a smooth opening a
the thin end (White-footed Mice mak
2 or 3 openings, Red Squirrels make a
ragged opening, and fox and gray
squirrels crush the nut).
Tracks: Similar to those of the Red
Squirrel, but slightly smaller. In snow
is nearly impossible to distinguish
tracks of the two species.

Habitat: Various deciduous forests such as beec
maple, oak-hickory, and, in the South,
live oak.

Range: Eastern U.S. (except for n New Englar
and s tip of Florida) east of Minnesota
Kansas, and e Texas.

The flying squirrels are the only
nocturnal tree squirrels. Although it i
active in all seasons, the Southern Flyi
Squirrel may remain in its nest in very
cold weather and will enter torpor in
times of extreme cold or food scarcity.
The state of torpor is not as deep as tru
hibernation, but the animal's body
temperature can drop to 22°F (−6°C),
and it may take up to 40 minutes to
wake. The flying squirrel does not trul
fly, but glides through the air, up to 8
yards (meters) or more, from the top o
one tree down to the trunk of another.
flies with its legs outstretched and the

fold of skin between foreleg and hindleg acting as a combination parachute and sail (or glider wing). While gliding, it can turn or change its angle of descent. Just before landing, it drops its tail and lifts its forequarters, slackening the flight skin, which then serves as an air brake. It lands very lightly on all four feet, and at once scurries around to the other side of the tree trunk, in case a predator has followed its flight. Agile and extremely surefooted aloft, it is relatively clumsy on the ground. The most carnivorous of the tree squirrels, the Southern Flying Squirrel feeds on nuts, acorns, seeds, berries, fungi, lichens, birds and their nestlings and eggs, some insects, and sometimes other vertebrates, including carrion. Hard parts and wings of larger insects are often discarded. Flying squirrels will gnaw bark from maple trees and drink the sap, and also eat moths that come to the sap to feed. Great quantities of nuts, acorns, and seeds are stored for winter use, in tree hollows, in their nests, in crotches or cracks in trees, and in the ground. Hickory nuts and acorns may be buried throughout the home range, adding to the general store of nuts buried by other species of squirrels. Southern Flying Squirrels may store up to 15,000 nuts in a season. They use their front incisors to pound the nuts into the ground or a crack in a tree. Woodpecker holes are favored nest sites, but the Southern Flying Squirrel may build a summer nest of leaves, twigs, and bark that is similar to that of gray or fox squirrels, but is only about 8 inches (200 mm) in diameter. Typical dens are dead tree stubs 8 to 20 feet (2.5–6 m) high that contain woodpecker holes, 1½ to 2 inches (40–50 mm) in diameter. The nest cavity is lined with shredded bark or, in the Deep South, Spanish moss and palmetto fibers. There is often a primary nest, plus many secondary nests used for temporary shelter. Some

dens are used exclusively for defecation
over time, humus can build up to 1½
feet (half a meter) deep. In winter,
several individuals may den together in
one tree hole, as their combined body
heat brings up the den temperature; as
many as 50 individuals have been found
in one nest in winter. Flying squirrels
know their home range very well, and
when abroad will hide in a hollow tree,
under loose bark, or another convenient
spot, such as an old bird or squirrel nest.
The Southern Flying Squirrel mates in
early spring. The female is receptive for
just one day. She usually mates with the
dominant male, and often a subordinate
as well. At about four weeks of age the
young resemble adults; at five weeks,
they exit the nest to take solid food.
Females of this species defend their
young vigorously, and will move them
to another nest if danger threatens. The
main calls of adults are faint and bird-
like notes, described as similar to those
of night-migrating birds. The young
produce squeaks, which include
ultrasonic components. One researcher
listened to ultrasonic sounds on a bat
detector of a female and its young as the
two became reunited after both hit a bat
net (only the young became entangled).
The Southern Flying Squirrel is more
aggressive than its northern counterpart.
Predators include owls and many
mammals, but the house cat is the
most dangerous.

POCKET GOPHERS
Family Geomyidae

This family of five genera and 35 species occurs only in Central and North America. Three genera and 18 species are found in the U.S. and Canada. Primarily as a result of pronounced adaptation to their almost completely subterranean existence, all the members of this family are very similar in structure and habits. Pocket gophers are among the continent's most specialized mammalian burrowers. They have thickset bodies with short necks, short fur, small ears and eyes, a naked or sparsely haired tail, and large, external, fur-lined cheek pouches—the "pockets" that give them their common name. These pouches, extending from cheek to shoulder, are often crammed with food or bedding material for nests. The animal empties the pouches by squeezing their contents forward with both forepaws; it then turns them inside out and cleans them. A special muscle pulls the pouches back into place. As with moles, the fur of a pocket gopher can lie either forward or backward, enabling the animal to move about in its burrow equally well in both directions. Because of their narrow pelvic girdles, pocket gophers can turn around in their burrows. The tail probably functions as a tactile organ. The lips of this animal close behind the large incisors, which serves to keep dirt from entering the mouth during underground gnawing; consequently the big yellow incisors are always on display. As with all rodents, the chisel-like incisors grow throughout the animal's life. They are coated with enamel only on the front, which ensures a sharp, beveled edge, as the back of each tooth wears down faster. With the constant gnawing on food, as well as the cutting of stems, roots, and tubers while burrowing, the incisors are continuously worn away. The three genera of pocket

gophers are distinguished by their upper incisors: The western pocket gophers (*Thomomys* species) have no conspicuous grooves down the middle of the teeth; the Yellow-faced Pocket Gopher (*Cratogeomys castanops*) has one groove; and the eastern pocket gophers (*Geomys* species, which range west to Wyoming and New Mexico) have two. There is pronounced sexual dimorphism in pocket gophers; females stop growing when they reach maturity, while males continue to grow throughout life.

Pocket gophers make two kinds of burrows: those near the surface for food gathering, and deeper ones for storage and shelter. The passages slant toward the ground surface (vertical shafts are made by moles). The animals use their long claws to dig the burrows, and their heavy incisors to cut through earth and roots and to pry rocks loose. The hindfeet push back the earth that accumulates under the animal's body. When dirt piles up in the burrow behind the pocket gopher, it turns and, like a bulldozer, pushes the dirt ahead with its head and forefeet. The excess soil is pushed out of a passage that slants to the surface, where it ends up a a mound; the passage is then plugged from below. Pocket gophers usually forage underground for roots and tuber or cut off stems belowground and pull plants into the burrow from below. What is not eaten immediately is stored in side chambers for later use. Because they derive their water from vegetation pocket gophers do not need to drink. These solitary animals do not hibernat but retreat to deep burrows in winter. spring, males find and enter a female burrow system, where mating occurs. Afterward, the males leave. There are usually one or two litters per year of 2 11 young each. Even before dispersing juveniles start to dig their own tunnels At two months, they leave home; at

three months they are sexually mature. Although pocket gophers damage crops, their tunneling aerates the soil, which helps both to conserve groundwater and to prevent erosion.

The different species of pocket gophers resemble each other closely. Their main differences are in skull and teeth characteristics not visible in the field. Even individuals of the same species vary in color, depending on the color of the soil in which they live. For these reasons, it is difficult to identify many of the species by external characteristics. Geographical range is probably the best means for identification. In the species accounts that follow, breeding and sign information is given for those species for which it is available.

141 Botta's Pocket Gopher
Thomomys bottae

Description: Usually *dark brown to grayish above,* with purplish cast on sides; slightly paler below. Tail tan to gray, essentially hairless. Variable coloration exhibited in some populations, ranging from white individuals in Imperial Valley of southern California to almost black animals in some coastal areas. White spotting under chin in Arizona. *Rounded ear* has similar-size *dark patch* behind it. As with all *Thomomys* species, upper incisors lack grooves. L 6⅝–10¾″ (167–273 mm); T 1⅝–3¾″ (42–97 mm); HF ⅞–1⅜″ (22–34 mm); E ⅛–¼″ (5–8 mm); Wt 2½–8⅞ oz (71–250 g).

Similar Species: All other western pocket gophers are very similar, distinguishable mainly by range.

Breeding: Most pocket gophers breed once a year, in spring, but this species often has multiple litters, and some females in irrigated fields in the Sacramento Valley evidently breed year-round. Litters average about 6 young. Gestation 18–19 days.

Sign:	Mounds of soil; in spring, "gopher cores": long coils of earth used to plug burrows in snow and left on ground after snow melts.
Habitat:	Varies: deserts to mountain meadows, in soils ranging from sand to clay, with loam preferred; in sw Oregon, sandy so in prairie, scrub, and dunes.
Range:	Extreme sw Oregon (Curry County) south through California, much of Nevada and Utah, sw Colorado, Arizona, New Mexico, and w Texas.

Botta's Pocket Gopher spends most of its time in underground burrows, whic may account for its ability to tolerate a wide variety of habitats. Its tunnels are extensive; one in Texas was 150 feet (4 m) long and very close to the surface, although the nest was more than 2 feet (600 mm) down. Side branches serve a refuse and latrine areas as well as food caches. This animal is solitary, living one to the burrow, and often fights if it meets another species member. Because this species breeds once a year in most of its range, its population remains in balance, even though the predation rat is low due to its protected underground environment. In Texas, however, breeding seems to occur throughout th year, with most young born in spring.

---

142 Camas Pocket Gopher
Thomomys bulbivorus

Description:	*The largest member of the genus;* male larger than female. Dark grayish brown L average: male 11¾" (300 mm), maximum 12⅞" (328 mm); T average: male 3½" (90 mm), female 3⅛" (81 mm); HF average: male 1⅝" (42 mm), female 1½" (39 mm).
Similar Species:	All other western pocket gophers are very similar, distinguishable mainly by range.
Breeding:	Reproductive season early April–early June; 1 litter per year of 4–9 young,

with 2″ (50 mm) newborns weighing about ¼ oz (6.1 g).

Sign: Mounds of soil; in spring, "gopher cores": long coils of earth used to plug burrows in snow and left on ground after snow melts.

Habitat: Sandy areas.

Range: Confined to Willamette Valley, south of Portland, in nw Oregon.

The habits of this pocket gopher, by far the largest in Oregon, are little known. Tunnels dug by the Camas Pocket Gopher can be more than 800 feet (240 m) long. There are several main tunnels, with additional inclined tunnels leading to the surface. Tunnels range from 2 to 5 inches (50–125 mm) in diameter, and are found from 4 inches to 3 feet (100–1,000 mm) below the surface, with deeper burrows in drier soil. The animal uses its teeth or claws to loosen the soil, which is pushed to one of the inclined passages and then out onto the surface. Lateral tunnels, which lead to storage chambers and such, are usually plugged with dirt from within; the gopher burrows through the plug to access them. Much new burrowing occurs when the soil is moist. This species cuts and stores roots of false dandelion *(Hypochaeris radicata),* vetch *(Vicia),* various trees, root crops (carrots, potatoes, etc.), and grasses. Roots and green leaves of a number of plants may be eaten immediately, or cut into pieces, transported in the pouches, and stored in chambers in the burrow system. Food is removed from the pouches with a sweeping motion of the foreclaws. The common and Latin species names of the Camas Pocket Gopher derive from its fondness for the sweet-tasting bulbs of the camas lily *(Camassia),* a plant that is uncommon in this animal's range. This pocket gopher often smells of the wild onion, leading to speculation that the animal has turned to that plant after having nearly extirpated the camas lily

in its range. Camas Pocket Gophers chatter or grind their teeth, and make "crooning" sounds when males and females are housed together. Great horned and barn owls have been known to eat this species, suggesting these pocket gophers are sometimes active aboveground at night.

Wyoming Pocket Gopher
Thomomys clusius

Description: Uniform brown with *distinct yellowish cast,* especially in young, with no light-colored patches around ears. Edge of ear fringed with white or whitish hair. Cheeks and top of head uniform in color. L 6⅜–7¼" (161–184 mm); HF ¾–⅞" (20–22 mm); Wt 1½–2½ o: (44–71.5 g).

Similar Species: Northern Pocket Gopher is larger and darker in color.

Sign: Mounds of soil.

Habitat: Meadows with good soil.

Range: Only in Sweetwater and Carbon counties, Wyoming.

The Wyoming Pocket Gopher was long considered a subspecies of the Northern Pocket Gopher, but recent studies have proved the two to be genetically distinct. The Wyoming's habits are assumed to be similar to those of the Northern.

Idaho Pocket Gopher
Thomomys idahoensis

Description: A small species, similar to Northern Pocket Gopher. Color varies, ranging from yellowish brown (with hairs tipped dark brown) to grayish brown or dark brown. *Color uniform above, but pales to light gray below.* Feet whitish. L 4⅝–6⅛ (117–156 mm); HF ⅞–1" (21–26 mm); Wt 1¾–3⅛ oz (51–88 g).

Similar Species: Northern, Mountain, and Townsend's pocket gophers are larger.

Sign: Mounds of soil; in spring, "gopher cores": long coils of earth used to plug burrows in snow and left on ground after snow melts.

Habitat: Meadows with good soil.

Range: Only in se Idaho and extreme sw Montana.

Little is known about the habits of the Idaho Pocket Gopher, but they are presumably similar to those of the Northern Pocket Gopher.

Western Pocket Gopher
Thomomys mazama

Description: Reddish brown or various shades of gray to black, depending on soil color. *Pointed ear, with dark patch behind that is 5 times size of ear.* L 7¼–9⅜" (183–239 mm); T 2⅛–3⅛" (53–81 mm); HF 1–1⅜" (25–35 mm); E ¼–⅜" (7–9 mm); Wt 1⅞–3⅜ oz (52–96 g).

Similar Species: Mountain Pocket Gopher, is duller tawny, russet, or brown. Very similar Northern Pocket Gopher occurs in different range.

Habitat: Prairies to mountain meadows.

Range: Western Washington and w Oregon, and south into n California.

Unlike most pocket gophers in its range, the Western spends a great deal of time aboveground, mostly at night, but also on warm, dark days. Sometimes it simply pokes its nose up through the ground to cut vegetation; at other times, it bends down vegetation to collect the seeds. It feeds heavily on grasses, false dandelions, garlic, and lupine. Owls, some larger mammals, and gopher snakes are frequent predators.

Mountain Pocket Gopher
Thomomys monticola

Description: *Tawny to russet or brown above;* slightly
buff to vividly golden below. *Snout
darker than face.* Pointed ear, with dark
patch behind that is about 3 times size
of ear. L 6⅝–10¾″ (167–273 mm);
T 1⅝–3¾″ (42–97 mm); HF ⅞–1⅛″
(22–30 mm); E ¼–⅜″ (7–9 mm);
Wt 2½–3⅛ oz (71–91 g).

Similar Species: Townsend's Pocket Gopher is larger.
Western Pocket Gopher is bright
reddish brown. Northern Pocket
Gopher has shorter, rounded ears.

Habitat: Open areas in coniferous forests.

Range: Northeastern California and extreme
wc Nevada.

The Mountain Pocket Gopher plugs its
underground burrows with soil to help
maintain suitable temperature and
moisture levels. The burrows are used
by many other animals, such as mice,
insects, and other invertebrates. In
winter, this pocket gopher builds a nest
aboveground, under the snow, and
burrows for plants.

143, 144 Northern Pocket Gopher
Thomomys talpoides

Description: *Color varies greatly: often rich brown* or
yellowish brown, but also grayish or
closely approaching local soil color.
White markings under chin. *Rounded
ear, with dark patch behind that is about
3 times size of ear.* L 6½–9⅛″ (165–233
mm); T 1⅝–3″ (40–75 mm); HF ¾–
1¼″ (20–31 mm); E less than ¼″
(7 mm); Wt 2¾–4⅝ oz (78–130 g).

Similar Species: While all western pocket gophers are
similar, Botta's Pocket Gopher in
Arizona has more white spots under
chin; Camas and Townsend's pocket
gophers are larger; and Western and
Mountain pocket gophers have longer,
more pointed ears.

Breeding: Reproductive season March–June.
1 litter per year in northern parts of
range; 2 or 3 in southern. Average of
3–7 young per litter. Gestation 18 days.

Sign: High, fan-shaped mounds; in spring,
"gopher cores": long coils of earth used
to plug burrows in snow and left on
ground after snow melts.

Habitat: Usually, good soil in meadows or along
streams; most often in mountains, but
also in lowlands.

Range: Southern British Columbia to s
Manitoba, and south to ne California
and n Nevada, through most of
Colorado to isolated portions of n
Arizona and n New Mexico; east to w
Nebraska and through most of North
and South Dakota.

The Northern Pocket Gopher seldom
appears aboveground; when it does, it
rarely ventures more than 2½ feet (750
mm) from a burrow entrance. Although
usually strictly terrestrial, one was
observed swimming across a Canadian
river nearly 300 feet (90 m) wide.
Except when seeking a mate, this pocket
gopher usually behaves ferociously
toward its own kind. Males are allowed
in burrows of females only in spring,
during the mating season. This species
is preyed upon by American Badgers,
weasels, and gopher snakes.

Townsend's Pocket Gopher
Thomomys townsendii

Description: *The largest pocket gopher in its range.*
Grayish brown, similar to color of local
soil. Male larger than female. L male
9⅞–13⅜" (251–340 mm), female
9–11⅜" (230–288 mm); T male 2⅝–
4⅜" (66–113 mm), female 2¼–3⅞"
(56–98 mm); HF male 1⅜–1⅝" (35–42
mm), female 1¼–1¾" (31–45 mm).

Similar Species: Botta's Pocket Gopher is very similar
but smaller.

Sign: Mounds of soil; in spring, "gopher cores": long coils of earth used to plug burrows in snow and left on ground after snow melts.

Habitat: Deep soils of river valleys.

Range: Isolated areas in ne California, n Nevada, e Oregon, w Idaho, and sc Montana.

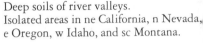

Little is known about Townsend's Pocket Gopher, but its habits are probably similar to those of Botta's Pocket Gopher

Southern Pocket Gopher
Thomomys umbrinus

Description: Usually some shade of *brown,* similar to color of local soil, varying from almost white to nearly black. In Arizona, often has a *dark band above from snout to base of tail,* which includes eyes and ears. L 5¼–10¾" (132–272 mm); T 1⅝–3⅞" (43–100 mm); HF ⅞–1½" (22–37 mm)

Similar Species: All other western pocket gophers are similar, but Botta's Pocket Gopher, which occurs with it in isolated colonies in Arizona and New Mexico, generally has no dark band above, and its sides have a purplish cast.

Breeding: Breeds in spring; 1 litter per year of 2–5 young.

Sign: Mounds of soil.

Habitat: In Arizona: oak belt in Huachuca Mountains; higher elevations, often in oak forests in Patagonia Mountains. In New Mexico: very shallow rocky soils in pine forests at high elevations in Animas Mountains.

Range: Mostly Mexico, but occurs north to extreme sw New Mexico and in mountains of se Arizona.

Where this pocket gopher occurs in our range, its habits presumably are similar to those of Botta's Pocket Gopher. The tunnels of the Southern Pocket Gopher are small, necessarily so to fit between the rocks in its habitat. This species feeds on agave in some areas.

Desert Pocket Gopher
Geomys arenarius

Description: Pale *gray-brown* above; *belly and feet white.* Like all *Geomys* species, has 2 distinct grooves on upper incisors. L 8¾–11″ (221–280 mm); T 2¼–3¾″ (58–95 mm); HF 1–1⅜″ (27–35 mm); Wt 5⅞–9¼ oz (165–264 g).

Similar Species: Botta's Pocket Gopher (and other *Thomomys* species) lacks groove on incisors. Yellow-faced Pocket Gopher has 1 groove. Texas Pocket Gopher is larger.

Breeding: Reproductive season may be prolonged, extending into summer. May exceed 1 litter per year; 4–6 young per litter.

Sign: Large, conspicuous mounds of excess soil pushed up from below.

Habitat: Sandy river bottoms, especially along irrigation ditches.

Range: South-central New Mexico and extreme w Texas.

The habits of this species are little known but presumably are similar to those of other species of *Geomys*. At times, it may greatly damage crops. Pocket mice and kangaroo rats often occupy its old burrow systems.

Attwater's Pocket Gopher
Geomys attwateri

Description: Short, fine hair; *light brown to black, usually paler below.* L 7½–9¼″ (192–235 mm); T 2–2¾″ (51–70 mm); HF 1–1⅛″ (25–28 mm).

Similar Species: The 3 species of pocket gophers in eastern Texas—Attwater's, Baird's, and Plains—are distinct genetically but very close morphologically. Baird's averages smallest, Attwater's is intermediate in size, and Plains is largest. Attwater's and Baird's, which occur together near College Station, Texas, occupy the same type of soil.

Breeding: 1 litter per year, of 1–3 young, usually born October–June. Some reproductively

active males may be found in any
month.

Sign: Large, conspicuous mounds of soil.

Habitat: Silty loam to silty clay-loam soils.

Range: Brazos River in Milam and Burleson
counties of east Texas, south along west
bank of Brazos River to Gulf Coast
(Matagorda County), southwest along
coast past Rockport (Aransas and San
Patricio counties), and northwest to
Atascosa County.

Attwater's Pocket Gopher appears to be
equally active throughout a 24-hour
period; about 60 percent of its time is
spent in its nest. This animal's complex
burrow systems contain many looping
tunnels. Average burrow depth is about
inches (180 mm), diameter 3 inches (80
mm), and length 300 feet (90 m).
Attwater's Pocket Gopher feeds on green
vegetation. Stems and tubers of poppy
mallow *(Callirhoe involucrata),* widow's-
tears *(Commelina erecta),* and Bermuda
grass *(Cynodon dactylon)* have been found
in burrows. Reingestion of defecated food
(coprophagy), which enables additional
absorption of nutrients, has been
observed in this species. There are many
possible predators on this species, but
only the Coyote has been documented.

Baird's Pocket Gopher
Geomys breviceps

Description: Short, fine hair; *light brown to black,
usually paler below.* L 7½–8¾" (192–22
mm); T 2⅛–2⅝" (54–67 mm); HF ⅞–
1⅛" (23–28 mm).

Similar Species: The 3 species of pocket gophers in
eastern Texas—Attwater's, Baird's, and
Plains—are distinct genetically but
very close morphologically. Baird's
averages smallest, Attwater's is
intermediate in size, and Plains is
largest. Attwater's and Baird's, which
occur together near College Station,
Texas, occupy the same type of soil.

Breeding: 1 or 2 litters per year, of 1–4 young each, born February–August. Those that produce 2 litters seem to do so in rapid succession.

Sign: Large, conspicuous mounds of excess soil. Large nest mounds of soil up to 6' (1.8 m) in diameter and 1–2' (300–600 mm) high.

Habitat: Silty loam to silty clay-loam soils.

Range: Eastern bank of Brazos River in c and se Texas, east into w Louisiana, and north into e Oklahoma and sw Arkansas.

The habits of Baird's Pocket Gopher are very similar to those of Attwater's Pocket Gopher. Burrows dug by Baird's average 2⅜ inches (60 mm) in diameter, and are 4 to 27 inches (100–680 mm) deep. They range in length from 180 to 600 feet (55–180 m). This species constructs large nest mounds; nests, often of Bermuda grass (*Cynodon dactylon*), may be relocated to higher ground in the mound during wet seasons, lower ground in dry seasons. This species feeds on roots, stems, and leaves of many green plants. Cellulose-digesting bacteria occur in the cecum (a pouch at the beginning of the large intestine) and large intestine, helping with digestion. Reingestion of fecal pellets has also been observed in this species. Baird's Pocket Gopher attracts a variety of predators, including kingsnakes, great horned owls, red-tailed hawks, Long-tailed Weasels, and Striped Skunks.

146 Plains Pocket Gopher
Geomys bursarius

Description: A relatively large pocket gopher; larger in North, smaller in South. Pale brown to black (in Illinois); slightly paler below. Little variation in color at any one locality, but color varies greatly over its range, matching the soil. *Long, sparsely haired tail. White feet.* L 7⅜–14"

(187–357 mm); T 2–4¼″ (51–107 mm); HF ⅞–1⅝″ (23–43 mm); Wt 4½–12½ oz (127–354 g).

Similar Species: Yellow-faced Pocket Gopher has only 1 groove on its upper incisors. Texas Pocket Gopher is usually larger. The 3 species of pocket gophers in eastern Texas—Attwater's, Baird's, and Plains— are distinct genetically but very close morphologically. Baird's averages smallest, Attwater's is intermediate in size, and Plains is largest.

Breeding: 1 litter per year of 1–7 young, born in spring after gestation of at least 51 days. Time of birth can vary greatly with latitude. Birth weight ⅛ oz (5 g); newborn length 1¾″ (46 mm).

Sign: Conspicuous mounds of excess dirt up to 1′ (300 mm) high, more than 2′ (600 mm) wide, often in a line, with fresh mounds indicating direction of excavation.

Habitat: Prairie areas with sandy loam or loam soils; pastures; lawns; sometimes plowed ground.

Range: Eastern North Dakota, Minnesota, and w Wisconsin south through c Illinois, nw Indiana, and much of Missouri and Arkansas to w Louisiana; west and south through s South Dakota, se Wyoming, e Colorado, e New Mexico, and ne two-thirds of Texas.

The only pocket gopher present over most of its range, the Plains Pocket Gopher is much more active in summer than in winter. Burrows are shallow in summer, usually within 1 foot (300 mm) of the surface, and deeper in winter, when dirt is pushed up into the snow, leaving earthen cores when the snow melts. A burrow in Wisconsin was described as having a main tunnel 200 feet (60 m) long, with many short lateral branches, most not reaching the surface. There was a single nest cavity in the center of the system, about 10 inches (245 mm) below the surface. Tunnels were about 6 inches (150 mm) below the

surface, but reached a depth of 38 inches (965 mm) in one place. The main tunnel was 3 inches (75 mm) wide and 4½ inches (115 mm) high. There were two food caches, and feces were found in plugged lateral tunnels. The Plains Pocket Gopher feeds primarily on roots, bulbs, and tender green plants, many of which are probably cut off at the roots and pulled into the burrow. Some plants are probably collected near the burrow entrances. Food is transported in the animal's cheek pouches, and much is stored in caches in the burrow system. The cheek pouches are filled very rapidly with the forepaws, using a wiping motion that forces the food into the open end of the mouth. To empty the pouches, the animal brings both forefeet from back to front against its cheeks, thus forcing the food out in a pile in front. In early spring, the male leaves his burrow to seek a female; after having mated, he returns to his solitary ways.

Jones' Pocket Gopher
Geomys knoxjonesi

Description: A *relatively large* pocket gopher. *Brown above,* but variable with soil color. Long, sparsely furred tail. White feet. Measurements unavailable.

Similar Species: Plains Pocket Gopher is very similar morphologically, but differs genetically.

Breeding: 1 litter per year.

Sign: Mounds of excess soil.

Habitat: Prairie areas with loam or sandy loam soils; open areas.

Range: Extreme se New Mexico and adjacent w Texas.

Jones' Pocket Gopher was recently deemed to be a species distinct from the Plains Pocket Gopher, which it closely resembles. Of 75 individuals from the area where the Plains and Jones' pocket gophers occur together and occasionally hybridize (known as the "hybrid zone"),

20 were the Plains Pocket Gopher, 14 were Jones' Pocket Gopher, and 41 had some intermediate characteristics. This ratio indicates that the two species are remaining distinct where they occur together, and that occasional but restricted hybridization occurs. If they were freely hybridizing, the original two species would have become one.

Texas Pocket Gopher
Geomys personatus

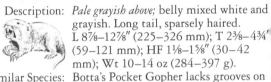

Description: *Pale grayish above;* belly mixed white and grayish. Long tail, sparsely haired. L 8⅞–12⅞" (225–326 mm); T 2⅜–4¾" (59–121 mm); HF 1⅛–1⅝" (30–42 mm); Wt 10–14 oz (284–397 g).

Similar Species: Botta's Pocket Gopher lacks grooves on incisors. Yellow-faced Pocket Gopher has only 1 groove. Attwater's, Baird's, and Plains pocket gophers are smaller. Jones' and Llano pocket gophers are browner.

Breeding: Reproductive season January–May; 1 or 2 litters of 2–4 young produced per year

Habitat: Sandy soils (not silt loam); areas of thorny brush such as mesquite, and in various grasses.

Range: Southern Texas.

The tunnels of this accomplished burrower are from 4 to 5 inches (100–125 mm) wide and can be over 100 feet (30 m) long. The Texas Pocket Gopher feeds mostly from within the burrow on roots and other plant parts, especially grasses, by cutting off plants at the base and pulling them inside. This animal reingests fecal pellets, which are about ¾ inch (19 mm) long and ¼ inch (7 mm) in diameter. Its activities sometimes cause erosion or interfere with cultivation. Marsh hawks and domestic cats are the only documented predators. Biting lice are characteristic and exceedingly host-specific parasites

of pocket gophers, and in this species, even different populations of gophers may have different species of lice.

---

145 Southeastern Pocket Gopher
Geomys pinetis

Description: Various shades of *brown,* according to soil color. Relatively *short, nearly naked tail.* L 9–13¼″ (229–335 mm); T 3–3¾″ (76–96 mm); HF 1⅛–1½″ (30–37 mm).

Similar Species: This is the only pocket gopher in its range.

Breeding: Breeds throughout year, but reproductive activity peaks February–March and June–August. Litters of 1–3 young; sometimes 2 litters per year. Newborn is 2″ (50 mm) long and weighs ¼ oz (5.8 g).

Sign: Mounds of excess soil.

Habitat: Open fields; pastures; open woods.

Range: Southeastern Alabama, s Georgia, and n Florida.

The Southeastern Pocket Gopher is active all year, yet stores great quantities of food in its extensive burrow system, which ranges from just below the ground surface to 2 feet (600 mm) deep. The animal uses its long claws to dig the burrow, and its heavy incisors to cut roots. When dirt accumulates in the burrow behind the pocket gopher, the animal turns and pushes it forward with its head and forefeet. Excess soil is pushed in this manner out of a lateral burrow, making a mound at the surface. The mounds made by this species are often very conspicuous, and sometimes number several hundred per acre (.4 ha). This pocket gopher feeds on roots, rhizomes, and green succulent plants that can be reached from a tunnel opening. In Florida, they often are called "salamanders," apparently an alteration of "soil mounder" spoken with a southern

accent. This animal can be a major pest in agricultural regions, attacking sweet potatoes especially, but also peanuts, sugarcane, and peas. Three separate races of Southeastern Pocket Gopher from Georgia were formerly recognized as distinct species, but are currently recognized as subspecies of the Southeastern: the Colonial Pocket Gopher *(G. p. colonus),* found along the coast in the extreme southeast; Sherman's Pocket Gopher *(G. p. fontanelus),* found in a small area in Chatham County, in the northeast; and the Cumberland Island Pocket Gopher *(G. p. cumberlandius),* from Cumberland Island.

Llano Pocket Gopher
Geomys texensis

Description: A *relatively large* pocket gopher. Brown above; shade varies with soil color. Measurements unavailable.

Similar Species: Jones' and Plains pocket gophers are very similar.

Sign: Mounds of excess soil.

Habitat: Prairie areas with sandy or sandy-loam soils.

Range: Isolated in portion of Edwards Plateau of Texas, including Kimble, McCulloch, Mason, San Saba, Llano, Gillespie, and Blanco counties.

Although most similar in appearance to Jones' Pocket Gopher, the Llano Pocket Gopher previously was considered a subspecies of the Plains Pocket Gopher. However, it was recently elevated to species status on genetic grounds.

147 Yellow-faced Pocket Gopher
Cratogeomys castanops

Description: Fur varies from *pale yellowish buff to dark reddish brown above* mixed with dark-tipped hairs. *Dark feet.* Hairless tail. 1 groove on upper incisors. L 8⅞–12⅝"

(226–320 mm); T 2¾–4⅛″ (70–105 mm); HF 1¼–1⅝″ (31–43 mm); Wt 7½–11⅝ oz (213–330 g).

Similar Species: Western pocket gophers have no grooves on incisors; eastern pocket gophers *(Geomys)* have 2 grooves.

Sign: Dirt mounds, often beneath bushes or cacti.

Habitat: Clay or sandy soils in open lands. Inhabits more rocky soils when other pocket gophers (*Geomys* species) are present. It is quite common in the lower Rio Grande Valley, and has replaced Botta's Pocket Gopher in several places as the soil has become drier.

Range: Southeastern Colorado and sw Kansas south through extreme w panhandle of Oklahoma, e New Mexico, and w Texas.

The burrows of the Yellow-faced Pocket Gopher are 3 to 4 inches (75–100 mm) wide and usually shallower than those of the *Geomys* species in the same area. Five burrows excavated ranged from 140 to 343 feet (42 to 104 m) long, with numerous short side burrows up to 10 feet (3 m) long. Typically, there is a shallow, foraging level and a deeper level that contains the nest and food chambers. One burrow system is occupied by one animal except during mating and rearing of the young. There is one functional nest per burrow system. Old nests are abandoned, and the burrows to them are plugged. Feces are scattered along runways, thrown out during mound building, or packed into plugs for exit openings or old lateral tunnels. Although these pocket gophers will harvest some plant material near the entrance, they do most foraging from underground, cutting off plants from below and pulling them down. Cut-up plant materials are then carried in the cheek pouches to food caches. Because this species feeds on plant parts, including bark from tree roots, and because of its large size and the size of its burrows, it is considered a pest in

cultivated areas. Even though mainly subterranean, these animals are preyed upon by hawks, owls, and mammalian predators. At one time farmers in the Davis Mountains of Texas killed great horned owls for fear they would eat their chickens, only to have pocket gophers invade their alfalfa. The owls were apparently preying on the pocket gophers, for once the owls were gone, the pocket gophers increased greatly in number.

POCKET MICE AND KANGAROO RATS
Family Heteromyidae

Heteromyids include 59 species in six genera and occur only in Central and North America. There are 38 species in North America north of Mexico, where they are found west of the Mississippi River, occupying plains, prairies, and deserts. Neither mice nor rats, and not closely related to any other North American rodents, these species are grouped in five genera: two genera of pocket mice (*Perognathus,* with 10 species, and *Chaetodipus,* nine species); and one genus each of kangaroo mice (*Microdipodops,* two species), kangaroo rats (*Dipodomys,* 16 species), and the Mexican Spiny Pocket Mouse (*Liomys,* one species). Members of this family are nocturnal, burrowing animals with external, fur-lined cheek pouches used for carrying food. In all but *Liomys* the incisors have grooves on the front face. The precise function of the grooves is uncertain, but they probably enhance chewing by giving greater surface area to the cutting edge and providing extra strength to the teeth.

Heteromyids dig underground burrows with chambers for sleeping, nesting, and food storage. Many of them build mounds of soil at the entrances that are often large and conspicuous. The entrances to and escape holes from the burrows tend to be small; some are plugged with soil or sand during the day to help maintain proper levels of temperature and moisture. While most heteromyids do not hibernate, some are inactive in cold or extreme heat. They subsist chiefly on seeds and some greens, which they store in underground chambers. From these, their bodies manufacture water; some species never drink. Mammalogists have determined, by comparing stomach contents with material in cheek pouches, that

heteromyids eat perishable items while in the field and store hard items. Heteromyids often form a major proportion of the desert community of small rodents; within a given area, there are often species of different sizes represented, which feed on different-size seeds. Heteromyid populations are subject to periodic fluctuations. The young are born in burrow nests, usually in late spring through early fall. A litter may comprise from one to eight offspring; there may be one to several litters per year, with fewer in the north. Vocalizations, very seldom heard, are thin, high-pitched squeaks. Predators include rattlesnakes, hawks, Coyotes, foxes, Bobcats, Cougars, weasels, Badgers, and skunks.

Kangaroo rat:
4-toed hindfoot

5-toed hindfoot

Kangaroo rats (*Dipodomys* species) are among our most beautiful rodents. The many species are remarkably alike in markings, but may be distinguished by their range, size, and intensity of coloration. Upperparts are pale yellow or tan to dark brown; the belly is white. The long tail, which aids in balancing, has dark stripes on the top and bottom, separated by pale side stripes. The color, width, and position of these stripes help in the identification of some species. The tail is often crested or tufted with long hairs along its terminal fifth. Most kangaroo rats have a white band running across the thigh and joining the white side stripe on the tail. Facial markings are usually black and white, though they are very pale in a few species. The forelegs are short. The hindfeet are large, and the hindlegs very long and powerful; the soles of the hindfeet are densely furred. Kangaroo rats are either five-toed, with a tiny fifth toe near the ankle, or four-toed, with the fifth toe lacking. Kangaroo rats travel by using their spring-like hindlegs, which has earned them their common name. When alarmed, they ca

make spectacular hops; some of the larger species can leap almost 9 feet (3 m). As they seldom forage more than 35 to 40 feet (10–12 m) from an escape hole, they can reach the comparative safety of a burrow in a few bounds. They also use their hindlegs when fighting, and they make a drumming sound with their hindfeet, which may serve to ward off predators, attract females, advertise a territory, or repel intruders. Kangaroo rats sometimes kick sand at objects, such as snakes, apparently to determine whether they are alive or dead. They often sand- or dust-bathe, presumably for cleansing or grooming, but also, in some species, as a way of marking territories.

Kangaroo mice *(Microdipodops),* smaller editions of kangaroo rats, also have long hindlegs for jumping, but their tails are quite different; they are broadest in the middle and lack a terminal crest or tuft. The sole of the hindfoot is densely furred, as with kangaroo rats.

Pocket mice *(Perognathus* and *Chaetodipus)* are smaller than kangaroo mice, look more mouse-like, and have only moderately long tails that are haired but not tufted or only slightly tufted. Pocket mice have shorter hindlegs than kangaroo mice, and are poor jumpers. Their pale yellowish to brown upperparts are separated from their white underparts by a buff-colored side stripe. The two pocket mice genera are very similar in appearance, but may be distinguished by the following characteristics: Members of the genus *Chaetodipus* have a naked hindfoot sole, and their fur is coarse, often with spiny bristles. Stiff, coarse hairs usually project across the front margin of the ear, and the front, inner base of the ear has an upright lobe. Species of the genus *Perognathus* have a sparse covering of short hairs on the back third to half of the sole of the

hindfoot, as well as relatively fine, soft fur; there are no spine-like bristles on the rump, or stiff, coarse hairs across the front margin of the ear. The front, inner base of the ear usually lacks an upright lobe. Pocket mice prefer sandy soil. The Mexican Spiny Pocket Mouse (*Liomys irroratus*) differs from other members of this family in lacking grooves on its upper incisors.

White-eared Pocket Mouse
Perognathus alticolus

Description: Soft-furred. *Olivaceous buff to brownish above;* white to whitish underparts; faint lines on sides. Tail pale; colored like upperparts at base, but dusky or black at tip; white below. Only *Perognathus* species with hindfoot more than ¾" (20 mm) long. *Ears have white or yellowish hairs outside and inside; lobe (antitragus) at base of ear.* L 6⅜–7⅛" (160–181 mm); T 2⅞–3¾" (72–97 mm); HF ⅞" (21–23 mm); E average ⅝" (17 mm).

Similar Species: Closely related to Great Basin Pocket Mouse, in which hair inside ears is buff rather than white or yellowish. No other *Perognathus* species has lobe (antitragus) at base of ear.

Sign: Small burrow openings with piles of sand, usually under a plant.

Habitat: Open grassy areas and dry bracken fern beds in yellow-pine forests and Joshua tree woodlands.

Range: Southern California.

The White-eared Pocket Mouse occurs as two isolated subspecies, both in California: *P. a. inexpectatus,* which occurs in the Tehachapi Mountains, in Kern County; and *P. a. alticolus,* which occurs in the San Bernardino Mountains in San Bernardino County. Little is known of the habits of this pocket mouse.

65 Arizona Pocket Mouse
Perognathus amplus

Description: Soft-furred. Color variable: *pinkish buff to pale ochraceous salmon above,* overlaid with black in varying degrees (some forms are nearly black). White or faintly buffy below. Buffy line on sides. Tail more than ¾ length of head and body. L 4⅞–6¾" (123–170 mm); T 2⅞–3¾" (72–95 mm); HF ⅝–⅞" (17–22 mm).

Similar Species: Little Pocket Mouse is smaller, with shorter tail.

Breeding: Mating season starts late February–early March; females have been found pregnant in April. 1–7 young per litter (average 3–5).

Sign: Small burrow openings with piles of sand, usually under a plant.

Habitat: Desert scrub; in greasewood, rabbit bush, ephedra, and shortgrass; sometimes in short junipers and in creosote bush flats.

Range: Much of Arizona.

The Arizona Pocket Mouse is primarily nocturnal, but is sometimes abroad in daylight. This species feeds almost entirely (95 percent) on seeds, with insects and green vegetation forming the remainder. Creosote bush seeds are the most important, followed by *Pectocarya,* heronbill, and plantain. The animals apparently find the seeds by digging at random in the sand. This pocket mouse most often occurs with Merriam's Kangaroo Rat, but is also associated with Harris' Antelope Squirrel, the Round-tailed Ground Squirrel, and the Desert Kangaroo Rat.

Olive-backed Pocket Mouse
Perognathus fasciatus

Description: Soft-furred. *Grizzled olive-gray above,* although may vary from dark olive to bright buffy. White below. Buff or yellowish line on side; buff patch

behind ear. Tail slightly less than half total length. Large hindfeet; small forefeet. L 4½–5⅝" (115–143 mm); T 2¼–2⅝" (57–67 mm); HF ⅝–¾" (16–18 mm); Wt ¼–⅜ oz (8.9–9.9 g).

Similar Species: Great Basin Pocket Mouse is much larger. Silky and Plains pocket mice are generally smaller, and yellowish above.

Breeding: Probably 1 litter per year, of 4–6 young born between June and August. Gestation about 30 days.

Sign: Small burrow openings with piles of sand, usually under a plant.

Habitat: Dry, sandy grasslands with little vegetation.

Range: Southeastern Alberta, s Saskatchewan, and sw Manitoba south to extreme ne Utah, n Colorado, and nw Nebraska.

Summer nests and storage chambers of the Olive-backed Pocket Mouse are 12 to 15 inches (300–375 mm) belowground, while those used for estivation or hibernation are as much as 6 feet (2 m) deep. While this species feeds on some insects, its primary food is seeds, including those of foxtail grass, bugseed, knotweed, Russian thistle, blue-eyed grass, and tumbleweed.

66 Plains Pocket Mouse
Perognathus flavescens

Description: Soft-furred. Yellowish buff sprinkled with black hairs above; white below. *No yellow patch behind ear.* L 4⅜–5⅛" (113–130 mm); T 1⅞–2⅜" (47–62 mm); HF ⅝–⅞" (15–21 mm); Wt ¼–⅜ oz (8–11 g).

Similar Species: Silky and Merriam's pocket mice have yellow patch behind ear.

Breeding: Breeds in spring; 2–6 young per litter.

Sign: Tiny burrows leading in all directions from under plants in sandy areas.

Habitat: Sandy plains with sparse vegetation, sand dunes, and shifting sands.

Range: Southeastern North Dakota, w and s Minnesota, and n Iowa southwest to e

Colorado and Texas panhandle. Apache Pocket Mouse subspecies is found in se Utah, sw Colorado, ne Arizona, and nw New Mexico.

As do many other heteromyids, this species varies greatly in color, generally resembling the color of the substrate on which it lives. During the day, the Plains Pocket Mouse closes the main burrow entrance but usually leaves open other, less conspicuous ones. Formerly considered a separate species, the larger Apache Pocket Mouse *(P. f. apache),* which has a distinct buff side stripe, was combined with this species when it was discovered that the two intergraded (bred and produced an intermediate form) where their ranges meet.

67 Silky Pocket Mouse
Perognathus flavus

Description: *Soft-furred. Pale yellowish above,* often with many black hairs; belly white. *Yellow patch behind ear;* white spot below ear. *Small hindfoot.* Juvenile dull gray. L 3⅞–4¾" (100–122 mm); T 1¾–2⅜" (44–60 mm); HF ⅝" (16–17 mm); Wt ¼ oz (6–9 g).

Similar Species: Plains Pocket Mouse lacks yellow patch behind ear. Arizona and Little pocket mice are larger, with larger hindfeet. Merriam's Pocket Mouse is very similar, but differs genetically and usually has a somewhat longer tail.

Breeding: In New Mexico, breeding activity peaks April–July and September–October. Usually ceases in winter, but some may breed in January and produce young in February. 1or 2 litters per year of 1–6 (average 3 or 4) young, born after gestation of 22–26 days.

Sign: Tiny burrow openings as wide as a finger, at the base of plants or in vertical banks.

Habitat: Prairies; sandy, gravelly, or rocky areas with sparse vegetation of various grasses and forbs.

Range: Southeastern Wyoming and w Nebraska
south to extreme se Utah, Arizona, New
Mexico, and w Texas.

The Silky Pocket Mouse is nocturnal.
It sometimes hibernates; otherwise it
may forage nightly throughout winter,
although its activity may be reduced.
The animal enters its small burrow
through a mound of dirt. The burrow
is usually not more than 4 inches
(100 mm) deep, and often has several
entrances from different levels on the
mound. Inside the mound there is a
central chamber from which a single
tunnel goes down into a second
chamber, larger than the first. The nest
is about 2½ inches (65 mm) in diameter
with one opening near the top of the
mound. Tunnels radiate from the nest
chamber and often contain small caches
of seeds. Additional seeds are stored
regardless of the number of caches
already existing. Burrows often are four
under cactus, yucca, or shrubs. Burrows
in a plowed field in Zacatecas, Mexico,
were 2 to 4 inches (50–100 mm) deep,
and extended up to about 3 feet (1 m)
long. Blind passages near the surface
enable the Silky Pocket Mouse to break
through to escape snakes and other
underground predators. This species
sometimes uses abandoned burrows of
other pocket mice, kangaroo rats, or
pocket gophers. It has been known to
inhabit the mounds of harvester ants,
presumably to obtain seeds gathered by
the ants. Like many other pocket mice,
it drinks no water, metabolizing it from
food. The latter is primarily seeds, but
this pocket mouse also eats other plant
material—including the small seeds
of thistle and millet, as well as several
grasses, wild sunflowers, amaranth
pigweed, and opuntia—along with
very few invertebrates. Predators include
gopher snakes, coachwhips, rattlesnakes,
owls, foxes, Coyotes, skunks, American
Badgers, and Ringtails.

San Joaquin Pocket Mouse
Perognathus inornatus

Description: *Soft-furred. Orange-buff above;* whitish
below. *Tail, nearly uniform light brown*
above and below, has small tuft;
averages longer than head and body
length. Ears may have pale spots at base.
L 5–6⅜″ (128–160 mm); T 2½–3⅛″
(63–78 mm); HF ¾–⅞″ (18–21 mm);
Wt ⅜–⅝ oz (13–17 g).

Similar Species: Little Pocket Mouse does not occur in
San Joaquin Valley. White-eared Pocket
Mouse has white ears, occurs in pine
zones.

Breeding: Reproductive season March–July, with
at least 2 litters per year of 4–6 young.

Habitat: Weedy or grassy areas on fine soil.

Range: Only in San Joaquin Valley, in wc
California.

Under adverse conditions of great heat
or cold, the San Joaquin Pocket Mouse
may become torpid and retire to its
burrow. The Latin species name *inornatus*
alludes to its uniform coloration, which
lacks "ornamental" markings. This
species tends to alternate sides when
sand-bathing, rubbing one side along
the ground, starting with the cheek, and
then the other side. Before bathing it
digs rapidly with its forepaws. It has
several vocalizations (growl, low grunt,
and squeal), and also engages in teeth-
chattering and foot-drumming.

74 Little Pocket Mouse
Perognathus longimembris

Description: Soft-furred. Grayish yellow or buff
above, interspersed with black hairs that
vary from paler to darker with color of
soil; underparts buff, brownish, or
white. *Tail uniformly pale brownish.
2 small white patches at base of ears.*
L 4¼–6″ (110–151 mm); T 2⅛–3⅜″
(53–86 mm); HF ⅝–¾″ (15–20 mm);
Wt ¼ oz (7–9 g).

Similar Species: San Joaquin Pocket Mouse occurs only
in San Joaquin Valley, where this specie
does not occur. White-eared, Long-
tailed, and Great Basin pocket mice are
larger, with longer hindfeet. Arizona
Pocket Mouse is generally larger. Silky
Pocket Mouse is generally smaller.

Breeding: Apparently reproductively active from
spring through fall. 1 or 2 litters per
year of 1–6 (average 4) young; gestatio
22–23 days.

Habitat: Gravelly soils in desert areas, often in
rolling terrain or in areas with ravines
or rock outcroppings. Associated plant
are rabbitbrush, tumble mustard
(Sisymbrium altissimum), blackbrush,
sand sagebrush *(Artemisia filifolia),*
Mormon tea, and yucca.

Range: Southeastern Oregon, Nevada, w and s
Utah, s California, and small isolated
areas in n, sc, and sw Arizona.

This seasonally active pocket mouse ma
hibernate for long periods under advers
conditions; in California, it is inactive
from October to January. The Little
Pocket Mouse can survive in the wild fo
three to five years, a considerable time
for such a small animal. In Nevada, the
Kit Fox is an important predator, but
presumably this mouse is eaten by all
common predators where it occurs.

Merriam's Pocket Mouse
Perognathus merriami

Description: *Very similar to Silky Pocket Mouse. Soft-
furred. Pale yellowish above,* often with
many black hairs; belly white. *Yellow
patch behind ear; white spot below ear. Sma
hindfoot.* Young very different from
adult; soft and silky, lacks black hairs.
Average measurements of 2 subspecies:
L 4½″ (116 mm); T 2″ (52 mm); HF ⅝″
(16 mm); Wt ¼–⅜ oz (6–10 g).

Similar Species: Silky Pocket Mouse is very similar, but
differs genetically. Plains Pocket Mous
lacks yellow patch behind ear. Arizona

and Little pocket mice are larger, with longer hindfeet.

Breeding: Breeding season April–November. Litter 3–6 young; more than 1 litter per year.

Sign: Tiny burrow openings at base of plants or in vertical banks.

Habitat: Shortgrass prairies and desert scrubland over most of range, but most common in open, arid brushland; also sandy, gravelly, or rocky areas with sparse vegetation.

Range: Western half of Texas; se New Mexico and extreme w Oklahoma.

Merriam's Pocket Mouse is primarily nocturnal, but occasionally is abroad in the daytime. The burrows of this pocket mouse are sometimes dug under logs or rocks or in vertical banks such as are found along roads, but most often at the base of a clump of grass or other plant. The burrow opening is ¾ to 1⅛ inches (20–30 mm) in diameter; the excess soil is scattered. Burrow openings are plugged during the day with soil that does not reach the surface; unoccupied burrows are plugged to the surface. This species maintains several burrows in its home range; a male might maintain six or seven, a female about five. Home ranges are small and do not overlap, indicating territoriality. A tunnel in a home burrow descends almost vertically for 6 to 8 inches (150–200 mm), then levels off in grade to the nest, which is about 18 to 24 inches (450–600 mm) in depth. There are often side tunnels for the deposit of fecal pellets. The home nest contains enlarged nest chambers 2¾ to 6 inches (70–150 mm) in diameter. Seeds and other food items are stored around the edge of the nests; nesting materials—dried grass, weed twigs, and seed husks—are lodged against one wall. Refuge burrows are blind tunnels used for storage, refuge, or temporary shelter. Merriam's Pocket Mouse digs in softer soils with its

forefeet; its hindfeet move the soil backward. In hard soils, the animal bite through the crust, then proceeds to dig as usual. Seeds of many species of grass and forbs are a primary food, but other vegetation and insects are consumed. It is probable that perishable foods are eaten in the field and hard parts returned to the burrows. Predators include foxes, owls, shrikes, and perhap grasshopper mice. The Silky Pocket Mouse was described as a species in 1855 and Merriam's Pocket Mouse in 1892. Morphological comparison led to the conclusion in 1972 that the two intergraded across the zone between them (breeding and producing an intermediate type), indicating that they were members of the same species. In 1991, however, genetic analysis of 28 populations across the Texas and New Mexico ranges of the animals re-established them as separate species.

68 Great Basin Pocket Mouse
Perognathus parvus

Description: A medium-size pocket mouse. *Soft-furred.* Pinkish buff or yellowish above, interspersed with blackish hairs; white or buff below; indistinct olive-greenish line on sides. *Long tail,* darker above, whitish below, *slightly crested toward tip.* Hair inside ear is buffy. L 5⅞–7¾" (148–198 mm); T 3–4¼" (77–107 mm); HF ¾–1" (19–27 mm); Wt ⅝–1⅛ oz (16.5–31 g).

Similar Species: Little Pocket Mouse is smaller. Long-tailed Pocket Mouse has tufted tail that is more distinctly crested along terminal third.

Breeding: Breeds spring–summer; producing 1–3 litters per year. First mates in April, after emerging from hibernation. Litter of 2–8 young each, usually born in May and August, after estimated gestation of 21–28 days.

Sign: Burrow systems with several openings and packed piles of soil near entrance.

Habitat: Arid, sparsely vegetated plains and brushy areas.

Range: South-central British Columbia south through e Washington, e Oregon, and s Idaho to ne and c California, Nevada, Utah, extreme se Wyoming, and extreme nw Arizona.

This pocket mouse is active from April through September, eating many kinds of insects and collecting seeds to be stored in its burrow, including those of Russian thistle, pigweed, wild mustard, and bitterbrush. It doesn't need to drink because it metabolizes water from its food. Summer nesting and storage burrows are shallow, but the animal digs a deep tunnel to a hibernation nest of dry vegetation in a chamber 3 to 6 feet (1–2 m) deep, where it spends the winter. Its many predators include snakes, owls, hawks, weasels, skunks, and foxes.

Yellow-eared Pocket Mouse
Perognathus xanthonotus

Description: A medium-size pocket mouse. Ocher-buff to cream-buff above; white below and on feet. *Hairs inside ears all yellowish.* Tail pale yellowish above, with slightly darker tinge on terminal fifth; white below. Tail has *tuft or crest, usually less than ⅝" (15 mm) long. Soles of hindfeet have some hairs near ankle.* Averages: L 6¾" (170 mm); T 3⅜" (85 mm); HF ⅞" (23 mm).

Similar Species: White-eared Pocket Mouse has white hairs inside ears. Great Basin Pocket Mouse has buffy hairs inside ears. Long-tailed Pocket Mouse has longer tail. Little and San Joaquin pocket mice are smaller, with shorter hindfeet. Other *Perognathus* species have hindfeet naked to the heel.

Habitat: Desert with buckbrush, manzanita,
           scrub oak, greasewood, and rabbitbrush
Range: Tehachapi Mountains, California.

Very little information is available on
the Yellow-eared Pocket Mouse or on its
habits and behaviors.

69 Bailey's Pocket Mouse
Chaetodipus baileyi

Description: *A large pocket mouse.* Fur coarse; grayish
               with yellow hairs interspersed; white
               below. Rump has black hairs but no
               spines or bristles. *Long tail tufted at tip.*
               Small white spot at base of ear. L 7⅞–9″
               (201–230 mm); T 4¼–5⅜″ (110–136
               mm); HF 1–1⅛″ (26–28 mm); Wt ⅞–
               1⅜ oz (24–38 g).

Similar Species: Most other pocket mice in its range
                   are smaller. Desert Pocket Mouse is
                   yellowish rather than grayish, and
                   usually smaller. Long-tailed Pocket
                   Mouse is smaller, and lacks yellow hairs
                   in its grayish fur. Rock Pocket Mouse
                   is smaller and has weak spines or
                   bristles on rump.

Breeding: Breeding is subject to environmental
            conditions, but peaks in spring and
            summer. Number of young averages
            3 or 4.

Habitat: Rocky slopes in desert and desert-
           grassland areas, often in transitional
           areas between rocky hillsides and
           desert flats.

Range: Extreme s California, s Arizona, and
         Hidalgo County, New Mexico.

Bailey's Pocket Mouse does not
hibernate. It is active all year, although
its winter activity may be reduced. It
feeds on a wide variety of seeds, as well
as other vegetation and insects. It will
carry some seeds in its cheek pouches
back to its burrow as a reserve for
winter. In Arizona, it often occurs with
Arizona and Desert pocket mice and
Merriam's Kangaroo Rat.

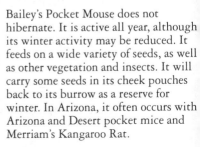

70 California Pocket Mouse
Chaetodipus californicus

Description: A large pocket mouse. *Brownish gray, with distinct white spines or bristles on rump, and brownish line on side.* Underparts yellowish white. *Tail* brownish above, whitish below, *with prominent tuft.* Hindfoot large. *Large ears;* long black or buffy hairs at front base of ear are nearly as long as ear. L 7½–9¼″ (190–235 mm); T 4–5⅝″ (103–143 mm); HF ⅞–1⅛″ (24–29 mm); E ⅜–½″ (9–14 mm); Wt average ¾ oz (23 g).

Similar Species: San Diego Pocket Mouse has smaller ears and is usually found at lower desert elevations. Spiny Pocket Mouse is grayish or yellowish buff, with side line faint or absent. Desert and Bailey's pocket mice lack spines or bristles on rump.

Breeding: 1 litter per year, of 2–5 young, born March–June; sometimes a second litter. Gestation 25 days.

Habitat: Coastal sage scrub communities on chaparral slopes.

Range: Southwestern California.

Along the coast of southern California, this is the most common pocket mouse. It enters a state of torpor in response to low temperatures and reduced food supplies.

71 San Diego Pocket Mouse
Chaetodipus fallax

Description: A fairly large pocket mouse. *Rough-haired.* Dark brown above, with *black spines on rump, white spines or bristles on hips;* underparts white, separated from upperparts by buff band. *Tail dark above with narrower light stripe below; crest about ½–⅝″ (12–16 mm) long at tip.* L 6⅞–7⅞″ (176–200 mm); T 3½–4⅝″ (88–118 mm); HF ⅞–1″ (21–26 mm); E ¼–⅜″ (7–9 mm).

Similar Species:	Bailey's and Desert pocket mice lack spines or bristles on rump. California Pocket Mouse has larger ears, and is found on brushy slopes rather than in low desert. Spiny Pocket Mouse is paler—yellowish or grayish.
Habitat:	Dry, open, sandy, weedy lowland areas.
Range:	Extreme sw California, in San Diego vicinity.

There is little information on the diet of this species, but it appears to eat almost exclusively grass seeds from summer through early winter, switching to seeds of shrubs and annual weeds the rest of the year. This pocket mouse carries the seeds to its burrow and deposits them in separate passages reserved for their storage.

Long-tailed Pocket Mouse
Chaetodipus formosus

Description:	*Soft-furred.* Gray or brown above; belly white, tipped with yellow. *Long tail,* grayish above, whitish below; has distinct *crest along terminal third and tufted tip. Large hindfoot.* Ears not clothed with white hairs. L 6¾–8¼″ (172–211 mm); T 3⅜–4⅝″ (86–118 mm); HF ⅞–1″ (21–26 mm); E more than ⅜″ (9 mm); Wt ½–⅞ oz (14–24 g).
Similar Species:	Desert Pocket Mouse is yellowish brown to yellowish gray. Bailey's Pocket Mouse is larger, with longer hindfoot. Great Basin Pocket Mouse has only slight crest on tail, lacks tuft, and has olive line on sides. White-eared Pocket Mouse has ears clothed with white hair and pale tail. Yellow-eared Pocket Mouse has yellowish hairs inside ears and pale yellowish tail. All other soft-furred pocket mice have smaller feet.
Breeding:	Reproductive season early April–early July. Litters average 5 young. Some females have more than 2 litters in a season.
Sign:	Small piles of sand at base of rocks.

Habitat: Usually rocky or gravelly ground, sometimes along riverbeds and sandy wastes in hard-packed sand; open mesquite.

Range: Nevada and w Utah south to s California and nw Arizona.

In extreme heat or cold, the Long-tailed Pocket Mouse becomes torpid and stays in its burrow. Its life span is four to five years. In Nevada, the Kit Fox is an important predator.

Hispid Pocket Mouse
Chaetodipus hispidus

Description: *A large pocket mouse.* Coarse fur; brownish mixed with yellowish above; buff below. No spines on rump. *Short tail* (less than half total length) dark brown above, whitish below; uncrested and sparsely haired, *not tufted.* Hindfoot large, with naked sole. L 7¾–8¾″ (198–223 mm); T 3½–4⅜″ (90–113 mm); HF 1–1⅛″ (25–28 mm); Wt 1–1⅝ oz (30–47 g).

Similar Species: Hispid Pocket Mouse is only *Chaetodipus* species with rump lacking spines or bristles, and with no crest on tail.

Breeding: In North, 1 or 2 litters per year of 2–9 young in spring and summer. In South, breeds throughout year.

Sign: Burrow openings the size of a quarter, usually plugged by day, surrounded by small mounds of dirt.

Habitat: Dry prairie areas with sparse or moderate (not dense) vegetation, with or without shrubs.

Range: Extreme sc North Dakota south through Texas; west to se Arizona; east to w Louisiana.

The Hispid Pocket Mouse is apparently active much of the year in the southern part of its range. In colder areas, it does not store fat for the winter or hibernate, but apparently subsists on stored seeds. Its burrows, which may be more than 15 inches (380 mm) belowground, are

simple when the animals are young; when they are older, the burrows contain several openings and branches for food storage and a maternity nest. This species eats seeds, including those of cactus, evening primrose, and wine-cup, but also grasshoppers and caterpillars. In Texas, major foods in winter include seeds of mesquite, sunflower *(Helianthus anuus),* cacti *(Mammillaria* and *Opuntia),* and sagebrush *(Artemisia ludoviciana).* In spring, important foods are mesquite blanketflower *(Gaillardia puchella), Opuntia,* and bluestem *(Andropogon halli).* Food varies with the seasons, and large quantities are stored when available. The Hispid's cheek pouches are large enough to contain sufficient food to sustain the animal for an entire day. Predators include snakes, owls, and some mammals. The western diamondback rattlesnake and the great horned owl feed heavily on this species.

Rock Pocket Mouse
Chaetodipus intermedius

Description: A medium-size pocket mouse. *Rough-haired.* Yellowish gray above (nearly black when on lava flows); whitish below. *Weak rump spines.* Long tail grayish above, white below; has both crest and tuft. Sole of hindfoot naked. L 6–7⅛" (152–180 mm); T 3¼–4" (83–103 mm); HF ¾–⅞" (19–24 mm); E less than ⅜" (9 mm); Wt ⅜–⅝ oz (12–18 g).

Similar Species: Nelson's Pocket Mouse is larger, with stronger spines on rump. Bailey's, Desert, and Hispid pocket mice lack rump spines.

Breeding: Season apparently begins February–March and continues for several months. Litters contain 3–6 young.

Sign: Tiny burrows under rocks with tiny trails leading from them.

Habitat: Rocky areas and lava flows in desert areas. Often in areas of large boulders.

Range: North-central and s Arizona, sw New Mexico, and w Texas.

The Rock Pocket Mouse's burrows are small and inconspicuous, often under rocks. Tiny trails lead from the burrows to feeding areas. Like other pocket mice, this species feeds on various weed seeds and often closes its burrow by day. It is preyed upon by owls, snakes, and carnivorous mammals.

Nelson's Pocket Mouse
Chaetodipus nelsoni

Description: A medium-size pocket mouse. *Rough-haired.* Grayish above, with *numerous distinct blackish spines or bristles on rump;* white below. *Tail longer than head and body,* slightly darker above than below; *crested toward tip* and tufted. L 7⅛–7⅝″ (182–193 mm); T 4⅛–4⅝″ (104–117 mm); HF average ⅞″ (22 mm); E less than ⅜″ (9 mm); Wt ½–⅝ oz (14–17 g).

Similar Species: Hispid and Desert pocket mice lack rump spines. Rock Pocket Mouse is smaller, with weak rump spines.

Breeding: Reproductive period February–July, peaking in March. Litter of 1–5 (average 3) young born after gestation of 30 days.

Sign: Burrow openings, 1⅛–1⅝″ (30–40 mm) wide, in banks or under shrubs.

Habitat: Rocky areas.

Range: Extreme se New Mexico and south into w Texas.

Although it is little known, Nelson's Pocket Mouse is often one of the most common pocket mice in its range. It is nocturnal, and usually runs rather than hops. It does not hibernate, and burrows in rocky areas among chino grass, sotol, and bear grass. Burrows may be found at the bases of thorny desert shrubs such as catclaw. One burrow excavated in Texas entered straight into a bank for 12 inches (300 mm), then branched.

The main branch sloped downward another 12 inches to a grass nest 2¾ inches (70 mm) high and 5⅛ inches (130 mm) wide. This species feeds on seeds, but also will eat other plant parts and insects. Seeds as well as green plant are carried in cheek pouches; honey mesquite, creosote bush, prickly pear, spurge *(Euphorbia),* and buckwheat *(Eriogonum)* seeds are among those that have been found in the pouches. Predator include snakes, especially the western diamondback rattlesnake, and owls.

---

72 Desert Pocket Mouse
Chaetodipus penicillatus

Description: Rough-haired. Yellowish brown to yellowish gray above, interspersed with black hairs; belly and underside o' tail white. *No spines on rump. Long tail* (over half total length) *crested and tufted.* Sole of hindfoot naked. L 6⅜–8½" (162–216 mm); T 3¼–5⅛" (83–129 mm); HF ⅞–1" (21–27 mm); Wt ½–¾ oz (15–23 g).

Similar Species: Nelson's and Rock pocket mice have rump spines or bristles; Hispid Pocket Mouse is generally larger, with shorter uncrested and untufted tail. Long-tailed Pocket Mouse is gray or brown.

Breeding: Reproductive period April–July. Litter of 2–8 young born after gestation of 23–26 days.

Habitat: Sandy deserts, among cactus or mesquite especially along streambeds or washes.

Range: Southern California, extreme s Nevada, w and s Arizona, s New Mexico, and w Texas.

During the heat of the day, the Desert Pocket Mouse closes its burrow and retires within; like other pocket mice, i is active at night. It feeds on weed and grass seeds, including mesquite, creosot bush, and broomweed, which it carries in its cheek pouches to be stored in side passages of its burrow. This species can

burrow its way into hardened soils by chewing through the hard crust. A baseball-size nest of a Desert Pocket Mouse was found in Arizona about 12 inches (300 mm) belowground.

73 Spiny Pocket Mouse
Chaetodipus spinatus

Description: *Rough-furred,* with distinct *white spines with brown tips on rump and flanks.* Grayish or yellowish buff above; whitish below; *line on sides faint or absent.* Tail has crest ½–1" (12–25 mm) long. Ears small. L 6½–8⅞" (164–225 mm); T 3½–5" (89–128 mm); HF ¾–1⅛" (20–28 mm); E ⅛–¼" (5–7 mm); Wt ⅜–⅝ oz (12.8–19.9 g).

Similar Species: *Perognathus* species of pocket mice, and Bailey's, Long-tailed, and Desert pocket mice lack spines or bristles on rump. California and San Diego pocket mice have definite line on sides.

Breeding: Probably averages 4 young per litter.

Habitat: Usually hot desert; in Nevada, found in seepage areas and near tamarisk or mesquite trees.

Range: Southern California; also at Granite Springs, in s Nevada.

Like other pocket mice, the Spiny Pocket Mouse stuffs its cheek pouches with food, mainly seeds and other plant parts, which it then stores in special chambers in its burrow.

97 Dark Kangaroo Mouse
Microdipodops megacephalus

Description: *Blackish or dark grayish above,* with hairs gray at base; white below. *Tail thickest in middle,* tapering at both ends; has *black tip,* no tuft. Incisors not grooved. Hindfoot has hair on sole. L 5⅞–7" (148–177 mm); T 2⅝–4" (68–103 mm); HF ⅞–1" (23–25 mm); Wt ⅜–⅝ oz (10–17 g).

Similar Species: Pale Kangaroo Mouse has light pinkish-cinnamon upperparts. Pocket mice and kangaroo mice have grooved incisors.

Breeding: Young produced throughout summer; most litters May–June, with 2–7 young.

Sign: Burrows 2–4' (600–1,200 mm) long and 1' (300 mm) deep open among brush; are closed by day and hard to find. *Scat:* Hard, dark, oblong; ⅛–¼" (3–7 mm) long.

Tracks: Prints are almost round, slightly longer than wide, about ½" (13 mm) long; forefeet usually print almost side by side, followed by hindfeet, also side by side. Trails may meander about colony area.

Habitat: Shad scale and sagebrush scrub.

Range: Southeastern Oregon, extreme ne California, much of Nevada, and nw Utah.

The Dark Kangaroo Mouse uses bipedal locomotion and will stand straight up when defending its nesting area. It is most active in the first two hours after sunset, but activity is greatly reduced if any moon is present. It is active from March through October, and probably hibernates. This species apparently stores food in caches in its burrow. It feeds heavily on the black seeds of desert star *(Mentzelia)* when available, and eats many other kinds of seeds as well; in addition, large numbers of insects have been found in its cheek pouches in summer. This species can exist without free water, but lack of fall and winter rains can adversely affect its reproduction. Predators are owls, foxes, American Badgers, and weasels.

98 Pale Kangaroo Mouse
Microdipodops pallidus

Description: *Light pinkish cinnamon* above; hairs white to base on belly and underside of tail. *Tail thickest in middle;* lacks tuft, distinct markings, and black tip. *Incisors not*

grooved. Hindfoot has hair on sole. L 5⅞–6⅞" (150–173 mm); T 2⅞–3⅞" (74–99 mm); HF 1" (25–27 mm); Wt ⅜–⅝ oz (10–17 g).

Similar Species: Dark Kangaroo Mouse has blackish or dark grayish upperparts and black-tipped tail. Pocket mice and kangaroo rats have grooved incisors.

Breeding: Reproductive season apparently spring–early fall; litters probably average 2–6 young.

Sign: Burrow entrances (usually at least 2) open in brushy areas, but are plugged with earth by day. Tracks and other signs are similar to those of Dark Kangaroo Mouse.

Habitat: Fine sand around scattered desert brush.

Range: Upper Sonoran Desert of w Nevada and adjacent Mono and Inyo counties, in California.

This species is primarily quadrupedal, but also uses bipedal locomotion. It does not climb. Some winter activity occurs, but the "kangaroo" tail stores fat, which serves as an energy source during estivation and hibernation, and also helps maintain balance during jumps. The simple burrows are 4 to 6 feet (1.2–1.8 m) long and 1 foot (.3 m) deep, in windblown sand; no food is stored in them, at least in summer. Various seeds, other vegetation, and insects form this animal's diet.

Agile Kangaroo Rat
Dipodomys agilis

Description: A medium-size kangaroo rat. Very dark brown above; white belly. Tail tufted, with broad, sharply demarcated white side stripes; *dark stripe on underside extends to tip. Hindfoot has 5 toes.* Soles of feet dark. Large ears. L 10½–12½" (265–319 mm); T 6⅛–8" (155–203 mm); HF 1⅝–1¾" (40–46 mm); E ½–⅝" (13–15 mm); Wt 1½–2⅝ oz (45–77 g).

Similar Species: Merriam's and Fresno kangaroo rats are smaller, with 4 toes on hindfoot. Giant Kangaroo Rat is larger. Panamint Kangaroo Rat's tail has dark stripe on underside, tapering to point near tip. Stephen's Kangaroo Rat is very similar, has dusky rather than dark soles; differs mostly in skull characteristics.

Habitat: Sand or gravel in brushy areas.

Range: Southwest California.

The Agile Kangaroo Rat is very similar in habits and characteristics to Heermann's Kangaroo Rat. The two may eventually be combined as the same species.

California Kangaroo Rat
Dipodomys californicus

Description: *A medium-size kangaroo rat with dark upperparts;* white belly. Tail distinctly white-tipped. *Hindfoot has 4 toes.* L 10¼–13⅜" (260–340 mm); T 6–8½" (152–217 mm); HF 1⅝–1⅞" (40–47 mm); E ⅜–⅝" (11–16 mm).

Similar Species: Smaller Merriam's Kangaroo Rat and larger Desert Kangaroo Rat are both lighter in color. Heermann's has 5 toes on hindfoot.

Breeding: Reproductive season February–September; most activity February–April. More than 1 litter of 2–4 young can be produced per year.

Sign: Burrows under shrubs or rocks. *Scat:* Fecal pellets, ⅛ × ¼" (3 × 6 mm) long, often evident along trails and in dusting areas. Pellets are tapered at the ends, slightly curved, and black.

Habitat: Desert with scattered shrubs or chaparral, with some open areas.

Range: South-central Oregon and n California mainly north of line extending approximately between Suisun Bay and Lake Tahoe and east of humid coastal regions to Cascade and Sierra Nevada foothills.

The California Kangaroo Rat becomes active soon after dark. Its locomotion is generally bipedal. Unlike some kangaroo rats, this species does not make long leaps when excited, but scampers away. It does not hibernate and may be active even at temperatures below freezing. This species usually lives in an environment with well-drained soil, where it can dig burrows, but sometimes it is found in areas of harder soil, where it burrows under rocks. It may also use the burrows of ground squirrels. This kangaroo rat feeds generally on seeds and berries, such as manzanita *(Arctostaphylus patula),* which is heavily used when present, but also on buckbrush *(Ceanothus cuneatus)* and many other plant species. It consumes green vegetation in spring and is often found with vegetation in its cheek pouches in summer, indicating that the food is being stored. It dust-bathes, which perhaps helps to remove parasites as well as fats and oils from its fur. Predators include foxes, Coyotes, owls, and snakes, but local climatic conditions may be more of a problem to this species than predators.

Gulf Coast Kangaroo Rat
Dipodomys compactus

Description: A medium-size kangaroo rat. *Paler than other kangaroo rats of Texas, with an orangish hue,* but has gray and red color phases. Tail, with short crest, has dusky brown stripe above; paler stripe below somewhat broken and not extending to tip. White belly and cheeks. *Hindfoot has 5 toes.* Average measurements for male: L 8½–9″ (216–230 mm); T 4⅜–5″ (111–126 mm); HF 1⅜–1¼″ (36–38 mm); Wt 1½–1⅝ oz (44–60 g).

Similar Species: Ord's, only similar kangaroo rat in range, has longer, more bushy tail with

longer crest and darker, less-broken stripe below.

Breeding: 2 records of 2 young each in July and late August.

Sign: Burrows under shrubs or rocks.
Scat: Fecal pellets, measuring ⅛ × ¼" (3 × 6 mm), often evident along trails and in sand-bathing areas. Pellets are tapered at ends, slightly curved, and black.

Habitat: Sparsely vegetated areas with sandy soils; often in dune areas, on side of dunes away from prevailing winds.

Range: Southern Texas, including Mustang and Padre islands, just offshore from Corpus Christi.

Where this species occurs with Ord's Kangaroo Rat, it lives in the more open areas, Ord's in the more brushy areas. Little is known of its biology, and some biologists do not agree that it is a separate species from Ord's.

102 Desert Kangaroo Rat
Dipodomys deserti

Description: One of the largest kangaroo rats. Yellowish buff above; white below. *Tail* crested, *with white tip preceded by dusky rather than black band.* Short dark cinnamon stripe on top of tail; *stripe on bottom of tail is absent or very pale,* only slightly darker than light sides of tail. *No dark facial markings. Hindfoot has 4 toes.* L 12–14¾" (305–377 mm); T 7⅛–8½" (180–215 mm); HF 2–2¼" (50–58 mm); E ½–⅝" (12–15 mm); Wt 2¾–4⅞ oz (80–138 g).

Similar Species: Banner-tailed Kangaroo Rat has black band of hairs preceding white tail tip and smaller hindfeet.

Breeding: 1 or 2 litters per year of 1–6 young, born from January to July; height of reproductive period in February. Gestation 29–32 days. Newborn weighs about ¹⁄₁₆ oz (3 g) and is about 2" (52 mm) long.

Sign: Well-worn trails often leading away
         from burrows.
Habitat: Areas of soft sand, such as dunes;
         creosote bush or shad scale scrub.
Range: Western and s Nevada south through se
         California and w Arizona.

The Desert Kangaroo Rat occupies the
most arid region of the North American
desert. When excited, it kicks sand and
drums the ground with its large
hindfeet; it also kicks sand at objects to
determine if they are alive. Like all
kangaroo rats, this animal is usually
abroad at night, when humidity is
highest, but it keeps to its burrow when
the moon is bright and it could easily
be spotted by predators. It is, however,
occasionally out during the day. This is
a solitary species: Only one individual
occupies a burrow, and drives other
individuals away. It moves soil by
pulling it between the hindfeet; the
hindlimbs may be used to push sand
backward or when the animal is
underground. The forelimbs or chest
may also be used to push sand. The
Desert Kangaroo Rat's burrow system
is up to 4½ feet (1.5 m) deep, with
entrances leading into a labyrinth of
passages that connect with several food
storage rooms and a nest chamber. The
nest is of grass and other plant material.
Mesquite pods and various kinds of
seeds are cached in the storage areas.
Food consists of many kinds of seeds,
but includes little green vegetation.
Kangaroo rats put seeds into their
mouth pouches very rapidly; their
forelimbs move together and can pocket
20 to 60 millet seeds per second. Where
the Desert Kangaroo Rat and Merriam's
both occur, the Desert stores larger
seeds. This kangaroo rat commonly
takes sand-baths that include side and
belly rubs. It first rapidly digs with its
forepaws, then anchors its forepaws in
the dirt to hold itself steady. The cheek
is lowered and the animal rapidly

extends and flexes the body with one side or the belly against the ground. The female Desert Kangaroo Rat sits on her hindlimbs while giving birth. The newborns are suspended by their umbilical cords and dragged across the ground until the cords break. The mother kicks sand on the newborns, perhaps to dry them. The adult color pattern appears at 21 days. This species drums more than other kangaroo rats. Predators include foxes, Coyotes, Bobcats, hawks, snakes, owls, and spotted skunks. Many individuals are killed on the highways.

Texas Kangaroo Rat
Dipodomys elator

Description: *A relatively large kangaroo rat.* Buff above; white below. Long, thick, *white-tipped tail. Hindfoot has 4 toes.* L 10¼–13⅝" (260–345 mm); T 6⅜–8⅛" (161–205 mm); HF 1⅝–1⅞" (42–49 mm); Wt 2¼–3⅛ oz (65–90 g).

Similar Species: Ord's Kangaroo Rat has 5 toes and no white tip on tail.

Breeding: Little information available, but 2–4 young are born; pregnant females and young have been found in the field May–November.

Sign: Burrow openings associated with small mounds of dirt among mesquite (*Prosopis glandulosa*).

Habitat: Scrub areas, especially in mesquite.

Range: Chattanooga, in Comanche County, sw Oklahoma, and 9 counties in nc Texas.

This species is active throughout the year. Burrows are dug in relatively firm clay-loam soils among mesquite and prickly pear plants to a depth of 3 feet (1 m). Burrow systems are about 2½ feet (.75 m) long, with several intertwined tunnels and a nest of shredded grass near the bottom. Tunnels with food caches occur throughout the system. The animal stuffs its cheek pouches with the seeds of grasses and other plants

(including storksbill, *Erodium circutarium*), and a few insect parts, then stores them in the burrow system. Dust-bathing areas are often found near the burrow openings. Texas Kangaroo Rats drum their hindfeet.

Big-eared Kangaroo Rat
Dipodomys elephantinus

Description: A *large kangaroo rat with large brownish ears.* Cinnamon-buff above; white below. *Tail long, heavily crested and tufted;* dark stripe under tail is narrower than white stripes on side of tail. *Hindfoot has 5 toes.* L 12–13¼″ (305–336 mm); T 7¼–8¼″ (183–210 mm); HF 1¾–2″ (44–50 mm); E ⅝–¾″ (16–18 mm); Wt 2¾–3⅛ oz (70–90 g).

Similar Species: Heermann's Kangaroo Rat is smaller and lighter-colored, with smaller ears. Desert Kangaroo Rat has smaller ears. Narrow-faced Kangaroo Rat is smaller and darker, with dark, pronounced facial markings.

Breeding: Young are apparently born in spring and summer.

Sign: Burrows under shrubs.

Habitat: Chaparral-covered slopes under rather heavy vegetation.

Range: Found only in San Benito and Monterey counties, California.

Very little is known about this species. It occurs together with Heermann's Kangaroo Rat, which is found in habitat that is more open. Some mammalogists believe it should be grouped as one species with the Narrow-faced Kangaroo Rat.

103 Heermann's Kangaroo Rat
Dipodomys heermanni

Description: A *medium-size kangaroo rat.* Brownish above; white below. Fur long and silky. *Tail uniformly dusky above and below,* with

typical white side stripes; crest varies
from slight and dusky to black with
well-developed white tip. *Hindfoot has
5 toes.* Ears dusky or nearly black.
Juvenile gray. L 9⅞–13⅜" (250–340
mm); T 6⅜–8½" (160–217 mm);
HF 1½–1¾" (38–46 mm); E ⅜–⅝"
(10–17 mm); Wt 1¾–3¼ oz (50–94 g)

Similar Species: Giant Kangaroo Rat is usually larger,
with longer hindfoot. Fresno and
California kangaroo rats have 4 toes on
hindfoot. Agile Kangaroo Rat is
generally darker. Big-eared Kangaroo
Rat has larger ears.

Breeding: Reproduction season February–October
peaking in April. 2–5 young born after
gestation of 30–32 days. Newborn
average weight ⅛ oz (3.7 g). Weaning
starts at about 17 days.

Sign: Burrows often under bushes; entrances
1⅝–3⅛" (40–80 mm) in diameter.

Tracks: When resting on hindlegs, base
of tail forms drag marks between
complete hindprints, 1½–2" (38–50
mm) long; when moving slowly, smaller
round foreprints made between or just
in front of hindprints; when making
long jumps, only fronts of hindfeet
touch ground, leaving small round
prints 1" (25 mm) across.

Habitat: Broad range of habitat types: open,
sloping terrain; grassland and woodland
in foothills; live oak and pines in
low valleys.

Range: Central and s California, south from line
between Suisun Bay and Lake Tahoe, to
slightly north of Point Conception, and
west from foothills of Sierra Nevada and
Tehachapi Mountains to Pacific Ocean.

Nocturnal, as is typical of kangaroo
rats, Heermann's is a fast-moving
species; it has been clocked running at
12 mph (20 km/h). Its burrow systems
may be 6 to 10 feet (2–3 m) long, with
two or three passages (there is usually a
nest in one) and with several blind
"escape" tunnels ending an inch (25
mm) below the surface. This species

often uses burrows of ground squirrels. The dirt excavated from burrows is sometimes used for dust-bathing, the sites for which are usually near burrow entrances. This kangaroo rat eats the seeds of many plants, as well as the green parts of prairie trefoil *(Lotus)*, *Dudleya*, *Lupine*, and bromegrass *(Bromus)*. It consumes green vegetation heavily in spring, eating soft or succulent foods while foraging and placing seeds in the cheek pouches for storage. Most of the food is gleaned from the ground, except when grasses are ripening; then the animals reach up on their hindlegs and cut off the heads. A California subspecies, the Morro Bay Kangaroo Rat *(D. h. morroensis)*, is listed as endangered by the U.S. government.

104 Giant Kangaroo Rat
Dipodomys ingens

Description: *The largest kangaroo rat.* Dusky above; white below. Dusky-tipped tail; dark brown stripe above and below; white side stripes. *Hindfoot has 5 toes.* L 12¼–13¾″ (311–348 mm); T 6¼–7¾″ (157–198 mm); HF 1¾–2⅛″ (46–55 mm); E ⅝–¾″ (15–19 mm); Wt 4⅝–6⅜ oz (131–180 g).

Similar Species: Other large kangaroo rats (Desert, Banner-tailed) have 4 toes on hindfoot. Heermann's and Fresno kangaroo rats have shorter hindfeet.

Breeding: Reproductive season January–May; 3–6 young per litter.

Sign: Two general types of burrows: one a vertical shaft with no apron of dirt, trails leading away, or dirt near the entrance; the other wider than high, slanting, and with a well-worn trail leading from the mouth.

Habitat: Open desert grasslands in valleys. This is often a dominant species in areas lacking brush.

Range: Southern California, in w San Joaquin Valley.

The burrow system of the Giant Kangaroo Rat has one to four entrances, a nest chamber, and special chambers in which large quantities of food are stored. This species does not store seeds in small pits in the walls of the tunnels, as most other kangaroo rats do. Near the entrances it makes tiny holes, which it fills with partially ripened seeds; when these are dry, the animal covers them with loose soil. At one site, 875 holes were dug by one individual. Some populations of this species make large stacks of seeds aboveground; there they are dried and later stored underground in large caches. A total of 24 caches was found in one burrow system, 12 old and 12 new. The new caches consisted mainly of storksbill *(Erodium)* and peppergrass *(Lepidium)* seeds. A typical burrow has two slanting shafts and one vertical shaft. Slanting shafts are about 3⅛ inches (78–81 mm) high and 3½ inches (88–89 mm) wide. Vertical shafts have openings about 2¼ inches (57–58 mm) in diameter. The total length of three systems ranged from 24 to 30 feet (7.3–9.1 m). This species is primarily a granivore, but feeds on green plant parts as well. It practices scent-marking and sand-bathing, which includes side and belly rubs. Both bipedal and quadrupedal locomotion are commonly used. This species has been heard drumming from within its burrow, and has been induced to drum by a person drumming on top of the burrow. The Giant Kangaroo Rat appears to exclude other species of kangaroo rats, and is often the only mammal in its habitat. Areas in which this species lives generally receive less than 15 inches (380 mm) of rain per year, mostly between April and November, and daytime temperatures exceed 100°F (37°C). Perennial grasses were probably the dominant plants in the original range of this species. The

Giant Kangaroo Rat is now often found in areas heavily grazed by sheep and cattle. Predators include barn owls and great-horned owls, Kit Foxes, Coyotes, and American Badgers. The Giant Kangaroo Rat was listed as endangered by the state of California in 1980 and by the federal government in 1987.

105 Merriam's Kangaroo Rat
Dipodomys merriami

Description: *One of the smallest kangaroo rats.* Light yellowish buff above; white below. Long tail with white side stripes that are wider than dark top and bottom stripes, and with *dusky tufted tip.* Dark line on either side of nose but not connected across it; facial markings paler than in most species. *Hindfoot has 4 toes.* L 8⅝–10¼" (220–260 mm); T 4⅞–6¼" (123–157 mm); HF 1⅜–1⅝" (36–41 mm); Wt 1⅜–1⅝ oz (38–47 g).

Similar Species: Among 4-toed species, Desert and Banner-tailed kangaroo rats are much larger, with longer hindfeet.

Breeding: Reproduction season January–August. Probably 2 or more litters per year of 1–5 young after gestation of 29 days.

Sign: Burrow openings with small mounds of soil near shrub bases; small, shallow, scooped-out sand-bathing spots.

Habitat: Sagebrush, shad scale, and creosote bush desert scrubs, in a great variety of soil types.

Range: Western Nevada, s California, sw Utah, nw and s Arizona, s New Mexico, and w Texas.

Merriam's Kangaroo Rat usually has one adult per burrow system; both male and female display territorial behavior. This species feeds mostly on seeds, especially of mesquite, creosote bush, purslane, grama grass, and ocotillo, but also consumes some green vegetation, especially from February through May

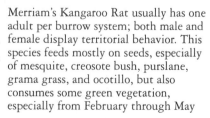

and in August. When Merriam's occurs
with another kangaroo rat, Merriam's
is often found in areas of hard soil and
the other on sand. This is especially
true when it is found with Ord's
Kangaroo Rat.

Chisel-toothed Kangaroo Rat
Dipodomys microps

Description: Buff to dusky above; whitish below.
Long tail with white side stripes narrower
than dark upper and lower stripes.
Hindfoot has 5 toes. *Lower incisors flat*
in front. L 9⅝–11⅝″ (244–297 mm);
T 5¼–6⅞″ (134–175 mm); HF 1½–
1¾″ (39–46 mm); E ⅜–½″ (9–12 mm
Wt 1⅞–2⅝ oz (55–75 g).

Similar Species: The other 5-toed kangaroo rats
occurring with this species—Ord's and
Panamint—and all other kangaroo rats
have lower incisors rounded in front.

Breeding: Mates February–March. 1–4 young
born after gestation of 30–34 days;
usually 1 litter per year. Newborn
weighs ⅛ oz (4 g).

Sign: Burrow entrances are often along banks
or other raised areas, scattered or
clustered in mounds up to 12″ (300
mm) high and 6–13′ (2–4 m) in
diameter.

Habitat: Sagebrush and shad scale scrub; piñon-
juniper woodlands.

Range: Southeastern Oregon, sw Idaho near
Murphy, e California, most of Nevada,
w Utah, and extreme nw Arizona.

The Chisel-toothed Kangaroo Rat is
active aboveground throughout the yea
Its daytime activities include resting
and the reingestion of fecal pellets. Its
main nocturnal activities outside the
burrow are foraging, caching food,
sand-bathing, and social interaction.
This species feeds primarily on green
vegetation and secondarily on seeds,
which predominate when leaves are no

available. Both leaves and seeds are stored in food caches buried in sand. The animal's chisel-shaped incisors are adapted for removing outer tissues of the leaves of shad scale *(Atriplex confertifolia)*. It also occasionally eats arthropods, such as insects and spiders, and subterranean fungi. Sand-bathing cleans and dries the animal, and also may act as olfactory communication by anointing the ground with scent from the face or anal area as the animal performs side or belly rubs. A male drums on sand with his hindfeet, evidently to attract a female from her burrow and arouse her. Males fight over females, rolling in the sand, growling, and sometimes leaping high above the ground. Life span is usually between four and five years.

106 Fresno Kangaroo Rat
Dipodomys nitratoides

Description: *One of the smallest species in the genus.* Uniform rusty brown or clay above, darkest on head; white below. *Upper and lower dark tail stripes are broader than white stripes on side of tail,* but meet along terminal one-third, thus interrupting white stripes. *Lacks white tuft on tip of tail. Hindfoot has 4 toes.* Lower incisors rounded in front. L 8¼–10″ (211–253 mm); T 4¾–6″ (120–152 mm); HF 1¼–1½″ (33–37 mm); E ½″ (12 mm); Wt average 1½ oz (44 g).

Similar Species: Giant, Panamint, Heermann's, and Chisel-toothed kangaroo rats all have 5 rather than 4 toes on hindfoot.

Breeding: Reproductive season is December–August or throughout year; gestation 32 days. Average litter has 2 or 3 young. Newborn weighs ⅛ oz (4 g); eyes open at 10 or 11 days.

Sign: Burrow openings usually at base of a low bush, with runways leading to clumps of vegetation. Burrow entrances 2⅜–3⅛″

(60–80 mm) in diameter. Relatively little dirt is actually excavated, thus no mound is formed. Burrow system may occupy a patch of grass 6–9' (2–3 m) across.

Habitat: Arid, often strongly alkaline, flat plains with sparse vegetation of grasses or sometimes orache *(Atriplex)*. Ponds occur during wet season; lack of drainage causes the alkalinity.

Range: Occurs only in San Joaquin and adjacent valleys in California.

A burrow system of this kangaroo rat may consist of one vertical entrance and several slanting ones; usually only two of the latter are in use at any one time. No dirt is excavated from the vertical shaft; it is dug from the inside and serves as an escape route. The Fresno Kangaroo Rat digs by standing on its hindlegs, which are placed far apart, and rapidly moving the forefeet, which shove the dirt back under the body. The dirt is then sent flying farther backward and away from the burrow by kicks of the hindfeet. As the burrow becomes deeper, the animal turns around and pushes dirt ahead and out of the burrow with the forefeet. Tunnels are 2 inches (50 mm) in diameter and about 8 to 10 inches (200–250 mm) below the surface. This species is primarily a granivore, feeding on seeds of *Erodium,* shepherd's purse *(Capsella bursa-pastoris)* and *Atriplex,* and, in spring, grasses *(Bromus, Festuca, Avena)* and forbs *(Erodium).* Seeds are cached in small pits in the walls of the burrow system; the presence of water may prohibit food-caching in the burrow during part of the year (seeds may sprout or mold). These animals perform tooth-chattering and foot-drumming; their vocalizations range from low grunts to growls and squeals. Their sand-bathing includes side and belly rubs. This species has been heavily disturbed by farming;

two subspecies (*D. n. nitratoides* and *D. n. exilis*) of California currently are listed as endangered by the federal government.

107 Ord's Kangaroo Rat
Dipodomys ordii

Description: Buffy, reddish, or blackish above; white below. *Tail crested but generally not white-tipped;* relatively short for the genus. *Dark tail stripe below extending to tip of tail.* Usually conspicuous white spots at base of ears and above eyes. *Hindfoot has 5 toes.* Lower incisors rounded in front. L 8¼–11⅛" (208–282 mm); T 3⅞–6⅜" (100–163 mm); HF average 1¾" (44–45 mm); Wt 1¾–3⅜ oz (50–96 g).

Similar Species: The only other 5-toed kangaroo rats occurring with Ord's are the Panamint Kangaroo Rat, which is larger and has a longer tail, and the Chisel-toothed Kangaroo Rat, which has flat-edged lower incisors. Merriam's Kangaroo Rat is similar, but has 4 toes on hindfoot.

Breeding: May produce young once or twice a year, or through entire year, depending on circumstances. Reproductive season arrives with appearance of rainfall and new vegetative growth. 1–6 young born after gestation of 28–30 days.

Sign: 3" (75 mm) burrow openings, often in banks or sand dunes, with small mounds outside; small, shallow, scooped-out dusting spots, as well as burrows, reveal an inhabited area; a tap at a burrow entrance may get a response of drumming, the occupant's foot thumping as an alarm signal. *Scat:* Brown or dark green, hard, oblong, very small (⅛–½"/3–13 mm long). *Tracks:* When moving slowly, all 4 feet touch ground and heel of hindfoot leaves complete print about 1½" (38 mm) long, somewhat triangular, much wider at front than rear; foreprints much smaller, round, and between hindprints;

when resting on ground, tail leaves a long, conspicuous drag mark. When hopping, heel of hindfoot is off ground so hindprints are shorter, little or no tail mark shows, and forefeet may or may not print. Straddle more than 2″ (50 mm) wide. Trails radiate and crisscross.

Habitat: Varies: including sandy waste areas, sand dunes, sometimes hard-packed soil.

Range: Southeastern Alberta, sw Saskatchewan s Idaho, sc Washington, e Oregon south to extreme ne California, Arizona, New Mexico, w Texas, and w Oklahoma.

Ord's Kangaroo Rat is active all winter in Texas, but farther north it is seldom seen aboveground in very cold weather. It spends its days in deep burrows in the sand, which it plugs to maintain stable temperature and humidity. Extra holes are dug throughout the home range as escape hatches. It eats the seeds of plants such as mesquite, tumbleweed, Russian thistle, sunflower, and sandbur as well as some green vegetation and subterranean fungi. Like other kangaroo rats, it eats perishable items while foraging, and puts seeds in its cheek pouches for later caching. Sometimes this species takes seeds from newly planted fields. Two individuals may skirmish, jumping into the air and striking out at each other with their feet. This kangaroo rat kicks sand into the face of an enemy, such as a rattlesnake; covers 6 to 8 feet (2–2.5 m) at a leap when speed is called for; drums by pounding the hindfeet; and takes sandbaths, which keep the fur from becoming matted. Ord's makes a sound similar to a bird's soft chirping. Rattlesnakes, owls, American Badgers skunks, foxes, weasels, and Coyotes are its predators. These animals may live for about seven and a half years. Canadian populations of Ord's Kangaroo Rat are currently under consideration for listing as threatened or endangered.

108 Panamint Kangaroo Rat
Dipodomys panamintinus

Description: Brownish gray above; cinnamon on sides; white below. *Well-crested tail has pale dusky stripes on top and bottom;* bottom stripe becomes indistinct; light side stripes meet on posterior third of tail; dark stripe on bottom tapers to point near tip. Tail about 40 percent longer than head and body length. Light cheek patches; *white spot behind ear.* Hindfoot has 5 toes. *Lower incisors rounded in front.* L 11¼–13⅛″ (285–334 mm); T 6⅛–8″ (156–202 mm); HF 1⅝–1⅞″ (42–48 mm); Wt 2 oz (57 g).

Similar Species: Chisel-toothed Kangaroo Rat has lower incisors flat in front. Stephen's has shorter hindfoot and white side tail stripes about half as wide as dark upper stripe. Ord's is smaller and has lower tail stripe that does not reach end of spine. Merriam's has 4 toes on hindfoot. Heermann's, light gray or whitish in color, has longer tail with little or no crest.

Breeding: Reproductive season February–May; 3 or 4 young born after gestation of 29 days.

Sign: Burrows in mounds accumulated around clumps of brush.

Habitat: Occurs in areas with widely scattered creosote bushes *(Larrea tridentata),* Joshua trees *(Yucca brevifolia),* and juniper trees.

Range: Extreme w Nevada south through scattered areas in s California, and in s Nevada near Searchlight.

Like other kangaroo rats, the Panamint does not hibernate, but it becomes inactive when the snow cover exceeds about 40 percent and temperatures drop below 23°F (−5°C). This species is primarily a granivore and metabolizes water from food. It probably stores seeds and other nonperishable plant parts, eating softer foods while foraging, although cheek pouches examined early in winter contained green grass shoots

and little dry material. Juniper fruits
have been found crammed into cheek
pouches of this animal. Like other
kangaroo rats, the Panamint has a
prominent oil-secreting gland on its back
between the shoulders and regularly
bathes in dust, which prevents its fur
from becoming matted with excess oil. I
drums its feet, and squeals and growls.

109 Banner-tailed Kangaroo Rat
Dipodomys spectabilis

Description: *One of the largest kangaroo rats.* Dark buff
above; white below. *Long tail has
prominent white tip preceded by black band;*
narrow white side stripes extend only
two-thirds length of tail; upper and
lower stripes grayish black to dusky.
Hindfoot has 4 toes. L 12¼–14⅜" (310–
365 mm); T 7⅛–8¼" (180–208 mm);
HF 1⅞–2" (47–51 mm); Wt 3⅜–4⅝ o
(98–132 g).

Similar Species: Banner-tailed is likely to be confused
only with Desert Kangaroo Rat, in
which bottom tail stripe is absent or
pale. Ord's and Merriam's kangaroo rats
are much smaller, with no white tip on
tail. Texas and other kangaroo rats are
smaller.

Breeding: Reproduction may occur throughout
year, with possible exception of
October–November. Gestation is 22–2
days; 1–3 litters per year. In North,
most births occur in April; in South,
most in December, June, and July.

Sign: Large, conspicuous mounds of earth and
vegetation up to 4' (1.2 m) high and 1⁵
(4.5 m) wide (although sometimes only
6"/150 mm high and 5'/1.5 m wide),
with several entrances opening to burrow
systems and trails leading from them.
Other signs resemble those of Ord's
Kangaroo Rat, but tracks are larger.

Habitat: Desert grasslands with scattered shrub
Scrub or brush-covered slopes, often
with creosote bush or acacia on hard or
gravelly soil.

Range: Southeastern Arizona, most of New Mexico, and w Texas.

This large kangaroo rat has a spectacularly white-tipped tail. The animal usually lives alone in its impressively mounded burrow system, which may have as many as a dozen openings to provide a convenient retreat from predators. Mound entrances are 4 to 6 inches (100–150 mm) in diameter, and contain numerous chambers and tunnels. Only one animal occupies a mound except when young are present, but a male may defend the mound of a female against another male. There are about 10 storage areas per mound. Most tunnels are within 20 inches (500 mm) of the surface, but usually one, particularly the nest burrow, goes deeper. Tunnels are about 3 inches (80 mm) high and 4¼ inches (110 mm) wide; storage areas are 6 to 10 inches (150–250 mm) wide. Old mounds are used for many years; one new mound appeared in three years among 287 existing mounds at one site, and six appeared among 105 in two years at another. About 40 to 95 percent of the mounds may be in use at any one time. The surface temperature above may reach 149°F (64°C), but deep in the mound the temperature remains at about 81°F (27°C). The Banner-tailed Kangaroo Rat eats green and succulent plants, and stores many types of seeds in side passages of its burrow. It will also eat some insect material and even rodents. Moonlight greatly decreases its activity. It takes sand-baths. As with other kangaroo rats, it does not hibernate or estivate, but will remain in its burrow during inclement weather. Banner-tails make a *peeee* sound lasting about a second, and also growl, squeak, squeal, and chuckle. They foot-drum in or near their mounds at night in response to neighbors or in case of a challenge to the mound. Foot-drumming

and sand-kicking, along with alert posture and avoidance, help them avoid snake predators. Other enemies include American Badgers, foxes, Bobcats, and great horned and barn owls.

110 Stephen's Kangaroo Rat
Dipodomys stephensi

Description: A *medium-size kangaroo rat.* White below. Many hairs in top and bottom *tail stripes* have white bases, giving stripes a *grizzled appearance.* Crested tail about 1½ times body length; white tail stripe about half as wide as dark dorsal stripe. Hindfoot has 5 toes; soles of feet dusky. L 10⅞–11¾" (277–300 mm); T 6½–7¼" (164–180 mm); HF 1½–1⅝" (39–43 mm); E ½–⅝" (13–16 mm); Wt 1⅛–1¾ oz (34–50 g).

Similar Species: Most similar to Agile and Panamint kangaroo rats. Agile has dark rather than dusky soles on hindfeet; many, rather than few, white hairs in tail tuft; tail has broad, sharply demarcated white stripes. Panamint, which occurs north of San Jacinto Valley, has larger hindfoot.

Breeding: Pregnant and lactating females have been found June–July. Litter averages 2 or 3 young.

Habitat: Sparsely vegetated habitats of sagebrush or annual grasses.

Range: Only in San Jacinto Valley and adjacent areas of w Riverside County; also sw San Bernardino and nw San Diego counties, California.

This species occurs with the Agile Kangaroo Rat, but Stephen's is more abundant in areas of sparse vegetation. The Deer Mouse is sometimes its only associate. In captivity, this species has built elaborate nests up to a week prior to giving birth. Barn and long-eared owls are its major predators. Agricultural development has limited Stephen's Kangaroo Rat to isolated populations. The species is listed as

endangered by the federal government, giving it complete protection.

Narrow-faced Kangaroo Rat
Dipodomys venustus

Description: *A medium-size kangaroo rat. Very dark-colored upperparts;* white below. Black nose. Hindfoot has 5 toes. *Ears are large* and nearly black, with pale areas at base and near fold. Individuals in south of range are lighter-colored. L 11½–13⅛" (293–332 mm); T 6⅞–8" (175–203 mm); HF 1¾–1⅞" (44–47 mm); E average ¾" (18.4 mm); Wt average 2⅞–3⅛ oz (82–88 g).

Similar Species: Occurs with or near ranges of Agile, Big-eared, and Heermann's kangaroo rats, all of which are much lighter-colored. Agile has much smaller ears and usually has shorter tail. Big-eared is larger, with brownish rather than blackish ears; its head is lighter-colored, with facial markings less pronounced. Heermann's has shorter ears.

Breeding: 1 or 2 litters yearly in spring or summer, each of 2–4 young.

Sign: Burrow openings, usually under plants. Surface food caches about 4" (100 mm) deep and 2" (50 mm) in diameter are distributed in a fan-like pattern around the burrow entrance.

Habitat: Chaparral or mixture of chaparral and oak or Digger pine from sea level to 5,800' (1,800 m) on sandy, well-drained, deep soils.

Range: Monterey and Santa Cruz areas of wc California.

In contrast to most other kangaroo rats, the Narrow-faced lives in a cool maritime habitat with annual rainfall over 30 inches (750 mm). Burrows are very simple and are located in open agricultural land. They consist of one or two main entrances, some auxiliary side passages that often end near the surface, and nest and food-storage cavities. One

typical burrow was 8 feet (2.4 m) long with a 4½-foot (1.4 m) side passage, five food caches, and a nest. Several supplementary burrows somewhat separate from the main system provide protection; these burrows are simple tunnels, often unbranched and without caches or nests. The main and auxiliary burrow entrances are usually under plants and usually plugged, while the supplementary burrow entrances are always open. Burrows are about 2 to 20 inches (50–500 mm) below the surface. When danger threatens, the animal breaks out of a side passage near the surface and goes to a supplementary burrow. The nest cavity is usually lower than the rest of the burrow and contains seed coats from stored foods, as well as live insects, such as crickets and dung beetles. This species does not burrow in orchards and cultivated areas, but will arrive and produce many burrows shortly after cultivation ceases; it often invades contour ditches or other earth structures that are not regularly cultivated. Such structures are sometimes undermined to the point of collapse. Seeds of annuals, especially golden aster *(Heterotheca grandiflora)* and bromegrass *(Bromus rigidus),* make up the contents of cheek pouches and food caches, which are on the surface as well as in tunnels. Large amounts of golden aster seed are stored

Mexican Spiny Pocket Mouse
Liomys irroratus

Description: *Upperparts grayish brown;* pale pink to buff lateral stripe usually present; whitish below. Coat rough because of flattened, grooved, and sharp-pointed *spiny hairs on back and rump.* Tail brownish above, whitish below; about half total length. *Incisors not grooved.* Juvenile gray. L 8–11⅝″ (202–295 mm); T 4⅛–6⅜″ (105–163 mm); HF 1–1¼″ (26–31 mm); Wt 1⅛–1¾ oz (34–50 g)

Similar Species: True pocket mice *(Perognathus* and *Chaetodipus)* have strongly grooved incisors.

Breeding: 2–8 young per litter. Births occur in most months, but peak of reproductive activity is August–November, when more than a third of females are pregnant.

Sign: Burrows, sometimes with small mounds closing entrances.

Habitat: In southern Texas, dense brush on ridges forming old banks of Rio Grande; often in prickly pear thickets. Also north of river, in Laguna Atascosa wildlife refuge, for example, and in many grassy dry scrub areas in lower Rio Grande Valley.

Range: Mainly Mexico, but extends north to extreme s Texas, in Cameron and Willacy counties near Brownsville.

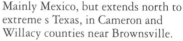

The Mexican Spiny Pocket Mouse feeds on seeds of shrubs, such as mesquite and hackberry, as well as other plant seeds. Unlike other species in the family, it and its more southern close relatives occasionally may drink water.

BEAVERS
Family Castoridae

There are two species in the family
Castoridae, the American Beaver and
the Eurasian Beaver *(Castor fiber),*
although many mammalogists consider
them to be the same species. (The
unrelated Mountain Beaver belongs to
another family.) Beavers are very large
rodents with large, flat tails. They often
build elaborate dams and lodges. Beavers
are hosts to a unique parasite that lives
on the exterior of their bodies—a highly
specialized beetle that parasitizes only
beavers. In addition, a huge number of
tiny mites, all members of the genus
Schizocarpus, occur on beavers, with
sometimes as many as nine or ten species
on one animal. Only one species,
S. mingaudi, has been found on both the
American Beaver and the Eurasian
Beaver, which suggests that *S. mingaudi*
was the ancestral species. Sixteen
additional species have been found on
the American Beaver and more than 25
others on the Eurasian Beaver. This
great difference in mite faunas could
suggest that the two beavers have been
separated for a long time.

220–223 **American Beaver**
Castor canadensis

Description: A very large, bulky rodent, with
rounded head and small, rounded ears.
Dark brown fur is fine and soft. *Scaly tail
large, black, horizontally flattened, and
paddle-shaped.* Large, black, webbed
hindfoot has 5 toes, with inner 2 nails
cleft. Eyes and ears small. Large, dark
orange incisors. L 3–4' (900–1,200
mm); T 11¾–17½" (300–440 mm);
HF 6⅛–8⅛" (156–205 mm);
Wt usually 44–60 lb (20–27 kg), but
sometimes up to 86 lb (39 kg).
Similar Species: Muskrats and Nutria are much smaller
and have slender tails.

Breeding: Mates late January–late February; 1–8 kits (usually 4 or 5) born after gestation of 4 months.

Sign: Alarm signal: Slaps tail on water loudly enough to be heard at a considerable distance. Dams of woven sticks, reeds, branches, and saplings, caulked with mud. Dome-like lodges in water, 6' (2 m) high or higher, up to 40' (12 m) wide. Scent mounds: heaps of mud, sticks, and sedges or grass, up to 1' (.3 m) high and 3' (1 m) wide, where beaver deposits scent from anal glands, apparently to mark family territory. Logs and twigs peeled where bark is eaten. Felled trees and gnawed tree trunks; gnawings at considerable heights made when beavers stand on surface of deep winter snow; successive gnawings, made when snow is at different levels, may produce totem-pole effect.
Scat: Seldom deposited on land; distinctive oval pellets, 1" (25 mm) long or longer and almost as thick, of coarse, sawdust-like material that decomposes quickly; may contain undigested pieces of bark.
Tracks: Distinctive when not obliterated by wide drag mark of tail. Usually only 3 or 4 of the 5 toes print, leaving wide, splay-toed track 3" (75 mm) long. Webbed hindfeet leave fan-shaped track often more than 5" (125 mm) wide at widest part, at least twice as long as forefeet; webbing usually shows in soft mud.

Habitat: Rivers, streams, marshes, lakes, and ponds.

Range: Most of Canada and U.S., except for most of Florida, much of Nevada, and s California.

Active throughout the year, the American Beaver is primarily nocturnal and most likely to be observed in the evening. Beavers living along a river generally make burrows with an underwater entrance in the riverbank;

these are known as bank beavers. Those in quiet streams, lakes, and ponds usually build dams and a lodge. The lodge has one or more underwater entrances; living quarters are in a hollow near the top. Wood chips on the floor absorb excess moisture, and a vent admits fresh air. The chief construction materials in the northern parts of the American Beaver's range—poplar, aspen, willow, birch, and maple—are also the preferred foods. To fell a tree, the beaver gnaws around it, biting out chips in a deep groove. Small trees 2 to 6 inches (50–150 mm) in diameter are usually selected, though occasionally larger ones as much as 33 inches (850 mm) thick are felled; a willow 5 inches (125 mm) thick can be cut down in three minutes. The beaver trims off branches, cuts them into convenient sizes (about 1 to 2 inches/25–50 mm thick and 6 feet/1.8 m long), and carries them in its mouth to the dam site. There it either eats the bark, turning the branches in its forefeet as humans eat an ear of corn, or stores them underwater for winter use by poking the ends into the muddy bottom of the pond or stream. Dam designs vary widely: To lessen water pressure in swift streams, dams may be bowed upstream; in times of flood, temporary spillways may be constructed. Dam repair is constant; the sound of running water stimulates the beaver to repair the dam. Well adapted to its highly aquatic life, the beaver swims, using its webbed hindfeet, at speeds up to 6 mph (10 km/h). The tail serves as a rudder, and the forefeet are held close to the chest, free to hold objects against the chest or to push aside debris. When the animal is submerged, valves close off the ears and nostrils; skin flaps seal the mouth, leaving the front incisors exposed for carrying branches; and clear membranes slide over the eyes protecting them from floating debris. A beaver can remain submerged for up

to 15 minutes before surfacing for air. When the animal is swimming, usually only the head is visible, whereas with muskrats, both head and back are partially above water. The beaver combs its fur with the two split nails on its hindfoot, and waterproofs it by applying castoreum, an oily secretion from scent glands near the anus. A thick layer of fat beneath the skin provides insulation from chilly water in winter. On land, the beaver is far less at ease than when in the water, and frequently interrupts its activity to sniff the air and look for signs of danger. Beavers are believed to pair for life. Kits are born well furred, with eyes open, and weighing about 1 pound (.5 kg). They may take to the water inside their lodge within a half hour and are skillful swimmers within a week; if tired, they may rest or be ferried upon the mother's back. On land, the mother often carries kits on her broad tail and sometimes walks erect and holds them in her forepaws. The young remain with their parents for two years, helping with housekeeping chores until they are driven away just before the birth of a new litter.

Great expanses of the U.S. and Canada were first explored by trappers and traders in search of beaver pelts, the single most valuable commodity in much of North America during the early 19th century. The fur was in constant demand for robes and coats, clothing trim, and top hats (sometimes called "beavers") that were fashionable in European capitals and urban areas of the eastern U.S. Some of America's great financial empires and real estate holdings were founded on profits from the trade in beaver fur. Unregulated trapping continued for so long—well into the 20th century in some areas—that the American Beaver disappeared from much of its original range. Now reestablished over most of the continent and protected from overexploitation, it

has become an agricultural pest in some regions, and it kills many trees, most of little value as timber. Its dams may block the upstream run of spawning salmon and flood stands of commercial timber, highways, and croplands, or change a farmer's pond or stream into a slough that will eventually become a meadow. However, the dams also help reduce erosion, and the ponds formed by the dams may create a favorable habitat for many forms of life: Insects lay eggs in them, fish feed on the insect larvae, and many kinds of waterfowl and mammals—including otters, Minks, Moose, and deer—come to feed and drink. The beaver's fine, soft fur is highly prized, and its meat is considered a delicacy by some residents of the far north. Aside from trappers, the otter is the beaver's most important enemy, though the Gray Wolf, Coyote, Common Red Fox, and Bobcat also prey upon it.

RATS AND MICE
Family Muridae

The rodents of the family Muridae form the largest, most successful, and most adaptable group of mammals in the world. No other family contains more species or more individuals, or occupies a greater geographic range. It currently includes 281 genera and 1,326 species worldwide, with 20 genera and 78 species in North America north of Mexico. These animals are found in every available habitat, and vary greatly in habits and form. Although there is no biological difference between rats and mice, rats are generally larger. The animals in this family range in size from the ¼-ounce (8 g) Northern Pygmy Mouse to the 4-pound (1.8 kg) Common Muskrat.

As currently recognized, the murid family in the U.S. and Canada includes three subfamilies: the Sigmodontinae (New World rats and mice), the Arvicolinae (voles and lemmings), and the Murinae (the three introduced Old World species of house rats and mice). Each of these is discussed in detail in its own section below.

NEW WORLD RATS AND MICE
Subfamily Sigmodontinae

Sigmodontinae, a subfamily of the family Muridae, contains 45 species of native rats and mice in nine genera. Sigmodontines generally have long tails (half the total length of the animal, or more) and large eyes and ears, and most have molariform teeth with two rows of well-developed cusps. As a group they are omnivorous and nocturnal. Most breed and produce litters several times per year, especially in warmer climates.

Three of the genera contain large individuals and are termed rats: the rice rats *(Oryzomys),* the cotton rats *(Sigmodon),* and the woodrats *(Neotoma).* Rice rats, generally found in marshy areas, look much like Old World rats, with brown fur above, buffy fur below, and medium-size eyes and ears. The cheek teeth have typical cusps in two rows. Cotton rats frequent grassy areas, and tend to replace voles in southern grasslands, but often occur in great numbers in thick vegetation around ponds or marshes. They have fairly large eyes and grizzled fur. Their ears are somewhat less prominent and their tails shorter and heavier than those of rice rats. The molariform teeth have transverse ridges. Woodrats, with prominent ears and eyes, and soft brownish or grayish fur with white bellies, resemble overgrown deer mice. They are found in a variety of habitats such as woods, swamps, rocky areas, and deserts, but not often in marshes or grasslands. The woodrat builds a large nest, usually in a bluff or brush pile, or in the desert under a cactus or other plant, and also caches food in large middens. Woodrats are known as pack or trade rats for their habit of collecting and hoarding all sorts of objects, especially shiny ones. A woodrat will often put down the twig it is holding to "trade" it for a shiny coin,

which it takes away to its cache. Woodrats have even been known to carry off mammalogists' traps (especially shiny new ones) to their stick nests.

The other genera in the subfamily Sigmodontinae contain smaller animals. The tiny harvest mice *(Reithrodontomys)* are distinguished by the deep vertical grooves in the fronts of their incisors. They live mainly in open grassy areas, feeding on grass seeds and building apple-size nests of shredded grass aboveground in vegetation. Deer mice *(Peromyscus),* often the most abundant mammals in a given habitat, have prominent ears and eyes, and usually have white bellies. They have internal cheek pouches used for carrying food, and most are good climbers. Many of the 17 species are so similar that they can be distinguished only by careful examination of their dental or skeletal characteristics. There are three species of grasshopper mice *(Onychomys),* all with short, white-tipped tails. The most predatory of the North American rodents, they eat mostly insects. The other three genera of sigmodontines, *Podomys, Ochrotomys,* and *Baiomys,* are each represented by one species in our range. The Florida Mouse is quite large, but is like the deer mice in appearance and has often been included in that genus. It has five tiny pads on the hindfeet rather than six, as the deer mice do. The Golden Mouse is also closely related to the deer mice, but is distinguished by its golden-cinnamon coloring. The tiny Northern Pygmy Mouse is the smallest rodent in North America, reaching a maximum weight of only 3/8 ounce (10 g).

Deer mice have been associated with two serious diseases in recent years: Lyme disease, a bacterial infection, and Hanta virus, a respiratory illness. Lyme disease is carried by deer ticks, tiny

ticks that when immature reside on deer mice. Hanta virus appeared in New Mexico in the early 1990s and caused a number of deaths; it is carried by the Deer Mouse and has also been found in many regions in several other *Peromyscus* species as well as in *Microtus* species (voles). One should avoid contact with deer mice and their droppings, especially in close quarters.

101   Key Rice Rat
Oryzomys argentatus

Description: Silvery gray. *Long, sparsely haired tail, showing scales;* gray above, paler below. L 9⅞–10¼″ (251–259 mm); T 4¾–5¼ (121–132 mm); Wt to 3 oz (84 g).

Similar Species: Marsh Rice Rat is more grayish brown than silvery.

Breeding: Probably several litters throughout the year of about 4–6 young.

Sign: Runways, cuttings of grass and other marsh plants, feeding platforms, and nests of shredded grass.

Habitat: Freshwater marshes.

Range: Found on Cudjoe, Big Torch, Middle Torch, Johnston, Little Pine, Raccoon, Saddlebunch, Summerland, and Water keys, Florida.

The Key Rice Rat was described in 1978 as a species distinct from the Marsh Rice Rat, based on differences in skull characteristics as well as fur coloration. However, many mammalogists do not accept this determination, and it may not stand the test of time. The Key Rice Rat is presently considered endangered because its range and habitat are so restricted.

Coues' Rice Rat
Oryzomys couesi

Description: Resembles Marsh Rice Rat, but is *distinctly larger, with longer tail. Yellowish*

buff or yellowish brown above; underparts buffy. Medium-size ears and eyes. Averages: L 10½″ (266 mm); T 5⅜″ (135 mm); HF 1¼″ (33 mm); E ⅝″ (15 mm); Wt 2⅜ oz (69 g).

Similar Species: Marsh Rice Rat, with longer, grayish-brown fur, is smaller and has a proportionally shorter tail. Cotton rats have longer, more grizzled fur and shorter tails. Norway Rat is larger, with a proportionally shorter, thicker tail. Black Rat's tail is longer and uniformly dark; young are difficult to distinguish from Coues'. Woodrats are larger and white below, with longer ears.

Sign: Extensive runways among marsh vegetation. Feeding platforms of plants bent over the water. Nests 1–3′ (.3–1 m) above water line in cattails and bulrushes, and built of same materials; sometimes made of leaves and small vines and placed in small trees.

Habitat: Mostly marshes; also drier areas among grasses or sedges.

Range: Only in Cameron, Willacy, Kennedy, and Hidalgo counties, in extreme s Texas.

Coues' and Marsh rice rats occur together in Texas, even in the same habitats. They show some genetic differences and remain distinct without hybridizing, thus fulfilling the criteria for separate species. Coues' Rice Rat feeds on marsh plants such as grasses, sedges, and cattails. It often sits on clumps of plants in water to feed, bending the plants over to form a feeding platform. Remnants of food items are dropped on such platforms.

---

100 Marsh Rice Rat
Oryzomys palustris

Description: Body and tail grayish brown above, pale or whitish buff below. *Long, sparsely furred tail,* showing scales. *Feet whitish.* Ears and eyes medium-size. L 7⅜–12″

(187–305 mm); T 3¼–6⅛″ (84–156 mm); HF 1⅛–1½″ (28–37 mm); Wt 1–2¾ oz (30–78 g).

Similar Species: Cotton rats have longer, more grizzled fur, and shorter tails. Norway Rat is larger, with proportionally shorter, thicker tail. Black Rat's tail is longer and uniformly dark. Woodrats are larger, pure white below, with longer ears. Coues' Rice Rat is larger, more brownish, with proportionally longer tail.

Breeding: Breeds throughout year; several litters of 4–6 young each; gestation 25 days. Young are weaned in 11–20 days.

Sign: Extensive runways among marsh vegetation; feeding platforms made of plants bent over the water, often surrounded by remnants of crabs; grapefruit-size nest of shredded grass and sedges, often in cattails and bulrushes above water line, but sometimes at base of shrub or under debris.

Tracks: In soft mud or dust, hindprint ⅜″ (10 mm) long, with 5 toes printing foreprint slightly smaller, with 4 toes printing. Walking stride, 2″ (50 mm), with hindprint directly behind, or slightly overlapping, foreprint. Those of other rice rats are similar.

Habitat: Mostly marshes; also drier areas among grasses or sedges.

Range: Mainly se U.S.: south from se Kansas, s Missouri, s Illinois, s Kentucky, and e North Carolina; south along East Coast from se Pennsylvania and s New Jersey west to e Texas (with populations in extreme s Texas).

The Marsh Rice Rat is a semiaquatic rodent. It swims underwater and dives with ease, foraging on the tender parts of aquatic plants. Its diet also includes crabs, fruits, insects, snails, and subterranean fungus *(Endogone)*. The nest is about the size of a large grapefruit, woven of grass and sedges. It is built in cattails or bulrushes, under

a shrub, on a high place on the ground under debris, or at the end of a shallow burrow. Barn owls, marsh hawks, and cottonmouth snakes are some of the Marsh Rice Rat's predators.

75 Fulvous Harvest Mouse
Reithrodontomys fulvescens

Description: Reddish brown interspersed with black above, shading to yellowish on sides; white below. *Tail more than half total length. Feet reddish above.* Grooved incisors. L 5¼–7⅞″ (134–200 mm); T 2⅞–4½″ (72–116 mm); HF ⅝–⅞″ (15–22 mm); Wt ½–1 oz (14–30 g).

Similar Species: Most other harvest mice have shorter tails. Western Harvest Mouse is less reddish yellow on sides.

Breeding: Breeding season peaks in March and July in Texas; several litters per year of 2–6 young each.

Sign: Burrows in arid areas; elsewhere, makes baseball-size nests of grasses and sedges, up to 4′ (120 cm) aboveground, with 1 or 2 small circular openings, plugged when the mouse is inside.

Habitat: Grassy or weedy areas, usually with shrubs; arid inland valleys. In regions where it overlaps range of Plains Harvest Mouse, this species chooses areas with less cover.

Range: Southeastern Arizona east to sw and e Texas, e Oklahoma, se Kansas, sw Missouri, w Arkansas, Louisiana, and w Mississippi.

Strictly nocturnal, the Fulvous Harvest Mouse is an excellent climber and spends much time aboveground. It eats a great variety of food, but eats little green vegetation. In one study, it fed predominately on insects in spring and summer and on seeds in fall and winter; in another, invertebrates dominated its diet throughout the year. The red-tailed hawk is a documented predator.

76 Eastern Harvest Mouse
Reithrodontomys humulis

Description: *Brownish above,* with darkish area in
middle of back; *dusky below.* Tail about
half total length; not sharply bicolored.
Grooved upper incisors. L 4¼–5⅞"
(107–150 mm); T 1¾–2⅝" (45–68 mm);
HF ⅝" (15–17 mm); Wt ⅜–½ oz
(10–15 g).

Similar Species: Fulvous Harvest Mouse has longer tail.
House Mouse has ungrooved incisors.

Breeding: Breeds spring–fall; several litters of
3–5 young each; gestation 21–22 days.
Young are weaned and independent
after 3 weeks.

Sign: Small (3"/7.5 cm), spherical nests of
finely ground grasses.

Habitat: Old fields; brushy areas; brier patches;
broom sedges; low areas.

Range: Southeastern U.S. from s Ohio, West
Virginia, Pennsylvania, Virginia, and
Delaware south through Kentucky,
Tennessee, nw Arkansas, Mississippi,
and Louisiana to e Texas.

The Eastern Harvest Mouse can become
locally abundant; there was one record
of 90 per acre (220 per ha). It feeds on
seeds and young sprouts, and probably
also on invertebrates, storing surplus
seeds in its nest or occasionally in an
extra cache nearby. The Eastern Harvest
Mouse can go for long periods without
water, probably obtaining what it needs
from food. This animal often builds its
nest in low herbaceous or woody
vegetation. Snakes, screech owls,
kestrels, shrikes, and weasels are likely
predators.

77 Western Harvest Mouse
Reithrodontomys megalotis

Description: *Brownish above, buff along sides; white
below.* Indistinct broad stripe down
spine. Tail length less than that of head
and body. Grooved incisors. L 4½–6¾"

(114–170 mm); T 2–3¾″ (50–96 mm); HF ½–¾″ (14–20 mm); Wt ⅜–¾ oz (9.1–21.9 g).

Similar Species: Very similar Plains Harvest Mouse has more distinct but narrow stripe down spine. Fulvous Harvest Mouse has longer tail. Salt-marsh Harvest Mouse is pinkish cinnamon or tawny below. House Mouse has ungrooved incisors.

Breeding: Breeds early spring–late autumn, with reduced activity in midsummer. Each of two captive females produced 14 litters in 1 year; 1 produced 58 young, the other 57, in litters of 2–6 individuals, but litters of up to 9 occur. Gestation 23–24 days; newborn weighs up to ¹⁄₁₆ oz (1–1.5 g).

Sign: Spherical nests of plant fibers, lined with finer downy material, about 5″ (125 mm) in diameter; 1 or more entrances near base.

Habitat: Early-stage dry weedy or grassy areas; sometimes forms large populations in rye or other cultivated grasses. Also deserts, salt marshes, and pine-oak forests.

Range: From extreme sw Canada through much of w U.S. southeast to sw Wisconsin, nw Indiana, ne Arkansas, and w Texas.

The Western Harvest Mouse, which is gradually expanding its range eastward, is nocturnal and is most active on dark nights. It frequently makes use of the ground runways of other rodents and is a nimble climber. Although primarily a seed-eater, in spring this mouse also eats new growth and in summer consumes many insects, especially caterpillars and grasshoppers. It stores surplus food, particularly seeds, in underground caches. Nests are usually in a protected place on the ground, such as under a bush or log or in a clump of vegetation; sometimes they are placed aboveground in a shrub. Predators include hawks, owls, snakes, and such mammals as foxes, coyotes, and weasels.

Plains Harvest Mouse
Reithrodontomys montanus

Description: *Brownish above with indistinct, narrow dark stripe down middle of back;* white below. *Tail usually less than one-half total length.* Grooved incisors. L 4¼–5⅝" (107–143 mm); T 1⅜–2½" (48–63 mm); HF ⅝" (14–17 mm); Wt ¼–⅜ oz (6–10 g).

Similar Species: Very similar Western Harvest Mouse has broader and less prominent dark stripe down middle of back. Fulvous Harvest Mouse has much longer tail. House Mouse has ungrooved incisors.

Breeding: Breeds much of the year; several litters of 1–9 young each; gestation 21 days. Young weigh less than 1/16 oz (1 g) at birth, are weaned at 2 weeks, and reach adult size in about 5 weeks.

Sign: Spherical nest with soft interior of thistle or milkweed down.

Habitat: Open grassy areas, including prairies and other types of grasslands.

Range: Central U.S. from w Wyoming and sw South Dakota south to se Arizona, New Mexico, and n Texas.

The Plains Harvest Mouse is both herbivorous and insectivorous, with grass seeds and grasshoppers making up the bulk of its diet. Nests are usually a few inches above the ground, but are also placed on the ground or in burrows. Three species of harvest mice were found in an abandoned field in Freestone County, Texas: Eastern, Plains, and Fulvous. Eastern and Plains were found in areas with less cover; Fulvous occurred throughout the area. Undoubtedly this species has many natural enemies, but only the barn owl has been documented

---

78 **Salt-marsh Harvest Mouse**
Reithrodontomys raviventris

Description: *Dark brown above; pinkish cinnamon or tawny below.* Tail similarly bicolored. Upper incisors grooved. L 4⅝–6⅞"

(118–175 mm); T 2¼–3¾″ (56–96 mm); HF ⅝–⅞″ (15–21 mm).

Similar Species: Western Harvest Mouse has a white belly.

Breeding: Reproductive season March through October or November; litters of 4 young.

Sign: Nest of grass lined with softer grass or down, in a clump of vegetation aboveground or in an old bird nest.

Habitat: Salt marshes.

Range: San Francisco Bay area of California.

The Salt-marsh Harvest Mouse is most active on moonlit nights. If cover is adequate, it will move into grasslands bordering the marshes. This mouse feeds mainly on seeds, including those of many grasses, and can drink seawater. It does not burrow. Owls, snakes, and many mammals prey upon it. The Salt-marsh Harvest Mouse was declared endangered in 1970 by the federal government. The species does well in glasswort, succulent plants that recently have been increasing in number around San Francisco.

Texas Mouse
Peromyscus attwateri

Description: Brown above with sides pinkish cinnamon; white below. *Tail half or more of total length, bicolored,* hairy with *prominent tuft at tip. Ankles dark or dusky.* Ears large. L 6¼–8⅝″ (160–218 mm); T 2¾–4⅜″ (68–111 mm); HF ⅞–1″ (24–27 mm); E ⅝″ (15–17mm); Wt ⅝–1⅜ oz (19–41 g).

Similar Species: White-footed Mouse's tail is less than half total length. White-ankled Mouse has white ankles, but is difficult to distinguish from Texas Mouse where the two occur together in Texas. Cotton Mouse is difficult to differentiate, but the two overlap only in e Oklahoma.

Breeding: Breeding may peak in spring, again in autumn; litters of 3–6 young.

Sign: Nest of grass and dry leaves in a crevice or rock pile.

Habitat: Close association with rocky areas: rock bluffs with red cedar or scrub woodland; high cliffs among oak stands.

Range: Central Texas east through e Oklahoma to se Kansas, sw Missouri, and nw Arkansas.

A good but cautious climber, the Texas Mouse uses its long, tufted tail for balance and support while scaling vertical cliff faces and climbing trees. This mouse feeds heavily on acorns, camel crickets, pine and grass seeds, and beetle larvae. In one study in Missouri, 60 percent of the volume of its food was seed fragments, 31 percent insects, 4.4 percent berries, and 1.6 percent green plant material. This animal travels over a home range of about half an acre (.2 ha). The Texas Mouse occurs with the White-ankled Mouse in west-central Texas, where the latter species occurs in rocky areas, especially on limestone ledges, while the former occupies a wider range of habitats. The Texas Mouse also occurs together with the White-footed Mouse and the Deer Mouse in Missouri, where the Deer Mouse is found in old field areas, the White-footed in hardwood forest, and the Texas Mouse in cedar glades. The Texas Mouse recently was recognized as a species separate from the Brush Mouse. Broad-winged hawks and coachwhip snakes are among the many predators of the Texas Mouse.

80 Brush Mouse
Peromyscus boylii

Description: Brownish or grayish above to buff or tawny on sides; white below. *Tail distinctly bicolored, hairy, equal to or longer than head and body,* with slightly tufted tip. Ankles dusky. Large ears. L 7⅛–9¾″ (180–238 mm); T 3⅝–4¾″ (91–123

mm); HF ¾–1″ (20–26 mm); E ¾″ (16–20 mm); Wt ¾–1¼ oz (22–36 g).

Similar Species: Deer and White-footed mice in range of this species have tails shorter than head and body. Piñon and Northern Rock mice have longer ears. Cactus and Canyon mice have smaller ears.

Breeding: Breeds throughout year, producing several litters of 1–6 young. First-year females may produce litters late in the year.

Sign: Nest of dried vegetation under brush piles or in rocky crevices.

Habitat: Varied. Arid to semiarid brushland, especially in rocky areas; abundant at times in oaks, junipers, piñons.

Range: Extreme s Oregon south through California and east to w and s Nevada, Utah, Arizona, Colorado, New Mexico, w Texas, and w Oklahoma.

A skilled climber, the Brush Mouse clambers about on cliffs and in trees and will often run up trees to avoid predators. It frequently occupies higher portions of the habitat, while other species, such as deer mice or cotton rats, occupy the lower areas. This mouse uses its long tail as a prop when climbing and as a balancing organ when on a limb. It will not jump more than about 2 feet (60 cm). Foods include conifer seeds, acorns, berries, and insects. The Brush Mouse eats cactus fruit extensively when this food is in season. In Kansas it eats acorns especially heavily. It also eats wheat, corn, and oat seeds when they are available. In California this species feeds heavily on acorns in winter and manzanita berries in summer, with cutworms and other insects making up half or more of its diet in spring. The animal stores seeds in a crevice or hollow, but apparently not in the nest. Brush Mice commonly associate with woodrats and have been found in their nests. Owls, coyotes, hawks, snakes, and many other animals undoubtedly prey on this species.

81 California Mouse
Peromyscus californicus

Description: *The largest North American mouse in its genus.* Yellowish brown or gray mixed with black above; whitish below, often with buff spot on breast. Tail usually not sharply bicolored. Feet white. Ears large. L 8⅝–11⅛" (220–285 mm); T 4⅝–6⅛" (117–156 mm); HF 1–1⅛" (25–29 mm); E ¾–1" (20–25 mm); Wt 1⅛–1⅞ oz (33.2–54.4 g).

Similar Species: All other deer mice are smaller.

Breeding: Up to 6 litters per year of 2 young each; gestation in laboratory 21–25 days.

Sign: Large nest of grass, weeds, and sticks, often containing as much as half a bushel of vegetative material.

Habitat: Brushy hillsides and ravines; chaparral.

Range: Southwestern California.

The California Mouse eats the fruits, seeds, and flowers of shrubs, as well as fungi, laurel seeds, berries, and insects, spiders, and other arthropods. It often lives in the house of a woodrat. Unlike most mice, this species is monogamous, and a male and female will share a nest for extended periods. Territorial, it actively defends its nest from members of its own sex. Weasels and barn owls are among its chief predators.

Canyon Mouse
Peromyscus crinitus

Description: Grayish or yellowish brown above; whitish below, the color varying among populations in relation to the color of the soil. *Tail more than half total length, thinly covered with long, soft hairs, somewhat tufted. Ears large.* L 6⅜–7½" (162–191 mm); T 3⅛–4⅝" (79–118 mm); HF ¾–1" (18–25 mm); E ⅝–⅞" (17–23 mm); Wt ⅜–¾ oz (12.6–23 g).

Similar Species: Cactus Mouse has sparsely furred tail, not tufted at tip. Piñon and Northern Rock mice have longer ears. Prairie form

of Deer Mouse usually has shorter tail where the two species overlap.

Breeding: At lower elevations, breeds in spring, and at higher elevations in spring and fall; generally 2 but up to 8 litters of 1–5 young per year; gestation 24–25 days in nonlactating females, 29–31 in lactating.

Habitat: Rocky canyons, from below sea level to over 10,000' (3,050 m).

Range: Southeastern Oregon and sw Idaho south through Nevada and Utah to s California, w and n Arizona, w Colorado, and nw New Mexico.

Little is known of the habits of this secretive mouse, which lives among the barren rocks and crevices lining canyon walls. The Canyon Mouse eats seeds, insects, and green vegetation.

82 Cactus Mouse
Peromyscus eremicus

Description: *Upper parts yellowish orange to cinnamon-buff;* white below. Dark populations occur on lava areas of Southwest. *Tail sparsely haired, indistinctly bicolored, longer than head and body. Ears relatively large,* nearly hairless. L 6¼–8⅝" (160–218 mm); T 3¼–5" (84–128 mm); HF ¾–⅞" (18–22 mm); E ½–¾" (13–20 mm); Wt ⅝–1⅜ oz (18–40 g).

Similar Species: Canyon Mouse's tail is somewhat tufted at tip; when the two species occur together, the Canyon Mouse occupies the rockier areas. Brush Mouse has long hairs toward tip of tail. White-ankled Mouse's tail is more clearly bicolored. Piñon Mouse has larger ears. California Mouse has well-furred tail. Merriam's Mouse is usually in shrubby lowland areas with deep soil.

Breeding: Reproductively active throughout year, but less so during hottest periods; average litter 2 or 3 young; gestation 21 days. Newborn weighs ⅟₁₆ oz (2.1–2.9 g).

Sign: Nests; leavings of food items.

Habitat: Deserts, especially rocky outcroppings
with cactus or yucca stands.

Range: Southern California, s Nevada, extreme
sw Utah, Arizona, s New Mexico, and
sw Texas.

Well adapted to desert living, the Cactus
Mouse tolerates higher temperatures
and needs less water than most other
North American deer mice. Often
climbing to forage, this species is known
to eat the fruit and flowers of shrubs
such as mesquite and hackberry, some
green vegetation, and insects. Seeds of
desert annuals constitute the major
portion of its diet. The Cactus Mouse
hoards food and can become torpid
during the day and during food shortage.
It may nest in clumps of cactus, among
rocks, or in the abandoned burrows of
other small mammals. This species
produces tooth-chattering sounds,
squeals, and single chits when it is
in aggressive situations, injured, or
defending the nest.

83 Cotton Mouse
Peromyscus gossypinus

Description: Reddish brown above; white below. *Tail
short-haired, usually bicolored, slightly less
than half total length.* 6 pads on sole of
foot, as in all deer mice. L 5⅝–8⅛"
(142–206 mm); T 2¾–4½" (71–116
mm); HF ¾–1" (20–26 mm); E ⅜–⅞"
(10–21 mm); Wt ¾–1⅝ oz (20–46 g).

Similar Species: White-footed Mouse is slightly smaller,
usually with smaller hindfoot. Florida
Mouse has 5 pads on sole of foot. Deer
Mouse is smaller. Oldfield Mouse is
smaller and lighter in color.

Breeding: Mates throughout year, with a decline in
activity during summer and a peak in
late autumn and early winter; several
litters of 1–7 young each. Newborn
weighs ¹⁄₁₆ oz (2.2 g).

Sign: Nests, elevated as much as 2–3' (60–90
cm) off the ground, are often on sandy

ridges along bayous, in palmetto scrub, or in or under logs or other protected places.

Habitat: Bottomland hardwood forest and other woodlands; swamps; brushlands; rocky areas; beaches.

Range: Southeastern U.S. from e Texas and se Oklahoma east to se Virginia, e North Carolina, e South Carolina, Georgia, and Florida.

A nocturnal rodent, the Cotton Mouse is omnivorous, eating many invertebrates as well as seeds, fruits, and nuts. A skillful climber, it runs up trees the way gray squirrels do and is a fairly strong swimmer. Both of these skills are useful adaptations for the southern swamps where this species is most abundant. Cotton and White-footed mice are very similar and can hybridize in the laboratory; in the field, however, hybridization is apparently rare. Cotton Mice have been known to invade buildings. A Florida subspecies is classified as endangered.

Osgood's Mouse
Peromyscus gratus

Description: *Grayish brown above;* whitish abdomen. Tail bicolored. Very large ears. Measurements unavailable

Similar Species: Piñon Mouse is almost identical, differing mainly genetically.

Habitat: Rocky areas with brushy ground cover.

Range: Southeastern Arizona and sw New Mexico.

Osgood's Mouse and the Piñon Mouse do not differ morphologically from each other and were originally classed as a single species. However, the two are genetically different and remain distinct where they occur together, and thus are considered separate species. The habits of Osgood's mouse are presumed to be similar to those of the Piñon Mouse.

84 White-footed Mouse
"Wood Mouse"
Peromyscus leucopus

Description: Occurs over a large geographic range
and range of habitats; *physical description
varies with location.* Body brownish or
reddish to grayish, often with darker
stripe down middle of back; white
below. *Tail* similarly bicolored, *nearly
half total length.* Large ears. Juvenile
gray above; white below. L 5⅛–8⅛"
(130–205 mm); T 2½–3⅞" (63–100
mm); HF ¾–⅞" (19–24 mm); E ½"
(13–14 mm); Wt ⅜–1½ oz (10–43 g).

Similar Species: Deer Mouse is often very similar and
can be difficult to distinguish.
Woodland form of Deer Mouse is
generally larger, with longer hindfeet
and tail. Prairie form is smaller, with
shorter tail, smaller hindfeet. Oldfield
Mouse is smaller, lighter colored.
Cotton Mouse is slightly larger, with
larger hindfoot.

Breeding: Breeding peaks in spring and fall; litter
of 3–5 young; gestation is a minimum
of 22 days.

Sign: Black cherry pits, acorns, or various
seeds stored in, around, and under logs
and tree trunks. Nests in any concealed
location, constructed of a variety of
materials, depending on availability:
Grasses, leaves, hair, feathers, milkweed
silk, shredded bark, moss, cotton, or
shredded cloth are common.
Tracks: In dust, hindprint ⅝" (15–17
mm) long, with 5 toes printing;
foreprint ¼" (6–8 mm) long and wide,
with 4 toes printing; straddle, 1 ⅜"
(34–36 mm), with foreprints printing
behind and between hindprints. In
mud, foreprints and hindprints each
approximately ¼" (6–8 mm) wide;
straddle 1½" (37–39 mm).

Habitat: Primarily wooded and brushy areas, but
also many cultivated and open habitats,
especially adjacent to woods.

Range: Eastern U.S.: from e Montana,
n Colorado, and c Arizona east to

mid-Maine and south to w North Carolina, n South Carolina, n Georgia, and n Alabama.

The White-footed Mouse is primarily nocturnal and active year-round, although it may remain in its nest during extremely cold weather. A few of these mice may hibernate. This species is semi-arboreal, often climbing in trees; the shrub and lower tree strata of its area are often part of its home range. The White-footed Mouse uses its tail for balance when climbing. Omnivorous, it feeds on nuts, seeds, and fruits, as well as beetles, caterpillars, and other insects. Two favorite foods are the centers of black cherry pits and jewelweed seeds, the latter coloring the stomach contents turquoise blue. The White-footed Mouse stores caches of nuts and seeds in autumn near the nest, often in a bird nest, the abandoned burrow of another small mammal, or a building. When a nest becomes soiled, the mouse abandons it and builds a new one in a different location. An alarmed individual will drum its forefeet rapidly.

85 Deer Mouse
Peromyscus maniculatus

Description: *Color varies* greatly with habitat and geographic area. Often grayish to reddish brown above; white below. *Tail distinctly bicolored* and short-haired. 2 forms in eastern part of range: woodland and prairie. Woodland form has much longer tail and larger feet, ears, and body than prairie form. Woodland form: L 4⅝–8¾″ (119–222 mm); T 1¾–4⅞″ (46–123 mm); HF ⅝–1″ (16–25 mm); E ¾–⅞″ (18–21 mm); Wt ⅜–1¼ oz (10–33 g). Prairie form: L 4⅛–6⅜″ (106–162 mm); T 1⅞–2⅝″ (48–68 mm); HF ½–¾″ (14–19 mm); E ½–⅝″ (12–16 mm); Wt ⅜–⅞ oz (12.2–25.6 g).

Similar Species: Because it is the most common species
in many small mammal communities
and is exceedingly variable, the Deer
Mouse can be difficult to distinguish
from other *Peromyscus* species. In East,
White-footed Mouse's tail is shorter
than that of woodland form (tail of
woodland form is more than half total
length), and its tail and hindfeet are
longer than those of prairie form (prairie
form's tail is less than half total length).
In West, Piñon and Northern Rock
mice have much longer ears. Most other
western species have shorter tails.

Breeding: Breeding season is variable, usually
during the period that provides the best
environment and food for raising young.
Several litters per year of 2–7 young
each; gestation 21–24 days.

Sign: Prairie form: small burrows in ground
or nests in raised areas, if available.
Woodland form: nests in hollow logs.
Western forms: nests in such protected
places as underground burrows, clumps of
vegetation, hollow limbs on or above the
ground, and rock crevices, among others.
Tracks: Similar to White-footed Mouse.

Habitat: Exceedingly variable: prairies and other
grasslands; brushy areas; woodlands.

Range: In West, s Yukon and Northwest
Territories to Mexico; in East, Hudson
Bay to Pennsylvania and s Appalachians
and across northern tier of states and
south into c Arkansas and e Texas.

The Deer Mouse occurs over a large
geographic area and range of habitats,
and is highly variable in appearance.
More than 100 subspecies have been
described. In the eastern portion of its
range, there are two primary forms: the
prairie and the woodland. The smaller
prairie form *(P. m. bairdii)* occurs
throughout much of the Midwest,
whereas the many woodland forms occur
in the Alleghenies and northward. It is
presumed that the prairie and woodland
forms in the eastern region form a
continuous series of interbreeding

populations whose habitats range from grasslands through brushy terrain to woodlands. The forms at the end of this continuum overlap in some areas, but they are separate morphologically and ecologically and do not interbreed. Thus they act as species, although they are not described as such and are referred to mainly as subspecies. This type of geographic pattern, called circular overlap, occurs occasionally but defies traditional taxonomic treatment, as it happens in a continuum, with interbreeding populations connecting the whole. Deer Mice are often highly arboreal. They feed on various foods, including seeds and nuts, small fruits and berries, insects, centipedes, and the subterranean fungus *Endogone.* The Deer Mouse caches food for winter use, routinely storing seeds and small nuts in hollow logs or other protected areas, but not as extensively as the White-footed Mouse. The most important foods of the prairie form include seeds of foxtail grass and wheat, among other sorts of seeds, as well as caterpillars and corn. The prairie form is common in cultivated areas and remains even during harvesting and plowing periods. It may have additional small refuge burrows as well as home burrows. The woodland form feeds on woodland nuts, seeds, and fruits as well as insects and other invertebrates. In the West, the Deer Mouse occurs in myriad habitats, feeding on the various seeds, fruits, nuts, caterpillars, and other insects available. Practically all predators of suitable size prey on this species, and since it is so common, it serves as a diet mainstay of many animals.

Black-eared Mouse
Peromyscus melanotis

Description: Similar to Deer Mouse. In winter, upperparts *ocher-tawny mixed with dusky,* especially in middle of back. Fairly

distinct ocher line on each side. *Ears dusky, edged with white.* Tail bicolored: sooty brown above, white below. In summer, less dusky overall. L 5¼–6⅞" (132–175 mm); T 2¼–3⅛" (58–81 mm); HF ⅝–⅞" (17–22 mm).

Similar Species: Deer Mouse often can be distinguished only by genetic analysis.

Breeding: Laboratory crosses between Black-eared Mice from Arizona and Mexico averaged 4 young per litter.

Habitat: Open grassy and brushy areas.

Range: Known from 3 mountaintop populations in Pima, Graham, and Cochise counties of s Arizona.

Certain populations of mice in southern Arizona are externally almost identical to the Deer Mouse but will not breed with that species. Some authorities believe them to be Deer Mice, but they appear to be the Black-eared Mouse, previously known only in Mexico. The two species evolved in a like manner because they inhabited similar environments. They probably eat similar foods and have many behaviors in common.

Merriam's Mouse
Peromyscus merriami

Description: Upperparts ocher-buff overlaid with dusky, producing *dull gray overall* appearance. Underparts whitish with buff or tawny wash. *Tail bicolored,* dusky above, whitish below; more than half total length of animal. L 7¾–8⅝" (197–218 mm); T 4–4⅝" (102–118 mm); HF ¾–⅞" (20–24 mm).

Similar Species: Cactus Mouse is almost impossible to distinguish in the field. Most other deer mice are much browner overall.

Breeding: Litters probably produced throughout much of the year; 2–4 young per litter.

Habitat: Low desert; mesquite flats and heavy, forest-like stands of mesquite.

Range: In U.S., Pima, Pinal, and Santa Cruz counties in extreme sc Arizona.

This species is most often found in mesquite stands, and it has been suggested that it might be better called the mesquite mouse. It probably eats various seeds and invertebrates.

Northern Rock Mouse
Peromyscus nasutus

Description: *A large mouse.* Brownish above; white below. *Tail more than half total length. Large ears.* L 7⅛–10¼" (180–260 mm); T 3⅝–5⅝" (91–145 mm); HF ⅞–1⅛" (22–28 mm); E ⅝–1" (17–28 mm); Wt ⅞–1⅛ oz (24–32 g).

Similar Species: Piñon Mouse is usually grayer, with shorter tail. Brush, Canyon, and Pocket mice have smaller ears.

Sign: Nests in crevices of cliffs; remnants of piñon nuts and juniper fruits.

Habitat: Rocks, canyons, and lava cliffs, generally with piñon and juniper trees.

Range: Southeastern Utah, w Colorado, e Arizona, New Mexico, and w Texas.

As its name suggests, the Northern Rock Mouse is very skillful at climbing about on rocky cliff faces. Much of its diet is piñon and juniper nuts and fruits, but in summer it feeds heavily on insects.

Columbian Mouse
Peromyscus oreas

Description: Very similar to Deer Mouse. *Pale coloration* overall. Long tail. L 7⅛–8¾" (181–224 mm); T 3⅝–4¾" (92–122 mm); HF ⅞–1" (21–25 mm); E ⅝–⅞" (16–22 mm).

Similar Species: Very similar to Deer Mouse where the two occur together. Columbian Mouse usually has a longer tail and longer hindfeet.

Breeding: Principal reproductive period from April to June, with complete cessation by end of July; average of 6 young per litter.

Sign: Nests in protected places; leavings of food items.
Habitat: Primarily wooded and brushy areas.
Range: Southwestern British Columbia, excluding Vancouver Island, through w Washington.

This species is closely related to the Deer Mouse, but because the two remain distinct where they occur together, they are considered separate species. The habits of the Columbian Mouse are presumed to be similar to those of the Deer Mouse. It is preyed upon by owls, hawks, coyotes, weasels snakes, and other animals.

White-ankled Mouse
Peromyscus pectoralis

Description: Gray to brown above; creamy white below. *Tail more than half total length,* thinly haired, and indistinctly bicolored. *Ankles white.* Ears small. L 7⅛–9⅛″ (180–232 mm); T 3¾–4⅞ (94–123 mm); HF ¾–⅞″ (20–23 mm E ½–⅝″ (14–17 mm); Wt ⅞–1⅜ oz (24–39 g).
Similar Species: Often hard to discern, the white ankle of this species cannot always be relied on as a distinguishing feature. Brush Mouse has larger ears and dusky ankle Canyon Mouse's tail has tufted tip. De Mouse has shorter tail. Texas Mouse h longer feet and dusky ankles.
Breeding: Capable of breeding year-round, but most active September–May, with pea October–March. Litter size varies fror 2 to 9 young.
Sign: Nests in rock piles and other protectec areas, including debris around human settlements.
Habitat: Rocky areas, especially in association with oak and juniper, catclaw, and grama-bluestem plants. When it occu with the Brush Mouse, the White-ankled Mouse is in the brush-covered

foothills, while the Brush Mouse is more often in the pine-oak stands of higher altitudes.

Range: Extreme se New Mexico, sw and c Texas, and extreme s Oklahoma.

Little is known of the habits of the White-ankled Mouse. Juniper berries, acorns, hackberries, cactus fruit, and various invertebrates are its known foods.

86 Oldfield Mouse
Peromyscus polionotus

Description: *Whitish to fawn above;* white below. *Tail short, bicolored.* L 4¾–6″ (122–153 mm); T 1⅝–2⅜″ (40–60 mm); HF ⅝–¾″ (15–19 mm); Wt ¼–½ oz (8–15 g).
Similar Species: White-footed, Cotton, and Florida mice are larger and darker, with longer tails.
Breeding: Reproduces year-round in South Carolina, producing several litters of 3 or 4 young each; gestation 23–24 days.
Sign: Mounds of sand at burrow entrances, plugged with sand when mouse is inside; tracks in sand.
Habitat: Old fields; beaches.
Range: Southeastern U.S. from Alabama, Georgia, and sw South Carolina south through n and e coastal Florida.

A burrowing species, the Oldfield Mouse constructs, at the far end of its burrow and above the nest, a branch tunnel extending upward and ending just below the ground's surface. If a predator starts digging into the burrow, the mouse will often "explode" through this escape hatch, thereby eluding the startled predator. Burrows are open at times, closed at others, especially if heavy rains are imminent. Seeds and insects form the bulk of this animal's diet, although it also eats blackberries and wild peas. Several subspecies of the Oldfield Mouse have been described that are geographically separated from

one another and are evolving in differen
directions. Five of these are classified as
endangered by the U.S. government.

Sitka Mouse
Peromyscus sitkensis

Description: Very similar to Deer Mouse, but larger.
Upper parts *brown, shading to darker
in middle of back;* white below. *Tail
bicolored.* L 8⅛–9″ (205–230 mm);
T 3¾–4½″ (97–116 mm); HF 1″
(25–27 mm); E ½–¾″ (14–18 mm).
Similar Species: Deer Mouse is very similar but smaller.
Sign: Nests in logs, under rocks, or in other
protected areas.
Habitat: Forested areas on islands off se Alaska.
Range: Forrester, Chichagof, Warren, Duke,
Coronation, and Baranof islands,
Alaska.

The Sitka Mouse is much larger than
the closely related Deer Mouse. It occurs
only on a few islands, some of which are
separated from the others by a great
distance. Intervening islands may be
occupied by Deer Mice, but no island is
known to have both species.

87 Piñon Mouse
Peromyscus truei

Description: Grayish brown above, whitish below.
Tail bicolored, tufted. Ears very large.
L 6¾–9⅛″ (171–231 mm); T 3–4⅞″
(76–123 mm); HF ⅞–1⅛″ (21–27 mm);
E ¾–1″ (18–26 mm); Wt ⅝–1⅛ oz
(19–31 g).
Similar Species: Osgood's Mouse is essentially identical
but differs genetically. Northern Rock
Mouse is more brownish, with longer
tail. Brush, Canyon, and Cactus mice
have smaller ears. California Mouse is
larger.
Breeding: Reproductive period April–September
(February–November in Arizona);
several litters per year of 3–6 young.

Sign: Small piles of piñon-nut and juniper-seed remnants on rocks and logs and around bases of trees.

Habitat: Piñon-juniper areas, preferably rocky cliffs and hillsides with ample brush.

Range: Southwestern Oregon south through California and east to Colorado, New Mexico, w Texas, and extreme w Oklahoma; also in nc Texas.

The Piñon Mouse is an agile climber and often forages in trees for the piñon nuts and juniper seeds that are its staple diet. It also eats other seeds, leafy plants, berries, and insects, and stores seeds and nuts in the nest. This mouse lives in hollow juniper trunks, or under rocks, in a nest made of grasses, leaves, and shredded juniper bark. Often there are small secondary nests of juniper bark scattered about the home range. Coyotes and Common Gray Foxes are among the many predators of the Piñon Mouse.

Florida Mouse
"Gopher Mouse"
Podomys floridanus

Description: *A large mouse.* Brownish above; white below. Nearly naked, *large hindfeet with 5 pads.* Large ears. L 7⅜–8¾" (186–221 mm); T 3⅛–3¾" (80–95 mm); HF ⅞–1⅛" (24–29 mm); Wt ⅞–1¾ oz (25–49 g).

Similar Species: All species of deer mice have 6 pads on hindfeet.

Breeding: Breeding peaks July–December and January–February; several litters per year of 2–4 young.

Sign: Burrow entrances; tracks in sand; remnants of nuts, seeds, and other foods.

Habitat: High, sandy ridges with abundant blackjack and turkey oak; scrub palmetto.

Range: Northern and e peninsular Florida; isolated population in Franklin County (in the panhandle).

The Florida or "Gopher" Mouse, the typical mouse of the driest Florida habitats, was previously in the genus *Peromyscus.* This omnivorous mouse feed on seeds and nuts, as well as insects and other invertebrates. Almost exclusively a burrow dweller, it seldom makes its own burrows, usually using that of a gopher tortoise or a pocket gopher. These large burrows provide good protection from climatic changes and predators. The typical burrow nest is simpler than the surface nests of deer mice that live aboveground. It consists of a platform composed of leaves, Spanish moss, and occasionally some feathers. Preyed upon by most common predators in its habitat, the Florida Mouse is often seen with a short tail, presumably the result of a narrow escap from a would-be captor.

88 Golden Mouse
Ochrotomys nuttalli

Description: *Golden cinnamon above;* white below, often tinged with yellowish brown. *Tail long.* Southern individuals generall smaller. L 5–7½" (127–190 mm); T 2⅝–3¾" (68–97 mm); HF ⅝–¾" (17–20 mm); E ⅜–1" (11–26 mm); Wt 2⅜–3¼ oz (68–93 g).

Similar Species: All species of deer mice are brownish rather than golden cinnamon above.

Breeding: Breeds mid-March to early October in e Tennessee (and presumably in most o its range), peaking in late spring and early autumn; breeds year-round in Louisiana, peaking late autumn–early winter; several litters per year, as little as 25 days apart, of 1–4 young. Newbor is 2" (51 mm) long, weighs ¹⁄₁₆ oz (2.7 g

Sign: Globe-shaped nest of leaves, pine needle grass, or Spanish moss, lined with finel shredded bark, grass, or fur, 5⅞–7⅞" (150–200 mm) long, 3⅞–4⅞" (100–125 mm) wide, and 3⅞–7⅞" (100–20 mm) high, with 1" (25 mm) opening a

end; positioned in tangled vegetation
from near the ground up to 30′ (9 m)
high.

Habitat: Greenbrier thickets; boulder-strewn
hemlock slopes; brushy hedgerows;
swamps.

Range: Extreme e Texas, se Oklahoma,
s Missouri, and s Illinois to Atlantic
Coast; s Virginia south to mid-
peninsular Florida.

This highly arboreal, gregarious mouse
often climbs trees to 30 feet (9 m) or
more, running about the high branches
with ease and using its long, prehensile
tail for balance and support. It appears
to favor dense, brushy undergrowth and
is often seen in honeysuckle, greenbrier,
or grape vines, or, in Florida, in Spanish
moss. The Golden Mouse has well-
developed internal cheek pouches that it
stuffs with foods for transport. Its diet
includes acorns and other types of seeds,
such as those of sumac, poison ivy,
greenbrier, and wild cherry, and many
invertebrates. Male and female Golden
Mice in captivity occupied a nest
together, but the male would not enter
the nest after the young were born.
Larger nests may provide quarters for
several individuals; up to eight have
been found in a single nest. This species
also makes smaller structures that serve
mainly as feeding shelters.

89 Northern Pygmy Mouse
Baiomys taylori

Description: *The smallest rodent in North America.* Body
and tail *grayish brown above; slightly paler
below. Tail nearly one-half total length.*
L 3⅜–4⅞″ (87–123 mm); T 1⅜–2⅛″
(34–53 mm); HF ½–⅝″ (12–15 mm);
Wt ¼–⅜ oz (7–10 g).

Similar Species: Harvest mice have grooved upper
incisors. Deer mice are generally white
below. House Mouse has longer, nearly
naked, unicolored tail.

Breeding: Breeds year-round, peaking late
fall—early spring; several litters of 1–5
young; gestation 20–23 days. Newborn
weighs less than 1/16 oz (1.1–1.2 g).

Sign: Ball-of-grass nests underground or in
low, tangled vegetation, usually with 2
entrances. Networks of tiny runways
beneath matted grass.

Habitat: Grassy or weedy areas; mixed desert
scrub and prickly pear associations;
prairies and savannas, especially with
dense ground cover. Also highway and
railroad rights-of-way.

Range: Extreme se Arizona and extreme sw
New Mexico; also e and s Texas.

The Northern Pygmy Mouse utilizes a
small home range, usually less than 100
feet (30 m) in diameter. For reasons not
yet known, populations occur in isolated
pockets, frequently separated by areas
seemingly suitable for habitation.
Individuals sometimes use two nests. In
captivity, both male and female help care
for the young. The Northern Pygmy
Mouse eats mainly vegetation, feeding
heavily on prickly pear cactus stems and
fruit when available. The fruit frequently
stains the mouse's mouth and urine red.
Grass blades and seeds, mesquite beans,
and granjero berries are common foods,
and snails may also be eaten. This mouse
produces high-pitched squeals similar to
those of grasshopper mice. Cotton rats
may outcompete Northern Pygmy Mice
for habitat. Predators include snakes
(rattlesnakes, cottonmouths, and
coachwhips) and barn and barred owls.

93 Mearns' Grasshopper Mouse
Onychomys arenicola

Description: A stocky mouse. *Grayish to pinkish
cinnamon above;* white below. *Very short,
thick, bicolored tail with white tip.* Ears
brown, with *conspicuous white tufts at base*
in front. L average 5½" (139 mm);
T 1⅞" (48 mm); HF ¾" (20 mm).

Similar Species: Very similar Southern Grasshopper
Mouse, best distinguished genetically, is
slightly larger, with drab brown back
fur, less conspicuous ear tufts, and a
larger white tip on tail. Northern
Grasshopper Mouse has shorter tail
on average.

Sign: Remnants of insect meals.

Habitat: Low deserts with creosote bush, yucca
and mesquite.

Range: Southeastern Arizona east to sc New
Mexico, extreme w Texas, and south
into Mexico.

Mearns' Grasshopper Mouse is closely
related to the Northern and Southern
grasshopper mice and exhibits physical
characteristics intermediate between the
two. It is currently recognized by some
mammalogists as a separate species
because it apparently remains distinct
where it occurs together with them. Its
habits and behaviors are also similar to
those of the Northern and Southern
grasshopper mice.

94, 95 Northern Grasshopper Mouse
Onychomys leucogaster

Description: *A heavy-bodied mouse.* 2 main color
phases above: *grayish* (nearly black
in northeastern part of range) and
cinnamon-buff; white below. Short, thick,
bicolored tail with white tip, usually less
than one-third total length. *Juvenile dark
gray.* L 5⅛–7½″ (130–190 mm);
T 1⅛–2⅜″ (29–62 mm); HF ⅝–1″
(17–25 mm); Wt ⅞–1⅞ oz (27–52 g).

Similar Species: Southern Grasshopper Mouse is smaller,
with longer tail. Deer mice have longer
tails lacking white tip.

Breeding: Breeds February–August, peaking
June–August; 3–6 litters per year of
1–6 young; gestation 26–37 days in
nonlactating females.

Sign: *Tracks:* In mud or dust, foreprints
and hindprints overlap partially or
completely, each about ⅝″ (15–17 mm)

long, ½" (12–14 mm) wide, with
hindprint sometimes slightly longer,
straddle 1¼" (31–33 mm).

Habitat: Low valleys; desert; prairies.

Range: Much of w North America from se
Washington, s Alberta, Saskatchewan,
and sw Manitoba south to ne California,
e Arizona, and w Texas.

Grasshopper mice are usually nocturnal
and are most active on moonless nights
or under heavy cloud cover. As their
name suggests, they feed heavily on
grasshoppers, but they also eat other
insects, especially beetles and their larvae,
as well as caterpillars, scorpions, and
spiders. They have even been known to
kill other species of mice, at least in
captivity. Plant material constitutes up
to a quarter of their diet—most notably
forbs, grasses, sedges, and seeds—with
the greatest amounts eaten in midwinter.
Grasshopper mice either dig burrows or
take over those abandoned by other
animals. They maintain a complex
system of burrows throughout their
rather large territories, including a nest
burrow, retreat burrows, cache burrows,
defecation burrows, and signpost
burrows. The nest burrow, sealed during
the day to retain moisture, is the center
of activity. Retreat burrows are 8 to 10
inches (200–250 mm) long and extend
into the ground at a 45-degree angle to
ensure speedy escapes from predators.
Cache burrows are used for storing
seeds, to be eaten when insects are not
available. Signpost burrows are short,
usually only 1 to 2 inches (25–50 mm)
long. Located at the edge of the territory
and marked with glandular secretions,
they are used to designate territorial
boundaries. The Northern Grasshopper
Mouse will actively defend its young,
even against humans. It has several calls,
one of which is high-pitched and
prolonged, and made in a wolf-like
fashion with raised nose and open
mouth. Another common call is a sharp,

bark-like alarm chirp, rapidly repeated. Coyotes, hawks, owls, snakes, and other animals prey on this mouse.

96 Southern Grasshopper Mouse
"Scorpion Mouse"
Onychomys torridus

Description: A stocky mouse. *Grayish or pinkish cinnamon above;* white below. Thick, short, *bicolored tail with white tip,* between a third and a half the total length of the animal. Juvenile gray. L 4⅝–6⅜″ (119–163 mm); T 1¼–2⅝″ (32–68 mm); HF ¾–⅞″ (18–23 mm); Wt ¾–1 oz (22–30 g).

Similar Species: Northern Grasshopper Mouse is larger, with shorter tail. Deer mice have longer, thinner tails lacking white tip. Mearns' Grasshopper Mouse is more reddish brown, with shorter tail.

Breeding: Most reproductive activity May–July; litters of 1–5 young; gestation 27–30 days.

Sign: Tracks; remnants of insect food.

Habitat: Low desert with creosote bush, mesquite, and yucca.

Range: Southern California, w and s Nevada, and extreme sw Utah, southeast through w and s Arizona, s New Mexico, and w Texas.

The Southern Grasshopper Mouse, like its relatives, is primarily nocturnal and is active throughout the year. The home range of the male extends up to 8 acres (3.2 ha), an unusually large area for a small rodent. Although this species eats small amounts of seeds, its diet consists almost entirely of animal material: scorpions, beetles, grasshoppers, and other small mammals, especially harvest and pocket mice. Like the large carnivores, grasshopper mice have developed efficient strategies for dispatching prey. When capturing certain beetles that produce a defensive secretion from the back of the abdomen, grasshopper mice hold the beetles in

their forepaws and jam the abdomen into the sand to avoid the secretion. They kill small mammals with a bite through the back of the neck. Before killing scorpions, they immobilize the deadly tail. The Southern Grasshopper Mouse either digs its own burrow or appropriates the burrow of another small mammal. The social unit includes one pair and its offspring per burrow system. The male and female both actively care for the young, although the male is excluded from the nest by the female for the first three days after birth. The highly territorial male employs a high-pitched, wolf-like call to ward off other males.

Arizona Cotton Rat
Sigmodon arizonae

Description: A large, grizzled brown rodent. Belly dark. Eyes medium-size; *ears partially concealed in fur.* Tail long. L 8–12½″ (202–317 mm); T 3⅜–5⅞″ (86–150 mm); HF 1⅛–1½″ (29–38 mm); E ¾–1″ (18–25 mm).

Similar Species: Nearly identical Hispid Cotton Rat is slightly smaller, with shorter hindfoot where the two species occur together. Tawny-bellied Cotton Rat has tawny belly. Yellow-nosed Cotton Rat has orange or buff nose. Woodrats are usually white below, with more prominent ears.

Breeding: Several litters throughout the year of up to 12 young each.

Sign: Runways; grass nests; cuttings of grasses and other plants. Tracks similar to those of Hispid Cotton Rat.

Habitat: Grassy and weedy fields.

Range: Southeastern California, extreme s Nevada, c Arizona, and south into Mexico.

Originally considered a subspecies of the Hispid, the Arizona Cotton Rat was established as a separate species when

genetic examination revealed it to have a markedly reduced number of chromosomes. In appearance, however, the two species are nearly identical, and presumably have similar habits.

Tawny-bellied Cotton Rat
Sigmodon fulviventer

Description:	Speckled, salt-and-pepper black and buff above; *tawny or buff below.* Tail uniformly brownish black. L 8¾–10⅝" (223–270 mm); T 3¾–4¼" (94–109 mm); HF 1–1⅜" (26–36 mm).
Similar Species:	Other cotton rats are gray below. Woodrats are usually white below and have more prominent ears.
Breeding:	Probably produces several litters throughout the year.
Sign:	Well-worn runways; nests of woven grass placed in runways among dense vegetation; grass and other cuttings. Tracks similar to those of Hispid Cotton Rat.
Habitat:	Grass and grass-shrub habitats; mesquite; piñon, juniper, and live oak woodlands.
Range:	Southeastern Arizona, sw and c New Mexico. A new subspecies has been described in extreme w Texas.

Populations of this grassland species now exist in scattered areas where domestic livestock do not graze. Little is known about its habits.

99 Hispid Cotton Rat
Sigmodon hispidus

Description:	*Dark brown or blackish, coarsely grizzled with cream above; grayish below.* Scaly, scantily haired *tail less than half total length;* dark above, slightly paler below. L 8⅛–14⅜" (207–365 mm); T 3–6½" (75–166 mm); HF 1⅛–1⅝" (28–41 mm); E ⅝–⅞" (16–24 mm); Wt 2¾–4¼ oz (80–120 g).

Similar Species: Rice rats generally have longer tails and paler fur on tops of hindfeet. Yellow-nosed Cotton Rat has orange or buff nose. Tawny-bellied Cotton Rat is tawny or buff below. Hindfoot of Arizona Cotton Rat averages larger.

Breeding: Several litters per year of 1–15 young each; gestation 27 days. Newborn weighs ¼ oz (7.2 g).

Sign: Ball- or cup-shaped nests of woven grass with a single entrance.
Tracks: In mud, foreprints and hindprints overlap; combined prints about ½" (12–14 mm) wide and long. Straddle to 1½" (40 mm); walking stride 1¼" (30 mm). Those of other cotton rats are similar.

Habitat: Grassy and weedy fields.

Range: Southeastern Arizona, New Mexico, and se Colorado east to Virginia and south through Florida. Isolated populations in se California and sw Arizona.

Cotton rats are active around the clock, but especially from dusk to dawn, often climbing about aboveground in vegetation. They primarily consume green vegetation, cutting grass plants into sections to reach the heads, and can cause great destruction to crops such as sugarcane and sweet potatoes. They also eat a few insects and at times reduce quail populations by eating the eggs and chicks. The Hispid Cotton Rat makes long, shallow runways with nest chambers. Populations in the northern parts of its range make larger and more tightly woven nests than those in more southerly areas, with the openings usually facing toward the southeast, thereby reducing cooling by northwest winds. The Hispid Cotton Rat is one of the most prolific of mammals, breeding as early as six weeks of age. Its enormous reproductive potential is kept in check by its many predators, including other mammals, birds, and reptiles.

Yellow-nosed Cotton Rat
Sigmodon ochrognathus

Description: Grizzled brownish above; grayish below. *Nose orange or buff. Tail bicolored:* very dark above; grayish buff below. L 9⅛–10¼" (233–260 mm); T 3⅞–4⅝" (100–117 mm); HF 1–1⅛" (25–29 mm); E ¾–⅞" (20–22 mm); Wt 1¾–4 oz (50–112 g).

Similar Species: Other cotton rats lack orange or buff nose.

Sign: Nests composed of grasses and agave fibers. Runways among vegetation, radiating from burrow entrances. Tracks similar to those of Hispid Cotton Rat.

Habitat: Occupies the driest habitat of any cotton rat: high grasslands and rocky slopes with patches of grass.

Range: Southeastern Arizona, extreme sw New Mexico, and sw Texas into n Mexico.

The Yellow-nosed Cotton Rat feeds on green vegetation. It makes its nests under piles of dead leaves or in the abandoned burrow of a Southern Pocket Gopher.

111 ## White-throated Woodrat
Neotoma albigula

Description: Brownish gray above; white or grayish below. Tail similarly colored. Feet white. *Hairs gray at base, except patch of all-white hairs on throat.* L 11⅛–15¾" (283–400 mm); T 3–7¼" (76–185 mm); HF 1⅛–1½" (30–39 mm); Wt 4¾–10⅜ oz (136–294 g).

Similar Species: Very similar Southern Plains Woodrat is steel-gray. Desert, Florida, and Mexican woodrats have throat hairs gray at base. Dusky-footed Woodrat has dusky hind ankles, tail darker brown above. Stephen's Woodrat has longer hairs at tip of tail.

Breeding: Breeds January–July in California; several litters of 2 or 3 young each; gestation at least 37 days.

Sign: Large house of sticks, cactus parts, and other debris under cactus or other plant.

Nest, inside house, 6–10″ (15–25 cm) in diameter. Well-worn trails leading from house.

Tracks: In mud, hindprint ¾″ (19 mm) long, with 5 toes printing; foreprint ½″ (13 mm) long, with 4 toes printing. Hindprints ahead of foreprints; distance between individual walking prints approximately 1¾–3″ (45–75 mm).

Habitat: Variable; often in arid situations, such as brushlands of dry plains and deserts. Often associated with prickly pear cactus and piñon-juniper; also dead trees.

Range: Extreme se California east to w Texas, extending north into se Utah and s Colorado.

The White-throated Woodrat is skilled at climbing spiny cacti, and cactus figures strongly in its diet, along with some juniper, yucca, and various other leafy plants. Highly adapted to desert living, this woodrat obtains its water from the food it eats, particularly prickly pear. It consumes very little grass or animal material and stores a small amount of food in its house, a large structure that contains the nest within it. The White-throated Woodrat generally chooses the base of a prickly pear or cholla cactus as the site for its house, using cactus needles to cover the entrance or entrances. Several houses may be present in a small clump of cacti, but no more than one adult resides in each. This species will also build in rocky crevices and sometimes adds an underground chamber if vegetation is scarce. Its dens are used by many other vertebrates and invertebrates. Like many woodrats, this species engages in hindfoot drumming, which is thought to express fear or serve as a means of communication. Woodrats are preyed upon by gopher snakes, rattlesnakes, Coyotes, great horned and probably other owls, Bobcats, Ringtails, and weasels.

112 Bushy-tailed Woodrat
"Mountain Pack Rat"
Neotoma cinerea

Description: Often brownish peppered with black hairs above, but varies from pale grayish to blackish; whitish below. *Tail squirrel-like,* bushy, and flattened from base to tip. L 11½–18½" (292–472 mm); T 4¾–9¼" (120–236 mm); Wt 5½–15½ oz (156–444 g).

Similar Species: All other woodrats have rounded, short-haired tails.

Breeding: Probably 1 or 2 litters per year of 2–6 young, usually born in spring or summer.

Sign: Stick houses concealed in crevices or abandoned buildings. Cup-shaped nest of fibrous plant materials, sometimes roofed over, is within house.
Scat: Piles of fecal pellets on ledges; black tar-like accretions of older fecal material and white calcareous encrustations from urine often in evidence on cliff faces.

Habitat: Rocky situations; coniferous forests.

Range: Southeastern Yukon and extreme sw Northwest Territories south into nw U.S. to n California and nw New Mexico, and east to sw North Dakota and w Nebraska.

The Bushy-tailed Woodrat is the original "pack rat," the species in which the trading habit is most pronounced. It has a strong preference for shiny objects and will drop whatever it may be carrying in favor of a coin or a spoon. Green vegetation is its preferred food, but the Bushy-tailed Woodrat also eats twigs, nuts, seeds, mushrooms, and some animal matter. In coniferous forests, this woodrat may build its house as high as 50 feet (15 m) up a tree, using its bushy tail for balance when climbing and jumping. In some areas, the house is used only for caching large quantities of dried vegetation, and the nest itself is concealed in a rocky crevice behind a barricade of sticks. The Bushy-tailed

Woodrat engages in hindfoot-drumming when alarmed. It will also drum when undisturbed, producing a slow, tapping sound. This rodent is heavily preyed upon by spotted owls and Bobcats.

Arizona Woodrat
Neotoma devia

Description: *Buff-gray above;* grayish, often mixed with buff, below. Tail bicolored. All hairs gray at base. *Hindfeet white.* L 10⅜–13″ (262–330 mm); T 4⅜–5⅞″ (111–149 mm); HF 1⅛–1¼″ (29–33 mm).

Similar Species: Very similar Desert Woodrat differs genetically. White-throated Woodrat's throat hair is white at base. Mexican Woodrat's tail is grayish to white below. Dusky-footed Woodrat is larger, with dusky ankles.

Sign: House at base of cactus or other desert plant.

Habitat: Desert; piñon-juniper areas.

Range: Western Arizona s and e of Colorado River.

Although very similar to the Desert Woodrat, the Arizona Woodrat is genetically distinct. The Colorado River forms the boundary between the two species, with the Desert Woodrat occurring to the north and west of the Colorado in western Arizona.

113 Florida Woodrat
"Eastern Woodrat"
Neotoma floridana

Description: *Grayish brown above;* white or grayish below. Bicolored tail is less than half total length. Throat hairs gray at base. L 12¼–17⅜″ (310–444 mm); T 5⅛–8″ (129–203 mm); HF 1½–1⅝″ (37–40 mm); Wt 7⅛–16 oz (200–455 g).

Similar Species: Very similar Allegheny Woodrat, distinguished by genetic and skull characteristics, is found north of

Tennessee River. Bushy-tailed Woodrat has flattened, squirrel-like tail. Southern Plains Woodrat is steel-gray above. White-throated and Stephen's woodrats have white throat.

Breeding: Breeds year-round.

Sign: Stick houses in crevices or caves in North; in brush, trees, palmetto, or holes in ground in South; piles of fecal pellets, bones, sticks, cedar, or other wood cuttings on ledges or around nests.

Habitat: Rocky cliffs, caves, tumbled boulders in s Illinois and elsewhere when available; Osage orange and other hedges and wooded low areas throughout South.

Range: Southern South Dakota and n Nebraska; e Colorado, s Nebraska, Kansas, and Missouri, south through Oklahoma and e Texas, and southeastward through southern states to mid-peninsular Florida.

The Florida Woodrat, once classified with the Allegheny Woodrat as the "Eastern Woodrat," feeds mostly on green vegetation, but also eats various fruits, nuts, fungi, ferns, and seeds. It eats food on-site for most of the summer, but in September and October it caches food in galleries in the top of its stick house, usually constructed in a protected location. When built in a cave, the house may be open at the top. The lodges are quite waterproof and well constructed, moderating temperature extremes. There is one woodrat per house, with the houses distributed over the available habitat; this tends to spread out the rats, reducing competition. In the South, this species often lives in hollow trees or holes in the ground, making large nests of sticks, leaves, or rubbish along the banks of streams, in dense tangles, or in trees. In Alabama, it builds bulky nests in Osage orange hedges. Young woodrats cling tenaciously to their mother's teats; if alarmed, the mother will drag the whole litter along as she flees. The Florida Woodrat has a host of predators,

including snakes, owls, weasels, and Bobcats, but only the snakes and sinuous weasels can get to a woodrat while it is inside its house.

114 Dusky-footed Woodrat
Neotoma fuscipes

Description: Buff-brown above, grayer on face; grayish to whitish below; belly often washed with tan. *Feet and ankles dusky;* toes and claws white. *Tail scantily haired; brown above, slightly paler below;* nearly half total length. L 13¼–18¾" (335–475 mm); T 6¼–8⅞" (158–227 mm); HF 1⅜–1½" (34–37 mm); E 1⅜–1⅞" (34–47 mm); Wt 8¼–9⅜ oz (233–267 g).

Similar Species: Desert and White-throated woodrats have white hindfeet. Desert Woodrat has a shorter tail, hindfoot, and ear, and its belly hairs are gray at base. Bushy-tailed Woodrat has flattened bushy tail. White-throated Woodrat has hairs on throat that are white to base.

Breeding: Breeds February–September; 1 litter per year of 1–4 young; gestation 33 days.

Sign: Sticks, bark, plant cuttings, and miscellaneous items in a conical pile below a bluff. Houses up to 8' (2.5 m) high and 8' (2.5 m) in diameter are often in or against a tree or shrub; there are openings where tree limbs protrude.

Habitat: Conifer and hardwood forests; chaparral.

Range: Western Oregon south through California.

The Dusky-footed Woodrat builds its house, or "lodge," on the ground in open areas or in a tree up to 50 feet (15 m) high when it dwells in woodlands. A house, often the result of work by several generations of woodrats, is usually occupied by a female and her young, although two females sometimes occupy a single nest. One woodrat occupies one or two houses, and may use a house for up to nine months. Several

rats may occur together in the same area. The diet of the Dusky-footed Woodrat is mostly green vegetation, but also includes fruit, nuts, seeds, and subterranean fungi. This animal often caches materials, such as numerous kinds of plants, in chambers inside its house. The cache chambers often provide homes for various frogs, small mammals, and invertebrates. The Dusky-footed Woodrat apparently establishes pair bonds for the mating season, but after mating, the male lives separately in a small tree nest. Teeth-chattering may occur when the woodrat is disturbed inside the house, and this species will rattle its tail against tree limbs outside. Predators are owls, Coyotes, weasels, skunks, house cats, and Bobcats.

115, 116 Desert Woodrat
Neotoma lepida

Description: *Buff-gray above;* grayish below, often washed with buff. Tail similarly bicolored. Hindfeet white. *All hairs gray at base.* L 8⅞–15⅛″ (225–383 mm); T 3¾–7⅜″ (95–188 mm); HF 1⅛–1⅝″ (28–41 mm); E 1–1¼″ (25–32 mm); Wt 3⅞–4¾ oz (109–136 g).

Similar Species: White-throated Woodrat's throat hair is white at base. Mexican Woodrat's tail is more whitish below. Dusky-footed Woodrat is larger, with dusky ankles. Very similar Arizona Woodrat is found only in w Arizona south of the Colorado River.

Sign: Stick and cactus-spine house at entrance of burrow.

Habitat: Desert; piñon-juniper areas.

Range: Southeastern Oregon, sw Idaho, and extreme ne California south through Nevada, Utah, and n Arizona, and east to w Colorado.

The Desert Woodrat is adept at moving about among spiny cacti without injuring

itself. Its diet is composed of spiny cacti along with yucca pods, piñon nuts bark, berries, seeds, and any available green vegetation. This rodent uses the house it constructs at the entrance of a burrow, often appropriated from a long-gone ground squirrel or kangaroo rat, both for protection and for food storage. The nest is placed underground in a deep, cool chamber. The Desert Woodrat produces a rattling sound when alarmed by vibrating its tail against dry vegetation.

Allegheny Woodrat
Neotoma magister

Description: Grayish brown above; white or grayish below. *Bicolored tail less than half total length.* L 13¾–17" (348–431 mm); T 5½–7½" (141–191 mm); HF 1½–1¾" (37–46 mm); Wt 13⅛–16 oz (370–455 g).

Similar Species: Very similar Florida Woodrat, separated genetically and by skull characteristics, occurs south of Tennessee River.

Breeding: Breeding season early spring to mid-fall; 3 or 4 litters per year of 1–6 (usually 2) young.

Sign: Stick houses in crevices, caves, or behind slabs. Piles of fecal pellets, bones, sticks, cedar, or other wood cuttings on ledges.

Habitat: Rocky cliffs, caves, and tumbled boulders.

Range: Much of Pennsylvania, s New York, and nw Connecticut south to Delaware, Maryland, Virginia, West Virginia, nw North Carolina, and Kentucky, and south to Tennessee River in Tennessee. Isolated populations on Ohio River bluffs in s Ohio and Indiana.

The Allegheny Woodrat was originally described as a separate species, then was combined with the Florida Woodrat as a single species called the "Eastern Woodrat." However, it is presently recognized as a species on the basis of recent genetic and morphological work.

It occurs north of the Tennessee River, while the Florida Woodrat lives to the south. A very agile animal, it inhabits rocky bluffs and cliffs. It feeds mostly on green vegetation, but also eats various fruits, nuts, fungi, ferns, and seeds. The Allegheny Woodrat constructs both middens of stored food and debris piles. Middens, containing only food items, are formed in fall along trails or on ledges; they are depleted during winter. In Indiana, they consist of the foliage of tree of heaven, and red cedar, samaras, and pokeberry fruit. Debris piles are usually located around den sites, apparently helping to protect the openings. The Allegheny Woodrat builds its stick house in a protected location, such as a cave, sometimes leaving it open at the top. The species seldom produces sounds except teeth-chattering and foot-drumming. Snakes, owls, weasels, and Bobcats all prey on woodrats, but only snakes and weasels can reach them once they are inside their houses.

117 Mexican Woodrat
Neotoma mexicana

Description: *Grayish to brownish above;* yellowish to white below. Populations that inhabit lava flows may be blackish above. *Tail distinctly bicolored,* black to dusky above, grayish to white below. *Throat hair gray at base.* L 11⅜–16⅝″ (290–417 mm); T 4⅛–8⅛″ (105–206 mm); HF 1¼–1⅝″ (31–41 mm); Wt 5–6½ oz (140–185 g).

Similar Species: White-throated and Southern Plains woodrats have throat hair white at base. Desert Woodrat, difficult to distinguish, is more grayish below, and its tail is less strongly bicolored.

Breeding: In Colorado, breeds March–May with litters April–June; in South, breeds longer, but few details available; litters of 1–3 young; gestation 33 days.

Reaches sexual maturity as early as 2 months of age.

Habitat: Rocky outcrops, slopes, and cliffs in mountains; most common in piñon-juniper and ponderosa pine forests, but may occupy a variety of other habitats.

Range: Southeastern Utah, c and s Colorado, w Arizona, most of New Mexico, and w Texas.

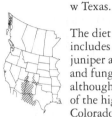

The diet of the Mexican Woodrat includes green plants, piñon nuts, juniper and other berries, nuts, acorns, and fungi. This animal also eats cacti, although cacti are not present in many of the higher areas it inhabits. In Colorado, for instance, it cures and stores great quantities of foliage. Rather than building a typical nest, the Mexican Woodrat often nests in a rocky crevice, fortifying it with a blockade of sticks, vegetation, and other debris, although collecting tendencies are not as pronounced as in some other species. In less rocky areas, its nests may be located in hollow trees, brush piles, or buildings. Like most woodrats, this species engages in foot-drumming, scent-marking by perineal dragging (lowering the rump while dragging itself along), and rolling on the back. Males, which have a scent gland only on the belly, also use abdominal rubbing for scent-marking.

118 Southern Plains Woodrat
Neotoma micropus

Description: *Steel-gray above; pale gray below.* Tail heavy, dark grayish above; white below. *Throat, chest, and feet white.* L 13⅛–16⅛″ (334–411 mm); T 4¾–7¼″ (120–185 mm); HF 1⅜–1¾″ (34–45 mm); E 1–1⅛″ (25–29 mm); Wt 7¼–11⅛ oz (204–317 g).

Similar Species: Other woodrats have browner coloring.

Breeding: Northern populations breed in early spring and bear 1 litter per year;

southern populations have a more prolonged breeding season and more litters. In both, litter of 2 or 3 young; gestation 30–39 days.

Sign: Large house may be 4–5' (1.2–1.5 m) high, is constructed of sticks, cactus parts, thorns, and other debris, usually near or under a cactus; up to 5 entrances, and thorns and cactus placed around them. Trails radiating from house are often visible. Nest a globular mass lined with fine grasses.

Habitat: Rocky outcrops and semi-arid brushlands; cactus, mesquite, and thornbush thickets.

Range: Southeastern Colorado, sw Kansas, w Oklahoma, and most of New Mexico east through w Texas.

The Southern Plains Woodrat is a vegetarian, obtaining all the water it needs from succulent plants. It feeds heavily on prickly pear leaves and the fruit of many cacti, along with the leaves of sotol, agave, and other plants, as well as mesquite pods, seeds, and nuts. This species usually builds its den under a shrub or cactus, such as mesquite, acacia, allthorn, yucca, or, most often, prickly pear. If vegetation is sparse, the rodent will dig a burrow in the earth below the house instead of constructing the usual surface runways. The den has a large central chamber used for the nest, with several side chambers for caching food. Trails lead from dens to feeding areas. The female often uses the same den throughout her adult life. Many other organisms live in the nests, including Desert Shrews and several species of assassin bug. Hawks, owls, roadrunners, foxes, Common Raccoons, Coyotes, Bobcats, and snakes are the main predators of this woodrat, which is the diet staple of the western diamondback rattlesnake in southern Texas.

Stephen's Woodrat
Neotoma stephensi

Description: A medium to small woodrat. Yellowish
to grayish buff above, but darkened by
dusky hairs. *Fur grayish at base; hairs on
throat white to base.* Underparts white or
creamy, often with pinkish-buff tinge.
Chest, abdominal regions, and inner
forelegs white; feet white with dusky
patch. Long hairs at tip of tail. L 10¾–
12¼" (274–312 mm); T 4½–5⅞" (115–
149 mm); HF 1⅛" (28–30 mm).

Similar Species: Desert Woodrat is slightly larger and
has gray hairs at throat. Only Bushy-
tailed Woodrat has bushier tail.

Breeding: Reproduction February–July; 1–5 litters
per year of 1 or 2 young each; gestation
31 days. Young are weaned at 35 days.

Sign: Houses at the bases of junipers or up to
6½' (2 m) high; often a mass of sticks,
sometimes a small amount of sticks,
bones, and cactus segments stuffed into
cracks in small cliff faces and shelf-like
rocky outcrops. Runways in rocky areas
often lined with moss.

Habitat: Rocky areas in piñon-juniper forest.
Sometimes in ponderosa pine, prickly
pear, or agave.

Range: Extreme sc Utah, n Arizona, and nw
New Mexico from Grant County
northward.

Throughout its range, Stephen's
Woodrat depends on juniper, using it
for food and shelter. It clips off a juniper
sprig and takes it to a midden just
inside the house to eat. Individuals
engage in foot-drumming, scent-
marking via perineal dragging and
abdominal rubbing, and also roll on
their backs. This species is presumably
fed upon by most predators in its area.

VOLES AND LEMMINGS
Subfamily Arvicolinae

Worldwide this subfamily comprises 26 genera and 143 species of rodents. Nine genera occur in North America north of Mexico; in this book we recognize 30 species in those genera.

Arvicoline rodents in our region fall into three groups: voles (including the bog lemmings), muskrats, and true lemmings. Members of this subfamily usually have rather inconspicuous ears and eyes, and stout bodies with short legs and tails, although muskrats have long tails. The molar teeth of arvicolines are modified with flattened crowns and ridges in "loop and triangle" patterns suitable for constant grinding of the fibrous grasses and leaves that make up the bulk of their diet. In most species, the molariform teeth and incisors grow throughout the animal's life; otherwise they would soon be worn down to the gums from the cellulose in the plant materials. The digestive tracts of voles and lemmings have a large cecum (a pouch at the beginning of the colon) that contains microflora; these help break down cellulose and other difficult-to-digest items. Many species in the subfamily are active day or night throughout the year.

Widespread and adaptable, voles (five genera, 24 species) live in a great variety of habitats. The red-backed voles of the genus *Clethrionomys* (three species) are reddish animals of moist woodland habitats. Members of the genus *Phenacomys* (three species) are usually rare and difficult to characterize. The "meadow voles" of the genus *Microtus* are often abundant in grassy locations; all are almost strictly herbivorous, feeding on green vegetation in warmer months, and roots and tubers in winter. In this guide, accounts are provided for

15 species of Microtus. In this we differ from the "Revised Checklist of North American Mammals" (Jones et al., 1992), which lists 17 Microtus species, including two that seem better classed as subspecies, at least until more data become available: the Insular Vole, which is closely related to the Singing Vole and occurs only on islands off the coast of Alaska; and the Beach Vole, which is closely related to the Meadow Vole and lives on an island off the Massachusetts coast.

The Sagebrush Vole, sole member of the genus *Lemmiscus,* is found in patches of grass in sagebrush areas of desert. Bog lemmings (two *Synaptomys* species), which are often rare and difficult to capture, are very similar to *Microtus* voles; they differ in having very short tails and very shallow grooves running the length of the front incisors. Bog lemmings are found in a variety of wooded and open (even very dry) habitats, seldom in bogs; they are called lemmings probably because of their very short tails.

Muskrats (one species each in the genera *Neofiber* and *Ondatra*) are large, long-tailed, aquatic rodents that are classed as voles but differ from them in form and habits. Larger than voles and lemmings, muskrats have long, laterally flattened (higher than wide) tails that serve as rudders when they swim. They live either in houses that they build of cattails or other vegetation in open marshes, or in burrows in banks along rivers, streams, or lakes. Their dens, within the house or burrow, are built above the high-water line, with several tunnels leading into the water, some usually below the water line. Vegetable matter is the chief food of muskrats, although they occasionally eat mussels, fish, and crayfish.

True lemmings (*Lemmus* and *Dicrostonyx* species, four of which are recognized here) are stouter than voles and have exceedingly short, thick, hairy tails. In North America, these rather bulky animals are restricted to tundra habitats in the far north. They undergo cyclical variations in population, increasing in number enormously when food is plentiful, then decreasing when the population outgrows the food supply. Such cycles are responsible for the mass migration of the Scandinavian Lemming (*Lemmus lemmus*) of Europe, which occurs every three or four years and is one of the most amazing phenomena of the natural world. When food supplies diminish, vast hordes of Scandinavian Lemmings leave their native regions to seek food in new areas, swarming through forests, towns, and cities, and across rivers and streams. They travel in one direction only and allow no obstacle to stop them. Many reach the sea, which represents merely another barrier to be conquered in their frenzied drive for food. They dive in, swim until they become exhausted, and eventually drown. The closely related Brown Lemming of northern North America undergoes similar population cycles, but individuals disperse separately after the population peaks rather than searching for food en masse.

As with the *Microtus* species, there is disagreement among mammalogists over the number of species in the genus *Dicrostonyx,* the collared lemmings. The "Revised Checklist of North American Mammals" now lists seven: the Peary Land Collared Lemming (*D. groenlandicus*), the Labrador Collared Lemming (*D. hudsonius*), and Richardson's Collared Lemming (*D. richardsoni*), which we accept as legitimate species; plus four others: Nelson's Collared Lemming (*D. nelsoni*),

the Bering Collared Lemming *(D. rubricatus),* the Unalaska Collared Lemming *(D. unalascensis),* and the St. Lawrence Island Collared Lemming *(D. exsul),* which we consider to be subspecies of the first three. For instance, the St. Lawrence Island Collared Lemming and the Unalaska Collared Lemming, which occur on Unalaska and Umnak islands in southwest Alaska, are island forms and perhaps best considered as subspecies of *D. groenlandicus,* with which we suspect they would interbreed if they occurred together in the field.

Western Red-backed Vole
Clethrionomys californicus

Description: *Chestnut brown,* or brown mixed with considerable black *above;* gradually lightening on sides and grading into buffy-gray belly. Indistinct reddish stripe on back. *Bicolored tail about half as long as head and body.* L 2 ½–5⅜″ (65–137 mm); T 1¾–2¼″ (45–56 mm); HF ¾–⅞″ (18–21 mm); E ⅜–½″ (10–12 mm).

Similar Species: Closely related Southern Red-backed Vole occurs to the north and east of the range of this species, and is redder, with more sharply bicolored tail. Meadow voles *(Microtus* species) lack reddish coloration. Red Tree Vole has much longer tail.

Breeding: Breeds February–November on slopes Cascade Range in n Oregon, year-round west of Cascades; 2–7 young per litter; gestation about 18 days.

Sign: Pieces of green vegetation among rocks.

Habitat: Mostly coniferous forests.

Range: From w Oregon south to nw California west of Cascade Range.

The Western Red-backed Vole lives in the woods and feeds on the green vegetation found there. It runs about on the forest floor, using the available

cover of logs, rocks, and moss, and usually does not form well-marked runways.

---

121 Southern Red-backed Vole
Clethrionomys gapperi

Description: *Rust-reddish above;* sides buff or grayish; gray to buff-white below. Tail short, slender, slightly bicolored. Gray phase sometimes occurs in Northeast. L 4¾–6¼" (120–158 mm); T 1⅛–2" (30–50 mm); HF ⅝–⅞" (16–21 mm); Wt ½–1½ oz (16–42 g).

Similar Species: Northern Red-backed Vole is brighter, with buffy or yellowish sides; tail is thicker, with thicker, more bristly fur. Western Red-backed Vole has indistinct reddish stripe on back. Meadow voles lack rust-reddish coloration. Red Tree Vole is arboreal and has very long tail for a vole. Woodland Vole can be distinguished from gray phase by softer, mole-like fur and much shorter tail.

Breeding: Sexually mature at 2–4 months; breeds late winter to late fall; several litters per year of 2–8 young, with larger litters at higher elevations and latitudes; gestation 17–19 days.

Sign: Runways and burrows on forest floor, under rocks and logs and in moss; pieces of cut green vegetation along runways. *Nest:* 3–4" (75–100 mm) in diameter; constructed of plant material in clumps, stumps, hollow logs, or old bird nests.

Habitat: Cool, damp forests; bogs and swamps.

Range: Southern tier of Canadian provinces south into Oregon, through entire Rocky Mountain system to Arizona and New Mexico; east to North and South Dakota, Minnesota, Wisconsin, upper half of s peninsular Michigan, New England south to Maryland, and Allegheny Mountain system to North Carolina.

Southern Red-backed Voles usually hop, but often run. They use natural runways

along rocks and logs, and are often seen running or hopping across a bed of moss or up an old tree stump. The home range is larger when the ground is covered with snow than when there is no snow. This species eats green herbaceous plants, such as false lily of the valley and goldthread, as well as bunchberry and various other ripe berries. Underground fungi comprise an important part of the diet, especially in fall. Individuals store bulbs, stems, tubers, and nuts for later use. The Southern Red-backed Vole uses the burrows of other animals when these are available, and does not construct elaborate burrow systems like those of the meadow voles. Male, female, and young may stay together in the nest, but the male usually moves out as the young get older. This species has population highs in late summer and fall, then decreases in number during winter. Maximum longevity is about 20 months, but most individuals do not live more than 10 to 12 months, and few survive two winters. Broad-winged hawks and Short-tailed Weasels are common predators, while Coyotes and Gray Wolves occasionally prey on this species. The gray phase that sometimes occurs in the Northeast can easily be mistaken for a vole of the genus *Microtus*.

Northern Red-backed Vole
Clethrionomys rutilus

Description: *Upperparts bright reddish; sides preponderantly buffy or yellowish.* Tail short, densely furred; yellowish below. Prominent *ears stand out from fur.* L 5¼–6⅛″ (133–154 mm); T 1⅛–1⅜″ (28–36 mm); HF ¾–⅞″ (18–21 mm); E ⅜–½″ (10–14 mm).

Similar Species: Southern Red-backed Vole is less brightly colored and lacks yellowish

color on sides; tail is longer, thinner, and more sparsely furred.

Breeding: Reproductive period late May–early September; gestation 17–19 days. A female could theoretically have 4 or 5 litters of 4–9 young per year; most probably have at least 2 litters per year; one female had 2 litters 18 days apart.

Sign: Pieces of cut vegetation among boulders and logs.

Habitat: Forest floor.

Range: Alaska and nw Canada.

The Northern Red-backed Vole is closely related to the Southern Red-backed Vole, and presumably has similar habits.

White-footed Vole
Phenacomys albipes

Description: Rich brown above; gray, often washed with brown, below. *Tail slender, sparsely haired; distinctly bicolored.* Tops of feet usually white. L 5⅞–7⅛″ (149–182 mm); T 2¼–3″ (57–75 mm); HF ¾–⅞″ (18–21 mm); Wt ⅝–1 oz (17–29 g).

Similar Species: Creeping Vole has shorter tail. Montane Vole is difficult to distinguish; inhabits high mountain meadows. Townsend's and Long-tailed voles usually are larger. Red-backed and Red Tree voles are reddish; Red Tree Vole has thicker, hairier tail.

Habitat: Along small streams, especially in areas with alder.

Range: Coastal Oregon to extreme nw California.

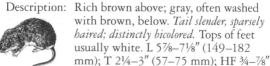

Probably the rarest of North American voles, the White-footed is a secretive species about whose habits little is known. Its small eyes and claw structure indicate that it is a burrower. The species feeds on green vegetation and roots. Major predators include weasels, Minks, skunks, and house cats.

Heather Vole
"Mountain Phenacomys"
Phenacomys intermedius

Description: A small vole. *Grizzled brown above; silver below.* Proportionally short tail (less than 1⅝"/42 mm) distinctly bicolored. *Feet white.* Yellow-faced individuals in some populations. L 5⅛–6" (130–153 mm); T 1–1⅝" (26–41 mm); HF ⅝–¾" (16–18 mm); Wt ½–1⅜ oz (15–41 g).

Similar Species: Long-tailed, White-footed, Red Tree, Yellow-cheeked, and Rock voles have longer tails. Montane Vole is difficult to distinguish; usually has longer tail and dusky or silvery-gray feet.

Breeding: Breeds May–August; up to 3 litters per year, each of 2–9 young; averages 4 or 5 young per litter; gestation 19–24 days. Male does not breed in first year.

Habitat: Broad range of habitats, usually with heather: most often coniferous forests; also open grassy heather and blueberry patches in scattered clearings on mountaintops.

Range: Southern tier of Canadian provinces, s Yukon, and Northwest Territories; south to n California, Idaho, w Montana, Wyoming, Utah, Colorado, and n New Mexico; also far ne Minnesota. Occurs in more isolated populations to the south.

The diet of the Heather Vole consists of plants such as bearberry, white heather, bear grass, lousewort, blueberry, and huckleberry, supplemented with twigs, bark, lichens, seeds, and occasionally underground fungi. In winter, this species nests on the ground surface in runways covered with snow; the rest of the year it nests in burrows up to 4 inches (100 mm) in diameter and 8 inches (200 mm) deep. Nests are built of such materials as fine, dry snow grass and lichens. Known predators are American Martens, Long-tailed Weasels, snowy and short-eared owls, and rough-legged and other hawks.

Red Tree Vole
"Tree Phenacomys"
Phenacomys longicaudus

Description: *Bright reddish above;* whitish below. *Blackish, well-haired tail very long for a vole.* L 6¼–8⅛" (158–206 mm); T 2⅜–3¾" (60–94 mm); HF ¾–⅞" (18–24 mm); Wt ⅞–1⅝ oz (25–47 g).

Similar Species: Red-backed voles have shorter tails and live on the ground. Meadow voles are grayish to brownish.

Breeding: Reproductive season February through September; ovulation induced by copulation; gestation typically 27–28½ days; 1–4 (most often 2) young per litter.

Sign: *Nest:* Bulky nest of twigs in Douglas fir trees.

Habitat: Primarily Douglas fir forests; also redwood or Sitka spruce, and areas with salal shrubs if ample Douglas fir not available.

Range: Coastal Oregon and nw California.

This highly specialized tree dweller rarely descends to the ground; its long tail is useful for balancing among the limbs. When leaping, a Red Tree Vole may spread its legs for balance and land uninjured on the ground; this ability seems to be learned, as young individuals seldom land on their feet. This species depends upon the Douglas fir for nest sites and food, although it also eats the needles of the grand or lowland white fir, Sitka spruce, and western hemlock. The animal may consume food where it finds it, but usually eats it at the nest. Douglas fir needles have resin ducts along each edge, which the vole bites off and discards or saves for inner nest material; the remainder is eaten, although young needles may be eaten entirely. The Red Tree Vole builds its nest 6 to 150 feet (2–46 m) aboveground, usually in a Douglas fir; it may also base the nest in an old squirrel, woodrat, or bird nest. It composes the nest of twigs from the

host tree, using additional twigs and food refuse to enlarge the nest continually, and urine and fecal pellets to help mold it together and secure it to the tree. Generations of voles may use the same nest; a large one may have several chambers with connecting passageways. One passage leads to the bottom of the nest; an escape exit is next to the trunk; other passages lead upward toward foraging areas. Sexes remain in separate nests unless breeding. Spotted owls are the main predator, along with saw-whet and long-eared owls. Populations of Red Tree Voles are widely separated and are disappearing due to logging and forest removal.

123, 124 California Vole
Microtus californicus

Description: *Grizzled brownish with scattered black hairs above;* gray below, with hairs often white-tipped. Relatively long, bicolored tail. *Feet pale.* L 6¼–8⅜″ (157–214 mm) T 1½–2⅝″ (39–68 mm); HF ¾–1″ (20–25 mm); Wt 1½–3½ oz (42–100 g

Similar Species: Montane Vole has dusky feet. Creeping Vole is not grizzled. Townsend's Vole has dusky feet and blackish tail. Long-tailed Vole has longer tail. Heather Vole is smaller. White-footed Vole is usually smaller.

Breeding: Major reproductive season is September–December, or several months after autumn rains; terminates with desiccation of vegetation, usually in June. In some years, there may be a minor reproductive period in autumn, with sporadic pregnancies the rest of the year. Several litters per year, each of 1–9 young; gestation 21 days. Young may be weaned as early as 14 days.

Sign: Runways and cuttings.

Habitat: Grassy meadows from sea level to mountains.

Range: Southwestern Oregon through much of California.

Like most voles, the California Vole is a burrower, but it also forms surface runways. This species feeds on grasses and other green vegetation when available; piles of cuttings are found along its runways. In winter, it eats mostly roots and other underground parts of plants. Like most other voles, this species considerably alters its habitat with its burrows, runways, and cuttings. Its main predators are hawks, owls, weasels, and snakes.

Gray-tailed Vole
Microtus canicaudus

Description: Fur on back yellowish brown or yellowish gray in summer, darker in winter. *Belly grayish white.* Feet gray. *Tail gray with black upper stripe.* L 5½–6⅝″ (140–168 mm); T ¾–1¾″ (20–45 mm); HF ⅝–⅞″ (15–21 mm).

Similar Species: Very similar Montane Vole, distinguished by skull and genetic characteristics, occurs east of the range of this species. Townsend's Vole has longer tail. Creeping Vole is much less robust and has smaller eyes.

Breeding: No information is available on reproduction in the wild. In captivity, 18-day-old females have mated successfully. Average litter has 5 young, fewer in very young animals; gestation 21–23 days.

Sign: Middens of cuttings, usually of grass, up to 6″ (150 mm) in diameter and 4″ (100 mm) deep.
Scat: Fecal pellets along runways.

Habitat: Now almost entirely agricultural grassy and legume fields; presumably, formerly in prairies maintained for centuries by Native Americans in Willamette Valley, Oregon.

Range: Extreme sc Washington and nc Oregon.

The Gray-tailed Vole's food presumably consists of green grasses and forbs and their roots. This animal constructs a

complex system of burrows and
runways; it nests under boards, bales,
or other items on the ground, or
sometimes underground. Fields
occupied by these voles are commonly
flooded in winter for several days at a
time. Voles may continue to live in the
flooded fields, using air trapped in the
burrows, though they sometimes have to
swim through flooded tunnels to enter
and exit. Gray-tailed Voles are less apt
to mate with familiar than unfamiliar
animals of the same species, perhaps
helping them to avoid inbreeding. This
species forms dominance hierarchies
("pecking orders"), as do many groups
of animals. Owls, hawks, skunks, foxes,
and house cats are common predators.
The Gray-tailed Vole formerly was
considered a subspecies
of the Montane Vole.

129 **Rock Vole**
"Yellownose Vole"
Microtus chrotorrhinus

Description: Yellowish brown above; gray below.
Relatively long tail slightly paler below.
Nose orangish or yellowish. L 5⅜–7¼″
(137–185 mm); T 1⅝–2½″ (42–64
mm); HF ¾–⅞″ (18–24 mm); Wt 1–1⅔
oz (30–48 g).

Similar Species: Yellow-cheeked Vole, the only other
North American vole with a yellow or
orangish snout, is considerably larger.
Meadow, Red-backed, and Woodland
voles lack yellowish nose.

Breeding: Reproductive period from early spring
to late autumn; 2 or 3 litters per year,
each of 1–7 young; gestation 17–19
days.

Sign: Scattered green food remnants among
rocks.
Scat: Piles of fecal pellets.

Habitat: Moist, rocky, woodland slopes;
mountains in southern part of range.

Range: Southeastern Ontario to Labrador, south
to extreme ne Minnesota, w Maine, and

e New York, and southward in scattered populations through Smoky Mountains; ne Pennsylvania.

Although often active during the day, the secretive Rock Vole is rarely seen. It occurs in scattered pockets of moist, rocky woodland, feeding mainly on green plants, including bunchberry, thread moss, grasses, and blueberry; it also eats some caterpillars and subterranean fungi. This species commonly has a specific latrine area.

125, 126 Long-tailed Vole
Microtus longicaudus

Description: A small vole with a *long, bicolored tail.* Grayish brown above; light grayish below. Feet off-white. L 6⅛–8¾" (155–221 mm); T 2–4½" (50–115 mm); HF ¾–1⅛" (20–29 mm); Wt ¾–3 oz (22–87 g).

Similar Species: Townsend's Vole has blackish, indistinctly bicolored tail. Most other voles have shorter tails.

Breeding: Breeds May–October (shorter season in North); 1 or 2 litters per female's lifetime (12 months); 2–8 young per litter. Newborn produces ultrasonic distress call.

Sign: Grass cuttings.

Habitat: Wide variety, with many different dominant plant species: dry, grassy areas far from water, mountain slopes, and alder and willow-sedge areas.

Range: Southeastern Alaska through Yukon and sw Northwest Territories (Mackenzie district) to California, Nevada, ne Arizona, and New Mexico; also w South Dakota.

The Long-tailed Vole feeds primarily on green plants when available, and on many underground fungi. It eats roots and bark when green vegetation is scarce. In winter, these voles spread out across the mountain slopes in search of

food; in summer, they usually retreat into grassy areas. This species does not form well-defined runways. Its life span is one year. Barn, great horned, long-eared, and short-eared owls, prairie falcons, weasels, and Martens are known to feed on this species. The Coronation Island Vole, formerly considered a separate species, is now classified as belonging to this species.

122 Mexican Vole
Microtus mexicanus

Description: *Grizzled cinnamon-brown above; buff to cinnamon below.* Relatively short tail. L 4¾–6″ (121–152 mm); T ⅞–1⅜″ (24–35 mm); HF ⅝–⅞″ (17–21 mm); Wt ⅞–1⅝ oz (26–48 g).

Similar Species: Montane Vole is white to gray below. Long-tailed Vole has longer tail.

Breeding: Reproductive activities occur throughout most of year, but primarily May–October; about 30–40 days between litters of 2–5 (average 3) young.

Sign: Numerous meandering runways through tall grass; piles of cuttings along runways.

Habitat: Clearings in yellow pine forests; arid grasslands; wet boggy situations.

Range: Extreme se Utah and extreme sw Colorado, south throughout e Arizona (including Hualpai Mountains) and sw New Mexico to extreme w Texas; also in wc Arizona.

The Mexican Vole is active mostly by day, and presumably feeds on grasses and forbs, like most other voles. Its globular nest, constructed of dried grass and forbs, is placed in a dense clump of vegetation, under a log or rock, in a depression on the ground, or in a chamber in its burrow. It is preyed upon by Common Gray Foxes, Bobcats, American Badgers, skunks, and Coyotes.

Singing Vole
"Alaska Vole"
Microtus miurus

Description: *Pale grayish or buff above;* gray below.
Very short tail, heavily furred. Feet gray or
buff; ear spot usually buff or tawny.
L 4–6⅜″ (101–161 mm); T ¾–1⅝″
(19–41 mm); HF ¾–⅞″ (18–21 mm);
Wt ¾–2⅛ oz (22–60 g).

Similar Species: Tundra, Yellow-cheeked, Long-tailed,
and Meadow voles are larger, with
longer tails.

Breeding: Reproductive period May–September;
may produce up to 3 litters per year,
each of 4–12 (average 8) young;
gestation 21 days.

Sign: Burrows, about 1″ (25 mm) in diameter,
with fresh dirt at entrance; surface
runways; stacks of cut vegetation around
willows and birches.

Habitat: Above timberline; subarctic tundra;
willow clumps along water; often drier
ground near water.

Range: Alaska, Yukon, and w Northwest
Territories (Mackenzie district).

The Singing Vole climbs well for a vole
and often forages in low bushes. It feeds
mainly on lupines, arctic locoweed,
horsetails, sedges, and the leaves and
twigs of willows. This species caches
tubers underground, and cuts and stacks
green vegetation on the surface where
willow or birch boughs hang low; these
piles sometimes contain a half-bushel of
material. Native peoples frequently
collect tubers from the vole's storage
chambers. The burrows, up to 3 feet (90
cm) long, lead to a large nest and
storage chamber, which may be 1 foot
(30 cm) long and is often within 2
inches (50 mm) of the surface. In
constructing the burrow, the animal
carries the excess soil in its mouth and
deposits it next to the burrow entrance.
This vole's common name refers to its
high-pitched trill, which usually is
sounded while the animal stands in its

burrow entrance. Colonial by nature,
the Singing Vole also gives warning
chirps when danger approaches. It is
probably preyed upon by most of the
common avian and mammalian
predators in its range. The Insular Vole,
sometimes classified as a separate species
(M. abbreviatus), is considered here as a
subspecies of the Singing Vole.

127 Montane Vole
"Mountain Vole," "Montana Vole"
Microtus montanus

Description: Grizzled brown to blackish above, often
with buff tint; white to gray below.
Moderately long, bicolored tail. Feet
dusky or silvery gray. L 5½–7½"
(140–192 mm); T 1¼–2¾" (31–69
mm); HF ¾–1" (18–25 mm); Wt 1¼–
3 oz (37–85 g).

Similar Species: Creeping Vole is usually smaller, not
grizzled. Meadow Vole is difficult to
distinguish; is usually darker brown, has
darker feet, and is not found in
mountains. Mexican Vole is buff to
cinnamon below. Prairie Vole is buff
below with shorter tail. California Vole
is difficult to distinguish, has pale feet,
and is found in lowlands. White-footed
Vole usually has white feet; occurs in
alder thickets along streams.

Breeding: Several litters, usually of 4–6 young,
produced from spring through fall.

Sign: Runways; cuttings of grass and forbs.
Scat: Fecal pellets in runways.

Habitat: High mountain meadows; valleys
associated with dry, grassy areas.

Range: South-central British Columbia, e
Washington, most of Oregon, ne
California east to se Montana, e
Wyoming, and w Colorado, extending
south into New Mexico and extreme ec
Arizona.

The little-known Montane Vole lives in
runway and burrow systems under
grassy cover. This grass, including

sedges and rushes, is the mainstay of its diet during most of the year. In summer, it eats leafy plants, although it does not forsake the cover of the grassy areas; it also eats some fungi. Populations undergo dramatic fluctuations from year to year, depending on food availability.

128 Prairie Vole
Microtus ochrogaster

Description: *Grizzled brown or yellowish brown above;* buff below. Relatively *short, bicolored tail.* Incisors ungrooved. L 5⅛–6¾″ (130–172 mm); T ⅞–1⅝″ (24–41 mm); HF ⅝–⅞″ (17–22 mm); Wt 1¼–1⅝ oz (37–48 g).

Similar Species: Meadow Vole has silvery-gray belly and longer tail. Woodland Vole and bog lemmings have much shorter tails.

Breeding: Reproduces throughout year; several litters per year, each of 3 or 4 young; gestation 20–23 days.

Sign: Runways and piles of grass cuttings.

Habitat: Dry grass prairie or mixed grassy-weedy situations.

Range: Southeastern Alberta, s Saskatchewan, and s Manitoba south to nw New Mexico, n Oklahoma, n Arkansas, Tennessee, and w West Virginia. There have been isolated populations in se Texas and sw Louisiana.

The diet of the Prairie Vole mainly consists of green vegetation and tubers from many different species of grasses and forbs, including bluegrass, bromegrass, curly dock, clover, and lespedeza, but also includes some insect material. This animal often caches food in underground chambers or tree stumps. It can damage orchards and other crops, and often competes with cotton rats for space, though not for food. When the Prairie Vole and the Meadow Vole occur together, the Prairie Vole inhabits the drier areas with less cover, the Meadow Vole the moist,

heavily vegetated areas, but either will occupy the other's habitats if the other species is absent. This species may be monogamous. Ultrasound produced by the young helps the adults find them Prairie Voles have indistinct cyclical population fluctuations, with numbers peaking, then declining, every three or four years. Most predators in the range feed upon this species.

130 Tundra Vole
Microtus oeconomus

Description: Grayish to brownish above, with *yellowish or reddish cast; white below. Tail bicolored:* dusky to black above, whitish to buff below; relatively short. L 6–8⅞' (153–225 mm); T 1⅜–2⅛" (36–54 mm) HF ¾–1" (19–25 mm); Wt ⅞–2¾ oz (25–80 g).

Similar Species: Meadow Vole usually lacks yellowish cast. Singing Vole is smaller and has shorter tail. Long-tailed Vole has longer tail. Yellow-cheeked Vole has yellowish or orangish nose.

Breeding: Reproductive period May–early September; 2 or 3 litters per year, each of 5–11 young.

Sign: Runways; cuttings of grasses and forbs; piles of feces near runways.

Habitat: Tundra areas around streams, lakes, and marshes; sometimes thick grass in drier areas.

Range: Alaska, w Yukon, and w Northwest Territories (Mackenzie district).

In summer, the Tundra Vole feeds primarily on green grasses and sedges, while storing rhizomes (especially knotweed and licorice root) and grass seeds for use at other times of the year. This burrowing species builds a nest of sedges and grasses within its burrows. Moving about on the ground surface, it sometimes uses frost cracks as extension of its runway system. This vole deposits droppings in specific areas off the main

runways. At one time, peoples of the far north trained dogs to find the Tundra Vole's caches of licorice root.

Creeping Vole
"Oregon Vole"
Microtus oregoni

Description: The smallest vole in its range. *Short-haired.* Brown above; silvery below. *Tiny eyes, only ⅛" (2 mm) in diameter. Short tail* indistinctly bicolored. Ears protrude from fur. L 4¾–6⅛" (120–156 mm); T 1⅛–2" (30–52 mm); HF ½–⅞" (14–21 mm); Wt ½–1⅛ oz (14–31 g).

Similar Species: California, Long-tailed, and Townsend's voles have longer tails, larger eyes, and longer fur. Montane Vole's hindfoot and tail average longer.

Breeding: Breeds March–November in northern part of range; 4 or 5 litters per year, each of 3 or 4 young; gestation 23 days.

Sign: Small, subsurface burrows appearing as raised ridges, similar to those made by a small mole.

Habitat: Coniferous forests, but clear-cut and grassy areas support more individuals; brushy and grassy areas, usually on drier upland slopes.

Range: Southwest British Columbia, south to w Washington, w Oregon, and nw California.

The diet of the Creeping Vole is mainly green vegetation, but also includes blueberries, other berries, and subterranean fungi. This species spends most of its time in shallow burrows, which can be so close to the surface that small ridges appear; however, burrows are not always apparent. Sometimes this vole uses the burrows of other species, such as moles. It usually constructs its grass nests underground, although occasionally it may place them in hollow logs or under pieces of bark or other debris. This species goes through major changes in population size, but

there is mixed evidence as to whether three- to four-year cycles occur. Barn owls, Bobcats, and Coyotes are predators.

---

131 Meadow Vole
"Field Mouse"
Microtus pennsylvanicus

Description: Color variable: from yellowish brown or reddish brown peppered with black, to blackish brown above; usually gray with *silver-tipped hair below. Long tail* dark above, paler below. Feet dark. L 5½–7⅝" (140–195 mm); T 1¼–2½" (33–64 mm); HF ¾–⅞" (18–24 mm); Wt ⅝–2½ oz (20–70 g).

Similar Species: Montane Vole, difficult to distinguish, is usually found in mountains. Prairie Vole is buff below, with shorter tail. Tundra Vole is yellower. Singing Vole has shorter tail, occurs above timberline. Yellow-cheeked and Rock voles have yellowish or orangish noses.

Breeding: Several litters of 1–11 young produced from spring through fall in North, year-round in South; up to 13 litters have been produced in a single season; gestation 21 days.

Sign: Grass cuttings, 1–1½" (25–40 mm) long, in piles along runways in dense vegetation.
Nest: Found under objects (hay bales or boards) or in clumps of grass.
Scat: Small, elongate fecal pellets, dark-colored.
Tracks: In light snow, hindprint ⅝" (16 mm) long, with 5 toes printing; foreprint ½" (13 mm) long, with 4 toes printing; hindprints ahead of foreprints, with distance between individual walking prints ½–⅞" (13–22 mm); straddle approximately 1½" (37 mm). Print patterns vary greatly, but most often show as alternating series of tracks very close together. Jumping distances between tracks range from 1¾ to 4¼" (45–110 mm).

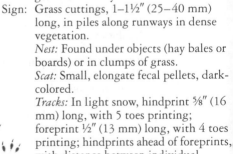

Habitat: Lush, grassy fields; also marshes, swamps, woodland glades, and mountaintops.

Range: Alaska (except for northern portions) and Canada south and east to n Washington, Idaho, Utah, New Mexico, Wyoming, Nebraska, n Missouri, n Illinois, Kentucky, ne Georgia, and South Carolina.

The Meadow Vole is active usually at night, occasionally during the day. It is less active during a full moon. The diet of this vole consists almost entirely of green vegetation and tubers, including many grasses, clover, and plantain. The animal produces grass cuttings as it reaches up and cuts off the stalk, pulls it down and cuts it again, until the seed heads are reached. The vole apparently consumes flowers, leaves, and all but the tough outer layer of the stalk, eating almost its own weight daily. The Meadow Vole constructs a system of surface runways and underground burrows. The spherical grass nest may be located in the burrows in summer or in a depression on the surface under matted vegetation; in winter, it is usually placed on the surface as long as there is snow cover for protection and insulation. A three- to four-year population cycle is well developed in this species. When alarmed, the Meadow Vole stamps its hindfeet like a rabbit. It uses vocalizations as a threat to other meadow voles. It is preyed on by house cats, foxes, coyotes, snakes, hawks, owls, and most other common predators; this widespread species is a mainstay in the diet of many carnivores. The Beach Vole, which is larger, more grizzled, and pale brown, is often considered a separate species *(M. breweri),* but is here considered an island subspecies of the Meadow Vole; it is the only vole found on Muskeget Island, Massachusetts.

132 Woodland Vole
"Pine Vole"
Microtus pinetorum

Description: Reddish brown above with *short, soft fur;* grayish, washed with buff below. Reddish-brown *tail short,* not much longer than hindfoot. Eyes and ears small. Upper incisors not grooved. Skull extremely similar to that of Prairie Vole. L 4⅛–5¾″ (105–145 mm); T ½–1⅛″ (12–29 mm); HF ⅝–¾″ (16–20 mm); Wt ⅝–1⅜ oz (19–39 g).

Similar Species: Prairie Vole has longer, more grizzled fur. Meadow Vole and voles of the genus *Clethrionomys* have longer tails. Bog lemmings have shallowly grooved upper incisors.

Breeding: Reproduces year-round in South, late March through August or September in North; 1–4 litters per year, each of 1–6 young; gestation 20–24 days.

Sign: Underground burrows about 1¼″ (30–35 mm) in diameter and usually in areas with soft soil; they can be found in woodlands by poking fingers into the ground.

Habitat: Deciduous woodlands with thick leaf mold or thick herbaceous ground cover; sometimes park-like grassy areas.

Range: Eastern U.S. west to c Iowa and c Texas; south from c Wisconsin and c New England through southern states, except for most coastal areas.

The Woodland Vole may be active any time of the day or night. The alternate common name "Pine Vole" and the Latin species name *pinetorum* are misleading, as this species is rarely found in or around pinewoods. Forbs, grasses, roots, and tubers are its dietary mainstays, but it also will eat seeds, fruits, bark, a few insects, and underground fungi. It stores large amounts of food in underground caches. This vole spends most of its time in tunnel systems one to several inches below the surface. It usually constructs

these burrows, digging with its forefeet and incisors and pushing back the dirt with its hindfeet. Vocalization is a harsh chatter, with one to five notes per call. The Woodland Vole is somewhat colonial, although colonies sometimes disband and disappear for no apparent reason. This species does not show population cycles, although it varies greatly in number through time. Its main predators are hawks, owls, foxes, and black snakes. It may be a problem in orchards, where it girdles young trees and can damage roots; however, since its burrows are so close to the surface, it usually can be controlled by cultivation.

Water Vole
Microtus richardsoni

Description: *A large vole.* Long fur, grayish brown to reddish brown above; grayish with white or silvery wash below. *Bicolored tail long.* 5 rather than the normal 6 plantar tubercles (small pads on soles of feet). L 7¾–10¾" (198–274 mm); T 2⅝–3⅞" (66–98 mm); HF 1–1⅜" (25–34 mm); Wt (overwintered adults) 2⅜–5¼ oz (68–150 g).

Similar Species: Other voles are smaller; most have shorter tails and shorter hindfeet (less than ⅞"/23 mm).

Breeding: Breeds from May or June through August or September; several litters per year, each of 2–10 young; minimum gestation 22 days; young may breed in their birth year.

Sign: Well-worn surface runways with exposed soil 24–33" (610–838 mm) wide; mounds of newly excavated soil; vegetation cuttings along runways or on the mounds.
Scat: Small groups (1–20) of elongated droppings.
Nest: About 4" (100 mm) high and 6" (150 mm) long, constructed of vegetation, and placed in cavities under rises or under logs or stumps.

Habitat: Along alpine and subalpine streams and
lakes, especially along clear, swift
streams with gravel bottoms.

Range: Southwestern and se British Columbia
and sw Alberta south through c and e
Washington to c and e Oregon, n Idaho,
nc Utah, w Wyoming, and w Montana.

This large, primarily nocturnal, semiaquatic vole is an excellent swimmer,
and will frequently enter the water to
avoid predators. In summer, its food
consists of various leafy plants,
including valerian, lousewort, lupine,
sneezeweed, grasses, sedges, and the
buds and twigs of willows; winter food
consists of roots and rhizomes of plants.
The Water Vole lives in small colonies
along high-elevation mountain streams,
constructing its burrows among sedges
or below alders or willows. Burrows are
large, up to 4 inches (100 mm) in
diameter; they are linked to the edge
of the water by wide, damp runways,
most of which are under the mat of root
and dead leaves on the ground surface.
There are also numerous openings to
the burrows along the banks of the
stream or lake. This species excavates
and reexcavates its burrows and nest
chambers throughout the summer; in
winter, it moves farther from water and
builds grass nests in its runways under
the snow cover. Few Water Voles survive
more than one winter. Occasional
population "explosions" have been
reported. Predators include American
Martens and Short-tailed Weasels.

Townsend's Vole
Microtus townsendii

Description: *Dark brown sprinkled with black above;*
grayish or grayish brown below. Long
tail blackish, indistinctly bicolored. *Feet*
dusky. Large ears project well above the
harsh fur. 6 plantar tubercles (small

pads on soles of feet). L 6⅝–9⅜″ (169–238 mm); T 1⅞–3⅜″ (48–85 mm); HF ¾–1″ (20–26 mm); Wt 1½–3⅝ oz (42–103 g).

Similar Species: California, Long-tailed, and Gray-tailed voles have more distinctly bicolored tails and lighter feet. Montane and Creeping voles are smaller, with proportionally shorter tails; Creeping Vole has smaller eyes. Water Vole has 5 rather than 6 plantar tubercles.

Breeding: Breeds from early spring through late summer or early fall; several litters of 1–9 young each; gestation 21–24 days.

Sign: Piles of grass cuttings in runways, which are often used by several generations and worn 1–2″ (25–50 mm) deep. *Scat:* Piles of droppings sometimes up to 7″ (180 mm) long and 3″ (75 mm) wide, often found at runway intersections.

Habitat: Marshes and moist, grassy areas, often in rank vegetation.

Range: Southeastern British Columbia and Vancouver Island south to nw California.

Townsend's Vole may be active day or night. Its varied diet includes velvet grass and other grasses, horsetail, alfalfa, clover, rushes, sedges, purple-eyed grass, and buttercups. Green food is still available in winter, but Townsend's Vole often stores and eats bulbs at that time. A good swimmer, this vole often constructs the entrances to its burrow system underwater. Summer and winter nests are constructed of grass. In summer, the nest is placed inside a hummock above water level; in winter, it is placed on dry ground away from water, which might freeze and prevent access. These animals use runways most of the year except when vegetation in summer is thick enough to completely conceal their bodies; then they move about at will under the cover of the vegetation. Owls and hawks are important predators, as well as house

cats, weasels, Coyotes, foxes, skunks, snakes, and the great blue heron. In winter, this vole occasionally causes damage in nurseries, killing saplings by eating their bark.

133 Yellow-cheeked Vole
"Taiga Vole"
Microtus xanthognathus

Description: *A large vole.* Dull brown above; gray below. *Tail long. Nose orangish.* Upper incisors of adults grooved. L 7⅜–8⅞" (186–226 mm); T 1¾–2⅛" (45–53 mm); HF ⅜–⅞" (9–23 mm); Wt 4–6 oz (113–170 g).

Similar Species: Rock Vole is smaller.

Breeding: 7–10 young per litter; gestation 21 days.

Sign: Burrows with 3–9' (1–3 m) wide dirt piles at entrances; wide, radiating runways extend 50–75 yd (45–70 m).

Habitat: Forests (especially spruce) bordering bogs or marshes.

Range: East-central Alaska, c Yukon, and s Northwest Territories (Mackenzie district) south to n Alberta, Saskatchewan, and Manitoba.

The Yellow-cheeked Vole, most active around dusk, is very vocal, chirping when alarmed. Grasses, horsetails, and lichens are important in its diet; it also eats blueberries. The species is colonial and undergoes major population fluctuations; a very large colony will often disappear within one year.

134 Sagebrush Vole
Lemmiscus curtatus

Description: *Pale gray above;* whitish to silvery or buff below and on feet. Bicolored, well-furred *tail usually less than 1" (25 mm) long.* Ears and nose buff. Back parts of soles of feet densely haired. L 4–5⅝"

(103–142 mm); T ⅝–1⅛″ (16–30 mm);
HF ½–¾″ (14–18 mm); Wt ⅝–1⅜ oz
(17–38 g).

Similar Species: Other voles are larger, darker-colored, with longer tails, and are rarely found in Sagebrush Vole's habitat.

Breeding: Breeds year-round in South, but shorter season in North; several litters of 1–13 young; gestation 25 days.

Sign: Burrow entrances, often under sagebrush clumps; runways between burrows; grass cuttings in runways. *Scat:* Greenish fecal pellets outside burrow entrances or along runways.

Habitat: Semi-arid prairies; sagebrush and bunchgrass usually dominant.

Range: Central Washington, s Alberta, and sw Manitoba south through e Oregon to Nevada, Utah, and ne Colorado.

The Sagebrush Vole is active throughout the day year-round, but exhibits greatest activity three hours before and three hours after dusk. This vole feeds on grasses and many other species of green plants in summer (usually not on the heads), and on bark and twigs of sage and various roots in winter. Bromegrass appears to be its mainstay in Oregon. This animal does not store food. Burrows are relatively short, often less than 2 feet (60 cm) long, and usually less than 12 inches (30 cm) deep, with a nest chamber 7 to 10 inches (18–25 cm) in diameter. The nest is constructed of shredded sagebrush bark lined with grass. Burrows in winter are usually under the snow. Runways may be 2½–3 inches (60–80 mm) wide. Sagebrush Voles are thought to dwell in colonies because their burrow entrances occur in clusters, generally of 8 to 30. Colonies vary greatly in size and density from year to year. This animal's main predators are owls, Long-tailed Weasels, and Bobcats, but a rattlesnake and a shrike have also been recorded.

Round-tailed Muskrat
"Water Rat," "Florida Water Rat"
Neofiber alleni

Description: *Large aquatic rodent.* Dense fur, dark brown above and below, with silky sheen when dry. *Tail long, round, and sparsely haired.* Hindfeet slightly webbed. Ears and eyes small. L 11¼–15 (285–381 mm); T 3¾–6⅝" (95–168 mm); HF 1½–2" (38–50 mm); Wt 5½–12⅜ oz (155–350 g).

Similar Species: Common Muskrat is larger, with laterally flattened tail.

Breeding: Young produced in any month; 4 or 5 litters per year, each of 1–4 young; gestation 26–29 days.

Sign: House, or lodge, 1–2' (30–60 cm) in diameter, of tightly woven grasses or sedges usually resting on a base of decayed vegetation, located in shallow water or on wet ground near water; similar to Common Muskrat's, but smaller and built without mud. Feeding platforms consisting of floating masses of cut vegetation. Also cattail or other plant cuttings in marshes.

Tracks: Smaller than Common Muskrat and longer in proportion to width, with more of sole printing; in mud, hindprint approximately 1½" (40 mm) long; foreprint to 1" (25 mm) long, printing slightly ahead of overlapping hindprint; distance between sets of prints 2½" (65 mm).

Habitat: Marshes.

Range: Southeastern Georgia south through peninsular Florida.

Although primarily nocturnal, the squeaky-voiced Round-tailed Muskrat sometimes active by day, and tends to be gregarious. Its known foods include maiden cane and other aquatic plants, such as sea purslane, pickerelweed, and arrowhead. It constructs feeding platforms, flat areas just above water level, essentially from the debris left over from eating, usually crisscrossed

pieces of cattail or aquatic plants. Plunge holes, sometimes covered with vegetation, lead directly into the water. Crayfish remains are common on feeding platforms, but they are left by rice rats, not the Round-tailed Muskrat. The house of this species has two entrances just below the surface of the water. Sometimes these animals construct houses in fields of broom sedge, a short distance from water; these are connected by a series of runways beneath the sedge. Occasionally this species constructs larger and more complex houses with several chambers, instead of the usual single chamber, and with several entrances. Abandoned houses are used by cotton and rice rats. During the breeding season, the Round-tailed Muskrat adds more material to the walls and floor of the house and fine grass to the nesting chamber. Nests and feeding platforms are sunning sites for snakes, such as the cottonmouth, a predator. Other predators include the marsh hawk and the barn owl.

216, 217 Common Muskrat
Ondatra zibethicus

Description: A large rodent. Dense, glossy fur, dark brown above, lighter on sides; finer, softer, and paler below to nearly white on throat. Small dark patch occasionally on chin. *Long tail* scaly, nearly naked, *laterally flattened (higher than wide) and tapering to a point.* Hindfeet partially webbed and larger than forefeet. Eyes and ears small. L 16⅛–24″ (409–620 mm); T 7⅛–12⅛″ (180–307 mm); HF 2½–3½″ (64–88 mm); Wt 1¼–4 lb (541–1,816 g).

Similar Species: Round-tailed Muskrat is smaller, with round tail; found only in se Georgia and peninsular Florida. Nutria is larger, with round tail. American Beaver is much larger, with very large, paddle-shaped tail.

Breeding: Breeds late winter through early September in North, year-round in South; 1–5 litters per year, each of 1–11 young; females often breed while still nursing; gestation 25–30 days.

Sign: Houses, or lodges, constructed of aquatic plants, especially cattails, up to 8' (2.4 m) in diameter and 5' (1.5 m) high, built atop of piles of roots, mud, or similar support in marshy areas, streams, or lakes, or along water banks. Also, burrows in stream or pond banks, with entrances above water line. Feeding platforms of cut vegetation in water, or on ice, slightly smaller than houses and marked by discarded or uneaten grass and reed cuttings. Other signs include floating blades of cattails, sedges, and similar vegetation near banks or feeding platforms; piles of clamshells sometimes at feeding sites ir freshwater areas; and scent posts along banks, composed of small mats of leave. and grass blades mixed with mud and musk secretions.

Tracks: Hindprint in mud 2–3" (50–75 mm) long, with all 5 toes printing; foreprint half as long, narrower, generally with only 4 toes printing (fift toe is vestigial). Hindprint can be ahead of or behind foreprint, sometimes overlapping. Distance between sets of walking prints approximately 3" (75 mm). Tail often leaves drag mark between prints; may alternately be lifte and dropped, or held up completely, leaving no mark.

Scat: Oval, generally ½" (12 mm) long, frequently in clusters on rocks, along banks, on logs in or next to water, and on feeding platforms.

Habitat: Fresh, brackish, or saltwater marshes, ponds, lakes, rivers, and canals.

Range: Most of Canada and U.S., except for Arctic regions, much of California, Southwest, Texas, and Florida.

The Common Muskrat is the largest rodent in its subfamily. Most active at dusk, at dawn, and at night, it may be seen at any time of day in all seasons, especially spring. An excellent swimmer, this aquatic rodent spends much of its time in water. Propelled along by its slightly webbed hindfeet and using its rudder-like tail for steering, the Common Muskrat can swim backward or forward with ease; it dislikes strong currents and avoids rocky areas. Its mouth closes behind protruding incisors, thus allowing it to chew underwater. It can remain submerged for long periods, and will travel great distances underwater. One individual was filmed underwater for 17 minutes, coming to the surface for air for 3 seconds, then submerging for another 10 minutes. The Common Muskrat eats mostly aquatic vegetation, such as cattails, sedges, rushes, water lilies, and pond weeds, along with some terrestrial plants. In some areas, this animal eats freshwater clams, along with crayfish, frogs, and fish. Ordinarily the muskrat tows food out to a feeding platform, which is littered with plant cuttings and other scattered food debris. Muskrat houses, or lodges, are similar to American Beaver lodges but much smaller. The muskrat adds to the house and feeding platform as long as they are used. The house usually shelters only one individual, although several may live together harmoniously except during the breeding season. The house is kept immaculately clean; fecal droppings are deposited on logs and rocks outside. Sometimes rather than build a house, the muskrat burrows into the bank along the water's edge and constructs a bank den with several entrances, usually below water level except when the water is low. While a house commonly contains one nesting chamber with one or more underwater

entrances, a bank den may have several chambers, each with one or more tunnels leading underwater. Scent posts covered with musky secretions from the perineal glands help muskrats identify each other by sex. Naked at birth, the young become furred about two weeks after birth, and can then swim and dive; in a month, they are weaned and are soon driven away by the mother. Droughts and flooding are common hazards faced by the Common Muskrat, leading to periodic population fluctuations. Overcrowding, especially when it occurs during fall or winter, causes fighting among individuals, forcing many to travel several miles overland to seek a new place to live. Common Raccoons, Minks, and humans are this rodent's major enemies (the first two open muskrat houses to capture the young), although many other animals also prey upon it. Until the decline of the fur industry, muskrat fur was considered extremely desirable because it is durable and waterproof. In the 1980s, nearly 10 million muskrats were trapped annually. Their flesh, sold as "marsh rabbit," provides good eating, although its popularity has declined. Muskrats often cause damage to dams or levees with their tunneling activities; they may also feed upon crops.

---

135 Brown Lemming
"Common Lemming," "Black-footed Lemming"
Lemmus sibiricus

Description: *Robust. Chestnut-brown above,* with grayish head; buff-gray below. *Tail stubby.* Feet silvery. Ears hidden in thick coarse fur. L 4¾–6⅝″ (122–168 mm); T ⅝–1″ (16–26 mm); HF ¾–⅞″ (18–2 mm); Wt 1⅝–4 oz (48–113 g).

Similar Species: Collared Lemming usually has black stripe down back, and is all white in

winter. Bog lemmings have grooved incisors. Most voles have longer tails.

Breeding: Mates spring–fall, sometimes through winter; usually 1–3 litters per year, each of 1–13 young.

Sign: *Nest:* Surface nests of woven balls of grass 6–8″ (150–200 mm) wide, among clumps of grass.
Tracks: Prints, ½″ (12 mm) long, look like miniature bear tracks; pattern irregular, like those of voles.

Habitat: Wet tundra and alpine meadows.

Range: Alaska and n tier of Canadian territories and provinces south into n British Columbia.

The Brown Lemming is closely related to the Scandinavian Lemming *(Lemmus lemmus),* famous for its mass migrations (see the Voles and Lemmings subfamily account). The North American animal feeds on grasses, along with sedges and leafy plants in summer, and bark and twigs of willow and birch in winter. This species uses runways and tunnels, the latter leading to two or three chambers about 6 inches (150 mm) in diameter. One chamber contains a nest of grass lined with fur; the other chambers are empty and probably used for resting and latrine areas. In winter, the Brown Lemming uses surface nests, frequently abandoning the old and building a new one. The Brown Lemming's home range is very small, from 3½ to 6 square yards (3–5 sq m). Populations are cyclic, peaking every three or four years, the highs usually produced when breeding continues through the winter. The following summer, overcrowding is severe, food supplies are depleted, and the lemmings become nervous and hyperactive. They then disperse separately, but do not migrate en masse like the Scandinavian species. Lemmings are preyed upon by the major avian and mammalian predators.

Northern Bog Lemming
Synaptomys borealis

Description: *Brown above;* grayish below. *Very short, bicolored tail.* Eyes small. Rust-colored hairs at base of ears; ears hidden in fur. *Upper incisors lightly grooved.* Males often have patch of white hair on flank. L 4⅝–5½" (118–140 mm); T ¾–1" (19–27 mm); HF ⅝–⅞" (16–22 mm); Wt ¾–1⅛ oz (23–34 g).

Similar Species: Southern Bog Lemming lacks rust-colored hairs at base of ears. Other voles have ungrooved incisors.

Breeding: More than 1 litter per year, between March and October; 2–6 young per litter; gestation 21–23 days.

Sign: Grass cuttings 1½" (40 mm) long; spherical grass nests 8" (200 mm) in diameter.
Scat: Green fecal pellets piled in runways.

Habitat: Bogs, spruce woods, alpine meadows, and tundra.

Range: Southern Alaska and British Columbia through s Yukon, s Northwest Territories (Mackenzie district), and eas across Canada; south in U.S. to extreme n Washington, n Idaho, and nw Montana.

This colonial lemming is active day and night year-round. Its diet includes sedges, grasses, and leafy plants. Like many of the voles, the Northern Bog Lemming uses surface runways and burrow systems. It places summer nests within burrows, and winter nests on the ground surface, under the snow.

136 **Southern Bog Lemming**
Synaptomys cooperi

Description: *Brown above;* silvery below. *Tail very short,* brownish above, lighter below. Ears and eyes inconspicuous. *Upper incisors shallowly grooved.* L 4⅝–6⅛" (118–154 mm); T ½–⅞" (13–24 mm);

HF ⅝–⅞″ (16–24 mm); Wt ¾–1¾ oz (21–50 g).

Similar Species: Northern Bog Lemming has rust-colored hair at base of ears. Other voles have ungrooved upper incisors.

Breeding: May breed year-round, but not in winter in some areas; 1–8 young per litter; gestation 23–26 days; one female in captivity produced 6 litters in 26 weeks.

Sign: Grass cuttings about 3″ (75 mm) long.
Nest: Globular grass nest, up to 7″ (175 mm) in diameter, with 2–4 entrances, among vegetation.
Scat: Bright green fecal pellets.

Habitat: Grassy meadows; sometimes ne forests; often dry broom sedge; seldom bogs.

Range: Southeastern Manitoba east to Newfoundland, south to Kansas, ne Arkansas, w North Carolina, and ne Virginia.

Although called a bog lemming, this species seldom occurs in bogs. Grass and clover form the bulk of its diet; it also sometimes eats fungi and algae. The Southern Bog Lemming lives in a system of subsurface runways and burrows about 6 inches (150 mm) below the ground; it also commonly uses the runways of other small mammals. Its globular grass nest may be placed in an underground chamber or aboveground. Populations of these animals fluctuate greatly; in some years they are abundant, while at other times they seem nonexistent in the same area. Although few records of predation are available, many mammals, birds of prey, and snakes feed upon this lemming.

Collared Lemming
"Varying Lemming," "Pied Lemming," "Banded Lemming"
Dicrostonyx groenlandicus

Description: Buff-gray above with black stripe down back; buff-gray to white below. *Pale or tawny ruff (or "collar") across throat. Tail*

short. All white in winter, with
elongated hardened footpad below
middle 2 foreclaws for digging in snow.
Small ears hidden in fur. L 5¼–6⅜"
(132–162 mm); T ⅜–¾" (10–20 mm);
HF ⅝–⅞" (15–22 mm); Wt 2–4 oz
(56–112 g).

Similar Species: Brown Lemming lacks black stripe
down back. Northern and Southern bog
lemmings have grooved upper incisors.
Most voles have longer tails.

Breeding: Breeds March–September, with several
litters of 1–7 young per year; both
parents care for young; gestation
probably 21 days.

Sign: Surface nests of balls of grass 6–8"
(150–200 mm) wide, usually among
rocks or in snowdrifts.

Habitat: Arctic tundra.

Range: Northern and w Alaska; n tier of
Canadian provinces east to Hudson Bay.

The Collared Lemming and the
Labrador Collared Lemming are the
only North American rodents that turn
white in winter. The Collared Lemming
is often found together with the Brown
Lemming in runways beneath the snow.
This species' tunnel systems include rest
and latrine areas, as well as nesting
chambers as deep as the permafrost line.
In summer, the Collared Lemming's diet
includes grasses, sedges, bearberry, and
cotton grass. Twigs and buds of willow
are eaten year-round. This lemming is a
major food source for Arctic carnivores,
with Arctic Foxes, wolves, Wolverines,
snowy owls, gulls, and jaegers among its
many predators.

---

137 Labrador Collared Lemming
Dicrostonyx hudsonius

Description: In summer, adults *nearly uniform buffy
gray* above, with narrow gray-black
stripe down back; *underparts gray, usually
strongly tinged with buffy. Pale or tawny
ruff across throat. Tail very short.* Very

small ears hidden in fur. *In winter, all white,* with elongated footpads for digging in snow. Young darker, more brownish. L average 5⅞″ (148 mm); T average ¾″ (20.3 mm); HF average ⅞″ (22.2 mm); Wt ⅞–2¼ oz (25–65 g).

Similar Species: Collared Lemming is very similar, but occurs to north and west of Hudson Bay. Brown Lemming lacks black stripe down back. Northern and Southern bog lemmings have grooved upper incisors. Most voles have longer tails.

Breeding: Reproductive period usually March through early September. 2 or 3 litters per year, each of 1–8 young; gestation 21 days.

Sign: Surface nests of balls of grass 6–8″ (150–200 mm) wide, usually among rocks or in snowdrifts.

Habitat: Arctic tundra.

Range: Labrador and n Quebec; also Belcher Islands, in Northwest Territories.

The habits of this species are probably similar to those of the Collared Lemming, which also turns white in winter. The forefoot has a small thumb with a rudimentary nail. The larger third and fourth claws develop a bulbous portion in winter, with partially split tips, giving the appearance of a double claw. The lower portion of this "claw" then grows to exceed the upper. As spring approaches, the forefoot returns to its normal appearance. Major predators are the Arctic Fox and the snowy owl.

Richardson's Collared Lemming
Dicrostonyx richardsoni

Description: *Adult buffy gray above, with black stripe down back;* underparts buffy gray. Pale or tawny ruff across throat. *Tail short.* Ears very short. L 3¾–5¾″ (95–145 mm); T ⅜–⅝″ (11–15 mm); HF ⅝–¾″ (15–18 mm); E ⅛–¼″ (3–6 mm); Wt ⅝–2⅜ oz (19–69 g).

Similar Species: Collared Lemming is very similar, differing primarily genetically. Brown Lemming lacks black stripe down back. Northern and Southern bog lemmings have grooved upper incisors. Most voles have longer tails.

Breeding: Breeds spring–fall; 2 or 3 litters per year.

Sign: Surface nests of grass, 6–8″ (150–200 mm) wide, usually among rocks or in snowdrifts.

Habitat: Arctic tundra.

Range: Southwestern Northwest Territories and n Manitoba.

The habits of Richardson's Collared Lemming are similar to those of the Collared Lemming. While this species' nests are usually found in clumps of grass or under stones, it will also make nests in Caribou carcasses. It feeds on green vegetation and other plant matter including blueberries.

OLD WORLD RATS AND MICE
Subfamily Murinae

The three species of Old World rats and mice accidentally introduced into the New World—the Norway Rat, the Black Rat, and the House Mouse—have long tails and large ears. Like most rodents, they are primarily nocturnal and are active throughout the year. They are distinguished from native New World rats and mice by their molariform teeth, which have three rows of cusps rather than two.

These animals are among the major scourges of mankind, damaging millions of dollars of goods each year and carrying diseases such as plague and typhus, and bacteria that can cause food poisoning. It has been estimated that disease-causing organisms borne by rats may have cost more lives in the past ten centuries than all the wars and battles ever fought. Today, however, these animals can also be said to have a beneficial aspect, for the white rat bred from the Norway Rat and the white mouse bred from the House Mouse play important roles in biological and medical laboratory research.

119 Norway Rat
"Common Rat," "Brown Rat," "Water Rat," "Sewer Rat"
Rattus norvegicus

Description: Brownish gray above; *grayish below. Scaly tail* slightly less than half total length, darker above than below. Small eyes. Prominent ears. L 12⅜–18⅛" (316–460 mm); T 4¾–8½" (122–215 mm); HF 1⅛–1¾" (30–45mm); E ⅝–1" (15–25 mm); Wt 6⅞–17 oz (195–485 g).

Similar Species: Black Rat has proportionally longer tail (more than half its total length). Woodrats have white underparts.

Breeding: Breeds year-round; like other rodents, sometimes mates within hours of givin birth; gestation 21–26 days; female ma bear up to 12 litters per year of 2–22 young (usually about 5 litters of 7–11 young). Young born hairless and blind, open eyes at 2 weeks; are weaned at 3– weeks.

Sign: Dirty smudges around holes in walls a other passageways; damaged goods; tunnels under boards or in riverbanks; pathways from tunnels to food supplie *Scat:* Dark brown, about ¾″ (20 mm) long.

Tracks: Hindprint to 2″ (50 mm) long, with 5 long, pointed toes printing; narrower foreprint less than half as lon with 4 toes printing. Sets of tracks in alternate pairs, with hindprints placed approximately 3″ (75 mm) ahead of foreprints.

Habitat: Farms, cities, and many types of huma dwellings; in summer, often cultivated fields.

Range: Southern Canada and entire continenta U.S.; Pacific Coast north to Alaska.

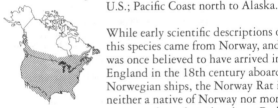

While early scientific descriptions of this species came from Norway, and it was once believed to have arrived in England in the 18th century aboard Norwegian ships, the Norway Rat is neither a native of Norway nor more common there than elsewhere. Probab originating in Central Asia, from the 16th to the 18th century it spread acrc Europe both overland and aboard trading vessels; it arrived in North America about 1776 in boxes of grain brought by the Hessian troops hired b Britain to fight the American colonists The Norway Rat makes a network of interconnecting tunnels 2 to 3 inches (50–75 mm) across, up to 1½ feet (45 mm) deep, and 6 feet (2 m) long. Such network contains one or more chambe for nesting or feeding, one or more ma entrances, and several escape exits. Th

rat digs by cutting roots with its incisors, freeing dirt, pushing it under its body with its forefeet and out behind with its hindfeet, then turning around and continuing to push the dirt out with its head and forefeet. Its vocalizations include squeaks, whistles, and chirps. This loosely colonial rat is a good climber and swimmer. Omnivorous, it feeds on meat, insects, wild plants, seeds, and stored grain, contaminating with its droppings what it does not eat. It will kill chickens and eat their eggs. Food shortages and unfavorable climates sometimes limit this rat's reproductive potential, resulting in fewer and smaller litters. When food is abundant, females may produce a dozen litters in a year. At two years, females stop breeding and males' reproductive powers diminish. Snakes, owls, hawks, skunks, weasels, Minks, and dogs are predators. The life span is about three years, but few Norway Rats live that long. If local populations become severely overcrowded, mass migrations may occur. In 1727, hordes of Norway Rats were observed crossing the Volga River in Russia; though millions drowned, many survived. The German nursery legend about the Pied Piper of Hamelin, who rid the town of rats by musically charming them into the Weser River, where they drowned, probably grew from observations of rat migrations. Rats are a major carrier of diseases such as typhus, spotted fever, tularemia, and bubonic plague, and their destructive powers are enormous. As well as eating grain and ruining property, rats have started fires by gnawing matches and caused floods by tunneling through dams. The white rats used in laboratories are specially bred albino strains of the Norway Rat.

120 **Black Rat**
"Roof Rat," "Ship Rat"
Rattus rattus

Description: Brownish or grayish above; underparts
grayish to whitish, but not white. *Scaly
sparsely haired tail* uniformly dark;
longer than half total length. Prominen
ears. L 12¾–17⅞" (325–455 mm);
T 6⅜–10⅛" (160–255 mm); HF 1⅛–
1⅝" (30–40 mm); E ⅝–1" (17–27 mm
Wt 4–12⅜ oz (115–350 g).

Similar Species: Norway Rat has tail proportionally
shorter (less than half total length). Ric
rats have tail darker above than below.
Woodrats have white underparts.

Breeding: Breeds year-round; several litters per
year, each of 2–8 young; gestation
21–26 days.

Sign: Similar to sign of Norway Rat.

Habitat: Mainly around seaports and buildings;
sometimes in natural habitats.

Range: Southern and coastal U.S.; inland in
West, as far north as w Nevada; east of
Rockies to e Arkansas, w Kentucky, n
Alabama, n Georgia, and most of Nort
Carolina and Virginia. Most abundant
South, along Atlantic Coast north to e
Maine, and along Pacific Coast to
extreme sw British Columbia.

The Black Rat occurs in a great many
varieties and races, or subspecies, of
which few are actually black, despite
the common name. Believed to have
come from Southeast Asia, this species
spread through Europe centuries ago,
long before the arrival of the Norway
Rat. It appeared in Central and South
America in the mid-16th century,
evidently carried there aboard Spanish
ships; it arrived in North America with
the early colonists at Jamestown in
1609, and gradually spread across the
continent. Formerly much more
common, it has often been displaced b
the slightly larger and more aggressive
Norway Rat; this may be because the

Black Rat does better in tropical climates and the Norway Rat in temperate climates, rather than because of overt competition. As Black Rats are far more common than Norway Rats on ships, they continue to be reintroduced at seaports. Excellent climbers, in the South they live in the upper stories of buildings; they also make nests in tangled vines and in trees. Omnivorous but partial to grain, the Black Rat does enormous damage in docks and warehouses, contaminating with its droppings what it does not eat. Like other rats, it carries a number of diseases, including bubonic plague, which is transmitted by its fleas. Snakes, owls, dogs, and cats are its chief predators.

79 House Mouse
Mus musculus

Description: *Grayish brown above; nearly as dark below.* Tail dusky above and below; nearly hairless; less than half the body length. *Ungrooved incisors.* L 5⅛–7¾" (130–198 mm); T 2½–4" (63–102 mm); HF ½–⅞" (14–21 mm); E ⅜–¾" (11–18 mm); Wt ⅝–¾ oz (18–23 g).

Similar Species: Deer mice (*Peromyscus* species) have white underparts. Harvest mice have grooved incisors.

Breeding: Gestation 18–21 days; several litters per year, each of 3–16 young; reproduces spring through fall in North, year-round in South.

Sign: Musky odor. In buildings: small dark droppings, damaged materials, holes in insulation, and shredded nesting material; in fields: small dark droppings, small holes in the ground.

Habitat: Buildings; areas with good ground cover, especially cultivated fields. Uncommon in undisturbed or natural habitats.

Range: Pacific Coast south from Alaska through w and s Canada and throughout all of continental U.S.

The House Mouse originated in Asia and spread throughout Europe many centuries ago. In the early 16th century, it arrived in Florida and Latin America on ships of the Spanish explorers and conquistadores, and about a century later came to the northern shores of North America along with English and French explorers, traders, and colonists. The House Mouse makes its own nest but lives in groups, sharing escape holes and common areas for eating, urinating and defecating. It takes turns grooming its fellows, especially on the head and back, where it is difficult for the animal to groom itself. If the population grows too dense, many females, particularly adolescents, become infertile. A highly migratory existence and rapid rate of reproduction enable the House Mouse to thrive; it takes advantage of situations not readily available to other species, including cultivated fields, which offer a rich if temporary habitat. As a crop develops, the mice move in and have several litters in quick succession, building large populations quickly; when the field is harvested or plowed, they move out. Many perish, many find other fields, and still others invade buildings. Sometimes these migrations assume plague proportions: In 1926–1927, an estimated 82,000 mice per acre (202,000 per ha) wreaked havoc in the Central Valley of California. In such densities, House Mice, though generally timid, have been known to run over people's feet and even to bite. In cultivated fields, some of their actions are beneficial, as they feed heavily on weed seeds, with foxtail grass a favorite, along with caterpillars and other insects; in houses, barns, and storage buildings, they are entirely destructive. These mice eat or their droppings contaminate large quantities of grain and other valuable foodstuffs. Their scientific name derives

from the Sanskrit *musha,* meaning "thief." They chew or shred anything chewable or shreddable, including furniture and wires, and sometimes start fires. They can scurry up rough vertical walls and even pipes; they gnaw holes in walls, floors, and baseboards. Like Black and Norway rats, House Mice can spread disease. In the wild, birds and mammals are predators. Centuries ago, cooked mouse meat was a folk remedy for colds, coughs, fits, and fevers, but it is not recommended today. The white mice used in research laboratories are albinos bred from this species.

JUMPING MICE
Family Dipodidae
Subfamily Zapodinae

Jumping mice are yellowish or reddish mice with very long tails, large hindfeet, and deeply grooved orange incisors. They are good runners and agile jumpers; their long tail helps them to maintain balance in the air. The three species of meadow jumping mice *(Zapus)* probably seldom jump more than 3 to 4 feet (1–1.2 m), but the Woodland Jumping Mouse *(Napaeozapus)* can make spectacular leaps of at least 6 to 8 feet (2–2.4 m). The seeds of grasses and other green plants, berries, insects, and fungi scratched from surface litter are the chief foods of these mice. Primarily nocturnal, they are long and deep hibernators, sleeping six to eight months of the year. During this period, they draw upon reserves of body fat for energy, as they do not store food.

The jumping mice had long been in their own family, the Zapodidae, but are now considered a subfamily of the family Dipodidae, which has 51 species worldwide. Zapodinae, the subfamily to which they belong, has five species worldwide, four in North America.

90 Meadow Jumping Mouse
Zapus hudsonius

Description: Brownish back; *yellowish sides;* white belly. *Long tail with tip usually not white.* 1 small molariform tooth precedes 3 large molariform teeth. L 7⅜–10⅛″ (187–255 mm); T 4¼–6⅛″ (108–155 mm); HF 1⅛–1⅜″ (28–35 mm); Wt ⅜–1 oz (13–28 g).

Similar Species: Woodland Jumping Mouse lacks a reduced first molariform tooth; tip of its tail is white. Western Jumping Mouse generally occurs west of the range of this species.

Breeding: Births peak in June, July, and August; most females produce 2 litters of 2–9 (usually 5 or 6) young per year; females born late the previous year probably produce 1 litter in July; gestation 19 days.

Sign: Match-length cuttings of grass with grass head parts on top; also grasses that have been "topped."

Habitat: Mainly moist fields; but also brush, brushy fields, marshes, stands of touch-me-not *(Impatiens),* and woods with thick vegetation, especially where Woodland Jumping Mouse does not occur.

Range: Southern Alaska and across most of southern tier of Canadian provinces; e Wyoming east through ne U.S. and west and south to ne Oklahoma and ne Georgia.

When startled from a hiding place, the Meadow Jumping Mouse may take a few long jumps of 2 to 3 feet (60–90 cm), then shorter ones, but generally it soon stops and remains motionless, which is its best means of eluding predators. By the end of October, nearly all individuals of this species have retired to hibernation nests of shredded grass in a protected place, such as under a board or clump of grass, or in a bank, mound, hollow log, or other raised area. Apparently many later-born, smaller mice, unable to accumulate adequate fat reserves, perish during hibernation. In spring, caterpillars, beetles, and other insects constitute about half of this rodent's diet. It feeds on the seeds of grasses and many other green plants as they ripen, either by cutting off grasses at the base, then pulling the stem down to reach the head, or by climbing a stalk, cutting off the head, and carrying it to the ground in its mouth. In summer and fall, the subterranean fungus *Endogone* forms about an eighth of this animal's diet. The Meadow Jumping Mouse stores no food, but

in the two weeks before entering
hibernation, it puts on about ¼ ounce
(6 g) of fat. Males emerge first, in late
April or early May; one to two weeks
later, females emerge and the first
mating takes place.

91 Western Jumping Mouse
Zapus princeps

Description: Yellow sides; dark band down middle of
back; belly white, sometimes tinged
yellow. *Long tail* darker above, whitish
below. *Very large hindfeet.* 4 upper
molariform teeth on each side, the first
reduced in size. L 8½–10¼″ (215–260
mm); T 5–6⅜″ (126–160 mm);
HF 1⅛–1⅜″ (28–34 mm); Wt ½–1⅜
oz (15–38 g).

Similar Species: Meadow Jumping Mouse is smaller and
occurs farther north and east. Pacific
Jumping Mouse generally occurs to the
west of this species.

Breeding: Mates soon after emergence from
hibernation; probably 2 or 3 litters per
year, each of 3–9 young, born
June–July, sometimes much later.
Newborn weighs less than ¹⁄₁₆ oz (1 g).

Sign: 1″ (25 mm) cuttings of grass in piles
with flower parts on top.

Habitat: Variable: primarily moist fields,
thickets, and woodlands, especially
where grasses, sedges, or other green
plant cover is dense; grassy edges of
streams, ponds, and lakes.

Range: Western North America from s Yukon,
all of British Columbia, s Alberta, s
Saskatchewan, and extreme sw
Manitoba south through w and ne
Montana, ne South Dakota, w Oregon,
w California, n Nevada, Utah, ec
Arizona, and c New Mexico.

Although it often runs on all four feet,
the Western Jumping Mouse may make
a series of jumps 3 to 5 feet (90–150 cm)
long when startled from its hiding
place, then "disappear" by remaining

motionless in a new one. It can swim
and climb. Between August and
October, the mouse fattens, then retires
to a winter burrow to hibernate, curled
into a ball with its long tail wrapped
around its body. It lives in burrows that
it digs itself or in those of other animals.
While it does not make runways, it will
use those of other animals. Although
primarily nocturnal, like other jumping
mice, this species has been observed in
the daytime foraging in salal thickets 1
foot (30 cm) or more off the ground.
The seeds of grass, dock, and many
other green plants, strawberries,
blueberries, blackberries, and
subterranean fungi are its chief foods.
Young are born in a spherical nest of
interwoven, broad-leaved grasses or, if in
a bog, of sphagnum moss, constructed
in a depression in the ground. Those
young born too late in the year to
acquire sufficient fat reserves probably
perish during hibernation. When
alarmed, this species may drum on the
ground with its tail. Owls and Bobcats
are its chief predators.

Pacific Jumping Mouse
Zapus trinotatus

Description: Yellow sides; dark band down middle of
back; belly white, sometimes tinged
yellow. *Long tail* darker above, whitish
below. *Very large hindfeet.* 4 moliform
teeth per side, the first reduced. L 8¾–
9⅜″ (221–238 mm); T 5⅛–5⅞″ (131–
149 mm); HF 1¼–1⅜″ (31–34 mm).

Similar Species: Deer mice have much shorter tails and
smaller hindfeet.

Breeding: Mates soon after emergence from
hibernation; gestation about 21 days; 1
or 2 litters per year.

Habitat: Primarily moist fields, thickets, and
woodlands, especially where grasses,
sedges, or other green plant cover is
dense; grassy edges of streams, ponds,
and lakes.

Range: Southwestern British Columbia south
         through nw California.

The habits of the Pacific Jumping
Mouse are presumably very similar to
those of the Western Jumping Mouse.
All three *Zapus* species look much alike
and live in similar habitats. All feed
heavily on underground fungi.

---

92 Woodland Jumping Mouse
Napaeozapus insignis

Description: Brightly colored: brownish back; *orange
              sides* with scattered dark hairs; white
              underparts. *Long tail with white tip.*
              Forefeet small; hindfeet long. 3 large
              molariform teeth. L 8–10½″ (204–256
              mm); T 4½–6⅜″ (115–160 mm);
              HF 1⅛–1¼″ (29–33 mm); Wt ⅝–⅞ oz
              (17–26 g).

Similar Species: The 3 species of meadow jumping mice
                  (*Zapus* species) generally lack white tail
                  tip, are more yellowish, and have a smal
                  molariform tooth preceding the 3 large
                  molars.

Breeding: Mates April–May; most females have 1
           litter per year of 2–7 young (usually 4 o
           5); often a second, smaller litter later,
           especially in the South; gestation 23–29
           days.

Habitat: Coniferous and hardwood forests in cool
          moist environments, especially with
          dense, green vegetation.

Range: Southeastern Canada from Manitoba to
        Labrador, south to ne Minnesota and e
        West Virginia; also in Allegheny
        Mountains to ne Georgia.

Although a good swimmer and a
superlative jumper, the Woodland
Jumping Mouse usually walks; when in
a hurry, it will make great leaps of 6 to 8
feet (2–2.4 m). It lives in a burrow,
either constructing its own or taking
over that of another small mammal. Its
consumption of the subterranean fungus
Endogone, which forms about a third of

its diet and provides water as well as food, also benefits the fungus; excretion of the fungal spores aids in their dispersal, and the mouse's digestive juices are probably essential for their germination. Seeds, caterpillars, beetle larvae, and berries are other major foods. This animal begins to put on a layer of fat about two weeks before hibernation, which starts in September or October. Emergence is in April or May, with males preceding females by a few days, and breeding soon follows. Young are born hairless and blind, and are weaned at 34 days. If frightened, the Woodland Jumping Mouse drums its tail on the ground. Predators include skunks, weasels, Minks, Bobcats, owls, rattlesnakes, and domestic cats.

NEW WORLD PORCUPINES
Family Erethizontidae

Worldwide there are four genera and 12 species in this family, which exists in North and South America. The outstanding feature of these large, heavy-set rodents is their body covering which is formed by spines with small overlapping barbs. Their feet are modified for life in the trees. The single North American species is described below.

230, 231 Common Porcupine
Erethizon dorsatum

Description: Large, chunky body, with *high-arching back,* short legs. *Long guard hairs* on fron half of body; black or brown in the East yellowish in the West. *Quills on rump an tail.* Feet have unique soles with small, pebbly-textured fleshy knobs and long, curved claws; 4 toes on forefeet, 5 toes on hindfeet. L 26–37" (648–930 mm); T 5⅞–11¾" (148–300 mm); HF 3⅜–4⅞" (86–124 mm); Wt 7¾–40 lb (3.5–18 kg).

Breeding: Mates October–November; 1 young born May–June after gestation of about 7 months.

Sign: Large, irregular patches of bark stripped from tree trunks and limbs with neatly gnawed edges, plentiful tooth marks; "niptwigs" often strewn on the ground. *Scat:* Similar to that of deer, but the pellets vary greatly in size and appearance, depending on food and season: in winter, rough-surfaced irregular pellets, sometimes connected when food is relatively soft; in summer, pellets softer, more elongate, often curved, sometimes segmented. Accumulations found at entrances to crevice shelter or cave among rocks and, in winter, at base of single tree where porcupine has fed for a long time.

Tracks: Distinctive; toe in, almost like a Badger's; pebbled knobs on soles leave stippled impression; long claw mark far ahead of oval main prints. Foreprint, including claw marks, about 2½" (65 mm) long; hindprint well over 3" (75 mm) long, usually but not always printing ahead of foreprint. Stride is short and waddling, with prints 5–6" (125–150 mm) apart; straddle up to 9" (230 mm) wide. In snow, feet may drag or shuffle, connecting prints. Trail occasionally blurred, as if swept by a small broom, as belly brushes ground and stiff, heavy tail swishes from side to side in waddling walk.

Habitat: Deciduous, coniferous, and mixed forests; also, in West, dry, scrubby areas with scattered trees.

Range: Most of Canada and w U.S. south to Mexico; in East, south to Wisconsin, n half of Michigan, and most of Pennsylvania, New York, and New England.

The solitary Common Porcupine is active year-round, though in bitter cold it may den up in a hole in a rocky bluff, sometimes with others of its species. Primarily nocturnal, it may also rest by day in a hollow tree or log, underground burrow, or treetop, for it is an excellent, if slow and deliberate, climber. Yet the animal occasionally falls; about 35 percent of museum skeletons examined showed healed fractures. On the ground, it has an unhurried, waddling walk, relying on its quills for protection against more agile predators, although it prefers to retreat or ascend a tree rather than confront an enemy. Long claws are one adaptation for climbing, as they hold on to crevices in bark. The stiff, backward-pointing quills of the underside of the tail help keep the animal from slipping back down a tree. The tail is repeatedly lifted and lowered as the animal descends, serving as an

exploratory organ. The Common
Porcupine has about 30,000 quills on its
body; these are modified hairs, solid at
tip and base, hollow for most of the
shaft, and loosely attached to a sheet of
voluntary muscles beneath the skin.
The porcupine's generic name means
"one who rises in anger," and while a
porcupine cannot throw its quills at an
enemy, when forced to fight it erects
them, lowers its head, and lashes out
with its tail. If the tail strikes the
enemy, the loosely rooted quills detach
easily and are driven forcefully into the
victim, whose body heat causes the
microscopic barbs on the end of each
quill to expand and become even more
firmly embedded. Wounds may fester,
or the quill, depending on where it
enters, may blind the victim or prevent
it from eating. The short tail quills are
the most dangerous and can be driven
deeply into the flesh; if they strike a
vital place, they can even cause death.
Cutting the end off the hollow quill
releases air pressure and allows it to be
more easily withdrawn. But the
Common Porcupine is not aggressive;
if left alone, it will not attack. A black
line runs up the middle of the tail and
expands on the lower back, and there is
white on the head. This contrasting,
black-and-white "warning" pattern is
not as obvious as that of a skunk, yet in
the same way it apparently
communicates to a potential adversary
that it should keep its distance. The
porcupine attempts to keep the black-
and-white warning coloration of its
backside toward potential enemies. If
attack is imminent, it gives a second
warning, tooth-chattering for up to half
a minute, which may be repeated many
times. In addition, the porcupine can
produce a strong, pungent odor, which
in confined quarters such as a porcupine
den can cause eyes and nose to water. If
all else fails, the porcupine erects its
quills. A few carnivores, notably the

Fisher, are adept at flipping a porcupine over to attack its wiry-haired but unquilled underside, but even a Fisher occasionally receives a fatal injury.

A strict vegetarian, the Common Porcupine feeds on leaves, twigs, and such green plants as skunk cabbage, lupines, and clover in spring; in winter, it chews through the rough outer bark of various trees, including pines, fir, cedar, and hemlock, to get at the inner bark (cambium), on which it then mainly subsists. Like many herbivores, the porcupine has bacteria in its digestive tract containing enzymes that help to digest the cellulose and other substances not sufficiently broken down by normal digestive enzymes. This animal has favored feeding trees that can be recognized by their cropped and stunted upper branches and bare wood. Another unmistakable sign of porcupines, often littering the ground under favorite trees, are "niptwigs," terminal branches of trees that have been cut off and their leaves or buds eaten. In the Catskill Mountains of New York State, sugar maples are popular with these animals, as are young beech trees (not canopy beeches), basswood, apple, and aspen. They also eat young ash leaves, acorns, and beechnuts when available. Porcupines gain weight in summer and lose it in fall. Fond of salt, the Common Porcupine has a great appetite for wooden tool handles that have absorbed human perspiration through use. The animal may kill trees by stripping away the bark, and its gnawing may damage buildings and furniture.

The Common Porcupine mates mainly in October and November; it is most vocal at this time, giving a variety of squeaks, groans, and grunts. The jocular answer to the question "How do Porcupines mate?" is "Carefully." In fact, mating occurs in the same fashion as with other mammals, but not until

the female is sufficiently aroused so that she relaxes her quills before raising her tail over her back and presenting herself. Males may fight over females, and courtship is elaborate. Prior to mating, the male squirts high-pressure jets of urine over the female. After a gestation of nearly seven months—an unusually long period for a rodent—the single young is born in May or June in a very precocious condition. Quills are well formed but not injurious to the mother, as the baby is born headfirst in a placental sac and its short quills are soft; they harden within half an hour. The life span of the Common Porcupine is seven to eight years. In addition to the Fisher, predators include the Mountain Lion, Bobcat, and Coyote. Porcupine quills, both natural and dyed, are used in brilliantly executed decorative quillwork by Native Americans, who also eat the animal's flesh.

NUTRIA
Family Myocastoridae

This family contains but a single species, the Nutria, or "Coypu." It is native to South America and the Caribbean, and has been introduced into North America.

218, 219 Nutria
"Coypu"
Myocastor coypus

Description: *Large aquatic rodent. Brown above;* somewhat paler below. Long, scaly, sparsely haired *rounded tail.* Muzzle and chin whitish. Ears and eyes small. Incisors dark orange, protruding beyond lips. Hindfeet longer than forefeet, with inner 4 toes webbed. Male larger than female. L 26–55" (67–140 cm); T 11¾–17¼" (30–44 cm); HF 4⅜–5½" (11–14 cm); Wt 5–25 lb (2.3–11.3 kg).

Similar Species: Common Muskrat is smaller, with vertically flattened tail (higher than wide). American Beaver is larger, with large, horizontally flattened tail.

Breeding: Breeds spring through fall in North, year-round in South; gestation about 130 days; litter of 1–11 young (usually 4–6).

Sign: Feeding platforms of aquatic vegetation and debris, 5–6' (1.5–2 m) across; well-established trails of flattened vegetation along marshy shores. At dusk, a chorus of pig-like grunts may be heard.
Scat: Elongated droppings found on feeding platforms or shoreline.
Tracks: Similar to those of Common Muskrat, but larger, with hindfeet showing webbing between inner 4 toes; easily confused with American Beaver's when the beaver's fifth toe webbing does not print.

Habitat: Marshes, ponds, and streams.

Range: Widely introduced, especially in Southeast, but also in Washington, n Oregon, scattered locations in the Great Plains, s New Jersey, and Maryland.

The generic name for this nocturnal aquatic rodent comes from the Greek words for "mouse" *(mys)* and "beaver" *(kastor)*. Although it often feeds on land when disturbed the Nutria returns to the water, often with a loud splash. It can remain submerged for several minutes and often floats just under the surface with only eyes and nose exposed. Its nest of plant materials is made either in a burrow dug in a riverbank, with an entrance aboveground or in shallow water; or in the burrow of another animal or in the house of an American Beaver or a muskrat. Feeding on almost any terrestrial or aquatic green plant, the Nutria also consumes some grain, sometimes dipping its food into water before eating. It may occupy feeding platforms to rest and to avoid terrestrial predators. Like the lagomorphs (rabbits hares, and pikas), the Nutria reingests fecal pellets in order to digest food more completely while at rest. Courtship features much chasing, fighting, and biting. The young are born fully haired and with eyes open. Within 24 hours, the well-developed young swim with their mother and nibble green plants. Introduced in Louisiana in the 1930s for their fur, many Nutrias escaped from captivity during the hurricane floodings of the 1940s, multiplied enormously in the wild, and became more important than muskrats to Louisiana's trapping industry. More than 1 million Nutrias were taken in Louisiana in 1984 and 1985. However, the fur industry collapsed in the late 1980s, and in 1991 only 134,000 Nutrias were taken. The long, coarse guard hairs are used in making felt for hats; the soft belly fur, resembling that of the American Beaver, is used for coats and linings. When populations are high, Nutrias may undermine stream banks, deplete wild vegetation, and raid rice and other crops.

CARNIVORES
Order Carnivora

Most of the larger predators of small vertebrates, as well as the domestic dog and domestic cat, are in the order Carnivora, which encompasses 11 families with 271 species worldwide. Eight families and 53 species occur in North America north of Mexico. The name of the order, which means "meat-eating," is somewhat misleading, for while many carnivores live mainly on freshly killed prey, a number of others are omnivorous, eating a great deal of vegetation. Carnivores generally live on land, but some, such as seals, sea lions, and the Walrus, spend part or nearly all of their time in water. Except for the European Badger *(Meles meles)*, no carnivores truly hibernate, though several remain in their dens for long periods during exceptionally cold weather.

Terrestrial (or land) carnivores, which include many of the animals most familiar to humans, vary greatly in size and appearance. North American terrestrial species range from the world's smallest land carnivore, the Least Weasel, which is less than 2 ounces (37–50 g) of sinuous energy, to the largest, a subspecies of the Grizzly Bear known as the Alaskan Brown Bear, which can reach a massive 1,700 pounds (770 kg). All North American carnivores have three pairs of relatively small incisors above and below, and large, strong canines. Most have one annual litter of offspring, which are born blind and require a relatively extended period of parental care. The dwindling of habitat and with it the decrease in prey species, as well as persecution by humans, have diminished the populations of many North American carnivores, such as the Mountain Lion, the Ocelot, and the Red Wolf. Some

of these animals are recognized as endangered or threatened, and receive some measure of federal protection or regulation.

The aquatic carnivores, seals, sea lions, and the Walrus, were formerly placed in their own order, Pinnipedia, meaning "fin feet." Recent studies suggest that they represent a single branch from the same ancestral group that gave rise to the other Carnivora, and they are now classed with them. There are 34 species of pinnipeds worldwide, although one species, the Caribbean Monk Seal, is assumed to be extinct. Of these, 15 species in three families occur in North America north of Mexico. Pinnipeds are well adapted for their aquatic life, with large sizes that enable them to maintain body heat, torpedo-like bodies, legs modified as flippers, and, in some species, nostrils and ears that close when submerged. In all but the adult Walrus, fine fur covers the skin and is molted at regular intervals. The fur provides insulation against the cold, as does a layer of blubber directly beneath the skin. This fat also increases buoyancy and acts as a reservoir of energy when the animals fast during molting and breeding periods. Skillful and graceful swimmers and divers, some aquatic carnivores can descend to almost 2,000 feet (600 m) and remain submerged for up to 45 minutes. Their heartbeats slow to about 10 percent of the normal rate, and their blood vessels constrict, sending the available oxygen mainly to the brain and heart, and reducing the amount of heat lost through the skin. These animals have five flat, elongated bones in the forefeet and hindfeet that are connected by webbing, which gives a fin-like appearance. Their tails are reduced or vestigial. Aquatic carnivores are clumsy on land and ice, to which they return to give birth and to molt. Delayed implantation occurs in most

species, extending the gestation period so that mating and the birth of young can occur when the herds assemble at the same time and place each year. There is usually a single pup, rarely two, which grows rapidly on its mother's rich milk (about 50 percent fat; cow's milk is approximately 3.5 percent fat). These animals usually have an adult coat near the end of the first summer, reach sexual maturity at two to five years of age, and may live up to 40 years. Although unlike terrestrial carnivores in appearance, these aquatic animals have similar dentition; they feed solely on flesh, mainly fish and aquatic invertebrates. An aquatic carnivore's age can be determined by counting the number of rings around the root of an extracted tooth, as one ring is added each year.

WOLVES, FOXES, AND THE COYOTE
Family Canidae

This family is represented worldwide by 34 species; nine species, including the domestic dog, reside in the U.S. and Canada. All North American canids have a dog-like appearance that is characterized by a lithe body; a long, narrow muzzle; erect, triangular ears; long, slender legs; and a bushy tail. Wolves are social animals, traveling in packs with a clearly established dominance hierarchy, while Coyotes, hunt in smaller groups or pairs, and foxes are solitary. Although their sight and hearing are keen, canids rely heavily on their acute sense of smell, both for scenting prey and "reading" scent posts marked with the urine or secretions of other animals. Most canids chase down their prey, but foxes also rely on stalking and pouncing. Nearly all species have one annual litter.

North American canids have had mixed success in modern times. As a result of years of persecution, wolves have decreased greatly in number. The Coyote however, has thrived alongside humans, increasing both in number and range. When a wolf is reported where none has been seen for at least a century, it is usually a Coyote or a dog, rarely a coydog (a dog-Coyote hybrid). The canid family has given rise to the domestic dog *(Canis familiaris),* which most mammalogists believe to be derived from the wolf, while others suggest it is descended from an extinct wild dog quite different from any living canid.

Domestic Dog
Canis familiaris

Description: Color and appearance vary with breed. *Tail arches upward.* "Coydog," a hybrid of Coyote and domestic dog, is usually

Coydog

fairly large and somewhat resembles
Coyote, but has relatively short, thick
snout, and lacks dark vertical line on
foreleg. Weights and measurements of
dogs vary considerably depending on
breed. L 14⅛–57″ (36–145 cm); T 5–
20″ (13–51 cm); Wt usually 2–175 lb
(1–79 kg); record is 330 lb (150 kg).

Similar Species: Most dogs, even feral individuals,
cannot be confused with any native
mammal, although larger, gray, long-
haired dogs such as malamutes and
German shepherds can be confused with
wolves and Coyotes. Gray Wolf is
usually much larger and runs with its
tail held horizontal, not arched upward.
Coyote runs with tail straight but
slanting downward, has longer, thinner
snout and dark line running up foreleg;
behavior of Coyote is different enough
from dogs that it is usually clearly
identifiable.

Breeding: Female enters heat twice a year, usually
late winter–early spring and fall; heat
lasts about 12 days. Litter of 2–10
young born after gestation of 63 days.

Sign: Tracks with 4 clawed toes; in zigzag
pattern.

Habitat: Most domestic dogs live in and around
buildings, but wild individuals often
use heavy vegetation or some natural
shelter.

Range: Throughout North America.

Domestic species usually do not appear
in field guides, but the domestic dog
has been present for a long time, and
some individuals have become feral.
The dog was apparently domesticated
from the wolf; the earliest known
record is from Iraq, some 12,000 years
ago. Feral dogs eat many different
items; some of the most important
are mice, rabbits, and other small
vertebrates. While dogs often prey on
domestic stock, wildlife species such
as the Coyote usually get the blame.
Feral dogs can make an impact on
native wildlife populations, and can

become dangerous to humans when they form packs. Although the domestic dog and the Coyote will mate, coydogs are not very common and usually do not breed.

280–282, 284   **Coyote**
Canis latrans

Description: *Grizzled gray* or orangish gray above, with buff underparts. Long, rusty or yellowish legs with dark vertical line on lower foreleg. *Bushy tail with black tip.* Ears prominent. Slender, pointed snout. Nose pad to 1″ (2.5 cm) wide. Ht 23–26″ (58–66 cm); L 3′5″–4′4″ (105–132 cm); T 11¾–15¼″ (30–39 cm); HF 7⅛–8⅝″ (18–22 cm); Wt 20–40 lb (9.1–18.2 kg); a very large individual may reach 55 lb (25 kg).

Similar Species: Gray and Red wolves are larger, with larger nose pads; both hold tail horizontal. "Coydogs," hybrids of Coyote and domestic dog, especially shepherd mixtures, are larger, usually lack dark vertical line on lower foreleg, and have relatively shorter and thicker snouts.

Breeding: Mates February–April; 1 litter of 1–19 young born April–May, in a crevice or underground burrow.

Sign: Long meandering, habitually used hunting trails, or runways. Tracks and scat most often seen where runways intersect or on a hillock or open spot, vantage points where Coyotes linger to watch for prey.
Dens: Favored sites are riverbanks, well-drained slopes, sides of canyons, and gulches. Den mouths usually 1–2′ (300–600 mm) wide, often marked by mound or fan of earth and tracks.
Scat: Typically canine; often full of hair and usually deposited on runway.
Tracks: Similar to domestic dog's, but in a nearly straight line; hindfeet usually come down in foreprints. 4 toes print, all with claws. Foreprint about 2⅜″ (62

mm) long, hindprint slightly smaller. Straddle about 6–8″ (150–200 mm). Stride 13″ (330 mm) when walking, 24″ (600 mm) when trotting, 30″ (750 mm) or more when running; much wider gaps signify leaps.

Habitat: In West, open plains; in East, brushy areas.

Range: Generally common throughout e and s Alaska, s and w Canada, and all of w U.S., but has extended its range into entire U.S.

The Coyote's scientific name means "barking dog"; its common name comes from *coyotl,* the name used by Mexico's Nahuatl Indians. The best runner among the canids, the Coyote cruises normally at 25 to 30 mph (40–50 km/h), getting up to 40 mph (65 km/h) for short distances, and can make 14-foot (4.25 m) leaps. Tagged Coyotes have been known to travel great distances, up to 400 miles (640 km). The Coyote runs with its tail down, unlike the domestic dog (tail up) or wolves (tail straight). In feeding, the Coyote is an opportunist, eating rabbits, mice, ground squirrels, pocket gophers, and other small mammals, as well as birds, frogs, toads, snakes, insects, and many kinds of fruit. Carrion from larger animals, especially deer, is an important food source in winter. The Coyote usually hunts singly, but may combine efforts with one or two others, running in relays to tire prey or waiting in ambush while others chase the quarry toward it. Sometimes an American Badger serves as an involuntary supplier of smaller animals: While it digs for rodents at one end of a burrow, a Coyote waits to pounce on any that emerge from an escape hole at the other end. The Coyote stalks like a pointer, "freezing" before it pounces, and chases down larger prey, such as Snowshoe Hares and cottontails. A strong swimmer, it does not hesitate to enter water after prey. Like most carnivores,

Coyotes have maternal dens for raising the young, but do not have permanent homes. The typical den is a wide-mouthed tunnel, 5 to 30 feet (1.5–9 m) long, terminating in an enlarged nesting chamber. The female may dig her own den, take over and enlarge a fox or badger burrow, or use a cave, log, or culvert. If the den area is disturbed, the female will move the pups to a new home. The Coyote may pair for several years or even for life, especially when populations are low. Its vocalizations are varied, but the most distinctive—given at dusk, dawn, or during the night—consists of a series of barks and yelps, followed by a prolonged howl and ending with short, sharp yaps. This call keeps the band alert to the locations of its members and helps to reunite them when separated. One call usually prompts other individuals to join in, resulting in the familiar chorus heard at night in the West (although noticed increasingly in the East, as Coyotes grow in number there). Barking alone, with no howling, seems to be a threat display employed in defense of a den or a kill. Although captive Coyotes have lived for 18 years, and one was known to live in the wild for 14½ years, most individuals probably live only 6 to 8 years. Predators once included Grizzly Bears, Black Bears, Mountain Lions, and wolves, but with declining populations of these animals, they are no longer a threat. Humans are the major enemy, purportedly killing Coyotes to protect livestock, as Coyotes are accused, often unjustly, of killing lambs, pigs, and poultry. In the 1970s and 1980s, Coyote pelts became quite valuable, but since the collapse of the fur industry, there has been little demand for them. Despite years of being trapped, shot, and poisoned, Coyotes have maintained their numbers and continue to increase in the East.

285–289 Gray Wolf
"Timber Wolf"
Canis lupus

Description: A *very large* canid, usually grizzled gray, but showing great variation in color, ranging from white to black. *Long, bushy tail with black tip. Nose pad 1¼" (33 mm) wide.* Long legs. Male larger than female. Ht 26–38" (66–97 cm); L 4'3"–6'9" (130–205 cm); T 13¾–19¾" (35–50 cm); HF 8⅝–12¼" (22–31 cm); Wt 57–130 lb (26–59 kg).

Similar Species: Coyote is much smaller, with smaller nose pad; holds tail at downward angle. Wolf carries tail straight out. Domestic dog's tail curves upward.

Breeding: Mates February–March; 1 litter of 1–11 young born April–June, after gestation of 63 days.

Sign: Heavily used trails where prey is plentiful.
Scent posts: Scraped patches of ground, rocks, or stumps marked with urine.
Den: Entrance 20 × 25" (500 × 650 mm), with burrow 4–30' (1.2–9 m) deep, often marked by fan or mound of earth and sometimes by bones or scraps of prey brought for pups.
Rest or "rendezvous" area: Usually a grassy expanse—dry marsh, old burn, or meadow with plenty of mice and a wide view—where pack gathers when den is vacated. It is marked by scat, tracks, well-worn trails, and sometimes diggings where food has been cached and later uncovered.
Scat: Resembles a large domestic dog's but often contains hair of prey.
Tracks: Similar to domestic dog's but larger. Foreprint 4¼–5⅛" (110–130 mm) long; hindprint slightly smaller. Walking stride 30" (750 mm). Hindfeet usually come down in forefeet prints (as in Coyote), producing tracks in straight line; dogs' prints show zigzag pattern. Wolf tracks are about 9–14" (230–360 mm) from one toe tip to the one behind it, Coyote's about 6–8" (150–200 mm).

Habitat: Open tundra and forests.
Range: Once most of North America, now only
Alaska, Canada, n Washington, n Idaho,
n Montana, Isle Royale National Park in
Lake Superior, and ne Minnesota.
Thought to be gone from Wisconsin by
1960 but had returned by 1981; 15
estimated present by 1986.

A social animal, the Gray Wolf lives in
packs of 2 to 15, usually 4 to 7, formed
primarily of family members and
relatives, although sometimes two packs
combine. The strongest male of a pack is
normally the leader. Within the pack
hierarchy, there are male and female
hierarchies. The alpha male is dominant
over the entire pack, both males and
females, and he can go anywhere he
wants and take anything he wants. The
beta male is second in the hierarchy, but
the alpha female may be dominant over
some of the lower-ranking males.
Dominance behaviors are open mouth
and bared teeth, hair raised along the
back, and ears erect and pointed forward.
The Gray Wolf runs with a bounding
gait with its tail held horizontal. Except
perhaps for the Caribou, this wolf
travels more often and for greater
distances than any other terrestrial
animal. A large pack's territory covers
100 to 260 square miles (260–675 sq
km), traveled at regular intervals over
such runways as animal trails, logging
roads, and frozen lakes. Territories may
overlap slightly, but packs usually avoid
one another. If food is adequate, a pack
may use the same range for many
generations. The pack works together
on a hunt, either chasing down its
victim, usually by slashing tendons in
the hindlegs, or forcing it to circle back
to waiting pack members. Usually
hunting at night, this animal feeds
primarily on large mammals, including
Moose, Caribou, and deer, but will also
catch smaller ones, and sometimes eats
berries, birds, fish, and insects. The wolf

ideally needs about 3¾ pounds (1.7 kg)
of food per day, but it can go for two
weeks or more without food. It gorges
itself when food is plentiful. Wolves
gain an advantage over large prey in
deep, crusted snow: The crust is able to
support wolves, but a Moose or deer
may break through and be unable to run
efficiently enough to escape. However,
a healthy deer can outrun wolves
unless hindered by the snow. Myth
notwithstanding, wolves do not often
attempt long chases, although
occasionally they may run for several
miles. They can gallop and bound over
short distances at speeds of more than
30 mph (50 km/h), but if they cannot
capture running prey within about
1,000 yards (meters), they usually
abandon the attempt. Wolves try to
surprise a prey animal and cut off its
retreat, or ambush it. When an animal
runs away, the wolf's instinct is to dash
after it, but it is soon apt to give up such
a chase unless the pursued creature stops
and starts intermittently. Wolves, like
many other carnivores, test prey: A
Moose that stands and fights often
persuades a pack to seek an easier
quarry, but one that first defends itself
and then runs—perhaps because it is
injured, sick, defective, or very young or
old—signals the possibility of its
defeat, and the pack often continues to
pursue it. In deep snow, wolves will
follow humans or dog teams in order to
take advantage of the tamped trail. The
Gray Wolf makes various whines, yelps,
growls, and barks, although not all
wolves are capable of barking. The most
common bark is short, harsh, and
uttered in a brief series. Howls, used
to keep the pack together, may be at a
constant pitch, may rise and fall, or
rise and break off abruptly, sounding
anywhere from dismal to beautiful and
haunting. The howl of one wolf seldom
lasts more than five seconds, but others
in the pack may take it up, producing a

chorus that continues until the animals tire or disperse. Many wolves answer a bark—even a poor human imitation of one—by howling. They produce a communal howl sometimes in early morning, but most often in the evening when it may stimulate the urge to hunt. It usually begins with a few sharp barks by one or more pack members, which are often followed by a low, rather querulous howl that, in turn, stimulate steadier, louder communal howling tha dies away after a few minutes, often ending with a few more barks before th pack goes off together to hunt. A wolf separated from its pack may give the "lonesome howl"—a shortened call tha rises in pitch and then dies away plaintively. If answered, the wolf switches to a "location," or "assembly," howl—deep, even, and often punctuated by barks. A wolf seldom calls when actually chasing prey; rather it stops to vocalize in order to maintain contact and sometimes calls to signal arrival at an ambush point toward whic other pack members will then attempt to drive the quarry.

The Gray Wolf normally does not use a shelter except as a maternity den. During a blizzard, it curls its tail over its paws and nose and soon becomes covered with snow, which provides insulation from the cold. Wolves matur in their second year, but usually do not breed until their third. They mate for life. The den is in an enlarged chamber without nesting material, and usually o high ground near water. There may be several entrances, up to 2 feet (30–60 cm) in diameter. The tunnel usually is about 4 to 17 feet (1.2–5.2 m) long, though it may be longer. The same den may be used for years, although the young may be moved between dens at times. All members of the pack help to care for the young. At three weeks, pup emerge to play near the den entrance guarded by an adult. The rest of the

pack leaves the den site in late afternoon or at dusk to hunt, usually returning the next morning with food. Pups jump and bite at the snouts and throats of returning hunters, stimulating them to regurgitate undigested meat, which the pups and their guardian devour. At about two months, the young are moved from the natal den to one of a series of "rendezvous sites," usually near water, where they play and learn to hunt. By late summer, they begin to hunt with the adults. Some juveniles leave the adults at one year, some at two years, when mature. The life span of the Gray Wolf is 10 to 18 years. Humans are the wolf's only important predator, persecuting the animal unnecessarily, even in national parks. It is considered endangered throughout the U.S., except in Minnesota, where it is classified as threatened. While humans have long feared wolves, there have been only three documented attacks by wolves on humans in North America, and none led to a fatality.

283 Red Wolf
Canis rufus

Description: Coat primarily gray, interspersed with blackish hairs; sometimes yellowish or reddish hairs, especially on legs and underparts. *Nose pad more than 1" (25 mm) wide.* Ht 15–16⅛" (38–41 cm); L 4'7–5'5" (140–165 cm); T 13½–16½" (34–42 cm); HF 8¼–9⅞" (21–25 cm); Wt 40–80 lb (18–36 kg).

Similar Species: Coyote is smaller, with smaller nose pad and hindfoot. Gray Wolf is larger.

Breeding: Mates February or March of third year; 1 litter of 2–10 young born April–early June; gestation about 2 months.

Sign: *Scat:* Typically canine, often full of hair. *Tracks:* Prints 4½" (115 mm) long. 4 toes with claws printing; outer toes smaller than middle ones.

Habitat: Prairies, brush, forested areas, coastal plains, swamps, and bayous.

Range: Originally se U.S. north to Missouri,
Indiana, and Virginia, and west to
Texas, but extirpated from all but sw
Louisiana and se Texas by 1900.
Reintroduced into Alligator River
National Wildlife Refuge, in Dare
County, North Carolina; also Bull
Island, South Carolina; Horn Island,
Mississippi; and St. Vincent Island,
Florida. Range map shows dispersion in
mid-20th century.

By 1980, the misnamed Red Wolf had
been completely eliminated from its
natural range. Persecution by humans,
the destruction of habitat, and
interbreeding with Coyotes and domestic
dogs contributed to its downfall. In the
early 1970s, the U.S. Fish and Wildlife
Service trapped a number of Red Wolves
from the wild and initiated a breeding
program, preventing the animal's
extinction. The wolf was introduced on
islands to allow natural matings to occur,
producing wolves for introductions
elsewhere. The first several pairs of Red
Wolves were introduced at the Alligator
River refuge in North Carolina in 1987;
by 1992 a total of 36 wolves had been
introduced there. In the spring of 1992,
at least 19 wolves were present in the
refuge, including six of the original 36;
the rest were pups from at least seven
litters that had been produced in the
wild by that time. As of January 1992,
there were 151 Red Wolves in existence:
26 in the wild, six on propagation
islands, and the rest in captivity. The
Red Wolf can exist alongside humans,
but is much less adaptable than the
Coyote. It is possible that the Coyote
has had success in the East because it
fills a niche formerly occupied by the
Red Wolf.
The Red Wolf is mainly nocturnal, but
tends to be more diurnal in winter. It
makes its den along stream banks, in
enlarged burrows of other mammals,

under stumps, or in culverts or hollow logs. The usual social unit is a mated pair, sometimes with an extra male, but these animals are very social and often form temporary packs. The howl of the Red Wolf is closer to that of the Coyote in sound than to that of the Gray Wolf. The Red Wolf pursues the White-tailed Deer, but also regularly eats rabbits and hares, and even smaller prey on occasion, such as small rodents and birds. At Alligator River, deer, raccoons, and Marsh Rabbits are the most important foods, followed by several species of mice.

290, 291	**Arctic Fox** *Alopex lagopus*
Description:	In summer, *bluish brown or grayish, with white underparts;* in winter, white or creamy white. Rare blue phase is dark blue-gray in summer, pale blue-gray in winter. Tail bushy. Nose pad, eyes, and claws black. Snout blunt. Ears rounded and short. Juvenile has white facial markings. Male larger than female. Ht 9⅞–11¾″ (25–30 cm); L 30–36″ (75–91 cm); T 10⅝–13¾″ (27–35 cm); HF 5⅛–6⅜″ (13–16 cm); E usually 2¼″ (5.6 cm); Wt 5½–8¾ lb (2.5–4 kg).
Similar Species:	Other foxes have larger, more pointed ears. Common Gray Fox has black "mane" on top of tail. Red Fox has white tail tip.
Breeding:	Mates February–May; 1 litter of 6–12 young born April–July; gestation 49–57 days.
Sign:	*Den:* Burrow in sandy bank on hillside with several entrances about 1′ (300 mm) wide. *Scat:* Small, narrow, roughly cylindrical, tapered at one end, similar to that of Red and Common Gray foxes; may be pinkish bleaching to white when crustaceans are in diet. *Tracks:* In straight line. Similar to those of Common Gray Fox. Foreprints and

hindprints 2–3½″ (50–90 mm) long. 4 clawed, closely spaced toes print; sometimes slightly blurred because of densely haired footpads, especially in winter.

Habitat: Tundra at edge of northern forests; in winter, great distances out on ice floes. Rare blue phase usually in areas without permanent snow cover, where its color would be disadvantageous.

Range: Western and n Alaska east across n Canada south to n and e Northwest Territories, ne Alberta, n Manitoba, and n Quebec; rarely farther south.

The adaptations of the Arctic Fox to its subzero habitat include a compact body with short legs and ears (heat loss occurs mostly through extremities), dense fur, and thickly haired footpads, which insulate against the cold and provide traction on ice. Winter fur develops in October: The coat thickens, and the new hairs are much lighter, providing camouflage against snow and ice. Individual Arctic Foxes are relatively solitary, but several will congregate around a large carcass or a dump. The Arctic Fox dens in a bank or hillside and in winter may tunnel into a snowbank. The female builds a summer nest in a new den with several entrances. Both parents care for the young until the family disperses in mid-August; the male brings food for the young and his mate and guards the family. In summer, when prey is abundant, the Arctic Fox gluts itself but still keeps hunting. It stores surplus food by clawing through the soil and deep-freezing it on the permafrost below, or caching it in a crevice or under a rock. In winter, it follows Polar Bears, eating leftovers from their kills and, if food is scarce, even their droppings. It also eats voles, ground squirrels, young hares, birds, bird eggs, fish, berries, occasionally young seals or sea lions, and, in winter, carrion. Populations of Arctic Foxes

summer

winter

peak about every four years, paralleling and following by roughly one year those of lemmings, a chief food. When prey is scarce and hunting territories are consequently expanded, young and old foxes unable to defend their territories may be forced to emigrate. They travel south up to several hundred miles; few return north, as many are trapped for their pelts. This fox's long-haired pelt, especially that of the rare blue phase, is highly valued and much sought. Arctic Foxes are raised commercially on islands off Alaska.

292, 293	**Kit Fox** "Swift Fox" *Vulpes velox*
Description:	*Yellowish buff above,* whitish below. *Tail with black tip* and often a black spot at upper base. *Feet light-colored.* Ears large, triangular. Dark spots below eye. Ht 11¾" (30 cm); L 24–31" (60–80 cm); T 9–11¾" (23–30 cm); E 3⅛ –3¼" (7.9–8.4 cm); Wt 3⅛–6 lb (1.4–2.7 kg).
Similar Species:	Red Fox has white tail tip and black feet. Common Gray Fox is larger, darker, and has black "mane" on top of tail.
Breeding:	Mates January–February; 1 litter of 3–5 young born March–April.
Sign:	*Den:* Underground, with 3 or 4 entrances 8″ (200 mm) wide; usually with mound of earth at entrance; sometimes scattered with small bones or scraps of prey. *Scat:* Small, irregular, cylindrical. *Tracks:* Similar to those of Common Gray Fox but usually less than 1½" (40 mm) long. All prints show 4 toes and claws.
Habitat:	Shortgrass prairies and other arid areas.
Range:	Southern Alberta, Saskatchewan, and Manitoba south through e Montana and Wyoming, ne Colorado, the Dakotas, Nebraska, w Kansas and Oklahoma, e New Mexico, and n Texas; s Oregon and sw Idaho south through Nevada and w Utah to s California and Arizona.

Once considered separate species, the
Kit Fox and the Swift Fox have been
combined as *Vulpes velox.* For short
distances, this species can run as fast as
25 mph (40 km/h), hence the name Swift
Fox. Solitary and mostly nocturnal, it eats
rabbits, ground squirrels, rats and mice,
birds, insects, grasses, and berries; in
winter, it caches food under snow. This
species usually mates for life. It excavates
its own den or enlarges an American
Badger or marmot den in open country.
The young are born in a chamber 3 feet
(1 m) belowground with no nesting
material. The calls of the Kit Fox include
a shrill yap, and several whines, growls,
and purrs. This animal tries to evade
predators such as domestic dogs and
Coyotes by entering a burrow or suddenly
changing the direction of its flight.

294, 295 **Red Fox**
Vulpes vulpes

Description: *Rusty reddish above;* white underparts,
chin, and throat. *Long, bushy tail with
white tip.* Prominent pointed ears. Backs
of ears, lower legs, and feet black. Color
variations include a black phase (almost
completely black), a silver phase (black
with silver-tipped hairs), a cross phase
(reddish brown with a dark cross across
shoulders), and intermediate phases; all
have white-tipped tail. Ht 15–16″
(38–41 cm); L 35–41″ (90–103 cm);
T 13¾–17″ (35–43 cm); HF 5¾–7″
(14.6–17.8 cm); E 3–3½″ (7.7–8.9 cm);
Wt 7⅞–15 lb (3.6–6.8 kg).

Similar Species: All other North American canids lack
conspicuously white-tipped tail.
Common Gray Fox, often confused with
this species, is reddish on backs and
outsides of ears, around neck, and on
sides of belly, but otherwise is gray and
lacks white tail tip.

Breeding: Mates January–early March; 1 litter of
1–10 kits born March–May in maternity
den; gestation 51–53 days.

Sign: *Den:* Maternity den, commonly an enlarged Woodchuck or American Badger den, usually in sparse ground cover on slight rise, with view of all approaches; may also be in stream bank, slope, or rock pile, or in hollow tree or log. Main entrance in earthen mound, typically up to 3′ (1 m) wide, slightly higher, with littered fan or mound of packed earth; 1–3 smaller, less-conspicuous escape holes. Den well marked with excavated earth, cache mounds where food is buried, holes where food has been dug up, and scraps of bones and feathers.

Scat: Similar to that of Common Gray Fox, but sometimes paler.

Tracks: Similar to those of Common Gray Fox, but usually slightly larger, with smaller toeprints. Foreprint about 2⅛″ (55 mm) long, hindprint slightly smaller, narrower, more pointed. Often blurred, especially in winter when feet are heavily haired, with lobes and toes less distinctly outlined than those of Common Gray Fox. In heavy snow, tail may brush out tracks.

Habitat: Varied: mixed cultivated and wooded areas, and brushlands.

Range: Most of Canada and U.S. except for far north, nw British Columbia, much of w U.S., and s Florida.

Regarded as the embodiment of cunning, the Red Fox is believed by many field observers merely to be extremely cautious and, like other canids, capable of learning from experience. Even when fairly common, it may be difficult to observe, as it is shy, nervous, and primarily nocturnal (though it may be abroad near dawn or dusk or on dark days). It eats whatever is available, feeding heavily in summer on vegetation, including corn, berries, apples, cherries, grapes, acorns, and grasses, and in winter on birds and mammals, including mice, rabbits, squirrels, and Woodchucks. Invertebrates such as grasshoppers,

crickets, caterpillars, beetles, and crayfish compose about one-fourth of its diet. The hearing of the Red Fox differs from that of most mammals in that it is most sensitive to low-frequency sounds. The fox listens, for example, for the underground digging, gnawing, and rustling of small mammals. When it hears such sounds, it frantically digs into the soil or snow to capture the animal. The Red Fox is cat-like in stalking its prey. It hunts larger quarry, such as rabbits, by moving in as close as possible, then attempting to run the prey down when it bolts. The Red Fox continues to hunt when full, caching excess food under snow, leaves, or soft dirt. It probably finds its caches by memory, aided by smell, although other animals sometimes find them first. An adult fox rarely retires to a den in winter. In the open, it curls into a ball, wrapping its bushy tail about its nose and foot pads, and at times may be completely blanketed with snow. Adults usually are solitary until the mating season, which begins (usually in late January or February) with nocturnal barking. The maternity den is established shortly after mating and abandoned by late August when families disperse. The female usually cleans out extra dens, to be used in case of disturbance, but the same one may be occupied for several years. Upon birth, most pups already show the white tail tip. When about one month old, the young play aboveground and feed on what is brought to them by their parents and sometimes by "helper" foxes, unbred females or female progeny that have not left the territory. Food is given to the first pup that begs for it, and some young may die in years when nourishment is scarce. At first, the mother predigests and regurgitates meat, but soon she brings live prey, enabling the kits to practice killing. Later the young begin to hunt with

the parents. The kits disperse at about seven months, males traveling away up to 150 miles (240 km) or more, females usually remaining closer. Adults also disperse, remaining solitary until the next breeding season. The adult Red Fox has few enemies other than humans and the automobile, but rabies, mange, and distemper are also problems. In the mid-18th century, Red Foxes were imported from England and released in New York, New Jersey, Maryland, Delaware, and Virginia by landowners who enjoyed hunting them with hounds. The Red Foxes in most of the U.S. are combined strains derived from the interbreeding of imported foxes with native races, which, encouraged by settlement, gradually expanded their range south from Canada. For years, unregulated trapping and bounty payments took a heavy toll on Red Foxes, but the collapse of the fur industry and the abolishment of most bounty payments have improved matters. With poultry farms made nearly predator-proof, farmers kill fewer foxes as well. The Red Fox in the U.S. may be expanding its range, although competition with the Coyote, which is also spreading farther afield, may have a restraining effect.

296, 297 Common Gray Fox
Urocyon cinereoargenteus

Description: *Grizzled gray above, reddish on lower sides, chest,* and back of head; throat and belly white. *Tail* similarly colored, but *has black "mane" on top* and black tip. Legs and feet rust-colored. Ears prominent. Ht 14⅛–15″ (36–38 cm); L 31–44″ (80–113 cm); T 8⅝–17⅜″ (22–44 cm); HF 3⅞–5⅞″ (10–15 cm); E 2¾–3¼″ (7–8 cm); Wt 7¼–13 lb (3.3–5.9 kg).

Similar Species: Red Fox has white tail tip. Kit Fox is smaller and has yellowish-buff fur with black tail tip.

Breeding: Mates January–April; 1 litter of 1–7 young born March–May; gestation 53 days.

Sign: Tree and scent posts marked with urine; noticeable on snow as spattered stains and melting.
Caches: Heaped or loosened dirt, moss, or turf. Dug-up cache holes are shallow and wide, since foxes seldom bury very small prey except near the den in whelping season.
Den: Entrance size varies considerably, as most dens are in natural cavities; snagged hair or a few telltale bone scraps occasionally mark entrance; rarely, conspicuous mounds like those of the Red Fox. Several auxiliary or escape dens nearby.
Scat: Small, narrow, roughly cylindrical, usually sharply tapered at one end; darker than Red Fox's, particularly where wild cherries abound.
Tracks: When in straight line, similar to those of a very large domestic cat, except that 4 nonretractile claws may show. Sharper than those of Red Fox, but often smaller with larger toes. Foreprint about 1½″ (37 mm) long; hindprints as long, slightly narrower. Hind heel pad may leave only a round dot if side portions fail to print. Fox digs in when running, leaving claw marks even in hard ground, where pads do not print.

Habitat: Varied; more often in wooded and brushy habitats than Red Fox.

Range: Throughout e U.S. east from e North and South Dakota, Nebraska, Kansas, and Oklahoma; in West, w Oregon, California, s Nevada, s Utah, Colorado, Arizona, New Mexico and most of Texas.

Although active primarily at twilight and at night, the Common Gray Fox is sometimes seen foraging by day in brush, thick foliage, or timber. The only American canid with true climbing ability, it occasionally forages in trees and often takes refuge in them, especially

in leaning or thickly branched ones.
The Common Gray Fox feeds heavily
on cottontail rabbits, mice, voles, other
small mammals, birds, insects, and
much plant material, including corn,
apples, persimmons, nuts, cherries,
grapes, pokeweed fruit, grass, and
blackberries. Grasshoppers and crickets
are often a very important part of the
diet in late summer and autumn.
Favored den sites include woodlands
and spaces among boulders on the
slopes of rocky ridges. This fox digs if
necessary, and it sometimes enlarges a
Woodchuck burrow, but it prefers to
den in clefts, small caves, rock piles,
hollow logs, and hollow trees, especially
oaks. Occupied in the mating season,
dens are seldom used the rest of the
year. The male Common Gray Fox helps
tend the young, but does not den with
them. The young are weaned at three
months and hunt for themselves at four
months, when they weigh about 7
pounds (3.2 kg). This fox growls, barks,
or yaps, but is less vocal than the Red
Fox. Other than humans, who shoot,
trap, and run over Common Gray Foxes,
this species has few enemies. Bobcats,
where abundant, and domestic dogs may
kill a few. Rabies and distemper are
important diseases.

BEARS
Family Ursidae

This family consists of nine species, distributed in the Americas and Eurasia The three species found in the U.S. and Canada— the Black Bear, the Grizzly Bear, and the Polar Bear—are the largest terrestrial carnivores. Ranging from 200 to 1,700 pounds (270–770 kg), they have powerful, densely furred bodies; small, rounded ears; and small eyes set close together. While their vision is poor, their sense of smell is keen. They have five claws on each foot and, like humans, walk on the entire sole with the heel touching the ground. Bears tend to be omnivorous, eating leaves, twigs, berries, fruit, and insects as well as small mammals. Commonly believed to hibernate, bears enter a protected area and sleep away the harshest part of winter, but their sleep is not deep and their temperature falls only a few degrees below normal. North American bears produce one litter every other year. Although the bears mate in late spring or early summer, at which time the eggs are fertilized, six or seven months may pass before the embryos become implanted in the uterine wall, after which they develop rapidly. The young are born while the female is in her den for the winter. When born, bears are the size of rats, generally weighing only ½ to 2 pounds (225–900 g), which makes the magnitude of their eventual growth greater than that of all other mammals except marsupials.

All North American bears can be dangerous in the following situations: when accompanied by cubs, when surprised by the sudden appearance of humans, when approached while feeding, guarding a kill, fishing, hungry, injured, or breeding, and when familiarity has diminished their fear of humans, as in some Canadian and U.S.

parks. Once distributed over much of North America, bears have been eliminated from most areas of human habitation.

336–339 Black Bear
Ursus americanus

Description: In the East, *nearly black;* in the West, *black to cinnamon,* with white blaze on chest. A "blue" phase occurs near Yakutat Bay, Alaska, and a nearly white population on Gribbell Island, British Columbia, and the neighboring mainland. *Snout tan* or grizzled; in profile *straight* or slightly convex. 3 pairs of upper incisors equal in size. Male much larger than female. Ht 3–3′5″ (90–105 cm); L 4′6″–6′2″(137–188 cm); T 3–7″ (7.7–17.7 cm); HF 9–14⅝″ (23–37 cm); Wt 203–587 lb (92–267 kg).

Similar Species: Grizzly Bear is usually larger, and has generally somewhat concave facial profile, muscular hump above shoulder region, longer foreclaws, and outer pair of upper incisors much larger than 2 inner pairs.

Breeding: Mates June–early July; litter of 1–5 (usually 2) young born January–early February; birth weight not much over 7 oz (200 g).

Sign: *Feeding signs:* Logs or stones turned over for insects; decayed stumps or logs torn apart for grubs; ground pawed up for roots; anthills or rodent burrows excavated; berry patches torn up; fruit-tree branches broken; rejected bits of carrion or large prey, such as pieces of skin, often with head or feet attached. *Trees:* Scarred with tooth marks, often as high as a bear can reach when standing on its hindlegs; higher, longer claw slashes, usually diagonal but sometimes vertical or horizontal. In spring, rub marks and snagged hair on furrowed or shaggy-barked trees used repeatedly and by several bears as shedding posts, to rub away loose hair and relieve itching.

Scat: Usually dark brown, roughly cylindrical, sometimes coiled, similar to that of domestic dog; often showing animal hair, insect parts, seeds, grasses, root fibers, or nutshells. Where bears have fed heavily on berries, scat may be liquid black mass.

Trails: Those used by generations of bears are well worn, undulating, and marked with depressions.

Tracks: Broad footprints; 5 toes print on all feet, although innermost, smallest toe may fail to register. Foreprints 4" (100 mm) long, 5" (125 mm) wide, turned in slightly at front. Hindprints 7–9" (180–230 mm) long, 5" (125 mm) wide. Individually prints (especially hindprints) look as if made by a flat-footed human in moccasins, except that large toe is outermost. In soft earth or mud, claw indentations usually visible just in front of toe marks. Bears have a shuffling gait; hind tracks and front tracks are paired, with hind track several inches before front track on same side. Stride about 1' (300 mm) long. Sometimes, when walking slowly, hindprints either partially or completely overlap foreprints; when running, hindfeet brought down well ahead, with gaps of 3' (900 mm) or more between complete sets of tracks.

Habitat: In East, primarily forests and swamps; in West, forests and wooded mountains seldom higher than 7,000' (2,100 m).

Range: Most of Alaska southeastward through Canada to n Minnesota, Wisconsin, and Michigan, and Maritimes south through New England, New York, Pennsylvania, and Appalachian Mountains to Florida; south on West Coast through n California; Rocky Mountain states to Mexico. Also in Arkansas and se Oklahoma.

Although primarily nocturnal, this uniquely North American bear may be seen at any time, day or night. It occupies a range usually of 8 to 10

square miles (20–25 sq km), although sometimes up to 15 square miles (40 sq km). The home range of the male is about double the size of that of the female. The Black Bear's walk is clumsy, but in its bounding trot it attains surprising speed, with bursts up to 30 mph (50 km/h). A powerful swimmer, it also climbs trees, either for protection or food. Although this animal is in the order Carnivora, most of its diet consists of vegetation, including twigs, buds, leaves, nuts, roots, fruit, corn, berries, and newly sprouted plants. In spring, the bear peels off tree bark to get at the inner, or cambium, layer. It rips open bee trees to feast on honey, honeycombs, bees, and larvae, and will tear apart rotting logs for grubs, beetles, crickets, and ants. A good fisher, the Black Bear often wades in streams or lakes, snagging fish with its jaws or pinning them with a paw. It rounds out its diet with small to medium-size mammals or other vertebrates. In the fall, the bear puts on a good supply of fat, then holes up for the winter in a sheltered place, such as a cave, crevice, hollow tree or log, under the roots of a fallen tree, or, in the Hudson Bay area, sometimes in a snowbank. Excrement is never found in the wintering den. The bear stops eating a few days before retiring, but then consumes roughage, such as leaves, pine needles, and bits of its own hair. These pass through the digestive system and form an anal plug, up to 1 foot (30 cm) long, which is voided when the bear emerges in the spring. Sows mate during their third year, with most producing one tiny cub the first winter, two on subsequent breedings. While the mother sleeps in the den, the almost naked newborns nestle into her fur. The mother often lies on her back or side to nurse, but sometimes sits on her haunches, with cubs perched on her lap, much like human infants; they may nurse for about a year. The female Black

Bear is not receptive to males while nursing. This bear is mainly solitary, except briefly during the mating season and when congregating to feed at dumps. Bears are often a problem around open dumps, becoming dangerous as they lose their fear of humans; occasionally people have been killed by them. Hunting Black Bears is a popular sport in some areas, both for the flesh (which must be well cooked because of trichinosis) and the hides, used for rugs. The helmets of Great Britain's Buckingham Palace guards are made of the Black Bear's fur. A recent threat to this species is the illegal killing and export of its gall bladders to Asia. The Black Bear is classified by the U.S. government as a threatened species in Texas, Louisiana, and Mississippi.

340–344 **Grizzly Bear**
"Brown Bear," "Alaskan Brown Bear," "Kodiak Bear"
Ursus arctos

Description: *Yellowish brown to dark bown,* often with white-tipped hairs, giving grizzled appearance. *Hump above shoulders.* Facial profile usually somewhat concave. Outer pair of incisors larger than inner 2. Claws of front feet nearly 4″ (10 cm) long. Ht about 4′3″ (130 cm); L 5′11″– 7′ (180–213 cm); T 3″ (7.6 cm); HF 10¼″ (26 cm); Wt 324–1,499 lb (147–680 kg); some individuals to 1,700 lb (700 kg).

Similar Species: Black Bear is smaller, lacks shoulder hump, and has straight or slightly convex facial profile; all 3 pairs of its upper incisors are equal in size.

Breeding: Mates late June–early July; litter of 1– young born January–March. Newborn weighs 1 lb (450 g).

Sign: *Feeding signs:* Shallowly dug depression and high, loose mound of branches, earth, or natural debris heaped over it may conceal cache of carrion or a kill.

Beware of this sign, for a bear will not be far away. Also overturned rocks, torn-up berry patches, and raggedly torn logs. Large, gaping pits indicate that Grizzlies have dug for rodents.

Trees and other signs: Girdled, bark-stripped, clawed, and bitten "bear trees," with largest tooth marks higher than 6' (2 m) and claw slashes perhaps twice as high (marks higher than those made by Black Bears). Hair tufts on trees, which may be polished from rubbing over several seasons. Wide, deep snowslide occasionally gouged by Grizzly sliding down short incline on its haunches.

Bed: Usually in thickets, oval depression about 1' (300 mm) deep, 3' (900 mm) wide, 4' (1,200 mm) long, matted with leaves, needles, or small boughs.

Scat: Usually cylindrical, often more than 2" (50 mm) wide, possibly showing animal hair, vegetation fibers, or husks. May be rounded, or massed in areas where vegetation is the primary food.

Trails: Trampled in tall grass and marked by deep depressions; may undulate.

Tracks: Shaped and placed like those of Black Bear but larger, and with long, relatively straight foreclaws farther ahead of toe pads; hindclaws register only occasionally. Hindprint of large Grizzly may be 10–12" (250–300 mm) long and 7–8" (175–200 mm) wide in front; foreprint often as wide, about half as long. In soft mud, tracks may be larger. Even on hard ground, Alaskan Brown Bears often leave bigger prints, with hind tracks more than 16" (400 mm) long, 10½" (265 mm) wide, and sunk 2" (50 mm) deep. Stride averages 24" (600 mm); may be 8–9' (2.4–2.75 m) during a bounding run.

Habitat: Somewhat open country, usually in mountainous areas; also along coasts and rivers.

Range: Alaska, Yukon, and Northwest Territories south through most of British Columbia and w Alberta to sc Nevada.

Primarily nocturnal, the great, shaggy
Grizzly moves with a low, clumsy walk,
swinging its head back and forth, but
when necessary it can lope as fast as a
horse. Grizzly cubs can climb, though
not as nimbly as Black Bear cubs, but
they lose their climbing ability during
their first year. Omnivorous, the Grizzly
Bear feeds on a wide variety of plant
material, including roots, sprouts,
leaves, berries, and fungi, as well as fish,
insects, large and small mammals, and
carrion. It is adept at catching fish with
a swift snap of its huge jaws, and
occasionally will pin a fish underwater
with its forepaws, then thrust its head
underwater to clasp the catch in its
teeth. It digs insects from rotting logs
and small mammals from their burrows,
sometimes tearing up much ground in
the process. It caches the remains of
larger mammals, such as Elk, Moose,
Mountain Goats, sheep, or livestock,
returning to the cache until all meat is
consumed. When salmon migrate
upstream to spawn, these normally
solitary bears congregate along rivers,
and vicious fights may erupt among
them. More often, they establish
dominance through size and threats,
spacing themselves out, with the
largest, most aggressive individuals
taking the choicest stations. In winter,
Grizzlies put on a layer of fat, as much
as 400 pounds (180 kg) worth, and
become lethargic. They den up in a
protected spot, such as a cave, crevice,
dead tree, or a hollow dug out under a
rock, and will return year after year to a
good den. Not true hibernators, they
can easily be awakened. While the
Grizzly normally avoids humans, it is
the most unpredictable and dangerous
of all bears. A Grizzly in captivity has
lived 47 years, but the life span in the
wild is 15 to 34 years. The Grizzly Bear
is considered a threatened species in the
lower 48 states. Once regarded as a
separate species, the Alaskan Brown

Bear *(U. a. middendorffi)* is a very large subspecies of Grizzly, usually 800 to 1,200 pounds (360–545 kg) at eight or nine years of age, and eventually reaching 1,700 pounds (770 kg), making it the world's largest terrestrial carnivore. This subspecies ranges over the coasts and islands of southern Alaska, including Kodiak Island, where it is known as the Kodiak Bear.

345–349 **Polar Bear**
Ursus maritimus

Description: *White fur;* black nose pad, lips, and eyes. Very small ears. Long neck. Relatively long legs. Male considerably larger than female. Ht about 4′ (120 cm); L 7–11′ (213–335 cm); T 3⅛–5⅛″ (8–13 cm); HF 13″ (33 cm); Wt 925–1,100 lb (420–500 kg).

Breeding: Mates April–May every other year (sometimes every third year); litter of 1–4 (usually 2) young born November–January in winter den, or in excavation in snowbank, ice ridge, or hillside. Birth weight about 2 lb (900 g).

Sign: Feeding signs: Partly eaten seal carcass, with no blubber remaining. Wheeling, calling gulls sometimes mark location of carcass, as may Arctic Foxes or their tracks in association with bear tracks.
Scat: Large, dark, cylindrical, like that of big Grizzly Bear.

Tracks: Similar to those of Grizzly, but rounder and blurred by hair. Hindprint 12–13″ (305–330 mm) long, usually at least 9″ (230 mm) wide. Claws, though sharp enough to grip ice or slippery snow when the bear runs, are short and seldom leave marks in front of prints. Any bear track on an ice floe is that of a Polar Bear.

Habitat: Broken ice packs at northern edge of continent, near North Pole; seldom far inland.

Range: Extreme n Canada and Alaska and Arctic Islands.

Unlike Black and Grizzly bears, which are primarily nocturnal, the Polar Bear may be active at any time of the day or year, searching for prey on long summer days and sometimes on long winter nights. Adaptations to its Arctic habitat include its fur color, which blends with the snowy environment and so provides useful camouflage for capturing prey; its large size, which helps maintain body temperature by reducing surface-heat loss; and its furred feet, which insulate against cold and provide traction on icy surfaces. Because the hairs of its waterproof coat are hollow, they are especially insulating and increase the bear's buoyancy when swimming. An excellent swimmer, it paddles at about 6½ mph (10 km/h) with the front feet only, hindfeet trailing—a trait unique among four-footed land animals—and can remain submerged for about two minutes. While swimming or treading water, it stretches its long neck for a better view, as it does on land. Owing to the scarcity of plants in its icy habitat, the Polar Bear is the most carnivorous North American bear, with canine teeth larger and molariform teeth sharper than those of other bears. An acute sense of smell enables the Polar Bear to locate prey even when it is hidden by snow drifts or ice. It stalks young seals and Walruses, and sometimes adult seals, often by swimming underwater to their ice floes. While hair seals are its staple, it also feeds on fish, birds, bird eggs, small mammals, dead animals (including whales), shellfish, crabs, starfish, and mushrooms, grasses, berries, and algae, when available. The Polar Bear hollows out a winter den in a protected snowbank, where it retires in a lethargic condition. Females den from November to March, during which time they give birth. Males den for much shorter periods, usually from late November to late January, but may be abroad occasionally at any time of the

year. Cubs remain with their mother about a year and a half, denning with her the winter after their birth. The lowlands of Hudson Bay and James Bay, one of the world's largest denning areas for Polar Bears, is the only known region where Polar Bears den in earth rather than in snow. They excavate caves in lake and stream banks and peat hummocks by digging down to the permafrost. This area is unusually far south for the species, and it is believed that they use the permafrost dens again in summer to cool off. No large denning areas have yet been found in Alaska; some Polar Bears in that region may winter in Siberia, drifting across to Alaska on ice floes in spring. Polar Bears are a source of food and hides for Native Americans. There have been several recorded cases of Polar Bears attacking humans.

EARED SEALS
Family Otariidae

Found nearly worldwide except in the
Arctic Ocean and Antarctica, the eared
seals, which include the fur seals and the
sea lions, have small but noticeable
external ears and long, slender bodies.
They have longer necks and longer,
suppler forelimbs than the hair seals
(family Phocidae). Of the 14 species in
this family, four are found in waters off
North America. These aquatic carnivores
are the least specialized for an aquatic
lifestyle and the most agile on land.
Their hind flippers can rotate forward
under the body, and their long fore
flippers turn out at the wrist at right
angles; all four limbs can be used to
move on land in either a somewhat dog-
like walk or a "gallop," in which the fore
flippers and hind flippers work together.
In the water these animals use their fore
flippers for propulsion, while their hind
flippers, held together, serve as a rudder.
The flippers are thick and hairless.
There are well-developed nails on the
middle three digits of the hind flippers.
All members of the family have a coarse
coat of guard hairs, and the fur seals also
have a dense underfur layer. Coat
coloration varies, but no eared seal has
stripes or distinct markings.

Although eared seals remain at sea most
of the year, they are highly gregarious
and form large herds on marine beaches
during the breeding season, which is
mainly in June and July. The much
larger bulls usually maintain harems of
3 to 40 cows. Eared seals mate shortly
after the cows give birth. The single pup
is born hairy, not woolly like newborn
hair seals, and usually begins swimming
at two weeks of age. Those in our region
are weaned in three to four months.

354, 355 Northern Fur Seal
"Alaska Fur Seal"
Callorhinus ursinus

Description: *Very large flippers.* Tiny tail. Small head
with short, pointed nose, large eyes, and
long whiskers. Bull has greatly enlarged
neck; is *blackish above, with massive
grayish shoulders;* reddish below. Female
is gray above, reddish below. Male much
larger than female, beginning at birth.
L male 6′3″–7′3″ (1.9–2.2 m), female
3′7″–4′7″ (1.1–1.4 m); Wt male
330–594 lb (150–270 kg), female
84–119 lb (38–54 kg).

Similar Species: Sea lions and Northern Elephant Seal
are larger. Harbor Seal is spotted.

Breeding: Mates mid-June to mid-July; yearlings
mate in August. 1 glossy black pup born
June–July of following year; birth
weight 10–12 lb (4.5–5.4 kg).

Habitat: At sea most of year; in summer, breeds
on rocky island beaches.

Range: In summer, Point Barrow, Alaska, on
Arctic Ocean; breeds on Pribilof and
Commander islands and vicinity in
Bering Sea, and on San Miguel Island,
off California. In winter, south to San
Diego, California.

The greatest traveler of the aquatic
carnivores, this oceangoing fur seal
sometimes migrates up to 6,200 miles
(10,000 km), returning to land only
during the breeding season. At sea, it is
not gregarious and is rarely seen in
groups of more than three. By day the
animal rests and preens, and may swim
slowly with a flipper waving in the air.
It rests in a "jug handle" position, lying
on its back with the hind flippers bent
up over the belly and held there by a
fore flipper. Small schooling fish of
about 50 species form two-thirds of its
diet, with 10 species of cephalopods,
mostly squid, forming the other third.
In addition, at least 30 species of marine
mammals and oceanic birds are eaten
occasionally. Feeding at night, when fish

swim closer to the surface to eat, this seal forages to depths averaging about 230 feet (70 m); the maximum depth documented was 755 feet (230 m). Dives last for an average of two and a half minutes. This seal's major rookery on the Pribilof Islands is enormous, with more than 1 million seals within a 30-mile (50-km) radius. Older bulls arrive first, in late May and June, and battle savagely to establish territories in the best places, near the water and with best access to cows. Females arrive from mid-June to mid-July and within two days give birth to a pup conceived the previous summer. The biggest bulls may form harems of 40 or more cows (some of these cows may be stolen by bulls farther from shore). The cow mates 8 to 10 days after the pup is born, usually with the bull in whose territory she has given birth. The bull does not feed during the breeding period and may lose up to 20 percent of his weight. After remaining with the pup one week, the cow goes on feeding forays several times weekly, often more than 100 miles (160 km) out to sea. She returns regularly and stays one or two days to nurse the pup with her rich milk. The pup is weaned by October or November. As many as 12 percent of the young die in their first month, often from hookworm which causes anemia. In August, female yearlings come ashore and breed with the bachelor bulls too young and small to maintain their own territories; these bulls will join the territory scramble when 9 or 10 years old. Bulls, battle-scarred and thin from their two-month fast during the breeding season, barely have strength to return to the water in August; most winter in the Gulf of Alaska or south of the Aleutians, but some migrate to Asian waters. Females and juveniles leave the beaches by November; a number of them winter as far south as southern California. This seal's Pribilof breeding grounds were

discovered by humans in 1786; by 1834, overhunting had greatly reduced the numbers of seals, which were prized for their fine, soft fur and their meat and blubber. The Northern Fur Seal was afforded some protection as early as 1835, but in 1988 the U.S. National Marine Fisheries Service designated the Pribilof Islands herd a depleted stock, and made it illegal for anyone to harvest these seals except native peoples who rely on them for subsistence.

353 Guadalupe Fur Seal
Arctocephalus townsendi

Description: Brownish gray above, *with silvery cast on yellowish-gray "mane" on nape;* brownish black below, with chest lighter in adult males. *Snout pointed,* rust-orange on sides. Flippers large. L male to 6′3″ (1.9 m), female to 4′7″ (1.4 m); Wt male to 350 lb (159 kg), female to 100 lb (45 kg).

Similar Species: Sea lions and Northern Elephant Seal are much larger and lack pointed snout. Harbor Seal is spotted. Northern Fur Seal lacks yellowish mane. Juvenile hard to distinguish from young California Sea Lion or young Northern Fur Seal.

Breeding: Breeds June–July; 1 pup born late June–July of following year.

Habitat: Rocky coastal islands.

Range: California's Channel Islands to Cedros Island, off Baja California (Mexico); recently known to breed only on Guadalupe Island, about 180 miles (290 km) west of Baja California.

The Guadalupe Fur Seal prefers a rocky cave for breeding. The bull apparently returns to the same territory year after year and has been known to remain there for a period of 35 to 122 days. While the bull will defend his territory, he occasionally oversees his harem from the water, unlike other eared seals. The vocalizations of this species are unique:

The male has a low bark when he is on his territory and a growl, roar, or cough when threatening other males, and the adult female bawls when interacting with her pup. Guadalupe Fur Seals mate 7 to 10 days after the female gives birth to a pup conceived the previous year. The pup, born three to six days after the female leaves the water, can swim shortly after birth, although it does so only in emergencies. The female makes series of two- to six-day foraging trips at sea, interspersing them with short visits ashore to suckle her pup. This seal's diet includes squid and lantern fish. The female and her pup may stay on the breeding territory until the next spring. The Guadalupe Fur Seal was nearly exterminated by seal hunters in the 1800s; by 1892 only seven individuals were known to exist. Although a fisherman sold two males to the San Diego Zoo in 1928, and one seal was seen on San Nicolas Island, off southern California, in 1949, very few were found until 1954, when 14 were sighted on Guadalupe Island, off Baja California. Mexico declared Guadalupe Island a seal sanctuary in 1975, and by 1984 the seal population numbered about 1,600 (including about 650 new pups). A bull established a territory on San Nicolas Island in 1988 and returned there each year through 1991. The Guadalupe Fur Seal is classified as a threatened species by the U.S. government.

352 Northern Sea Lion
"Steller Sea Lion"
Eumetopias jubatus

Description: *The largest eared seal* (male larger than any bear). *Bull buff above, reddish brown below;* has dark brown flippers, and massive neck and forequarters. *Cow uniformly brown,* one-third size of male, more cylindrical in shape. Prominent forehead of male gives *snout and face an*

otter-like appearance. L male 8′10″–10′6″ (2.7–3.2 m), female 6′3″–7′3″ (1.9–2.2 m); Wt male to 2,200 lb (1,000 kg), female 600–800 lb (272–365 kg).

Similar Species: Northern Elephant Seal is larger, with relatively small flippers; male has large proboscis. Fur seals and California Sea Lion are smaller, the latter browner and usually barking.

Breeding: Breeds May–August; 1 dark brown to black young born late May–June of following year; birth weight 40–45 lb (18–20 kg).

Habitat: Rocky shores and coastal waters along them.

Range: Pacific Coast, from s Alaska and Aleutian and Pribilof islands south to s California.

The Northern Sea Lion usually stays in the water during poor weather, but otherwise spends much time "hauled out" on rocky shores; it will dive into the sea if a boat approaches. This sea lion eats fish, including blackfish, rockfish, greenling, and, more rarely, commercially valuable species, such as salmon, squid, clams, and crabs. In Alaska it favors walleye pollack, in Canada, herring, rockfish, squid, cod, and octopus. The Northern Sea Lion usually feeds at night in water less than 600 feet (180 m) deep and within 10 to 15 miles (15–25 km) of shore. In Oregon, this species often swims up rivers and feeds upon lamprey and salmon. Northern Sea Lions are believed to range widely during the nonbreeding season. Some females routinely travel 250 miles (400 km) south of their haul-out on six-week foraging trips. The female Northern Sea Lion mates in her third year. The male is sexually mature at six to seven years, but he is not large enough to compete for cows until he is 9 or 10 years old. During the breeding season, bachelor bulls and nonbreeding females herd separately from breeding colonies. The breeding bull, fasting for

one to two months during the mating season, often fights fiercely to defend his territory and his harem of 10 to 30 cows. The female arrives on territory in May or June and gives birth about three days later to a pup bred the previous summer. Within two weeks of the birth she mates again. About nine days after giving birth, the female starts making foraging trips of one to three days, with intermittent nursing bouts that last about a day. These foraging trips lengthen with time. Although the pup can swim on the day it is born, it lacks enough body fat to insulate it against the cold water. After a month, however, some young accompany their mothers on short forays. Most pups are weaned before the end of their first year, but a few continue to nurse until they are two or even three years old. The young and sometimes the adults frolic in the water and pups may play "king of the mountain" on rocks along the beach, gaining swimming and climbing experience. Adult seals have a deep, bellowing roar; pups have a lamb-like bleat. Males leave the rookeries in August, females and pups in early fall, at which time most animals from southern populations migrate north. Northern Sea Lions are declining in number for unknown reasons. Whereas in 1985 there were 67,000 individuals, in 1989 only 25,000 were counted from the central Aleutian Islands to the central Gulf of Alaska. The U.S. and Canadian governments list this species as threatened, affording it complete protection in their waters; however, a multi-year permit granted to commercial fisheries in 1988 allows them to take up to 1,350 individuals annually without penalty. The Northern Sea Lion's rookeries are scattered off the west coast of North America. There are small rookeries off the California coast on Año Nuevo, Farallon, and Cape St.

George islands, and off the Oregon coast on Oxford and Rogue islands. About 30 pups are born per year on Farallon Island, and several hundred on Oxford and Rogue islands. The largest rookeries are on a number of islands off the tip of Vancouver Island, British Columbia, at Cape St. James and North Danger Rocks, where there are about 4,000 to 5,000 individuals. Scattered rookeries in the Gulf of Alaska and on the Aleutian Islands produce numerous pups; about 2,200 young were born on Marmot Island, in the gulf, in 1989. The northernmost rookery is in Prince William Sound at Seal Rocks. This species is sometimes referred to as the "Steller Sea Lion," after the German naturalist Wilhelm Steller, who first described it in 1741, calling it a "lion of the sea" because of its golden, leonine eyes and its bellowing roar. Its predators are killer whales and sharks.

350, 351 California Sea Lion
Zalophus californianus

Description: *A slender seal. Buff to brown; appears black when wet.* Male dark brown, female "blond." Male's head paler with age; hair on its forehead much lighter than that on rest of body. *Male has high forehead.* L male 6'7"–8'2" (2–2.5 m), female 4'11"–6'7" (1.5–2 m); Wt male 440–860 lb (200–390 kg), female 100–220 lb (45–100 kg).

Similar Species: Northern Sea Lion is larger, paler; male has less prominent forehead. Guadalupe Fur Seal is smaller.

Breeding: Breeds June–July. 1 blue-eyed pup usually born in late June of following year; weighs about 35 lb (16 kg).

Habitat: Usually islands: sandy or rocky beaches; occasionally caves protected by cliffs.

Range: Pacific Coast from Vancouver south to Baja California and Gulf of California; breeds on islands, south from California

Channel into Gulf of California, mostly on San Miguel (largest herd), Santa Barbara, San Nicolas, and San Clemente islands of southern California.

The trained seal of zoo and circus, the highly gregarious California Sea Lion often indulges in playful antics in the wild, throwing objects and catching them on its nose and cavorting in the water. The fastest aquatic carnivore, it can swim up to 25 mph (40 km/h) when pressed, often "porpoising" along at the surface. It can descend to 450 feet (137 m) and stay submerged for 20 minutes, using sonar for underwater navigation and finding prey. Spending much of the day sleeping on islands, the California Sea Lion hunts primarily at night, feeding at depths of 85 to 240 feet (26–74 m) on squid, octopus, abalone, and more than 50 species of fish, usually Pacific whiting, juvenile rockfish, Pacific and jack mackerel, and market squid, which it eats by snapping off the head and swallowing the remainder. The bull is among the most vocal of all mammals, continually giving a honking bark while defending his territories. The cow makes a quavering wail to summon her pup, and barks and growls in aggressive interaction with other cows. The pup recognizes its mother's voice and responds with a lamb-like bleat. The sexes remain segregated outside the breeding season, males traveling north in August or September as far as British Columbia while females and young remain in the waters off rookeries. The breeding male establishes a territory in May, June, or July in southern California and defends it against neighboring males. The males lunge at one another, each trying to bite his adversary's front flippers. During the breeding season, the fasting bull maintains a territory but does not attempt to keep a specific harem. He may, however, patrol waters where females give birth and nurse their

pups. The female hauls out in May or June and gives birth about four to five days later to a pup conceived the spring before. The mother nurses the pup for about eight days, then begins foraging trips, which at first last two days, then three to four days. Between trips, the mother returns for 30 to 70 hours to nurse. This pattern continues until weaning, which in California rookeries takes place at four to eight months. Although a pup may nurse for eight months, it also eats fish. While learning to swim, it often rests on its mother's back. California Sea Lions mate three to four weeks after the females have given birth. They usually mate in the water or at water's edge as the female returns or departs on a foraging trip. The female breeds at four years, but the male must wait several more years before he is big enough to compete for territories. Killer whales and great white, hammerhead, and blue sharks occasionally prey on the California Sea Lion. Once killed for the oil rendered from its blubber, and also for its meat, which was used for dog food, this species is now fully protected by law in Canada and the U.S. However, human activities continue to endanger these mammals, whose numbers are much reduced. Many sea lions drown after becoming entangled in gill nets; others die when netting or other plastic materials get caught around their necks, causing deep wounds as the animals grow. In the late 1980s, California Sea Lions congregated in winter in the Lake Washington ship canal at Chittenden Locks, near downtown Seattle, where they consumed steelhead trout, a species that is itself considered endangered. Attempts have been made to move the animals from the Seattle area, but they often return, sometimes traveling from several hundred miles away. There were about 80,000 California Sea Lions in the U.S. in 1989, and a similar number in Mexico.

WALRUS
Family Odobenidae

The family Odobenidae contains one species, the Walrus. Fossils indicate that the Walrus evolved from an ancestral eared seal perhaps 7 million years ago. Like the eared seals, the Walrus has hind flippers that can rotate forward for terrestrial locomotion, and the bulls are polygamous; unlike them, the Walrus has no external ears and has nails on all five toes. The Walrus is distinguished by its tusks, which grow throughout the individual's life; an enamel layer coats the tusks when they erupt, but soon wears off, leaving pure ivory.

356, 357 **Walrus**
Odobenus rosmarus

Description: *Upper canines form large tusks,* up to 40″ (1 m) long in Pacific bull, 14″ (35 cm) in Atlantic bull; cow's tusks shorter, more curved. Nearly hairless body may be yellowish, reddish, pink, or brown to almost white. *Muzzle covered with about 400 bristles up to 12″ (30 cm) long.* No external tail. Short, round head. Fore flippers lack nails. Atlantic race: L male 8′2″–11′10″ (2.5–3.6 m); female 7′7″–9′6″ (2.3–2.9 m); Wt male 1,650 lb (750 kg), female 1,250 lb (570 kg). Pacific race larger: Wt male 3,300 lb (1,500 kg), female 2,200 lb (1,000 kg).

Similar Species: No other aquatic carnivore has tusks.

Breeding: Mates January–February; 1 calf born mid-April to mid-June of following year. Newborn weighs 100–150 lb (45–68 kg).

Habitat: Along continental shelf of northern seas, especially edge of pack ice; usually in water less than 60′ (18 m) deep.

Range: Atlantic race: Arctic seas around Greenland south to Hudson Bay. Pacific race migratory: generally in Chukchi Sea off ne Siberia in summer, Bering Sea off sw Alaska in winter.

A sociable animal, the Walrus gathers in a mixed herd of up to 2,000 bulls, cows, and calves when feeding and migrating, as well as when hauled out on ice floes, although the sexes sometimes segregate during the nonbreeding season. An ungainly creature, the Walrus spends more time out of the water than other aquatic carnivores, sunbathing and resting on ice and beaches for long periods. Individuals in the herd compete for the best basking spots. This animal will dive immediately if it scents humans. It sometimes sleeps at sea, hanging vertically in the water, held up by a pair of inflatable air sacs in the neck. Today the Walrus spends most of its time on pack ice, but some evidence indicates that before disturbance by humans, it lived more on islands and rocky shores. An excellent swimmer, this mammal uses its hind flippers alternately for propulsion. It can travel about 15 mph (25 km/h) and is able to dive to 300 feet (90 m) and remain submerged for almost half an hour. During deep dives, blood flows from the skin to the internal organs, leaving the skin pale; after the animal surfaces, its heartbeat is unusually rapid at first, with the increased blood flow gradually restoring the skin to its usual color. Walruses use their tusks to defend themselves against predators, including local people in kayaks. Tusk size is important in helping bulls establish dominance. Tusks are also used as grappling hooks to help the animal haul out onto ice or remain anchored to it. The Walrus generally feeds in early morning, foraging along the sandy sea bottom using its tusks as sled runners and the highly sensitive bristles on its face to search for prey. Mollusks, especially clams, and crustaceans are its chief foods. The animal uses its dome-shaped mouth like a vacuum cleaner to suck the meat from the shells, which are then discarded. A Walrus can eat as

many as 3,000 to 6,000 clams at a single feeding. Softshell crabs, shrimp, worms, and a few fish supplement the diet. An adult requires about 100 pounds (45 kg) of food daily, but may go without food for a week at times. The bull sometimes preys on seals, grabbing a victim with his fore flippers and stabbing it with his tusks. Walruses have been known to eat small whales, probably ones that are already dead. These polygamous animals court at sea and are thought to mate underwater. The bull does not form a harem, but performs an elaborate courtship display, adding a sound that resembles pealing church bells to his ordinary bellows and grunts. Males remain in the water about 23 to 33 feet (7–10 m) apart, apparently defending their territories, and may engage in courtship battles in which tusks are sometimes broken. Unlike the seals, the female Walrus does not mate in the same year she gives birth, but waits until the following year. The Pacific female is sexually mature at 8 years, with the first birth occurring at 10. The male is mature at 10, but apparently is unable to compete successfully for mates until he is about 15. Walruses live up to 40 years. The female Walrus usually nurses in the water, in an upright position, with the young hanging upside down. The young remains with its mother for about two years, nursing most of the time, until its tusks emerge and it can forage for itself. The calf often rides on its mother's back while she swims, holding on with its flippers. A cow is fiercely protective and will charge a Polar Bear to defend her calf. If a Walrus is attacked, neighboring animals will come to its defense, and injured herd mates will be helped from the water onto ice.

The Pacific Walrus is migratory, riding pack ice whenever possible, although it will swim if its chunk of ice begins going in the wrong direction. This race

spends the summer on the open pack ice of the Bering Sea, beginning northward movement in April and returning south in the fall; during the autumn months it frequents shallow water, keeping air holes open by butting ice with its head. A number of Pacific Walruses do not migrate northward, instead passing the summer on Round Island in Bristol Bay, Alaska. The Walrus has been hunted by northern peoples since at least the ninth century and is about as important to the local populace as the Bison was to the Native Americans of the Great Plains. Almost every part of the animal is used: Flesh provides food for humans and their dogs; skins are transformed into boat covers and leather, intestines into rain gear, and bones into tools; blubber is rendered for the oil, which is burned as fuel; the ivory tusks, previously made into sled runners, are now mainly used for carvings (scrimshaw), a source of considerable revenue; and even the bristles serve as toothpicks. Polar Bears, killer whales, and humans are the Walrus's sole enemies. Walrus herds were greatly reduced by hunting, but today many of the populations are recuperating. The Marine Mammal Protection Act of 1972 fully protects the Walrus from commercial hunting except by native peoples.

Hair Seals
Family Phocidae

The hair seals, sometimes called earless
seals or true seals, are the most
abundant and widespread of the aquatic
carnivores, with 19 species found
throughout most of the world's seas and
in some freshwater lakes, including 10
species in North American waters. The
hind flippers of these species are
permanently turned backward, and thus
are almost exclusively for aquatic use.
The fore flippers are smaller than those
of the eared seals or the Walrus. Both
front flippers and hind flippers are
haired and have claws on each of the five
digits. These seals have no external ear,
only a small orifice. Their fur is stiff
and, in several genera, distinctly
marked, but they generally lack a soft
underfur. In most species, the newborn
has a woolly, often white coat called the
lanugo that is usually shed within a
month and replaced with an adult coat
better adapted to a cold climate. Hair
seals move on land in a slow, clumsy
fashion, propelled solely by muscular
contraction of the body; whenever
possible, they roll or slide on ice. In
water they swim easily, moving their
hind flippers up and down and using
their fore flippers for steering. They
often appear upright at sea, treading
water with the fore flippers. Most
species of hair seals are gregarious,
gathering in small groups, and most are
monogamous, pairing up during the
breeding season. Members of this family
are believed to have evolved from otter-
like mustelid ancestors.

Spotted Seal
Phoca largha

Description: Variable coloration; *usually dark gray
mantle on silvery or brownish-yellow
background.* Numerous dark, *irregular*

spots mostly on back and sides; spotted creamy white below. Male larger than female. L male to 5′7″ (1.7 m), female to 5′3″ (1.6 m); Wt 180–270 lb (82–123 kg).

Similar Species: Very similar Harbor Seal is usually found on islands or mainland, rather than on ice floes and pack ice. Northern Elephant Seal is larger, lacks spots, and has enlarged snout. Gray Seal is larger, with less dog-like face; sometimes has splotches. Ringed Seal has streaks in a marbled pattern as well as spots.

Breeding: Mates in May; 1 white pup born March–April of following year; length at birth 31–37″ (77–92 cm), weight 15–26 lb (7–12 kg).

Habitat: Ice floes in Bering Sea.

Range: Southern edge of pack ice in Bering Sea from Alaska to Siberia.

The Spotted Seal is most frequently encountered between February and May on the ice floe stretching from Alaska to Siberia in the Bering Sea, usually within about 15 miles (25 km) of the southern or front edge of the pack ice. It is found in particularly high numbers in Alaska's outer Bristol Bay in April, and is rarely encountered more than 60 miles (100 km) from the pack ice, on which it mates, gives birth, and nurses its young. The Spotted Seal can dive to at least 980 feet (300 m); a pup can dive to 260 feet (80 m) a month after weaning. This carnivore feeds on many types of fish and invertebrates, but especially on crustaceans and octopus. Among the fish it consumes are walleye pollack, capelin, Arctic cod, saffron cod, sand launce herring, and rainbow smelt. Spotted Seals are annually monogamous and seasonally territorial, forming pairs on the pack ice in March. The female is sexually mature at three to four years. The male is thought to join the female about 10 days before she gives birth to a pup from the previous year's mating, and remains with her until they mate.

The family groups are spaced about 500 to 600 feet (150–200 m) apart. Each pair has a white pup by April. They mate a month later, after the pup is weaned, by which time it has shed its white lanugo. The pup stays on the ice floe for a few weeks after weaning, and is abandoned by its mother about the time it starts swimming. The seals stay with the ice as it breaks up in May and June, forming small herds. In late spring and early summer they move north and toward the coast. In summer, they haul out on St. Matthew and other nearby islands in the north-central Bering Sea. In late summer and fall, they enter the estuaries and embayments and replace the Ringed Seals, which by then have moved farther north. About 2,000 Spotted Seals occur in Kasegaluk Lagoon, near Point Lay, Alaska, from July through freeze-up. A Spotted Seal can live to about 35 years of age. Killer whales, Walruses, Northern Sea Lions, and large sharks are predators. First described as a species in 1811, the Spotted Seal was treated for many years as a subspecies of the Harbor Seal, until 1970, when it was again elevated to species status. Spotted and Harbor seals often occur together in the southern Bering Sea, and may even haul out in the same areas, but they form separate groups and do not interbreed, as their reproductive periods are about two months out of sync.

358 Harbor Seal
"Hair Seal," "Leopard Seal"
Phoca vitulina

Description: Highly variable in coloration; *two main color types: white, light gray, yellowish gray, or brownish; with dark spots: and black, gray, or brown with light rings.* Most often yellowish gray or brownish with dark spots; spotted creamy white below. Sometimes colored with an orange, rust

or green cast (found on one-fifth of the animals in San Francisco Bay), possibly due to algal growth. Nostrils close together, V-shaped. Male larger than female. L male 4'7"–6'2" (1.4–1.9 m), female 3'11"–5'7" (1.2–1.7 m); T 3⅜–4½" (8.7–11.5 cm); Wt male to 308 lb (140 kg), female to 175 lb (80 kg).

Similar Species: Very similar Spotted Seal also occurs in se Bering Sea and is best identified by habitat, preferring pack ice rather than islands or mainland. Northern Elephant Seal is larger and lacks spots. Gray Seal is often splotched, and young are very similar to Harbor Seal. Gray Seal has less dog-like face; eyes are farther from nostrils; and nostrils, when viewed head-on, are farther apart, appearing as a W rather than a V. Ringed Seal has streaks in a marbled pattern as well as spots.

Breeding: Breeding season varies; 1 pup born March–August, with adult fur; weighs 18–26 lb (8–12 kg).

Habitat: Coastal waters, mouths of rivers, beaches, and rocky shores; some northern populations permanently inland in freshwater lakes.

Range: In West: s Arctic from Yukon and n Alaska south along California coast. In East: s Greenland and Hudson Bay coasts south to Carolinas.

The gregarious Harbor Seal spends much time basking on beaches and rocky shores, sometimes alone but usually with several other individuals, and occasionally in groups numbering in the thousands. Whether alone or in a group, it stays alert to danger; a single seal spends much more time watching for danger than does one that is part of a group. At the first sign of trouble, it gives an alarm bark and dives into the water. The Harbor Seal can dive to 1,460 feet (446 m) deep and remain submerged for 27 minutes. It feeds mostly on fish, including rockfish, herring, cod, mackerel, flounder, and salmon, but it eats some mollusks

(about 5 percent of the diet), such as squid, clams, and octopus, and sometimes crayfish, crab, and shrimp. Some seals learn to steal fish from nets, often damaging them and incurring the wrath of commercial fishermen. This species feeds when the tide comes in, sometimes ascending rivers with it, and in the spring may follow fish runs up a river for hundreds of miles, returning to coastal waters in the fall. The animal hauls out at low tide, sleeping high and dry until the next rising tide unless disturbed. The Harbor Seal becomes sexually mature at three to seven years of age. Pups are usually born with adult fur, having shed the lanugo before birth. They are often born in the intertidal zone and can follow the mother on land or into the water within five minutes of birth. The young's vocalizations keep the mother and young in contact. A pup often stays on the mother's back while she dives. A female may lose more than 35 percent of her weight during lactation, which lasts about four weeks. She mates within a few days after weaning the young. An individual Harbor Seal may live for 30 or more years. Polar Bears, Northern Sea Lions, killer whales, and sharks are the main predators aside from humans; golden eagles sometimes prey on pups.

359   **Ringed Seal**
    Pusa hispida

Description: *The smallest aquatic carnivore.* Color varies: often brownish to bluish black above, streaked and marbled with black on back; often has *irregular light rings on sides;* whitish to yellowish below with scattered dark spots. Male slightly larger than female. L 4′–5′5″ (1.22–1.65 m); Wt 110–250 lb (50–113 kg).

Similar Species: Harbor and Spotted seals usually have spots rather than ringed pattern.

Breeding: Mates April–early May; gestation about 11½ months. 1 pup born March–April of following year with lanugo and full set of adult teeth; weighs 8–10 lb (3.6–4.5 kg) at birth.

Habitat: Land-fast ice; seldom on shifting ice of open sea.

Range: Arctic seas from Point Barrow, Alaska, east to Labrador and Newfoundland.

The often solitary Ringed Seal spends most of the year under thick ice, using the strong claws of its fore flippers to dig and maintain breathing holes, which often pock the edges of the ice where it lives. Juveniles frequently move to open water near the edge of the ice. This hair seal dives to 300 feet (90 m) and can stay submerged up to 18 minutes, but usually surfaces in about three minutes. Feeding mostly on crustaceans, shrimp, and small fish, in deep water it eats larger zooplankton, such as amphipods. Adult Ringed Seals form territories at freeze-up where they remain all winter, maintaining their own breathing holes. Young seals tend to disperse; some older seals may remain in the same place year-round. It is thought that territories are maintained by vocalizations, which are varied and are used throughout the year and at any time of day or night. The female Ringed Seal establishes her territory around a pupping den or birth lair, which she digs with her claws, usually under overlying snow near breathing holes, although occasionally a natural snow cave is used. The birth lair, unique among seals, is usually about 7 feet by 10 feet (2 × 3 m) and 2 feet (60 cm) high, with a 12-inch opening into the water. These seals also use resting chambers, and several individuals may share a multi-chambered lair. The Ringed Seal pup, helpless at birth, is left in the lair while the mother forages. The pup loses its lanugo three to eight weeks after birth; it is nursed for two months. Ringed Seals mate the month after the

female gives birth. During the mating season, the monogamous bull has a strong odor, resembling that of gasoline, and both sexes fast, losing about 30 to 50 percent of their body weight. At this time, they bask frequently on the ice near their air holes, raising their heads every minute or so to scan the horizon for the Polar Bears that prey upon them. The female usually does not breed until six to eight years of age. Pregnancy rates are very low—28 percent in some populations—apparently because of the presence of PCBs and DDT. The life span of this species is up to 40 years. Besides the Polar Bear, which sometimes digs up dens, predators include the Walrus, killer whales, and sharks; Arctic Foxes take pups in the den. Native peoples who live along the coasts kill great numbers of Ringed Seals, using them for food, clothing, tools, and oil, which is burned for light and heat.

---

364 **Gray Seal**
     Halichoerus grypus

Description: A large seal. Variable coloration; *grayish to almost black* above, somewhat paler below. Male usually dark with lighter splotches; female light with dark splotches. Head squarish; *snout long;* nostrils form W shape. Male larger than female; has wrinkled neck. L male to 9′10″ (3 m), female to 7′7″ (2.3 m); Wt male to 770 lb (350 kg), female to 440 lb (200 kg).

Similar Species: Harbor Seal very similar, especially when young; is smaller, with rounded head, more dog-like face, and eyes closer to nostrils; in frontal view, nostrils are close together, appearing as a V. Bearded Seal has prominent beard. Hooded Seal has inflatable pouch above nose.

Breeding: Breeding season varies with locality; in Canada, 1 pup weighing about 30 lb (14 kg) and covered with creamy white

lanugo born December–February. Total
gestation about 11 months.

Habitat: Waters along rocky coasts and islands.

Range: Labrador south to New England; into
Gulf of St. Lawrence. Breeds on Sable,
Basque, and Camp islands (east of Nova
Scotia), Amet Island (Northumberland
Strait), and Deadman Island (sw of
Magdalens); also on pack ice between
Prince Edward and Cape Breton islands
and in Northumberland Strait. A few
may breed on Muskeget, Tuckernuck,
and Monomoy islands in Nantucket
Sound, Massachusetts.

The highly gregarious Gray Seal does
not undertake well-defined, long-
distance migrations, although the pups
disperse quite widely. The species dives
to 475 feet (145 m) and can remain
submerged for 20 minutes. It gathers in
groups to feed on such bottom-dwelling
fish as pollack, haddock, capelin, cod,
flounder, and whiting, as well as squid
and octopus. Newly weaned seals often
eat crab and shrimp. Adults vocalize
with threatening hoots, barks, and
hisses; pups with a shrill yap. The Gray
Seal maneuvers better on land than
other members of its family and may
crawl far inland to breed and give birth.
Remote islands usually have the largest
rookeries. Both sexes of the Gray Seal, a
polygamous species, fast throughout the
four- to six-week breeding period. The
bull does not defend a territory, but
does attempt, with vocal threats and
fighting, to gain access to a group of
females that have gathered for pupping
and nursing. Male displays include
open-mouthed threats and "hooting,"
and the exchange of long, quavering
calls. Fighting between males may
result in neck wounds. The seals mate
toward the end of the nursing period of
the previous year's pup. The Gray Seal
pup is weaned at two to three weeks, at
which time the mother abandons it. A
weaned pup fasts for up to a month

before beginning to feed at sea. The pup sheds its white baby fur for its adult coat four to five weeks after birth, then enters the water. It becomes a wanderer at this time, and may disperse quite far away. The Gray Seal's life span in the wild is up to 30 years for a bull, 40 for a cow. Its predators are probably sharks and killer whales. Unlike many other seals, a Gray Seal bull cut off from water by a seal hunter will fiercely attack the hunter. Once quite rare, the Gray Seal was protected, and populations have been increasing rapidly in the past few years; the world population is now estimated at nearly 200,000 individuals, with 85,000 to 115,000 in the western Atlantic. This species is unpopular with fishermen because it destroys fishing gear and consumes quantities of game fish, which has led to a bounty system and a controlled harvest. In addition, the Gray Seal is the primary host of the codworm (or sealworm), whose larvae invade fish, reducing the commercial value of the catch.

---

360 Harp Seal
"Saddleback Seal"
Pagophilus groenlandicus

Description: Male yellowish white to grayish above, with *harp or horseshoe-shaped black saddle on back;* silvery with scattered small spots below. Female less distinctly marked, or markings broken up into irregular spots. Black or dark brown face. Tail dark. Harp develops over various molts, not fully formed until age 7 for male, age 12 for female. Male slightly larger than female. L 4'7"–6'7" (1.4–2 m); Wt to 400 lb (182 kg).

Similar Species: Ribbon Seal is darker overall, with creamy-white bands.

Breeding: Breeds February–March; 1 pup born late February to mid-March of following year; acquires white lanugo within 3 days; birth weight about 12 lb (5.4 kg).

Habitat: Drifting pack ice; occasionally streams.
Range: Arctic seas from n Hudson Bay and w
coast of Greenland south along Labrador
into Gulf of St. Lawrence, west to
mouth of Mackenzie River in Northwest
Territories.

The Harp Seal can dive to depths of 900
feet (275 m) and remain submerged for
up to 15 minutes. Its chief foods are
small fish, especially such schooling
kinds as capelin and herring, as well as
cod and some crustaceans; young pups
feed on even smaller items, especially
krill and various invertebrates. Annual
oceanic migrations of this species cover
about 6,000 miles (9,600 km); these
migrations can probably be attributed to
the abundance and movements of Arctic
cod. In April the Harp Seal begins an
annual molt, lasting about a month, in
which the coat and even large pieces of
skin peel off. After the molt, the Harp
Seal in our range—the western Atlantic
population—moves north via the Davis
Strait to Baffin Bay and then to Thule
and the Canadian Archipelago. The seal
leaves the high Arctic in September
when the bays begin to freeze; it moves
along the eastern coast of Baffin Island,
from Hudson Bay eastward through
the Hudson Strait, and south along
the Labrador coast to the Gulf of St.
Lawrence, the eastern coast of
Newfoundland, or the Grand Banks,
with the first migrants reaching
Labrador and Newfoundland in
December. January and February are
mostly spent feeding. In February,
adults move to the edge of the pack ice
for breeding. Some gather near the
Magdalen Islands, others on pack ice off
southern Greenland. The Harp Seal is
highly gregarious at this time, with the
density of seals reaching 5,200 per
square mile (2,000 per sq km). The
noise here (and during migration) is
tremendous and has been likened to
that of a barnyard. After being nursed

for only two weeks, the baby seal born from the previous year's mating has grown to 90 to 100 pounds (40–45 kg) and is then abandoned. For the next couple of weeks, it does not feed; its weight drops to about 50 pounds (23 kg), and the lanugo is molted, replaced by a silvery coat with small dark flecks. In later molts the coat becomes spotted, then "spotted harp," until finally, years later, the harp is fully formed. Many pup perish in the first few weeks, but the res learn to gather food for themselves and will migrate north with the herd. Meanwhile, the bulls, which have a musky odor during breeding season, have congregated in the water between ice floes, where they court cows by swimming about furiously, calling underwater, and emitting bubbles. The seals mate in the water. Female Harp Seals are sexually mature at five to eight years, males at eight years. The annual fast, observed by both sexes during the breeding period, causes the formation of rings on the canine teeth, which can be examined to determine a seal's age. The Harp Seal's life span may be up to 30 years, but animals older than 20 are rare. This seal is the mainstay of Newfoundland sealing, and kills of more than 300,000 per year caused a dramatic decline in populations by the early 1960s. The present harvest quota of 186,000 seals per year—shared by about 9,000 Newfoundlanders—should allow the herd to be maintained. This species has drawn much public attention because of the sealers' clubbing white-furred "lanugos" for their coats.

---

361 **Ribbon Seal**
Histriophoca fasciata

Description: Male dark brown with *creamy white rings around neck, rump, and base of front flipper.* Female gray, with less-distinct rings.

L to 6′ (180 cm); Wt male to 200 lb (90 kg), female to 176 lb (80 kg).

Similar Species: Harbor Seal has spots; lacks rings or has inconspicuous ones. Harp Seal is light-colored, with dark harp-shaped saddle on back.

Breeding: Breeds late April through mid-May; 1 calf born April–early May of following year, with eyes open and body covered with thick lanugo.

Habitat: Open sea and Arctic ice floes.

Range: Bering Sea from Alaska to Aleutian Islands; 1 taken near Morro Bay, California, in 1962.

As it seldom occurs near shore, this rare seal seems even less common than it is. Usually solitary, the Ribbon Seal molts, rests, and gives birth on pack ice, and swims in the open waters of the Bering Sea when the ice is gone. Among seals, the Ribbon Seal's way of moving on ice is unique: It slithers along, alternating front flippers and throwing its head and body from side to side to stretch its flippers forward. Various fish, including pollack, capelin, eelpouts, sculpin, and polar cod, as well as octopus and shrimp compose the diet of this species. The Ribbon Seal can remain submerged for almost half an hour. Two underwater vocalizations have been recorded, described as "low-intensity downward frequency sweeps" and "short broadband puffing noises." The function of these sounds is not known, but they may be for mating or territorial purposes. During the reproductive and molting season in spring, adult Ribbon Seals fast, losing 40 to 50 pounds (18–23 kg). Young are born on pack ice; their thick, white birth fur lasts four to five weeks, then is replaced by a bluish fur coat. Finally, the "ribbon" coat, similar to that of the adults, grows in during the second year. The lactating female parent leaves her pup on the ice floe while she feeds; the pup is weaned at one month,

when it weighs about 65 pounds (30 kg). Ribbon Seals can live at least 30 years, but most probably live only 20. Likely predators include killer whales, Walruses, and sharks.

362 Bearded Seal
Erignathus barbatus

Description: *A large seal,* with proportionally small head. Uniformly grayish to yellowish. *Bearded with tufts of long, flat bristles at sides of snout.* Fore flippers squared off, with third digit longer than others. Female slightly larger than male. L 6'11"–7'11" (2.1–2.4 m); Wt 440–550 lb (200–250 kg), maximum 935 lb (425 kg).

Similar Species: Other seals lack beard.

Breeding: Mates in May. 1 pup born April–May of following year, with eyes open and with bluish or brownish woolly coat; birth weight about 70 lb (32 kg).

Habitat: Arctic and subarctic continental shelf, in relatively shallow water up to 500' (150 m) deep; moves seasonally with drift ice. Rarely hauls out on land.

Range: Northern coastal waters and shallow seas from Alaska to Labrador, including Hudson Bay.

The nonmigratory Bearded Seal is generally solitary outside the breeding season. Like the Ringed Seal, this species uses the claws of its fore flippers to keep breathing holes open in ice. The Bearded Seal eats bottom-dwelling fish, especially cod, flounder, and sculpin, but also feeds heavily upon whelks, clams, crabs, and octopus, which it locates and digs out with its sensitive bristles and strong claws. The maximum depth it can dive to is about 650 feet (200 m). This species is exceptionally vocal, uttering eerie, melodious songs, which can last more than a minute, under the ice in spring. The seal dives in a slow, loose spiral, releasing bubbles as it

sings; it then resurfaces in the middle of its area of activity. Presumably related to courtship and territory, the song can be heard in the air at short distances, but can be heard much more clearly by placing a paddle in the water and pressing the other end against the ear. During the breeding season, Bearded Seals may congregate in groups of up to 50. At this time, although bulls do not form harems, they may fight each other; they give a long warble to woo cows. The female gives birth only on the surface of the ice. The pup is weaned after nursing for 12 to 18 days, at which time it weighs about 200 pounds (90 kg), most of it blubber; it is then abandoned by the mother. Females are sexually mature at six years, males at seven, but this slow-growing species does not attain full size until 10 years of age. The seal is at its fattest from late fall to early spring. Its life span may exceed 25 years. Polar Bears, humans, and an occasional Walrus or killer whale are its chief predators. The Bearded Seal has little commercial value, but 2,000 to 4,000 are taken annually by native people for their flesh; their rough hides are used for nonskid boot soles.

363 Hooded Seal
"Bladdernose Seal"
Cystophora cristata

Description: *Prominent inflatable hood on head of adult male;* when inflated, resembles two footballs fastened together end to end, stretching from nostrils to forehead; when deflated, appears wrinkled. Male can extrude *red membranous, balloon-like organ* from hood. Body steel-gray above, often with irregular whitish or brownish blotches; paler below. Female paler, with less distinct markings. Claws light-colored and strong. Juvenile bluish gray above. L male 6'7"–9'10" (2–3 m), female 5'11"–7'10" (1.8–2.4 m);

Wt male 660–880 lb (300–400 kg), female to 660 lb (300 kg).

Similar Species: No other species has distinctive hood.

Breeding: Breeds late March–early April; single pup born late March–early April of following year. Pup slate-blue above, with a black head, lighter below; birth weight about 50 lb (23 kg).

Habitat: Edge of Arctic pack ice; in deep water.

Range: Gulf of St. Lawrence to Greenland waters; not Hudson Bay.

Little is known of the feeding habits of this deep-diving seal, but it probably eats mostly mussels, starfish, squid, octopus, and such fish as herring and cod. It is thought to feed in deep water, at depths greater than 650 feet (200 m) During courtship, or when disturbed or threatened, the Hooded Seal bull inflates its hood and a red, membranous "balloon," which greatly increase the apparent size of its head and may make the seal more formidable to an enemy. The hood is actually an enlargement of the nasal cavity; the seal extrudes the red "balloon" by closing one nostril and blowing into the hood, inflating the balloon mainly from the other nostril. Essentially solitary except during the breeding and molting seasons and when migrating, Hooded Seal individuals are generally well spaced when they occur in small groups on the same ice floe. The young are born on winter ice northeast of Newfoundland with a slate-blue coat the white lanugo having molted before birth. The entire reproductive season of this seal—birth, nursing, weaning, mating—occupies only two to two and a half weeks, at which time the adults fast. The Hooded Seal has the shortest lactation time of any mammal: only fou days. During this period the pup gains weight at a rate of 15 pounds (7 kg) per 24 hours. When the pup is about two weeks old, the mother abandons it to mate. A bull may mate with one female or may visit several. After mating, in

groups segregated by sex, the Hooded Seal follows the retreating ice through Davis Strait to waters east of Greenland, where it is common, and hauls out onto the ice to molt. When molting, from late June to early July, it fasts again, and its weight drops significantly. Then it slowly migrates southward, feeding far out to sea, and finally winters off the Grand Banks of Newfoundland. Hooded and Harp seals often migrate together but do not intermingle. Pups migrate by the same route as the adults, but a month later. Members of this highly migratory species have strayed as far south as Florida. Polar Bears kill many Hooded Seal pups, and the killer whale probably feeds on this species also. The bluish-gray juveniles, known as "blue-backs" or "bluemen," are prized in the fur trade.

West Indian Monk Seal
"Caribbean Monk Seal"
Monachus tropicalis

Description: *Uniformly brownish gray above; yellow or yellowish white below.* Hind flippers with first and fifth toes longest, giving V-shaped outline. Nails well developed on front toes, vestigial on hindtoes. Pup born jet-black, with eyes open. Male larger than female. L male about 7'7" (2.3 m); Wt 150–300 lb (68–137 kg).

Habitat: Tropical seas off sandy beaches.

Range: Previously Gulf of Mexico; now considered to be extinct. Map shows historic range.

The West Indian Monk Seal, last seen in Jamaican waters in 1952, was quite common when first encountered by Columbus in 1494. Living on the shores of islands with no terrestrial predators, the species had no fear of people; consequently, groups of them sunning on the beaches were easily approached and slaughtered for meat and oil. Its

habitat since completely dominated by humans, this seal is presumed to be extinct. Suspected sightings should be reported to conservation authorities immediately.

365 Northern Elephant Seal
Mirounga angustirostris

Description: *A very large seal; both sexes have large snout.* Snout droops over muzzle in adult male; when inflated during mating season, curves down and can reach back into mouth. Body brown or gray above, lighter below. Chest of male broad, calloused, scarred. Hind flippers have 2 lobes, reduced claws. L male to 13' (4 m), female to 10' (3 m); Wt male to 4,400 lb (2,000 kg), female to 1,320 lb (600 kg).

Similar Species: No other species has similarly large snout. Sea lions and fur seals are much smaller and can rotate hind flippers forward on land.

Breeding: Mates January–March, peaking mid-February; 1 black-furred pup born December–March; birth weight about 65 lb (30 kg).

Habitat: Temperate seas; subtropical sandy beaches for breeding and molting.

Range: Pacific Coast from Gulf of Alaska south to Baja California, Mexico; colonies have been established in U.S. waters on Santa Barbara, San Nicolas, San Miguel, Santa Rosa, Año Nuevo, and Farallon islands, and on mainland coast at Point Año Nuevo and Point Reyes.

The Northern Elephant Seal feeds mostly on squid, octopus, and such fish as ratfish, hagfish, small sharks, and other deepwater marine life. A deep feeder, it can remain submerged for up to 80 minutes, often remaining at the surface for four minutes or so before diving again. The deepest dive recorded for an air-breathing vertebrate was 5,187 feet (1,581 m), by a Northern

Elephant Seal in 1989. Polygamous but not territorial, this seal is limited by its enormous bulk to breeding on sandy beaches. The bulls arrive on the beaches first, in early December, and begin spectacular fights for territories on which to attract females. Inflating their huge snouts, they rear and threaten each other, the distended nasal pouches causing their snorts and bellows to resonate as much as a mile (1.6 km) away. The main vocal threat is a loud clapping sound made as the male raises up on his flippers, his elongate nose dangling in his mouth. Visual and vocal threats usually suffice, but if not, the bulls lunge at each other with their large canine teeth. By the time the females arrive, most disputes have been settled. About six days after hauling out, a cow gives birth. Of all mammals, the Northern Elephant Seal has the milk richest in fat (54.5 percent) and lowest in water content (32.8 percent); after nursing for a month, the calf weighs 400 pounds (180 kg). The female protects her own pup, and will chase away pups not belonging to her if they stray too far from their own mothers. The biggest danger to a pup is the possibility of starvation if it becomes separated from its mother. A pup that attempts to suckle at a strange cow may be killed, although some orphans are adopted by cows that have lost their own pups. After a pup is weaned, by which time it has grown a silver-gray coat, the mother abandons it to mate. During the breeding season the bull fasts, but the cow feeds periodically. Some females can mate at two years, but most do not mate until their third or fourth year. Bulls are sexually mature at six to seven years, but many do not gain access to females until they are 9 or 10. After mating, the female leaves the rookery and feeds at sea for about two and a half months, then returns to shore to molt in late May. The weaned young remain on the

beach until May, living off their fat and practicing swimming nightly. The male departs a little later than the female and remains at sea for several months, regaining part of the 50 percent of its body weight lost during the territorial and mating activities. The males return to shore in late June and July. Both sexes fast while they undergo a molt, during which hair and skin fall off in large patches. After the molt, the seals swim out to sea, where they remain until December, feeding.

Killer whales prey upon the young, but healthy adults seem relatively immune to predation. In the 1890s, the Northern Elephant Seal was nearly exterminated by the whaling industry for the oil rendered from its great rolls of blubber. In 1892, a tiny colony of fewer than 20 animals was discovered on Guadalupe Island, off Baja California. It was protected, and the species made a spectacular comeback; the herd now numbers about 115,000 seals, which breed on offshore islands from Baja California north to San Francisco. The Northern Elephant Seal recently has been forced onto mainland beaches owing to the increase in its population. The first birth on the mainland occurred in 1975, and by 1979, almost 100 cows had given birth in a new rookery at Point Año Nuevo, near Santa Cruz, California.

RACCOONS AND THEIR KIN
Family Procyonidae

The procyonid family, with six genera and 18 species worldwide, is a diverse group, ranging from the lesser pandas *(Ailurus)* of Asia to the olingos *(Bassaricyon)* of Central and South America. Most species occur in tropical parts of the New World; only three species are found in the U.S. and Canada. North American procyonids are characterized by their long tails with dark and light banding. Their cheek teeth are somewhat blunted rather than sharp, indicating that these carnivores have adapted to a wide variety of foods, including vegetable matter. They all have five clawed toes on each foot. Two of the three, the Common Raccoon and the White-nosed Coati, walk flat on the soles of their feet, as do bears and humans. The Ringtail, which walks on its toes, has semi-retractile claws. These mammals are good climbers and will den in hollow trees when these are available; when not, the animals may use ground burrows or caves. The Common Raccoon and the Ringtail are both nocturnal, while the White-nosed Coati is abroad during the day. These animals tend to be social, often remaining together as family groups and, in the case of the White-nosed Coati, in larger bands.

---

243 Ringtail
"Miner's Cat," "Civet Cat," "Cacomistle"
Bassariscus astutus

Description: Body cat-like; face somewhat fox-like. *Yellowish gray* above; whitish buff below. *Very long, bushy tail with 14–16 bands,* alternating black and white, ending with black at tip; black bands do not meet on underside. Relatively large ears and eyes. *No black mask;* white or pale eye ring. 5 toes on each foot; claws

partially retractile. L 24–32″ (616–811 mm); T 12¼–17¼″ (310–438 mm); HF 2¼–3⅛″ (55–78 mm); Wt 1⅞–2½ lb (870–1,100 g).

Similar Species: White-nosed Coati is much larger and browner, with thin, indistinctly banded tail. Common Raccoon is larger, with black mask and shorter tail.

Breeding: In Texas, mates early April; 1 litter of 2–4 young born late May–early June, sometimes in a nest.

Sign: *Den:* In cliffs, between or under rocks, or in a hollow tree, stump, or log. Tree holes, often small, with gnawed rim.

Scat: Usually elongated and cylindrical, but with great variation in size and shape; in dry habitat, tends to crumble. *Tracks:* Not easily found in this animal's dry and often rocky habitat. Prints 1–2¾″ (25–70 mm) long, 2″ (50 mm) wide, cat-like; foreprint and hindprint similar. 5 toes on each foot, with no claws showing, no long heel print.

Habitat: Varies: usually rocky situations, such as jumbles of boulders, canyons, talus slopes, and rock piles; less commonly, wooded areas with hollow trees; sometimes around buildings.

Range: Southwestern Oregon, California, s Nevada, s Utah, w Colorado, and s Kansas south through Arizona, New Mexico, Oklahoma, and Texas.

In a narrow den often padded with moss, grass, or leaves, the Ringtail sleeps by day, lying on its side, its back (summer), or with its tail wrapped about its curled body (winter). It grooms itself upon awakening, scratching with a hindleg, licking its fur, and using its moistened forepaws to clean ears, cheeks, and nose. The Ringtail can leap like a squirrel, and its extraordinarily sharp claws permit it to climb walls or trees. By night, this carnivore ambushes its prey, pouncing and forcing the animal down with its forepaws, then delivering a fatal bite to the neck. It generally begins to eat by devouring its

victim's head. The Ringtail's varied diet includes grasshoppers, crickets, spiders, centipedes, and scorpions; snakes, lizards, toads, and frogs; small birds; small mammals such as rats, mice, squirrels, and rabbits, as well as carrion; and fruit such as persimmons, juniper berries, hackberries, and mistletoe. Young Ringtails are white-haired, fuzzy, and stubby-tailed at birth, but they soon acquire adult coloration and longer tails. Three to four weeks after the birth of the young, the male joins his mate in bringing food to the den. Ringtails hunt independently at about four months of age and disperse in late fall. They squeak when young, but can bark, scream, and snarl in adulthood. When threatened or fighting, the Ringtail screams and secretes a foul-smelling fluid from the anal glands, earning it the name "Civet Cat." This is an allusion to the African carnivore *Civettictis civetta,* which produces a musky substance called civet that is used in perfumes. The name "Cacomistle" derives from *tlacomiztli,* which in the language of Mexico's Nahuatl Indians means "half mountain lion." Better mousers than house cats, Ringtails were once placed in frontier mines to control rodents; hence the name "Miner's Cat." The chief predators of this animal are the Bobcat and the great horned owl. Neither the fur nor the meat of the Ringtail is considered valuable, but the animal is sometimes killed by humans because of its habit of raiding henhouses.

239–242 **Common Raccoon**
Procyon lotor

Description: Usually gray-brown or orange-brown above, with much black; grayish below. Face has *black mask* outlined in white. *Tail bushy, with 4–6 alternating black and brown or brownish-gray rings.* Ears are

relatively small. L 24–37" (603–950
mm); T 7½–16" (192–406 mm);
HF 3¼–5⅜" (83–138 mm); Wt 12–
48 lb (5.4–21.6 kg).

Similar Species: White-nosed Coati has long, thin,
indistinctly banded tail, and much less
prominent mask. Ringtail lacks mask
and has longer tail.

Breeding: Mates January–March; litter of 1–8
young born April–May after gestation
of 63 days. Birth weight 2 oz (60 g).

Sign: Crayfish leavings along shores, streams,
or ponds; broken stalks, shredded husk,
scattered kernels, and gnawed cob ends
in cornfields.

Den: Usually a hollow tree, sometimes
with scratched or torn bark; may have
scat accumulated at base. Fissures,
burrows, old chimneys or buildings, and
other protected spots may also be used.

Scat: Droppings inconsistent in shape
but generally cylindrical, uniform in
diameter, about 2" (50 mm) long;
granular, varying from black to reddish
and sometimes bleached white. Scat
resembles that of Virginia Opossum and
skunks, but is often deposited on logs,
large tree limbs, or stones crossing a
stream.

Tracks: Hindprint 3¼–4¼" (82–108
mm) long, much longer than wide;
resembles a miniature human footprint
with abnormally long toes. Foreprint
shorter, 3" (75 mm) long and almost as
wide; claws show on all 5 toes. Tracks
are relatively large for animal's size
because Common Raccoon is flat-footed
like bears and humans. Stride 6–20"
(15–50 cm), averaging 14" (35 mm).
When walking, animal's left hindfoot
print is almost beside right forefoot.
When running, makes many short,
lumbering bounds, bringing hindfeet
down ahead of forefeet in a pattern
similar to that of squirrel tracks.

Habitat: Various wooded and wetland habitats;
common along wooded streams. Often
found in cities and suburbs as well as in
rural areas.

Range: Southern Canada through most of U.S. except for portions of Rocky Mountains, c Nevada, Utah, and Arizona.

Native only to the Americas, the Common Raccoon is nocturnal and solitary except when breeding or caring for its young. An accomplished climber, it can ascend a tree of any size and is able to come down backward or forward. Few animals can descend a tree headfirst; the raccoon does this by rotating the hindfoot 180 degrees. On the ground this animal usually walks, but it can run and is a good swimmer. During very cold spells, the raccoon may sleep for several days or even a month or more at a time, but it does not hibernate. It may be out during warmer periods in winter, and sometimes even forages then, but it does not need to feed, as it stores a third or so of its body weight as fat and can survive the entire winter without eating. Omnivorous, the Common Raccoon eats grapes, nuts, berries, pawpaw, and black cherry; grubs, grasshoppers, and crickets; voles, deer mice, squirrels, and other small mammals; and bird eggs and nestlings. It spends most nights foraging along streams and may raid Common Muskrat houses to eat the young and to prey on rice rats nesting in the muskrat's walls (afterward perhaps taking the house as its den). The raccoon swims in woodland streams, prowling for crayfish, frogs, worms, fish, dragonfly larvae, clams, turtles, and turtle eggs; climbs trees to cut or knock down acorns; and, in residential areas, tips over or climbs into garbage cans. The Common Raccoon's nimble fingers, almost as deft as a monkey's, can easily turn doorknobs and open refrigerators. (In fact, the animal's common name is derived from *aroughcoune,* an Algonquin Indian word meaning "he scratches with his hands.") If water is conveniently close, this animal sometimes appears to wash its food, a trait reflected in its

species name, *lotor,* which means
"washer." The raccoon's objective,
however, is not to clean the food but to
knead and tear at it, feeling for inedible
matter that should be discarded.
Normally this is done with food found
in the water. The Common Raccoon
uses its den for bearing young, for
winter sleep, and for temporary shelter.
Communal denning is common—up to
23 raccoons have been reported in a
single den—but usually only one adult
male is present. During the day in
summer, the Common Raccoon may
simply sleep on top of a log, in a nest,
or on a clump of vegetation. Although
Common Raccoons are sedentary, males
travel miles in search of mates. After
mating, the male may remain with a
female a week or so before leaving to
seek another mate. The female is
lethargic during pregnancy; she prefers
to make a leaf nest in a large, hollow
tree, but may also use a protected place
such as a culvert, cave, rock cleft,
Woodchuck den, or space under a wind-
thrown tree. Young are born in spring
and open their eyes at about three
weeks; they clamber about the den
mouth at seven or eight weeks, and are
weaned by late summer. At first the
mother carries them about by the nape
of the neck, as a cat carries kittens, but
she soon leads them on cautious foraging
expeditions, boosting them up trees
when threatened and attacking predators
ferociously if cornered. Some young
disperse in autumn; others may remain
in the den until the female drives them
out upon expecting a new litter, as den
space is limited. This creature's
vocalizations are varied and include
purrs, whimpers, snarls, growls, hisses,
screams, and whinnies. Upon meeting,
two raccoons whose territories overlap
growl, lower their heads, bare their
teeth, and flatten their ears; the fur on
the back of their necks and shoulders
stands on end. Usually both animals

back off without coming to blows.
The Common Raccoon harbors a
nematode (roundworm), *Baylisascaris
procyonis,* which although harmless to
the raccoon, is very dangerous and often
fatal to woodrats and probably to many
other mammals, including humans.
Transmission is through the organism's
very tiny eggs, found in soil and dung,
which become much more infective
with time. In recent years, raccoons have
been carriers of rabies, especially in the
eastern U.S. For this reason, they should
not be encouraged to feed on porches,
and the animal's dung should not be
left around buildings where humans
and pets can come into contact with it.
Foxes, Bobcats, Coyotes, owls, and other
predators undoubtedly kill many young
raccoons, but the automobile, disease,
and accidents probably are more
important causes of death. In some
regions, "coon" hunting is a popular
sport in late autumn, when raccoons
are very active, fattening themselves for
winter. Such hunting expeditions
involve dogs trailing the raccoon until
it is treed, at which point the hunters
shoot the animal. Sometimes, however,
instead of climbing a tree, the raccoon
leads hounds to a stream or lake. A dog
that swims well can easily overtake a
Common Raccoon in the water, but the
raccoon, a furious fighter, can defeat a
single dog. Raccoon pelts were valuable
until the fur industry declined; interest
in the animal's fur probably peaked
during the 1920s, when owning a
coonskin coat was a collegiate craze.

---

244 White-nosed Coati
Nasua narica

Description: Dark brown above. Long, pointed
snout with facial mask; white toward tip
and around eye, sometimes with black
or dark brown patches on upper part.
Rather thin tail with 6–7 indistinct bands.

Feet dark. Ears small. Male much larger
than female. L 33–53" (850–1,340 mm)
T 16½–27" (420–680 mm); HF 3¾–
4¾" (95–122 mm); Wt 16½–27 lb
(7.5–12.2 kg).

Similar Species: Common Raccoon is stouter, and has
shorter, bushy, distinctly banded tail.
Ringtail is smaller, with much larger
ears and bushy, distinctly banded tail.

Breeding: Mates January–March; 1 litter of 4–6
young in spring; gestation about 77
days. Born in maternity den in rocky
niche or similar shelter, young have
darker coats than adults.

Sign: *Scat:* Usually cylindrical; similar to
Common Raccoon's and as variable.
Tracks: Foreprints and hindprints 3" (75
mm) long, 2" (50 mm) wide. 5 toes
print; claws show on foreprints only. As
coati does not walk quite as flat-footed
as Common Raccoon, less of hind heel
pad registers, and prints are shorter.

Habitat: Mountain forests, usually near water;
also rocky, wooded canyons.

Range: Southeastern quarter of Arizona, sw
New Mexico, Big Bend and Brownsville
areas of s Texas. Abundant in Huachuca,
Patagonia, and Tumacacori mountains of
se Arizona.

Active by day (very seldom at night),
the White-nosed Coati holds its long
tail high and nearly erect, except for the
curled tip. It swims well and is an
excellent climber, using its long tail to
keep its balance on branches. When
startled, a coati generally climbs a tree,
but when treed by humans, it descends
rapidly and may attack and injure dogs.
During the hottest part of the day, it
may nap in a shady spot. The White-
nosed Coati spends the night in trees,
ascending toward dusk. Because it
inhabits warm latitudes, it remains
active in winter and needs no den for
warmth. During the day the coati
forages for food, frequently pausing for
grooming sessions, during which it
combs its fur with its teeth or the claws

of either forefeet or hindfeet. Its diet includes invertebrates and lizards found in soil surface litter. Because coatis are extremely fond of fruit, including that of the manzanita, juniper, and prickly pear, a troop may ignore customary foods and visit a fruit-bearing tree daily until it is stripped. More gregarious than other members of the raccoon family, coatis are fairly conspicuous, traveling about in troops of 4 to 25 females and their young. The young engage in constant noisy play, chasing one another up and down trees. Males are solitary except during the breeding season. As is usual among social animals, the coati is very vocal (much more so than the Common Raccoon), issuing snorts, grunts, screams, whines, and chatters. The White-nosed Coati is abundant in the Huachuca Mountains of Arizona, where it has been blamed for damage to orchards and the deaths of chickens and dogs. Although a record of this species dates from 1892 at Fort Huachuca, its abundance in the area may be a relatively recent phenomenon. Hawks and eagles have been seen capturing coatis.

WEASELS, SKUNKS, AND THEIR KIN
Family Mustelidae

The family Mustelidae includes some
65 species in 25 genera worldwide.
Nine genera, with 16 species, are
represented in North America north
of Mexico, including the arboreal
American Marten, the aquatic otters,
the burrowing American Badger, and
the scent-spraying skunks. Mustelids,
which vary greatly in appearance and
habits, are solitary, primarily nocturnal
and active throughout the year. Most ar
relatively small animals, with long, low
slung bodies, short legs, short, rounded
ears, a thick, silky coat that makes then
valuable fur-bearers, and paired anal
scent glands, which in skunks are highl
developed for defense and can spray
their powerful secretion accurately as fa
as 15 feet (5 m). In many other musteli
species, the secretions, while often
pungent, are used more as social and
sexual signals. In their reproductive
cycles, many mustelids exhibit delayed
implantation, with the fertilized eggs
remaining free in the uterus for an
extended period (180 to 200 days in the
Western Spotted Skunk) instead of
immediately implanting in the uterine
wall. This delay is advantageous,
allowing the animals to mate in summe
or autumn and bear young in spring,
when food is plentiful and conditions fo
growth and survival are optimal. While
skunks are rather slow-moving, boldly
patterned animals, their close kin the
weasels *(Mustela)* are furtive hunters.
Weasels move so sinuously that they
seem to flow over rocks and logs,
slipping their supple bodies through th
smallest burrows or crannies in their
relentless pursuit of prey. Like skunks,
weasels have paired scent glands that
open into the anus, but they use the
foul-smelling secretions in marking
territories more than in defense.

251 American Marten
"Pine Marten," "American Sable"
Martes americana

Description: *Weasel-like. Brownish,* varying from dark brown to blond, with paler head and underparts, darker legs, *orange or buff throat patch.* Long, *bushy tail;* pointed snout; small ears. Male larger than female. L 19¼–27" (490–682 mm); T 5⅜–9⅜" (135–240 mm); HF 2¾–3⅞" (70–98 mm); Wt 1–3½ lb (.4–1.6 kg).

Similar Species: Much larger Fisher lacks orange throat patch. Mink is darker, has shorter tail white spotting on chin, and lacks orange throat patch.

Breeding: Mates in midsummer; implantation delayed for 6–8 months until midwinter, then stimulated by increasing day length. Litter of 2–5 young born mid-March to mid-April in a hollow tree, on the ground, or in an underground den. Blind and naked at birth, the young are weaned at 6 weeks.

Sign: *Scat:* Resembles that of Mink, but often contains bits of fruit or nuts. Scat stations, where droppings are left repeatedly.

Tracks: Typical mustelid trail of twin prints, like that of Mink but slightly larger; very difficult to differentiate from Fisher's. Prints 1⅝–1⅞" (40–47 mm) wide. Straddle 2½–3" (64–75 mm), to 6" (150 mm) in snow. Walking stride, 8⅞" (225 mm) for males, 6" (150 mm) for females; more than doubles when animal is running.

Habitat: Forests, particularly coniferous, with numerous dead trunks, branches, and leaves that provide cover for rodents, such as its principal food, red-backed voles.

Range: Most of Canada; in West, south to n California and through Rocky Mountains; in East, to n New York and n New England.

The American Marten is active in early morning, late afternoon, at night, and on overcast days, traversing a home range of 5 to 15 square miles (13–39 sq km). In daytime and during inclement weather, it uses a tree hollow, fallen log, rock den, or squirrel's nest as a resting site. Its small size allows it to use a woodpecker hole in a tree, where it constructs a nest of leaves or grass. An inquisitive animal, it can be coaxed from its den by mouse-like squeaks. The marten spends most of its time foraging on the ground for rodents. Its large feet serve as snowshoes in soft snow. Red-backed voles are the diet staple throughout its range, but the marten also takes other small rodents, especially other voles, Red Squirrels (often captured in winter while they are sleeping in their middens), and flying squirrels, as well as rabbits, reptiles, and birds. Its varied diet also includes carrion, insects, earthworms, eggs, berries, conifer seeds, and honey. The marten sometimes buries surplus meat, though generally it is a poor digger. Martens breed the year after birth or in the second year. The male holds the female by the nape and may drag her around before actual copulation occurs. Females may accept more than one male. Both sexes establish scent posts by dragging the scent glands on their abdomens over logs and branches. Usually martens avoid each other; if two meet, they bare their teeth and snarl. Their repertoire of vocalizations also includes huffs, pants, chuckles, growls, screams, whines, and eeps. The marten has few enemies other than humans. It is easy to trap, and its valuable pelt has led to its extirpation in many areas; in other spots, lumbering has destroyed its habitat and reduced populations. The American Marten is now protected and making a comeback in many localities.

252 Fisher
Martes pennanti

Description: Long body; looks stockier than most other mustelids because of its long fur. *Dark brown above and below.* Broad *head has grayish cast;* pointed snout; small ears. *Bushy tail.* Male larger than female. L 31–41″ (790–1,033 mm); T 11¾–16⅝″ (300–422 mm); HF 3½–5⅝″ (89–143 mm); Wt 3–18 lb (1.4–8.2 kg).

Similar Species: American Marten is smaller, with orange throat patch. Mink is much smaller, with white spotting on chin.

Breeding: Mates March–April; implantation delayed 10–11 months; 1–6 young born the following spring; total gestation nearly 1 year. Young born blind, usually in a nest in a large, hollow tree; weaned at 3–4 months.

Sign: Porcupine kills.
Scat: 4–6″ (100–150 mm) long, dark, roughly cylindrical, often segmented; may show fur, bone, berries, or nuts. Scat with Common Porcupine quills almost a sure sign of Fisher.

Tracks: Similar to Mink's and American Marten's but larger. Prints wider than long, with claws showing; 2″ wide on dirt, more than 2⅝″ (50–67 mm) on snow. Runs with forelimbs moving in sync, one slightly ahead of other. Forepaws leave ground just before hindpaws arrive, resulting in typical mustelid twin prints, or sets of 2 prints, 1 slightly ahead of the other. Tracks may end abruptly at base of tree.

Habitat: Mature, dense, coniferous or mixed coniferous-hardwood forest with closed canopy.

Range: Southern tier of Canadian provinces south to n California and Rocky Mountains to Utah; in East, to n New York and New England.

A solitary animal of dense forests, the Fisher is primarily nocturnal like most mustelids, but is sometimes abroad by

day. A good climber and swimmer, it travels a home range of 50 to 150 square miles (130–400 sq km), wandering ever farther in winter when food is scarce. It moves about the home range a great deal (the male more than the female), following well-established trails, running along fallen logs, moving among branches from tree to tree, and taking refuge in temporary dens in crevices, underbrush, or holes dug in snow. The Fisher also uses a maternity den, located in a hollow tree. Snowshoe Hares and Common Porcupines are the Fisher's main prey. Porcupines are well armed, and occasionally a Fisher is injured or even killed by the quills. The Fisher begins with repeated attacks at its victim's face; the porcupine rotates to keep its back to the circling Fisher, sometimes putting its face against a tree or other object for protection. The Fisher may climb the tree, then come down headfirst to force the porcupine away from the tree. The Fisher makes the kill when the porcupine has been weakened by facial wounds. It feeds on the internal organs first, consuming the entire animal over two or three days. Although an individual Fisher does not kill a large number of porcupines, this prey is very important because it supplies so much food, providing the Fisher with energy from stored fat for a month. The Snowshoe Hare is the Fisher's usual victim, but squirrels, mice, and chipmunks also fall prey to this animal, usually by a bite to the back of the neck. Carrion, fruit, and other plant material round out its diet. The Fisher's common name is a misnomer and its origin is unknown, for this animal seldom feeds on fish; the name was probably due to confusion with the Mink, an avid fisher. Female Fishers first mate at one year. Copulation lasts about an hour. The male apparently does not participate in raising the young. If disturbed, the Fisher hunches its back

like a cat and may hiss, growl, snarl, or spit. Because of its reputation as a valuable fur-bearer, with female skins especially prized, and the ease with which it can be trapped, the Fisher has been extirpated in many areas. Loss of habitat has also depleted populations, as this animal requires extensive wilderness.

245 Short-tailed Weasel
"Ermine," "Stoat"
Mustela erminea

Description: Elongated body. *Dark brown above; white below.* Tail brown with black tip. Legs short; *feet white.* In northern part of range in winter, fur entirely white except for black tail tip. Male almost twice as large as female. L 7½–13½" (190–344 mm); T 1⅝–3½" (42–90 mm); HF 1⅛–1¾" (28–43 mm); Wt 1¾–6⅜ oz (45–182 g).

Similar Species: Least Weasel is smaller, with shorter tail that lacks black tip. Long-tailed Weasel is usually larger, with longer tail and brownish feet in summer. Black-footed Ferret is yellowish brown, with dark or black mask around eyes.

Breeding: Mates June–July; egg development stops after 2 weeks; implantation delayed 9–10 months; active gestation about 1 month. 1 litter of 4–9 young born blind, with fine hair, in April, in a protected area, such as under a log, rock pile, or tree stump; eyes of young open at 35 days.

Sign: *Scat:* Similar to Long-tailed Weasel's but slightly smaller.
Tracks: Similar to Long-tailed Weasel's but usually slightly smaller.

Habitat: Varies: open woodland, brushy areas, grasslands, wetlands, and farmland.

Range: Most of Canada south to n California, w Colorado, and n New Mexico in West; to n Iowa, Michigan, Pennsylvania, and Maryland in East.

The Short-tailed Weasel, active day or night, may be fairly common at times but is seldom seen. Although it hunts mainly on the ground, often running on fallen logs, it can climb trees and occasionally pursues prey into water. This carnivore kills all that is available and stores the excess. Mice, particularly meadow voles, are its main food, but its diet also includes shrews, baby rabbits, birds, frogs, lizards, snakes, and many kinds of insects. This weasel captures prey several times its own weight, such as a young cottontail. Because of the danger in hunting much larger animals, however, it does so only if an excellent opportunity arises, or in times of scarcity. After a rapid dash, it pounces on its victim with all four feet, biting through the neck near the base of the skull. Weasels lick blood from their quarry before eating it, which is perhaps the source of the myth that they suck blood from prey. The Short-tailed Weasel's den, which is sometimes appropriated from a chipmunk, may be found in or beneath a log, stump, roots, brush pile, or stone wall. The den usually has several entrances and contains a nest of vegetation mixed with hair and feathers of prey. Rather than curling into a ball the way most animals do, a weasel curls into a disk, thereby exposing more skin surface to the air; the insulation material in the nest compensates for this. Male and female Short-tailed Weasels remain separate most of the year, though male territories encompass those of several females. Territories are expanded when food is scarce. The weasel keeps several nests throughout the territory, using them when it is in the vicinity. The male Short-tailed Weasel does not mature and mate until its second year, but the female is sexually mature in June of her birth year, when spontaneous ovulation begins. Ovulation occurs monthly until the female is bred, often in July. The

female remains passive while the male drags her around by the scruff of the neck during copulation. After the female gives birth, her annual estrus begins when her young are about six weeks old. At that time, the male breeds with the female adult and with her female progeny, who may not have their eyes open yet. Most likely, this male is not the newborn females' father, because Short-tailed Weasels are not monogamous. Cared for by both parents, the young begin to kill prey at 10 to 12 weeks of age. Vocalizations of the Short-tailed Weasel include grunts, hisses, chatters, and a shrill call note. Its predators are hawks, owls, house cats, and other carnivorous mammals and snakes. Although the names "Ermine" and "Stoat" are sometimes used to describe the species in its white winter fur and its brown summer coat, respectively, these names are also given to the Long-tailed Weasel. Ermine was highly valued before the decline of the fur industry, and the black-tipped tails are the traditional trim on the robes of royalty.

247, 248 **Long-tailed Weasel**
Mustela frenata

Description: A long-bodied, short-legged animal. *Brown above; white to deep yellow below. Tail long, often equal to head and body length;* brown with black tip. *Feet brownish.* In Southwest, white on face. During winter in northern latitudes, fur entirely white (but often stained with yellow on hips) except for black tail tip. Male weighs about twice as much as female. L 11–22″ (280–550 mm); T 3⅛–6⅜″ (80–160 mm); HF 1⅛–2″ (29–51 mm); Wt 3–9⅜ oz (85–267 g).

Similar Species: Short-tailed Weasel is smaller, with shorter tail and white feet. Black-footed Ferret is yellowish brown, with dark or black mask around eyes. Mink is larger

and darker, with bushy tail and white spotting on throat and chin.

Breeding: Mates midsummer; development proceeds for about 15 days to blastula stage; implantation delayed until following spring; total gestation 205–314 days. 4–8 young born blind, nearly naked, in April or May; birth weight about ⅟₁₆ oz (3 g).

Sign: In deep snow: drag marks; holes where weasel has plunged under snow. Cache of dead mice or voles under log or in burrow.
Scat: Dark brown or black, long, slender and segmented, often tapering at one end; frequently contains hair or bits of bone; deposited on rocks, logs, or stumps.

Tracks: Hindprints ¾″ (18 mm) wide, 1″ (25 mm) long or more, usually with only 4 of 5 toes printing. Foreprints slightly wider than hindprints, but approximately half as long. Hindfeet usually placed in or near foreprints, but sometimes side by side, or with 1 slightly ahead. Straddle 3″ (75 mm). Stride varies as weasels run and bound, often alternating long and short leaps: when carrying prey or stalking, 12″ (300 mm); when running, 20″ (500 mm).

Habitat: Varies: forested, brushy, and open areas, including farmland, preferably near water.

Range: Southern British Columbia, Alberta, Manitoba, and Saskatchewan south through most of U.S. except se California, se Nevada, and much of Arizona. Range extends southward to Bolivia.

Like other weasels, the Long-tailed was once thought to be strictly nocturnal, but in fact is often out by day, since voles, a primary food item, are mainly diurnal. The most widespread carnivore in the Western Hemisphere, it preys largely on mice and voles, while also taking rabbits, chipmunks, shrews, rats, birds, and poultry, and the occasional

insect or earthworm. When hunting, it follows a zigzag pattern, moving from burrow to burrow. It does not suck blood, contrary to myth, although it will often lap it from a wound. The Long-tailed Weasel will attack animals its own size and larger, but is very careful to attempt to take larger prey only when opportunity presents itself, or when smaller quarry is scarce. When attacking, the weasel rushes in toward the prey, grabs the victim at the base of the skull, and curls its body around it while grasping it with its forelimbs. It eats the head and thorax first, and caches the portion not eaten. Weasels occasionally go on killing sprees, as instinct dictates that they procure food when available and then store it. For this reason, a weasel in a chicken yard is disastrous. Like other weasels, the Long-tailed runs by a series of bounds, with its back humped at each bound and its tail trailing backward. It makes its dens in the abandoned burrows of other mammals, often chipmunks, and also ground squirrels, moles, or pocket gophers. Within the den it constructs a nest, primarily of hair from prey. The maternity den may also be in the burrow of another small mammal, or under a stump in a gully. The Long-tailed Weasel uses a variety of vocalizations: It may screech and squeal, utter a rapid trill, and purr when content. During the mating season, females give a reedy, twittering call. When the animal is alarmed, enraged, or excited by the mating urge, the anal glands release a powerful malodorous musk. A weasel may drag its rump on the ground, presumably to leave a scent that informs other weasels of its sex and perhaps even its identity. The Long-tailed Weasel male matures during its second summer, the female at three to four months of age. Only the female brings food to the young, which disperse at seven to eight weeks, when the males are already larger

than their mother. In the northern
part of its range, the Long-tailed Weasel
turns white in winter. The time of the
molt is governed by the length of
daylight, and the change is gradual;
weasels are piebald during transition.
In the southern part of its range, within
a 600-mile-wide (1,000-km-wide)
transcontinental belt, some individuals
molt to white, while others remain
brown. In Pennsylvania, for instance,
less than half of the weasels become
white, and south of the Maryland-
Pennsylvania border none do. The
color change is evidently genetically
determined: If a northern weasel is
captured and taken south, it will
still turn white in winter, while a
southern weasel transported north
will remain brown. Weasels are
valuable in controlling populations
of rodents, including rats. Hawks, owls,
cats, foxes, and snakes prey upon the
Long-tailed Weasel. This animal has
also been taken by trappers, although
its pelt is not considered especially
valuable.

249   Black-footed Ferret
Mustela nigripes

Description: Mink-like in shape. *Yellowish brown or
buff above,* with brownish wash on back;
slightly paler below. *Tip of tail and feet
dark or black.* Dark or black *mask around
eyes,* with *face white* above and below
mask. Male larger than female.
L 19¾–23″ (500–573 mm); T 4½–5½″
(114–139 mm); HF 2–2⅞″ (51–73
mm); Wt 18¾–22 oz (535–633 g).
Similar Species: Weasels are brown above, white below.
Mink is uniformly dark brown to black.
Weasels and Mink lack dark mask.
Breeding: Mates in spring; implantation is not
delayed. 1 litter per year of 3–5 young
born late spring–early summer, after
gestation of about 41 days. Both sexes
sexually mature at 1 year.

Sign: Fresh, untamped earth at entrance of prairie dog burrow often indicates occupancy by Black-footed Ferret. (Prairie dogs tamp down mounds of excavated earth.)

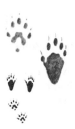

Scat: Approximately 3″ (75 mm) long, usually segmented when it contains much hair.

Tracks: Similar to Mink's, but longer and narrower, with more of heel printing. Hindprint approximately 2″ (50 mm) long, 1¼″ (37 mm) wide. Foreprint nearly as long, and about half as wide. Both prints show 5 toes, with claw marks visible.

Habitat: Arid prairies; prairie dog towns.

Range: Formerly Great Plains from s Alberta and s Saskatchewan south to w Oklahoma, nw Texas, New Mexico, and nw Arizona. Currently Montana, South Dakota, Wyoming, and Nebraska; scattered populations elsewhere.

The Black-footed Ferret is the only wild ferret in existence, and the rarest North American mammal. The U.S. Fish and Wildlife Service lists it as one of the ten most endangered species in North America. Its range originally extended as far north as Alaska, but the slaughter of prairie dogs—its primary food—has reduced its range considerably and driven this animal to near extinction. The Black-footed Ferret often lives in prairie dog towns, surrounded by its favorite source of food. It sometimes sits erect, looking for a burrow to raid. A prairie dog that catches sight of it will dart quickly underground. The ferret, keeping its body close to the ground, stalks the prairie dog's burrow and peers into the entrance. If the prey is near the surface, the ferret lunges for it. Otherwise, it slithers down the tunnel. It quickly kills its prey the way weasels do, with a bite to the base of the skull. After eating, it buries any remains. Having entered its victim's home, the ferret will sometimes poke its head from

the burrow to scan its territory. If it
decides to take up residence in the
burrow, it enlarges the entrance and
builds additional living chambers.
When prairie dogs are scarce, the ferret
will eat other rodents, including mice,
gophers, and ground squirrels, as well
as birds, eggs, and small reptiles.
Attempts are being made to restore
ferret populations in the West, notably
in South Dakota, where Wind Cave
National Park offers the best opportunity
to see this species in the wild.

246 Least Weasel
Mustela nivalis

Description: *A tiny weasel. Brown above, white below.
Tail very short, brown. Feet white. All
white in winter in North. L 6¾–8⅛″
(172–206 mm); T ⅞–1¼″ (24–38 mm);
HF ¾–⅞″ (19–23 mm); Wt 1¼–1¾ oz
(37–50 g).

Similar Species: Short-tailed Weasel and Long-tailed
Weasel both are larger and have much
longer, black-tipped tails. Black-footed
Ferret is yellowish brown, with dark or
black mask around eyes.

Breeding: Mates year-round. Ovulation is induced
by copulation; no delayed implantation.
Gestation 34–37 days. Up to 3 litters
per year of 1–6 young, usually born in
spring and late summer, often in
abandoned burrow of another animal.
Young weigh less than ¹⁄₁₆ oz (1.4 g) at
birth; they are weaned at 6–7 weeks.

Sign: *Scat:* Similar to that of larger weasels
but smaller.
Tracks: Similar to those of larger weasels
but much smaller. Straddle 1¼–1¾″
(30–45 mm); leaps occasionally 2′
(600 mm).

Habitat: Grassy and brushy fields, marsh areas;
also deep forests of high Allegheny
Mountains.

Range: Most of Canada south in Midwest to ne
Montana, Nebraska, Iowa, n Illinois, n
Indiana, Ohio, Pennsylvania, West

Virginia, and through s Appalachian Mountains.

The smallest living carnivore, the Least Weasel is seldom seen and rarely trapped, and does not appear to be common in any part of its range. Previously known as *Mustela rixosa,* it is now considered to be the same species as the European Least Weasel *(Mustela nivalis).* Like other weasels, it turns white in the northern part of its range, with some individuals remaining brown all winter from southern Michigan southward. White individuals have been seen, however, in Pennsylvania, Ohio, Indiana, and other areas in the middle part of its range. The Least Weasel may be abroad day or night, any time of the year. It moves over its home range of less than 2 acres (.8 ha) in search of prey, investigating every hole or crack and frequently standing on its hindfeet to scan for predators or prey. It may have several temporary dens scattered over its range. It will den in the abandoned burrow of another small mammal, such as a mouse, gopher, or ground squirrel, adding a lining of mouse hair to the rodent's grass nest. This species can run very fast, up to 6 mph (10 km/h). It feeds almost entirely on meadow voles, chasing them along their runways, and is small enough to chase mice inside their burrows. Pouncing on its prey, the weasel wraps its legs around the victim and kills it with a swift bite at the base of the skull. Occasionally it eats shrews, moles, birds (including eggs and nestlings), and insects. Despite legend, weasels do not suck blood, but do often lick blood from their prey. Weasels will kill more than they can eat when prey is available, caching the excess in the side passages in their dens. The Least Weasel consumes about 40 percent of its own weight per day. Male Least Weasels are sexually mature at eight months and females at four months, though they

seldom breed in their birth year. Males
and females remain separate except
during the breeding season. Like other
mustelids, the Least Weasel has the
characteristic anal glands used for
defense and for marking territory. When
disturbed, it gives a shrill squeaking
call, and it may hiss when threatened.
Foxes, cats, hawks, and owls are the
Least Weasel's chief predators.

250 Mink
Mustela vison

Description: Sleek-bodied, with *lustrous chocolate-
brown to black fur; white spotting on chin
and throat.* Tail long, somewhat bushy.
Male larger than female. L 19⅜–28″
(491–720 mm); T 6¼–7⅝″ (158–194
mm); HF 2¼–3″ (57–75 mm);
Wt 1½–3½ lb (700–1,600 g).

Similar Species: American Marten has longer tail and
orange or buff throat patch. Weasels
have white underparts and much
narrower tails. Black-footed Ferret is
yellowish brown, with dark or black
mask around eyes.

Breeding: Mates January–April; ovulation induced
by mating. Implantation occurs at 9–46
days, depending on how late it is in the
season. Litter of 1–10 young born blind
and naked in fur-lined nest in April or
May. Young weigh ¼–⅜ oz (8–10 g) at
birth, and are weaned at 5–6 weeks.

Sign: Hole in snow where Mink has plunged
after prey. Trough in snow, similar to
otter slide but smaller.
Den: Openings in stream banks, 4″ (100
mm) wide.
Scat: Dark brown or black, roughly
cylindrical, 5–6″ (127–152 mm) long,
sometimes segmented; often with bits of
fur or bone. Usually deposited on rocks,
logs, beaver lodges, and near den.
Tracks: Fairly round, 1¼–1⅝″ (30–40
mm) wide, more than 2″ (50 mm) in
snow. A clear print may show heel pad,
all 5 slightly webbed toes separately,

and semi-retractile claws. Hindfeet 2¼″ (55 mm) long in mud, 3½″ (90 mm) in snow, and placed nearly in prints of forefeet. Trail of twin prints 12–25″ (300–650 mm) apart, depending on animal's size and speed.

Habitat: Along rivers, creeks, lakes, ponds, and marshes.

Range: Most of U.S. and Canada except Arizona, s California, s and c Utah, s New Mexico, and w Texas.

Able to dive to a depth of more than 16 feet (5 m), the Mink is an accomplished swimmer and spends much time hunting in ponds and streams. Often out at night, it adapts its hunting time to prey availability. The Mink marks its hunting territory with a fetid discharge from its anal glands, which is at least as malodorous as a skunk's, although it does not carry as far. The home range of a male Mink encompasses the ranges of several females. While the Mink's preferred prey is the Common Muskrat, it also takes rabbits, mice, chipmunks, fish, snakes, frogs, young snapping turtles, and marsh-dwelling birds; occasionally it raids a poultry house. Like weasels, the Mink kills by biting its victims on the neck. It eats where it kills, or carries the prey by the neck into its den, where it caches any surplus. The Mink dens in a protected place near water, using a muskrat burrow, an American Beaver den, or a hollow log, or digging its own den in a stream bank. All dens are temporary, as the Mink moves frequently. If angered or alarmed, the animal may hiss, snarl, or screech, and discharge its anal glands (as it does when trapped). It produces a purring sound when content. Males mate with several females but eventually live with one. Young remain with their mother until the family disperses in early fall. Minks of both sexes are hostile to intruders, and males fight viciously with one another. Foxes, Bobcats, Lynxes,

alligators, and great horned owls are
known predators. Mink pelts were
highly valued before the decline of the
fur industry. Most commercial pelts
come from Minks raised on ranches; the
range of fur colors reflects selective
breeding.

---

257 Wolverine
"Glutton," "Skunk Bear"
Gulo gulo

Description: *Bulky,* somewhat bear-like. *Dark brown,*
with *broad yellowish bands* from
shoulders to over hips, meeting at base
of tail. Light patches in front of ears.
Male larger female. L 31–44″
(800–1,125 mm); T 6¾–10¼″ (171–
260 mm); HF 6½–8″ (165–205 mm);
Wt 18–42 lb (8.2–19.1 kg).

Similar Species: American Badger is smaller and more
grizzled, with white cheeks and white
stripe on forehead.

Breeding: Mates April–September; implantation
delayed until December–March. In two
recorded cases, total gestation was 215
and 271 days, with active gestation only
30–40 days. Litter of 2–5 young born
early spring in protected area, such as
thicket or rock crevice.

Sign: *Scat:* Often more than 5″ (125 mm)
long, more or less cylindrical, usually
tapering at one or both ends, semi-
segmented; often showing hair or bone.
Tracks: Sometimes all 5 toes and semi-
retractile claws print; small toe often
does not print. Foreprint 4–7″ (100–
175 mm) long, varying with size of
animal or condition of snow, and about
as wide; heel pad often showing 2 lobes,
a wide lobe in front of smaller round
lobe, which does not always print.
Hindprint similar to foreprint. Stride
extremely variable; straddle 7–8″ (175–
200 mm). Wolf and dog tracks are
similar but have only 4 toes and much
different lobing.

Habitat: Forest and tundra.

Range: Northern Canada south to nw
Washington; populations remain in
Alaska and w Canada; increasingly rare
in e Canada; spotty distribution in w
U.S., but increasing in California,
Colorado, Montana, Oregon, and
Washington.

Primarily nocturnal but active at any
time, the Wolverine is nonmigratory
and does not hibernate. Alternating
periods of activity and rest every three
or four hours, a male traverses a vast
home range of more than 1,000 square
miles (2,600 sq km), sharing it with two
or three females. A Wolverine can cover
great distances at a slow lope, swims
capably, and climbs quickly, often
pouncing on prey from a tree. Its
eyesight is poor, but its senses of smell
and hearing are excellent. Very powerful
for its size, the ferocious Wolverine is
capable of driving even a bear or
Mountain Lion from its kill. It eats
anything it can find, including Moose or
Elk slowed down in heavy snow,
beavers, deer, Porcupines, birds, and
squirrels, as well as eggs, roots, and
berries; it also consumes much carrion,
trailing Caribou herds and eating the
remains of wolf kills. The Wolverine
also follows trap lines, eating bait,
trapped animals, and cached food. It has
been known to raid cabins, marking
everything it cannot eat with musk,
urine, or droppings. The Wolverine was
once popularly called the "Glutton" (the
species name *gulo* means "glutton"), but
its truly voracious appetite is probably
an adaptation for survival where food is
often scarce. Careless about concealing
its own food caches, the Wolverine
marks them with a foul-smelling musk
that repels other carnivores. Its den,
which may contain leaves or grass, is
under an uprooted tree or in a crevice,
thicket, or other protected place. An
extended mating season increases the
probability that these solitary and

sparsely distributed animals will find a
mate. Young remain with the mother for
two years. Wolverine fur is used to line
or trim parka hoods, as the oils in it
make it frost-resistant.

---

258 American Badger
Taxidea taxus

Description: *Flattish body, wider than high, with short,
bowed legs. Shaggy coat grizzled gray to
brown. Short, bushy, yellowish tail.* Face
dark brown or black with *white cheeks;
narrow white stripe* runs from above nose
over head to nape. Snout pointed and
slightly upturned. Ears small. Feet dark
with *large foreclaws.* Male larger than
female. L 20–34″ (521–870 mm);
T 3⅞–6¼″ (98–157 mm); HF 3½–
5⅛″ (89–130 mm); Wt 7⅞–25 lb
(3.6–11.4 kg).

Similar Species: Wolverine is larger, bulkier, and
browner, and lacks facial stripes.

Breeding: Mates July–August; implantation
delayed until February. Litter of 1–5
young born March–April. Well furred
but blind at birth, the young are weaned
by June.

Sign: *Den:* Burrow with 8–12″ (200–300 mm)
elliptical entrance to accommodate
animal's flattish shape; surrounded by
large mound of earth scattered with
bones, fur, rattlesnake rattles, and
droppings. In vicinity of burrow, other
elliptical holes dug when foraging;
burrows of prairie dogs with openings
enlarged to capture occupants.
Scat: Cylindrical, usually segmented,
often showing bits of bone and fur;
definitive only in association with other
American Badger signs.
Tracks: Turned in sharply. Foreprint 2″
(50 mm) wide and long, even though
little heel pad shows; longer when claw
tips show. Hindprint narrower than
foreprint, 2″ (50 mm) long. Gait
variable, with hindprint printing before
or behind forefoot. Stride 6–12″

(150–300 mm); straddle 5–7″ (125–175 mm), wider in snow.

Habitat: Open plains and prairies, farmland, and sometimes edges of woods.

Range: Western U.S. east to e Texas, Oklahoma, n Missouri, n Illinois, n Indiana, and n Ohio; north to se British Columbia, much of Alberta, s Manitoba, and s Saskatchewan. Range appears to be increasing.

This powerful burrower is basically nocturnal but is often active by day, waddling about and occasionally moving at a clumsy trot. Its home range varies from about 590 to 4,200 acres (240–1,700 ha). The home range of the male is larger and encompasses the ranges of several females. Although primarily terrestrial, the American Badger swims and even dives, and on hot days sprawls in shallow water to cool off. It buries its droppings and cleans itself frequently, swallowing loose hair licked from its coat. Although the badger does not hibernate, it may become torpid in the coldest part of winter, remaining in a nest chamber deep within its burrow for several days or weeks. Adapted to feed mainly on small burrowing mammals, especially ground squirrels, pocket gophers, rats, and mice, which it usually digs out of the ground, a badger can destroy the burrows of an entire ground squirrel colony. This carnivore often forages by visiting its abandoned dens, as potential prey will frequently use them, and it may dig up hibernating animals. Occasionally a badger will dig itself into an inhabited burrow and await the occupant's return. A Coyote sometimes watches attentively as the badger digs for prey and may steal the rodent as it pops from an escape exit, but the Coyote and badger do not truly hunt together. Fond of rattlesnake meat, the badger is evidently unharmed by the venom unless the snake strikes its nose. The

American Badger also eats invertebrates
birds, reptiles, and carrion; it buries
surplus meat for future use. Burrows
or dens are central to the badger's
existence; it uses its many dens for
sleeping, giving birth, and food storage.
An active badger uses different dens
nearly every day, except when young
are present. Records show that one
badger dug a new burrow each day in
summer, reused one den for several
days in autumn, and used a single den
through most of the winter. Dens have
one entrance, with a pile of dirt just
outside that may serve as a latrine area.
The entrance may be plugged during
the coldest part of winter. Depending on
use, dens may vary in length and depth;
not all dens have a nest. Maternal dens
are more complex than other dens,
with the main tunnel branching and
rejoining, presumably so the badgers
can pass one another. Maternal dens
also have dead-end tunnels, pockets,
and chambers. Except during mating
and rearing of young, the American
Badger is solitary. Males and most
females do not mate until the second
year, though a few females become
pregnant as juveniles (4 to 5 months).
The young disperse in late summer.
Because the badger is a formidable
adversary, few animals will attack it. Its
thick fur, loose, tough hide, and heavy
neck muscles protect it as it bites, claws,
and exudes (not sprays) a skunk-like
musk, all the while snarling, squealing,
growling, and hissing. Despite such
ferocity, it seldom picks a fight,
preferring to retreat if necessary. A poor
runner, the badger will back into a
nearby burrow and face its tormentor
with its sharp teeth and strong claws;
once inside, it plugs the entrance hole.
If no burrow is available as refuge, it
may dig one, showering dirt in the face
of its attacker, and excavating so quickly
that it can outpace a person digging

with a shovel. The American Badger probably got its common name from the white, badge-like mark on its forehead. In Europe it was once considered sport to bait badgers, and the verb "to badger," meaning to tease, annoy, or persecute, derives from that cruel practice. Badger hair is used to make paintbrushes, and the coarse bristles were once used in shaving brushes. Although their excavation activity can pose hazards for hoofed animals, badgers are valuable in controlling rodent populations in agricultural areas.

253 Western Spotted Skunk
Spilogale gracilis

Description: A small skunk. Black, with horizontal white stripes on neck and shoulders; *irregular vertical stripes and elongated spots on sides.* White spots on top of head, between eyes. Tail has white tip. L 13⅝–18¾" (345–473 mm); T 4¼–7" (110–178 mm); HF 1½–2⅛" (37–55 mm); Wt 1–1½ lb (2.2–3.3 kg).

Similar Species: Eastern Spotted Skunk is best distinguished by range. Other skunks are larger, with horizontal stripes or bands only.

Breeding: Mates late September–October (as early as July in southern parts of its range); implantation delayed until around March, followed by 50–60 days of development. Female first mates at 4–5 months of age.

Sign: *Scat:* Dark, small, ⅞" (22 mm) diameter, irregularly cylindrical.
Tracks: Like Striped Skunk's but smaller. Hindprint 1¼" (32 mm) long; heel pad shows more definite lobing. Stride very irregular, unlike that of other skunks.

Habitat: Mixed woodlands and open areas, scrub, and farmland.

Range: From sw British Columbia, Idaho, and sw Wyoming south to Mexico and w Texas.

The closely related and similarly patterned Eastern and Western Spotted skunks have alternately been considered the same and separate species by different mammalogists. The two do not occur together over much of their ranges. They overlap geographically, apparently without interbreeding, in Wyoming and perhaps in Oklahoma, but their relationship in other areas is unknown. If they are capable of interbreeding, the two would likely be treated as separate subspecies of a single species.

Eastern Spotted Skunk
Spilogale putorius

Description: A small skunk, virtually identical to Western Spotted Skunk. Black with horizontal white stripes on neck and shoulders, *irregular vertical stripes and elongated spots on sides.* White spots on top of head, between eyes. Tail with white tip. L 13½–22″ (343–563 mm); T 2¾–8⅝″ (68–219 mm); HF 1¼–2¼″ (33–56 mm); Wt 27–35 oz (784–999 g).

Similar Species: Western Spotted Skunk is best distinguished by range. Other skunks are larger, with horizontal stripes or bands only.

Breeding: Breeds late winter in northern parts of range. No delayed implantation; gestation 50–60 days. 1 litter of 2–6 young born April–May, in a Woodchuck burrow or hollow log, under a foundation or in another protected place.

Sign: *Scat:* Dark, small, ⅞″ (22 mm) diameter, irregularly cylindrical.
Tracks: Like Striped Skunk's but smaller. Hindprint 1¼″ (32 mm) long; heel pad shows more definite lobing. Stride very irregular, unlike that of other skunks.

Habitat: Mixed woodlands and open areas, scrub and farmland.

Range: In Midwest: e Wyoming, Minnesota, and Dakotas east to Wisconsin, and

south to Texas and Louisiana. In East: Mississippi northeast to s Pennsylvania and South Carolina, and east to Florida.

Faster and more agile than the larger skunks, the Eastern Spotted Skunk is also a good climber, ascending trees to flee predators and occasionally to forage. This species is more social than other skunks, and several individuals may share a den in winter. Highly carnivorous, the Eastern Spotted Skunk feeds mainly on small mammals, but also eats grubs and other insects, as well as corn, grapes, and mulberries. Except when rearing the young, this skunk does not occupy a particular territory, but rather moves about and dens wherever convenient. Maternity dens are established in burrows of other animals, hollow logs, brush piles, or other protected places. The female breeds for the first time at 9 or 10 months. The young are born blind and furred in spring; they achieve adult coloration in early summer. Males do not participate in the rearing of the young. Although most larger carnivores will kill and eat this skunk if they can do so without being sprayed, they usually back off when the skunk starts its unique threat display. If a predator refuses to retreat when the skunk raises its tail, the skunk turns its back, stands on its forefeet, raises its tail again, spreads its hindfeet, and sprays, often for a distance of 12 feet (3.5 m). The great horned owl, the Eastern Spotted Skunk's chief predator, can strike from above without warning and carry off a young skunk before its mother can spray. Other predators are the Coyote, the domestic dog, and perhaps the barred owl, but humans are the main enemy of skunks, often killing them casually out of fear, or running over them with automobiles. The fur of the Eastern Spotted Skunk is the finest and silkiest of the skunk furs, and pelts were once considered valuable.

255 Hooded Skunk
Mephitis macroura

Description: Usually *black except for 1 or 2 widely separated narrow white stripes along upper side;* in one phase, back and tail are white with black hairs interspersed. Hair on back of neck often forms ruff. L 22–31″ (558–790 mm); T 10⅞–17⅛″ (275–435 mm); HF 2¼–2⅞″ (58–73 mm).

Similar Species: Striped Skunk usually has 2 broad whit stripes close together, and shorter tail. Common Hog-nosed Skunk has back and tail white without black hairs; top of snout is naked.

Breeding: 1 litter of 3–5 young born May–early June.

Habitat: Rocky ledges; tangled vegetation along streams.

Range: Southeastern Arizona, sw New Mexico, and extreme w Texas.

The habits of the Hooded Skunk are probably similar to those of the Striped Skunk. This species apparently feeds on insects; the stomach of one individual from New Mexico contained beetle fragments. The common name refers to the ruff of long white hair that often forms at the back of the neck.

254 Striped Skunk
Mephitis mephitis

Description: Typically *black with 2 broad white stripes on back meeting in cap on head and shoulders;* thin white stripe down center of face. Bushy black tail, often with white tip or fringe. Species coloration varies from mostly black to mostly white. Male larger than female. L 20–31″ (522–800 mm); T 7¼–15½″ (184–393 mm); HF 2¼–3½″ (57–90 mm); Wt 6–14 lb (2.7–6.3 kg).

Similar Species: Common Hog-nosed Skunk has white back and tail, snout naked on top, and no white facial stripe. Hooded Skunk

has longer tail and is usually mostly black with narrow white stripes (in white-backed phase, black hairs are interspersed).

Breeding: Mates February–April; implantation delayed about 19 days; total gestation 62–66 days. Litter of 4–7 young born in mid-May; young are blind at birth, their very fine hair clearly marked with black-and-white pattern.

Sign: Strong odor if skunk has sprayed recently. Small pits in ground or patches of clawed-up earth.
Den: Burrow, in any protected place, with up to 5 entrances, each about 8″ (200 mm) in diameter. Entrances well hidden, but sometimes marked with nesting material and snagged hairs. Burrow has 1–3 chambers, 12–15″ (300–380 mm) in diameter; one chamber has nest of dried grass.
Scat: Varies (as with most omnivores): generally dark, cylindrical, sometimes segmented, varying in size.
Tracks: 5 toes print when clear; sometimes claws show. Foreprints 1–1¾″ (25–44 mm) long, slightly wider. Hindprints 1¼–2″ (32–50 mm) long, slightly narrower, broader at front; somewhat flat-footed. Stride 4–6″ (100–150 mm). Because skunks shuffle and waddle, tracks are closer than in other mustelids, and foreprints and hindprints usually do not overlap. When animal is running, stride is longer, and hindfeet print ahead of forefeet. Trail undulates slightly because of waddling walk.

Habitat: Desert, woodlands, grassy plains, and suburbs.

Range: Most of U.S.; s tier of Canadian provinces.

Whereas most mammals have evolved coloration that blends with their environment, the Striped Skunk, like other skunks, is boldly colored, advertising to potential enemies that it is not to be bothered. Its anal glands

hold about a tablespoon of a fetid, oily, yellowish musk, enough for five or six jets of spray—although one is usually enough. When threatened, the Striped Skunk will face the intruder, arch and elevate its tail, erect the tail hairs, chatter its teeth, and stomp the ground with the front feet. This usually causes the intruder to retreat, but if it remains, the skunk will twist its back around, raise its tail straight up, evert its anal nipples, and spray scent 10 to 15 feet (3–5 m). The mist may reach three times as far, and the smell may carry a mile. Spray in the eyes causes intense pain and fleeting loss of vision. Sudden movement, noise, or a close approach can trigger the spray, and the Striped Skunk can spray even when held aloft by the tail. Ammonia or tomato juice can be used to remove the odor; carbolic soap and water are best for washing skin. The Striped Skunk is primarily nocturnal and does not hibernate, although during extremely cold weather it may become temporarily dormant. The animal's temperature drops only from about 98.6° to 87.8° F (37° to 31° C), rather than down to the temperature of its den. The Striped Skunk is an omnivore, feeding heavily on a wide variety of animal food in spring and summer, including insects and grubs, small mammals, the eggs of ground-nesting birds, and amphibians. Some of the more important invertebrate foods consumed are beetles and their larvae, grasshoppers and crickets, earthworms, butterfly and moth larvae, spiders, snails, ants, bees and wasps, and crayfish. This skunk eats fruits in season, such as wild cherries, ground cherries, blackberries, blueberries, and many others. In the fall, the animal gorges itself to fatten up in preparation for the lean winter months. The Striped Skunk usually dens in a burrow that has been abandoned by another animal, although it may also dig its own or use a

protected place, such as a hollow log, crevice, or the space beneath a building. Maternal and wintertime dens are underground; other dens are often aboveground. The young are weaned at six to seven weeks, at which time their scent has developed but is not yet very potent. Mother and offspring begin to hunt together at about this time. A mother skunk is fiercely protective of her young, and at the approach of an intruder she will posture and spray if necessary. The procession of a mother skunk followed by her young in single file is an amusing sight. The skunk is not a social animal, although in winter several skunks will sometimes occupy one den. The only serious predator of the skunk is the great horned owl. The Striped Skunk is currently the chief carrier of rabies in the U.S. Unusual behavior can be a sign that the animal is infected. As skunks are usually out at night, one active in the daytime could well be rabid. Mothballs sprinkled on the ground discourage skunks from digging up lawns for insects and visiting homes or campsites, since they and many other small animals are repelled by the smell of camphor. Pelts of the Striped Skunk are not highly valued, but the musk, once its odor is removed, is used as a perfume base because of its clinging quality.

256 Common Hog-nosed Skunk
"Rooter Skunk"
Conepatus mesoleucus

Description: *Top of head, back, and tail white;* lower portions black. *Long snout, naked on top,* with broad nose pad. Foreclaws enlarged. Male larger than female. L 20–36" (513–900 mm); T 6⅞–16⅛" (174–410 mm); HF 2½–3½" (65–90 mm); Wt 5–10 lb (2.3–4.5 kg).

Similar Species: Striped Skunks and most Hooded Skunks have black tails. Hooded Skunk

phase with white tail and back has
scattered black hairs.

Breeding: Mates in March; litter of 2–4 young
born April–May in den in rocky crevice.
Young are weaned in August.

Sign: Extensive patches of ground torn up and
pitted by rooting.
Scat: Similar to Striped Skunk's.
Tracks: Similar to Striped Skunk's, but
toeprints in forefeet are longer, often
longer than heel pad.

Habitat: Foothills and brushy areas.

Range: Southern Arizona, much of New
Mexico, se Colorado, w Oklahoma, and
nw and s Texas.

Although primarily nocturnal like other
skunks, the Common Hog-nosed Skunk
may forage by day in winter. Its broad
nose pad is an adaptation for rooting for
the insects that are its chief food, which
explains its alternate name, "Rooter
Skunk." This species also eats reptiles,
arachnids, mollusks, small mammals,
and vegetation. It makes its dens in
rocky crevices. The Eastern Hog-nosed
Skunk *(C. leuconotus),* found north of
Mexico only in extreme southern Texas,
is considered a separate species, although
it may prove to be a subspecies. The two
skunks do not occur together, although
some mammalogists believe that they
probably have the potential to intergrade
(merge gradually with one another by
interbreeding).

259, 260 **Northern River Otter**
Lutra canadensis

Description: A large aquatic mustelid with elongated
body and broad, flattened head. *Dark
brown above (looks black when wet),* with
paler belly. Throat often silver gray. Ears
and eyes small. Prominent, whitish
whiskers. *Long tail thick at base, gradually
tapering to a point. Feet webbed.* Male
larger than female. L 35–52" (889–
1,313 mm); T 11⅞–20" (300–507

mm); HF 3⅞–5¾" (100–146 mm); Wt 11–30 lb (5–13.6 kg).

Similar Species: Sea Otter is yellowish or grayish on head and back of neck.

Breeding: Mates in early spring, just after birth of litter; implantation delayed; total gestation 8–9½ months. Litter of 1–6 young born blind and fully furred, in March or April.

Sign: Rough trough through loose snow; plunge holes in snow.
Slides: The most obvious and best-known evidence of otters; may be of ice and snow or mud. Riverbank slides 8" (200 mm) wide, much wider with heavy use. Snowslides 12" (300 mm) wide or wider, up to 25' (7.5 m) long; often on flat ground, sometimes pitted with blurred prints where otter has given itself a push for momentum.
Rolling places: Areas of flattened vegetation up to 6½' (2 m) wide, with twisted tufts of grass marked with musk and sometimes droppings.
Haul-outs: Trails from the water, often containing crayfish parts and droppings.
Scat: Irregular, sometimes short, rounded segments, sometimes flattened masses, containing fish bones, scales, or crayfish parts; when fresh, often greenish and slimy. Scat most often found on banks of stream or pond, on logs, or on rocks in water.
Trail: Meandering, about 8" (200 mm) wide, between neighboring bodies of water or other favored spots, such as rolling areas or slides. Trail may show sidling walk.
Tracks: 3¼" (80 mm) wide or more; often show only heel pad and claws. Toes fan out widely, but webbing rarely prints, except in mud. Running stride 12–24" (300–600 mm).

Habitat: Primarily along rivers, ponds, and lakes in wooded areas, but otters will roam far from water.

Range: Alaska and most of Canada south to n California and n Utah; in East, from Newfoundland south to Florida;

extirpated from most areas of Midwest, but currently being reintroduced into some areas.

The Northern River Otter is active by day if not disturbed by human activity. Well adapted to its aquatic life, it has a streamlined body, rudder-like tail, and ears and nostrils that are valved to keep out water. The animal swims rapidly both underwater and on the surface, moving like a flexible torpedo, either forward or backward, with astonishing grace and power. To observe its surroundings, it raises its head high and treads water. A river otter can remain submerged for several minutes and can dive to a depth of 55 feet (17 m), swimming as far as ¼ mile (.4 km) underwater if necessary. Also at ease on land, the river otter will lope along, then slide, and it also runs fairly well. River otters are among the most playful of animals. A lone river otter often amuses itself by rolling about, sliding, diving, or "body surfing" along on a rapid current. In family groups, otters take turns sliding and will frolic together in the water. A river otter makes the most of a snowslide by running to get speed, then leaping onto the snow or ice with its forelegs folded close to its body for a streamlined toboggan ride. The Northern River Otter feeds mainly on fish, often caught in a quick broadside snap, but also eats small mammals such as mice, as well as terrestrial and aquatic invertebrates. It may capture fish by pursuit or by digging into the sand and lying in wait. Some anglers suspect the Northern River Otter of depleting game fish stocks, particularly trout. While it will eat game fish, it more often eats the slower-moving suckers, chubs, daces, darters, and catfish, as well as schooling fish such as the bluegill, which are caught more easily. A pair of river otters may work together to drive a school of

fish into an inlet, where they can be
easily caught. The otter can manipulate
items in its forepaws and carries large
catches to land to be eaten. It often digs
its permanent den in banks, establishing
underwater and aboveground entrances.
Inside it constructs a nest of sticks,
grass, reeds, and leaves. This species
rests under roots or overhangs, in hollow
logs, burrows of other animals, or beaver
lodges, which if heavily used by otters
may also contain some nesting materials.
The river otter's vocalizations include a
whistle, probably used to communicate
over distances, and a shrill, chattering
call, emitted during the mating season.
Otters chuckle softly to siblings or mates,
apparently as a sign of affection, and also
chirp, grunt, snort, and growl. The
male river otter presumably mates with
one or more females that have home
ranges within his territory. The female
establishes the natal den shortly before
giving birth. Weaned at four months,
the young disperse in fall or winter
before the arrival of the next litter. The
male, evicted while the young are small,
returns to help care for them when they
are half-grown. While sociable most of
the year, during the breeding season
competing males may battle. The
Northern River Otter's fur is durable,
thick, and beautiful, and excessive
trapping in the past has greatly
diminished the animals in number.
More recently, water and air pollution,
including mercury fallout, have taken
a toll on otter populations. Some river
otters, however, may be developing a
tolerance to certain toxic substances, and
their populations are slowly increasing.

261, 262 **Sea Otter**
Enhydra lutris

Description: *Dark brown; head and back of neck
yellowish or grayish.* Old males may have
white heads. Fairly short tail, thick at

base, gradually tapering. Feet webbed; *hindfeet flipper-like.* Male somewhat larger than female. L 30–71" (760–1,810 mm); T 10¼–14¼" (260–360 mm); HF 5⅞–8¾" (150–222 mm); Wt 25–80 lb (11.4–36.3 kg).

Similar Species: Northern River Otter is smaller, with longer tail and silver-gray throat and belly; lives mainly in freshwater.

Breeding: 1 (rarely 2) pups born in winter or spring, with fur and teeth, and eyes open. Gestation 6½–9 months.

Sign: A loud, rapid tapping: the sound of shellfish being cracked open on stones. *Scat:* When fresh, massed or cylindrical, thick, 4–5" (100–125 mm) long, containing bits of shellfish or shell; crumbles easily. Occasionally found on beaches.
Tracks: Rarely seen. Hindprint fan-shaped, 6" (150 mm) long, almost equally wide at front. Foreprint smaller, roundish.

Habitat: Coastal waters within a mile (1.5 km) of shore; especially rocky shallows with kelp beds and abundant shellfish.

Range: Pacific Coast from California to Alaska.

Highly aquatic, the Sea Otter eats, sleeps, mates, and even gives birth at sea, and can remain submerged for four to five minutes. Flipper-like hindlegs make the species clumsy on land; it takes to the beach only to wait out storms. By day, the Sea Otter feeds while floating on its back, sculling with its tail. If in a hurry, it swims on its belly, using its feet and tail like a Northern River Otter. At night, the Sea Otter wraps strands of kelp about its body to secure its position in the kelp beds where it sleeps. Unlike other oceangoing mammals, the Sea Otter has no insulating blubber; air trapped in its fine fur keeps it warm as well as buoyant. If the fur is damaged in an oil spill or by other pollution, the otter can die from exposure or cold. The Sea Otter is a greater user of tools than any other

mammal outside the primate family. When it dives for food, it also brings up a small rock. It then floats on its back, places the rock on its chest, and cracks the shell against it. Abalone, sea urchins, crabs, mussels, and fish are the chief foods of this species. A playful animal, the Sea Otter may interrupt a meal to dive and frolic underwater; it consorts amiably with seals and sea lions, sometimes touching noses. The Sea Otter watches for danger by standing in the water and shading its eyes with both forefeet; if it spots such predators as sharks and killer whales, it hides in kelp beds. Sea Otter pups are weaned at one year, but may remain with their mother even after she has a new pup. The mother floats on her back to let her offspring nurse, nap, and play on her chest. When alarmed, she may tuck her pup under a foreleg and dive for safety. Once an abundant species, the Sea Otter was so heavily hunted for its highly prized pelt that by 1911, when an international treaty forbade its massacre, it had nearly become extinct. The animal was not seen in California for many years, but in the spring of 1938, a herd appeared in the sea south of Carmel. Today the population there is perhaps 1,000, and even larger herds are found off the coast of southern Alaska and the Aleutian Islands. There is an ongoing controversy between fishermen and conservationists concerning the Sea Otter. Conservationists want to keep this animal on the U.S. government list of endangered species to ensure its protection, but fishermen want to control the Sea Otter in order to limit damage to abalone populations.

Cats
Family Felidae

Cats are native to most parts of the world except Australia and New Zealand. There are 18 genera and 36 species of cats worldwide. Counting the domestic cat, eight species are found in North America north of Mexico, although only the Mountain Lion, Lynx, Bobcat, and domestic cat are present in significant numbers. The short-tailed cats, the Bobcat and Lynx, are in the genus *Lynx,* while the rest have long tails and are included in the genera *Felis* and *Panthera.*

Cats are well equipped for hunting. All have long, sleek bodies, powerful legs, short heads with relatively small, rounded ears, and eyes that face forward, providing the binocular vision and depth perception so crucial to locating prey. Their night vision is superb; the pupils, contracted into vertical slits by day, expand in the dark almost to fill the eye, while a layer of cells behind the retina absorbs even the dimmest light. Cats are one of the few types of mammals that have color vision. Their highly sensitive whiskers provide tactile information. Along with weasels, cats are among the most carnivorous of mammals, and they are usually at the top of the food chain. They have few enemies other than humans. Their molariform teeth have well-developed shearing edges, and the canines are enlarged. Rough tongues, with which they groom their fur, can also rasp meat from bones of prey. The five toes on the forefoot and four on the hindfoot have retractile claws, which are usually withdrawn so as not to become blunted but are extended to slash prey. Soft footpads surrounded by fur permit stealthy stalking. Cats can swim though most do not like getting wet, and they climb well. All North American cats can purr, and the Mountain Lion makes

a variety of loud sounds, but the only
member of the family that roars is the
Jaguar, which is rare in the United
States. Most cats are tree-dwellers and
agile climbers that live in remote
wooded areas. They mark out territory
with feces as well as urine and tree
scratches. Most are highly secretive,
nocturnal, and solitary except during
the mating season. North American
species generally mate in winter or
spring. Copulation usually stimulates
ovulation in the female. Gestation is 50
to 110 days; most cats have one annual
litter of one to six young. The kittens,
born blind and helpless, receive
extended parental training in the ways
of the wild.

Domestic Cat
"House Cat," "Feral Cat"
Felis catus

Description: Too familiar to need much description.
A small cat; tail about half total length.
Color and pattern vary greatly in 30
domestic breeds. Averages: L 30″ (75
cm); T 15″ (38 cm); Wt 6½–13 lb
(3–6 kg).

Similar Species: Lynx and Bobcat have short tails. Long-
tailed cats are much larger.

Breeding: 2 litters per year of 1–8 young; births at
any time of year, but most often March–
August; gestation 62 days.

Sign: *Tracks:* Nearly round, showing no claw
marks; about 9″ (22–23 cm) apart.
Tracks usually in a straight line, with
the smaller hindfeet placed in the
foreprints.

Habitat: Usually around buildings; hunts in open
fields.

Range: Throughout North America.

The house cat was domesticated from
Felis silvestris, found in Eurasia and
Africa. It is usually not included in field
guides, but domesticated cats do often
become feral, or at least hunt afield, and

here we describe the behavior of the feral individual. Most free-ranging feral cats live in or around houses or other buildings and are fed, but truly feral individuals presumably raise their young in burrows or other protected areas. Whether fed or not, the feral cat spends much time hunting, and may have a major effect on native wildlife. Feral cats have even been known to eliminate endemic island species. When stalking prey, the feral cat waits for an extended period for the right moment, then pounces. Meadow and other voles are important food items, but the species also captures a good number of other small mammals and birds. The male feral cat urine-marks his territory, which is larger than that of the female and encompasses the territories of several females. Young cats can fend for themselves by one month of age and are weaned by two months. While the dog is the traditional enemy of this species, a cat often can hold its own against one, though a large dog can kill an adult cat by shaking it. The automobile is the most significant enemy of the feral cat.

263–266 **Mountain Lion**
"Cougar," "Puma," "Panther," "Catamount"
Felis concolor

Description: *A large, unspotted cat with a relatively small head and a long, dark-tipped tail. Pale brown to tawny above;* white overlaid with buff below. Dark spot at base of whiskers. Ears short and rounded, with dark backs. Legs long and heavy; feet large. Juvenile buff with black spots. L 6–9′ (1.5–2.75 m); T 21–37″ (53–92 cm); HF 8¾–12″ (22–31 cm); Wt 75–275 lb (34–125 kg).

Similar Species: Jaguar is spotted. Jaguarundi is much smaller and shorter-legged.

Breeding: No fixed mating season; 1–6 young usually born in midsummer every other

year. Gestation 82–98 days. Newborn weighs about 14 oz (400 g).

Sign: Scratches or gashes on trees used as scratching posts, longer and higher than those left by Bobcat or Lynx. Remains of a kill, often conspicuous, to which the cat may return; may be loosely covered with branches, leaves, and litter.
Scrapes and scent posts: Piles of dirt in home range kicked up by hindfeet; 6–18″ (15–45 cm) across, 1–2″ (3–5 cm) high. A more conspicuous scrape may be a scent post, where a male Mountain Lion has loosely piled leaves or debris and urinated on the pile to mark his territory.
Scat: Usually copious; varies from masses to irregular cylinders and pellets; frequently contains traces of hair or bone scraps. Sometimes covered with earth, but often left exposed or partly exposed as a scent post; if covered, scratchings on the ground probably indicate general direction of movement, as Mountain Lions habitually face their line of travel as they scratch.
Tracks: Prints quite round, usually with all 4 lower toes showing but no claw marks, as claws are retracted. Foreprint 3¼–4″ (80–100 mm) long; hindprint slightly smaller. Lobed heel pad has single scalloped edge at front, double scalloped edge at rear. Tracks usually in a fairly straight line, staggered in pairs, with hindfoot track close to or overlapping forefoot track, but seldom registering precisely within it. Straddle 8–10″ (200–250 mm); length of stride 12–28″ (300–700 mm). Longer gaps indicate bounding, when all feet come down close together. In snow, prints slightly larger, sometimes blurred by thicker winter fur, and elongated by foot drag marks; in deep snow, tail may drag and leave trace between prints.

Habitat: Originally varied; now generally mountainous, semi-arid terrain; subtropical and tropical forests and swamps.

Range: Western North America from British
Columbia and s Alberta south through
w Wyoming to California, w Texas and
s Texas. Despite numerous reports
throughout U.S., Everglades area of s
Florida, which contains perhaps 45–75
individuals, has only viable population
east of Mississippi River.

The most widely distributed cat in the
Americas (found from Canada to
Argentina), the Mountain Lion is a
solitary, strongly territorial hunting
species. Unlike most cats, it hunts day
or night, although it is generally active
by day only in undisturbed areas,
choosing to hunt at night in populated
areas to avoid humans. A good climber
and excellent jumper, able to leap more
than 20 feet (6 m), this animal swims
only when necessary. It feeds primarily
on large mammals, especially deer, but
also eats Coyotes, porcupines, beavers,
mice, marmots, hares, raccoons, birds,
and even grasshoppers. Sometimes it
waits for passing game, but more often
it travels widely after prey; a male may
cover up to 25 miles (40 km) in one
night. It can outrun a deer, but only for
short distances. After locating large prey
by scent or sound, it usually slinks
forward slowly and silently, with belly
low to the ground and legs tensed to
leap. It tries to stalk within 30 feet (9 m)
before running from its hiding place
and leaping onto its victim's back,
keeping its hindlegs on the ground for
support, control, and stability. The
Mountain Lion kills its prey by biting
into the back of the victim's neck.
Where deer abound, an adult Mountain
Lion may kill an average of one per
week. (This is often beneficial to the
deer herd, helping to keep it from
overpopulating.) This carnivore covers
the meat it does not eat immediately
with leaves, sticks, and like material for
later use, and may visit the cache several
times. Usually silent, the Mountain

Lion can produce many kinds of calls, including screams, hisses, and growls. It also utters a shrill, piercing whistle, evidently an alarm, when it has been treed or cornered; a female uses this whistle to signal her cubs. The Mountain Lion's bloodcurdling mating call has been likened to a woman's scream. The male has a large home range that does not overlap with that of another male; the female has a smaller one that may overlap with those of other females and may be enclosed by that of a male. The home range of a male (and sometimes of a female) is marked by "scrapes," piles of dirt kicked up by the hindfeet. The Mountain Lion breeds at two and a half years, then generally every other year thereafter. The young are born in a maternity den that is lined with a small amount of moss or other vegetation and located in a rock shelter, crevice, pile of rocks, thicket, cave, or other protected place. The newborn cubs, heavily spotted for the first three months of life, are raised only by the female. At about three months, the young are weaned and begin hunting with the mother. Mother and young, who remain together for about a year and a half, communicate by licking, rubbing, and vocalizing. The young produce loud chirping whistles. A female Mountain Lion can breed until at least 12 years of age, a male to at least 20. These animals pair only during the breeding season, when for about two weeks male and female hunt together and sleep side by side.

Because the Mountain Lion requires isolated or undisturbed game-rich wilderness, it has declined or been extirpated in much of the habitat where it once thrived early in the 20th century. This is due not only to persecution by humans, but also to the disappearance of its main food, the White-tailed Deer, over much of its range. In recent years, there have been a few sightings of

animals or tracks in Canada's Maritime Provinces and in upper New England, New York State, and elsewhere in the East, but most reports turn out to be false. The last documented records for Pennsylvania were in 1871 (except one taken near Edinboro in 1967); for Virginia, in 1882; and for West Virginia, in 1887. There are relatively recent reliable records in Alabama, including Tuscaloosa County, where one was shot in 1956, and Clarke County, where plaster casts were made from tracks in 1961 and 1966. There may still be a remnant population in Louisiana and extreme southeastern Arkansas, as there are reliable records from Caddo Parish, Louisiana, in 1965; from Madison Parish, Louisiana, in 1971; and from Ashley County, Arkansas, in 1969. In 1971, a specimen was taken near Pikeville, Tennessee. Radio-tracking is being used to study the behavior of Mountain Lions in Florida, and an office has been established to investigate reports of sightings in the southern Appalachians. Occasionally Mountain Lions have been known to injure or even kill people, usually children, but they tend to avoid humans unless cornered or extremely hungry. There have been rare, unexplained killing orgies, when an individual has slaughtered several deer or a flock of domestic sheep in one night. For many years, the species was pursued by bounty hunters and persecuted as a threat to livestock; currently it is fully protected where rare (such as the eastern U.S.), and classified as a game animal where abundant.

---

268 **Ocelot**
Felis pardalis

Description: A medium-size, slim cat. Grayish to tawny or gold, heavily marked with *black-bordered brown spots that tend to form*

lines. Spots include rosettes, rings, speckles, slashes, and bars. Tail fairly long, but shorter than hindleg. L 3'1"–4'6" (92–137 cm); T 10⅝–16" (27–40 cm); HF 5¼–7" (13–18 cm); Wt 20–40 lb (9.1–18.2 kg).

Similar Species: Jaguar is much larger and marked almost entirely with rosettes. Margay is smaller, with similar coloration and markings; has longer tail, longer than hindleg.

Breeding: 1 litter of 2–4 young born after gestation of 70 days; births fall–winter in Texas.

Sign: Scratchings and other signs similar to Bobcat's.
Scat: Similar to Bobcat's.
Tracks: Similar to Bobcat's, but slightly larger and wider: 2–2½" (50–62 mm) long, about equally wide and with forefoot larger than hindfoot. Front and rear edges of print of heel pad less scalloped than Bobcat's and sometimes even convex.

Habitat: Forested or brushy areas and dense chaparral.

Range: Southern Oklahoma, sw Arkansas, w Louisiana, Texas, and extreme se Arizona. The Santa Ana National Wildlife Refuge, along the Rio Grande in s Texas, offers the best opportunities for seeing the Ocelot in the wild.

The Ocelot climbs well and silently, and sometimes even catches birds perched in trees. Its principal foods include mice, rats, rabbits, birds, snakes, lizards, fish, frogs, and young or small domestic animals. A good swimmer, it occasionally hunts along streams. Ocelots are usually solitary but sometimes travel and hunt in pairs, probably as mates, maintaining contact and signaling each other with meows like those of domestic cats. The Ocelot population has declined drastically because of habitat loss, efforts at eradication in order to protect small livestock and poultry, and, more

importantly, the trade in exotic furs and pets. Mother Ocelots have often been killed as humans captured their kittens, which, while affectionate and easily tamed when young, become unpredictable and sometimes dangerous when mature. The Ocelot is now fully protected in the U.S., and the trafficking of skins and the selling of live Ocelots as pets are banned.

269 Margay
Felis wiedii

Description: Ocelot-like, but smaller. Ground color grayish or yellowish to tawny; belly white with brown spots. *Tail is proportionally long and fairly bushy.* Spots on sides brown and irregular in shape, often with dark buff centers. *4 dark brownish stripes on back, 1 on neck.* L 31–51" (80–130 cm); T 13–20" (33–51 cm); HF 3½–5⅛" (9–13 cm); Wt to 22 lb (10 kg).

Similar Species: Ocelot is larger, with similar color and markings, and shorter, narrower tail. Jaguar is much larger, with rosette-shaped spots.

Breeding: Mates year-round; litter of 1–2 young.

Sign: *Tracks:* Similar to Ocelot's, but forefeet and hindfeet same size.

Habitat: Forested areas.

Range: Rare in U.S.; only known sighting occurred at Eagle Pass in Maverick County, Texas, in 1850s. Occurs from Mexico to Argentina.

Skilled at climbing, the Margay is the only North American cat known to go down a tree headfirst. This arboreal acrobat can hang from a bough by one unusually long-clawed hindfoot. Little else is known of the habits of this nocturnal species, although they are believed to be similar to those of the Ocelot. The Margay is classified as an endangered species by the U.S. government.

270 Jaguarundi
Felis yagouaroundi

Description: A small, unspotted cat. *3 distinct coat colors: black, gray (entirely grayish), and reddish (white or pale below).* Long body and tail; legs short. L 3'–4'6" (89–137 cm); T 13–24" (33–61 cm); HF 4¾–6" (12–15 cm); Wt 15–18 lb (6.7–8.1 kg).

Similar Species: Mountain Lion is much larger. Other long-tailed cats are spotted.

Breeding: Mates any time of year, but most births occur in spring and late summer; 2 annual litters, generally of 2 or 3 young. Gestation 63–70 days.

Sign: *Tracks:* Foreprint 1½–1¾" (3.7–4.5 cm) long; hindprint almost as large. Typical cat tracks: 4 toeprints, no claws.

Habitat: Brushy thickets; areas with cactus, catclaw, mesquite, and other spiny plants; forests and swampy areas.

Range: Rare in U.S.: only extreme se Arizona and extreme se Texas. Also Mexico south to n Argentina.

The habits of the elongated and elegant Jaguarundi are little known. This species is occasionally active during the day, especially in the morning. It swims well, crossing rivers when necessary, and probably preys on fish and other aquatic species. Stalking and ambushing less than most cats, the Jaguarundi is an excellent runner, sprinting after and overtaking even the fastest quarry. Although this cat will climb, it spends most of its time on the ground, preying on birds (including poultry), reptiles, and small to medium-size mammals such as rats, mice, and rabbits. The diet may include a little fruit at times. A solitary animal, the Jaguarundi pairs only for breeding. It makes its den in thickets, brush, or under fallen trees. Kittens are born with distinct light spots, which soon disappear. All three coat colors may be displayed within the same litter. In the U.S., the Jaguarundi is listed as an endangered species.

267 Jaguar
Panthera onca

Description: A large, heavy-bodied, big-headed cat.
Yellowish to tawny, *spotted with black
rosettes or rings* in horizontal rows along
the back and sides; most rings are tan
inside, with 1 or 2 black spots. Legs,
head, and tail have smaller, solid spots,
usually giving way to incomplete bands
near the end of the tail. L 5′2″–7′11″
(1.57–2.42 m); T 17–27″ (43–67 cm);
HF 9–12″ (22–30 cm); Wt 119–300 lb
(54–136 kg).

Similar Species: Mountain Lion is unspotted. Ocelot is
smaller. Margay is much smaller and
lacks rosettes.

Breeding: Mates December–January in northern
part of range; 2–4 young born April–
May in den in cave, rock shelter, dense
thorn thicket, under tree roots, or in a
similar shelter.

Sign: *Scat:* Similar to Mountain Lion's.
Tracks: Very difficult to distinguish
from those of average-size Mountain
Lion. Foreprint 4–4½″ (100–115 mm)
long, about equally wide; hindprint
slightly smaller.

Habitat: Brush, primarily in forest; forested
areas, jungles, swamps, and arid
mountainous scrub.

Range: Very rare in U.S.: s Arizona, New
Mexico, and s Texas. Also from Mexico
to Brazil and n Patagonia.

The biggest and most powerful North
American cat, the Jaguar is the only one
that roars. It moves over a large home
range with a diameter of 3 to 15 miles
(5–25 km) where prey is abundant,
larger where prey is scarce. This cat
hunts mostly on the ground, but climbs
well and sometimes ambushes prey by
leaping from tree limbs or ledges. It is,
however, less arboreal than the Mountain
Lion, and its tail, less necessary for
balancing, is proportionally shorter.
Unlike most cats, the Jaguar is very
fond of water, delighting in playing in

shallow pools on hot days and without hesitation entering ponds and streams for fish and other aquatic life. The Jaguar's diet includes deer, peccaries, rabbits, large ground birds, sea turtle eggs (which it digs up along the coast), and livestock. This species has incredibly powerful jaws and is much stronger than the Mountain Lion; it can haul a full-grown cow for a mile (1.5 km) or more. The Jaguar is solitary except when breeding and rearing young. While some mated pairs remain together, it is more common for male and female to separate after a year, when the young disperse. The kittens are born with heavily spotted fur and with their eyes closed. At first, the male keeps his distance, but soon he brings food for his nursing mate and later for the young. The name "Jaguar" derives from an American Indian word meaning "the killer that takes its prey in a single bound." In pre-Columbian Mexico, Guatemala, and Peru, the Jaguar was worshiped as a god. Although a great traveler, the Jaguar has been sighted only rarely in the U.S. since the 1940s, and is classified as endangered by the federal government. Individuals occasionally wander north from Mexico.

271–275 **Lynx**
Lynx lynx

Description: A medium-size cat. Light gray, with scattered pale brown to blackish hairs; underparts cinnamon-brownish. *Short tail wholly tipped with black. Long black ear tufts. Large, pale cheek ruffs,* whitish with black barring, forming a double-pointed beard at throat. Feet very large and well furred. Male larger than female. L 30–42" (74–107 cm); T 2–5½" (5–14 cm); HF 7–13" (18–33 cm); Wt 11–40 lb (5.1–18.1 kg).

Similar Species: Bobcat is browner, with indistinct spotting and dark bars on forelegs; legs

are shorter; tail is black above but not
below; ear has smaller tuft.

Breeding: Mates February–March; 1 litter of 1–6
young born April–early June; gestation
63–70 days.

Sign: Scratching posts and kill caches
resemble Bobcat's. Lynx creates scent
posts by urinating on trees and stumps.
Scat: Similar to Bobcat's.
Tracks: Foreprint 3–4¼" (75–110 mm)
long, almost as wide; hindprint slightly
smaller; both with 4 toes, no claws
showing. Because of well-furred paws,
prints are much larger and rounder than
Bobcat's; prints are especially large
when toe pads spread and blur in
powdery snow. Straddle usually less
than 7" (175 mm). Normal stride
14–16" (350–400 mm), but may have
long gaps, as Lynx occasionally leaps as
if practicing pounce.

Habitat: Deep, coniferous forest interspersed
with rocky areas, bogs, swamps, or
thickets.

Range: Much of Canada and Alaska south into
much of Washington, n Oregon, n
Idaho, and extreme nw Montana. Also
Rocky Mountain areas of Wyoming and
n Colorado. Nearly eradicated in e U.S.
in the 20th century, but still occurs in n
New England; a few apparently still
remain in n Wisconsin, n Michigan, and
extreme n New York.

By day, the Lynx rests under a ledge, the
roots of a fallen tree, or a low branch. It
frequently climbs trees and sometimes
rests in them, waiting to leap down on
passing prey. The Lynx's long ear tufts
serve as sensitive antennae, enhancing
its hearing, while its big feet help make
it a powerful swimmer. The animal's
thick fur permits silent stalking and
speed through soft snow, in which
some animals may flounder—although
not the well-named Snowshoe Hare,
the Lynx's chief quarry, which makes
up about three-fourths of its diet. The
Lynx also eats birds, meadow voles, the

remains of dead Moose and Caribou, and occasionally small, winter-weakened deer, Caribou, or sheep, especially when the snow is deep. This cat will cache meat, particularly a large kill, by scantily covering it with snow or ground litter. Usually silent, during the mating season the Lynx may shriek or utter a scream that ends in a prolonged wail. The Lynx establishes and maintains a home range for several years. It is a solitary animal, associating with the opposite sex only during mating, at which time several males may follow a female. Females with young are tolerated within a male's home range. Kittens are born streaked and spotted, and remain with the mother through the first winter; they begin foraging with the mother at about two months, and are weaned at three months. Young Lynx bury their scat; adults do not. Lynx often urinate on stumps or bushes, which is suggestive of territorial marking, but little is known of territoriality in this cat. The populations of this species are characteristically cyclic, peaking about every nine to ten years, parallel to the population cycle of the Snowshoe Hare. Although the Lynx occasionally preys on domestic animals in remote areas, it usually poses no threat to humans or livestock. Its main natural predators are wolves and the Mountain Lion, but humans, who destroy its habitat and value its long, silky fur, are its chief enemy today.

276–279 **Bobcat**
Lynx rufus

Description: Tawny (grayer in winter), with *indistinct dark spotting. Short, stubby tail, with 2 or 3 black bars and black tip above; pale or white below.* Upper legs have dark or black horizontal bars. Face has thin, often broken black lines radiating onto broad cheek ruff. Ears slightly tufted. Male

larger than female. L 28–49" (71–125 cm); T 4–7" (10–17 cm); HF 5⅛–8¾" (13–22 cm); Wt 14–29 lb (6.4–13 kg).

Similar Species: Lynx has tail tip black above and below, larger feet, longer legs, more pronounced ear tufts, and longer, grayer fur without indistinct spotting.

Breeding: Mates February–March; 1 (occasionally 2 in South) litter of 1–7 (usually 2 or 3) young born late April–early May.

Sign: Scent posts, established by urinating, visible only on snow and identifiable only by tracks. Tree trunks used as scratching posts, with low claw marks. Food cache usually covered somewhat haphazardly and scantily with ground litter.
Scat: Similar to a domestic dog's. Often buried, but sometimes merely covered with dirt scraped about and accompanied by scratch marks on ground.
Tracks: Foreprints and hindprints about same size, 2" (50 mm) long, slightly longer than wide, with 4 toes, no claw marks. If clearly outlined, heel pad can be distinguished from canine print: domestic dog's or Coyote's is lobed only at rear; Bobcat's is lobed at rear and concave at front, giving print scalloped front and rear edges. Trail very narrow, sometimes as if made by a 2-legged animal, because hindprints are set on, close to, or overlapping foreprints; 9–13" (228–330 mm) between prints. This manner of walking may be an adaptation to stalking: Hunting as it travels, the cat looks for spots to place its forefeet noiselessly, then brings down its hindfeet on the same spots.

Habitat: Primarily scrubby country or broken forests—hardwood, coniferous, or mixed; also swamps, farmland, and rocky or brushy arid lands.

Range: Spotty distribution from coast to coast, and from s Canada into Mexico. Probably most plentiful in Far West, from Idaho, Utah, and Nevada to Pacific Coast and from Washington to Baja California. Scarce or absent in much of Midwest.

Found only in North America, where it
is the most common wildcat, the Bobcat
gets its common name from its stubby,
or "bobbed," tail. The animal spends
less time in trees than the Lynx, resting
by day in a rock cleft, thicket, or other
hiding place, but is also an expert
climber. Sometimes it rests on a boulder
or a low tree branch, its mottled fur
providing excellent camouflage; if hard-
pressed, it will swim. The Bobcat's
home range varies in size with sex,
season, and prey distribution and
abundance. It marks its range using
urine (in large or small amounts that it
may cover up), feces (also sometimes
covered up), anal gland scent, and
scrapes and scratches. It uses the same
hunting pathways repeatedly to prey
mostly on the Snowshoe Hare (in the
northern U.S.) and cottontails (in the
eastern U.S.), but also on mice,
squirrels, Woodchucks, Virginia
Opossums, moles, shrews, Common
Raccoons, foxes, domestic cats, birds,
reptiles, Common Porcupines, and even
skunks. The Bobcat, like many larger
predators, can fast for some time when
food is not available, but eats heavily
when it is. The animal consumes small
prey immediately, but caches and
revisits larger kills. The Bobcat and the
Lynx are capable of killing prey as large
as deer, but they seldom do so except in
deep snow, when food is scarce, or when
fawns are available. The Bobcat hunts
small prey by waiting for victims
motionlessly and then pouncing;
pursues medium-size animals from a
hunting bed or lookout, attacking by
stalking and then rushing, or by simply
rushing; and seeks large prey such as
deer when they are bedded down. After
a rush, the Bobcat will bite at the
throat, base of skull, and chest.
Occasionally the species preys upon
livestock, especially poultry. When food
is scarce the Bobcat will eat carrion,
usually animals killed by cars or by

hunters. This wildcat has a natal or maternal den and other auxiliary or shelter dens in less-visited portions of its home range. The natal den, with a nest of leaves or other dry vegetation, is often in a cave or rock shelter, if available, but can be in a hollow log, under a fallen tree, or in another protected place. Brush piles, rock ledges, stumps, hollow logs, or similar protected places serve as auxiliary dens.

Like the Lynx, the Bobcat is a solitary animal, the sexes coming together only for mating. The Bobcat generally does not mate until its second year. Males are sexually active all year, but most females are in heat in February or March. More than one male may be attracted to a female; the female and dominant male may mate several times after a series of chases and "ambushes." The other males remain apart during matings, but the female may mate with them later. The young are well furred and spotted at birth. They begin exploring at one month and are weaned at two. By fall they are hunting on their own, but remain with the mother for nearly a year. The various calls of the Bobcat sound much like those of the domestic cat, although its scream is piercing. When threatened, the animal utters a short, sudden, and resonant "cough-bark." It yowls loudest and most often during the breeding season. Humans (who hunt Bobcats with hounds in some areas) and the automobile are this animal's worst enemies, but predators such as foxes, owls, and adult male Bobcats may attack the young. Populations are stable in many northern states and are reviving in other states where intensive trapping formerly decimated the species. In some states, such as New Jersey, the Bobcat is being reestablished.

MANATEES AND SEA COWS
Order Sirenia
Family Trichechidae

There are four species of sirenians in two
families: the Dugong (Dugongidae),
found in the Indian Ocean, and three
manatees (Trichechidae), one each in the
rivers of central West Africa, the
Amazon Basin of South America, and
the shores of the Caribbean, including
Florida. The name of the order comes
from the supposed resemblance of its
members to the mermaids, or sirens, of
ancient myth. There is some thought
that the story was inspired by the sight
of a female sirenian cradling an infant in
her arms and pressing it to one of her
two breasts. However, sirenians nurse
underwater while the mother is in a
horizontal position.

Sirenians are large, cylindrically shaped
aquatic mammals that presently occur
in coastal waters or rivers in tropical
regions of the world. Their front legs are
quite dexterous, and they have a highly
specialized feeding apparatus. The
skeleton is heavy, an adaptation that
probably helps them to remain
submerged. There is no distinct neck,
and a fold of skin separates the head
from the body. Forelegs are rounded
flippers; hindlegs are absent. The tail is
a horizontally flattened, paddle-shaped
fluke. Although embryos are well
haired, adults are nearly hairless except
for thick bristles on the snout. Nostrils
are valved, closing when the animals
submerge. Incisors and canines are
absent in adults; when a front molar
wears away, it is replaced by a molar
moving forward from the rear, as with
elephants, and its roots are resorbed.
These highly evolved animals have their
closest evolutionary ties to the terrestrial
hoofed mammals, particularly the
elephants. Sirenians are nonruminant
herbivores, and their stomachs have

several chambers, as do those of most herbivorous hoofed mammals. The hindgut has abundant microflora, which help digest cellulose and other tough plant materials.

Another sirenian, known as Steller's Sea Cow *(Hydrodamalis gigas),* was present in the Bering Sea in historic times. In 1741, a Russian expedition led by Captain Vitus Bering and including the German naturalist Wilhelm Steller was stranded on Bering Island in the Commander Islands of the western Bering Sea. They found these huge sea cows to be abundant in the shallow bays and inlets around Bering and Copper islands, where they fed incessantly on the kelp and algae. A female measured 24 feet 8 inches (7.5 m) long and was estimated to weigh about 8,800 pounds (4,000 kg). This species apparently had been hunted by primitive peoples down to this last population. Members of the expedition and others visiting there later killed the animals for food and for the hides, and the species was believed to be extinct by 1768.

366, 367 **Manatee**
"West Indian Manatee"
Trichechus manatus

Description: *Massive, torpedo-shaped,* nearly hairless
aquatic mammal. Grayish to blackish
when wet. *Tail broad, flattened, paddle-
shaped.* Broad head with *upper lip deeply
cleft* and bearing stiff bristles. Front legs
like large flippers, with 3 nails at end;
hindlegs absent. No external ears.
L 9'10"–11'6" (3–3.5 m), maximum
about 13' (4 m); Wt 1,100 lb (500 kg),
maximum about 3,500 lb (1,600 kg).

Similar Species: Whales and dolphins have a much more
fish-like form.

Breeding: Mating season variable; 1 offspring
produced every 2–3 years; gestation
12–14 months.

Habitat: Shallow coastal waters, bays, rivers, and
lakes.

Range: Gulf and Atlantic coastal waters of se
U.S. north to Beaufort, North Carolina.
Florida's waters have largest remaining
populations. Mainly a summer migrant
north of Florida.

The Manatee is a primarily nocturnal
and moderately social animal that
congregates in warm water in winter. It
can swim quite rapidly for a short
distance, but its normal cruising speed
is only 1½ to 4 mph (2.4–6.4 km/h).
Propelling itself with undulations of the
hind end of its streamlined body, the
Manatee uses its flippers and tail mainly
for steering and stabilization. When at
rest, the animal either hangs in the
water, partially supporting itself on
submerged vegetation, or lies on the
bottom. It can remain submerged for up
to 24 minutes, but about four minutes is
the normal length of submergence. The
Manatee browses on aquatic vegetation,
particularly water hyacinth and hydrilla,
but also on a number of other species,
which it grasps in its lips and bristles,
using its flippers to hold loose grass
blades. An adult consumes 60 to 100

pounds (27–45 kg) of food per day. The Manatee sometimes swims far up on the beach to get plants on shore, and at times may even eat acorns. In Florida, the Manatee performs a valuable service by consuming quantities of water hyacinth, which chokes many waterways. The sounds the Manatee produces include squeals, chirp-squeaks, and a high-pitched scream when frightened; the female gives an alarm call to her young. Several males may be attracted to a female in estrus, and she may mate with several of them. Born underwater, the young Manatee is immediately brought to the surface on its mother's back; after about 45 minutes, it is gradually immersed again. Nursing takes place underwater and may continue for one to two years. Manatees cannot survive water colder than 46°F (8°C) and in winter move upriver to warm lakes or to the heated discharge from power plants. Large sharks, alligators, crocodiles, and killer whales are possible predators, at least on the young. As this animal is slow and incautious, many individuals are injured by boat propellers (in some Manatee habitats, boating is now prohibited or speeds regulated). Other dangers include too much human activity, which may drive the Manatee from areas of good browse, and the release of warm water from power plants, which may lure the animal to areas of insufficient browse. The Manatee has been hunted for its meat, which is partly why it is now an endangered species and fully protected in the U.S. It is still heavily hunted in Mexico and Central America.

ODD-TOED HOOFED MAMMALS
Order Perissodactyla
Family Equidae

Ungulates, or hoofed mammals, have
elongated feet and reduced numbers of
toes. The animals walk on the tips of
their toes, which end in thick, hard
hooves. Perissodactyls are hoofed
mammals with an uneven number of
toes (usually one or three) on each foot;
when they have three toes, the center of
gravity passes through the middle toe.
This order, which has three families and
18 species worldwide, includes horses,
rhinoceroses, and tapirs. There are no
living native perissodactyls in North
America (they disappeared about 10,000
years ago), but domestic horses have
been introduced and have become
established in the wild in the U.S. in
10 western states, and also on several
islands off the East Coast.

Equidae, the horse family, has seven
species, all in the genus *Equus.* Horses
are large herbivores with high-crowned
molars for grinding plant material.
On the foot, only the third digit is
functional, forming the hoof. There is
little information on feral horses in the
western states, but a great deal has been
accumulated on the populations—often
known as Assateague Ponies—that
inhabit several East Coast islands.

300–302 Feral Horse
"Assateague Pony"
Equus caballus

Description: Needs little description, as it is virtually identical to its progenitor, the domestic horse. Very large animal with *elongate snout, mane, long tail, and large, semi-circular, uncloven hooves.* Western race larger than Assateague (East Coast) race. Ht western 14–15 hands (4'8"–5'/1.42–1.52 m); Assateague rarely over 13 hands (4'4"/1.32 m); Wt western male 795–860 lb (360–390 kg), western female 595–750 lb (270–340 kg); Assateague Wt unavailable.

Breeding: Reproductive season generally late spring–early autumn. Gestation about 11 months, with estrus following birth by 7–11 days. Often reproduces in alternate years, especially when food is scarce. 1 (rarely 2) young born fully haired with eyes open.

Sign: *Tracks:* Hoofprint is large semi-circle with triangular notch behind.

Habitat: Variable: densely forested, mountainous terrain, brushlands, dry ridges, and swamps.

Range: In West, Oregon, California, Nevada (largest population), Idaho, Wyoming, Montana, Utah, Colorado, Arizona, and New Mexico. In East, on Assateague Island National Seashore and in Chincoteague National Wildlife Refuge off coast of Maryland and Virginia. Also some feral horses on Shackleford Island, off North Carolina, and on Cumberland Island, off Georgia.

The horse was domesticated in southern Ukraine about 5,000 years ago, and domestic horses have been introduced all over the world. Today there are feral populations on the East Coast and in the western U.S. The feral horse spends about 80 percent of the day grazing and 20 percent resting; at night it grazes about 50 percent of the time. The horse is entirely herbivorous, eating a great

variety of grasses and forbs in summer. In winter, it becomes a browser, in the West feeding heavily on brushy species such as saltbush, rabbit bush, sagebrush, and greasewood. About 80 percent of the diet of the Assateague Ponies of the East is grasses, particularly salt-marsh cordgrass, followed by American beach grass, American three-square rush, giant reed phragmites, and various woody plants in winter. Intestinal microfauna help facilitate digestion of cellulose. The animal drinks water from freshwater pools.

The feral horse exhibits several vocalizations: snorts, which indicate danger and are used mostly by the stallion; neighs, which are a distress call, used mostly by the mare; nickers, for communication and courtship; squeals, used by the female when the male sniffs her genitalia, or by the male as a sign of aggression; and screams, the aggressive call of the male. The Feral Horse also has facial expressions. One is the greeting, in which the head is extended to touch another horse's muzzle and lips; the greeting can change to the threat, in which the ears are directed backward, with the mouth possibly open as well. Another characteristic expression is the flehmen, in which the neck is extended and the upper lip curled, exposing the teeth; this expression is used by a stallion during pre-copulatory activity with a mare.

There are two types of social groups among wild horses: territorial and harem. Territorial groups consist of males and females, and can change in number and sex ratio at any time. Harem groups are made up of one or two dominant males with five or six mares. These groups stay together even if the stallion is lost or replaced by another stallion; a mature mare rarely changes harem group. The dominant individual in a harem group is usually

a mature male; if there is currently no dominant male, it may be the dominant female who rules. The dominant individual leads the herd to forage or to water. A dominant stallion has exclusive rights to the mares in the harem; he may herd them by biting their necks, flanks, or hindquarters if they move too slowly. The stallion needs to defend his harem against other stallions trying to raid it. In confrontations between stallions, the two stare at each other; then they both defecate, smell the feces, and again stare. Either combatant may leave; if one does not, the two move toward each other, necks arched and tails high. They sniff each other, emit screams, and then may fight, standing side by side, kicking, and attempting to knock the other off balance, until one or the other leaves or is beaten. A young male wards off attack by exhibiting a type of submissive behavior known as champing: The young male moves toward the stallion, facing him nose to nose, but the ears are in the upright (non-threatening) position. A solitary horse is usually an adolescent male, a male who has matured and is trying to establish his own harem, or an old stallion who has lost his harem.

The mare becomes sexually active at about three years of age; she separates from the band when about to give birth. Shortly after it is born, the newborn runs and swims with the mother; it is returned to the band in a few hours. The foal, which nurses four to seven times per hour, is protected by its own mother and other members of the band.

Nursing decreases until weaning occurs, around the time a new foal is produced. The young horse then disperses to a different area.

Mutual grooming occurs between members of a herd; it consists of using the incisor teeth to groom the neck, withers, and base of the mane. The horse also self-grooms by rubbing some part

of its body against a tree or fence post, or by rolling in dusting areas or in water or mud. Mutual grooming helps to maintain social ties.

Wild ponies on Assateague Island have no important enemies except humans and biting flies. Horses sometimes walk through thick brush or enter water to rid themselves of the flies. The origin of the Assateague Pony is not clear, but it has been present in its current range for at least three centuries. Originally this pony was a solid color, but many are now brown and white as a result of interbreeding with other ponies released in the region early in the 20th century. It is thought that the original Assateague animals were full-size, and that the small size is not genetically based, but possibly related to diet. The population size of the herd on the Assateague Island National Seashore is maintained by seashore personnel at 150 animals by birth-control measures, but otherwise is not manipulated. The Chincoteague animals are managed by the Chincoteague National Wildlife Refuge and are owned by the Chincoteague Volunteer Fire Department, which sells foals and adds horses to its herd.

EVEN-TOED HOOFED MAMMALS
Order Artiodactyla

The even-toed hoofed mammals, part of
the larger group known as ungulates
(hoofed mammals), are represented
worldwide by 10 families and some 220
species. Most members of the five
families that exist in our range have an
even number of toes (two or four) on
each foot. In animals with four toes, the
main weight-bearing axis passes
between the third and fourth toes,
which form a hoof, so that the animal
literally walks on its toes. The other
toes, called dewclaws, are much smaller
and higher on the leg, touching the
ground only in soft mud or snow. Except
for the omnivorous feral pig and the
Collared Peccary, North American
members of this order are herbivorous
and have a cartilaginous pad instead of
incisors at the front of the upper jaw.
Their molariform teeth are adapted for
nipping or tearing off and grinding
vegetation. Most have a four-chambered
stomach and ruminate, or chew their
cud. In this process, food is hastily
swallowed and stored temporarily in the
largest compartment (the rumen); it
then passes to the second stomach (the
reticulum), where it is shaped into
pellets (the cud). While the animal is at
rest, the cud is returned to the mouth
and slowly chewed to pulp and
swallowed. It then passes through the
four chambers—rumen, reticulum,
omasum, and abomasum, the last of
which is the true digestive stomach.
This complex process permits an animal
to feed quickly, thus reducing its
exposure to predators in open country,
and afterwards chew its cud at leisure in
the relative safety of a concealed spot.

OLD WORLD SWINE
Family Suidae

There are eight species of Old World
Swine in five genera, all in Eurasia and
Africa except for the domestic pig *(Sus
scrofa),* which has been introduced
worldwide. Members of this family are
characterized by their long, pointed
heads and barrel-like, stocky bodies. The
most familiar member of this family, the
domestic pig, is derived from the wild
pig. Since their introduction in North
America, domestic pigs have escaped and
formed feral populations. In addition, the
European Wild Boar, which is the same
species as the wild pig and the domestic
pig, has been introduced in many areas
as big game. Pigs have simple stomachs
and do not chew a cud. Their canine
teeth continue to grow throughout life.

---

298 **Feral Pig**
"Wild Pig," "Wild Boar," "Feral Hog"
includes European Wild Boar,
domestic pig
Sus scrofa

Description: A medium-size hoofed mammal with a
*truncate, flexible, yet tough cartilaginous
snout disc like that of the domestic pig.* Coat
usually coarser and denser than that of
domestic pig; has dense undercoat in
winter. Extremely variable in color: *most
often black,* but may be brown, gray, or
black and white. *Tail moderately long;*
lightly haired; *hangs straight* (never
coiled). Upper tusks (modified canines),
usually 3–5″ (7.5–12.5 cm), but up to 9″
(23 cm) long, curl out and up along
sides of mouth. Lower canines smaller;
turn out slightly; rise outside mouth
and point back toward eyes. Young have
pale longitudinal stripes on body until
6 weeks of age. Ht to 3′ (90 cm);
L 4′4″–6′ (1.32–1.82 m); T to 12″ (30
cm); Wt male 165–440 lb (75–200 kg);
female 77–330 lb (35–150 kg).

Similar Species: Collared Peccary is much smaller, more uniformly and thickly coated; grayish, usually with light collar over shoulders; has vestigial tail and upper tusks that point down. Larger domestic pig has much rounder body, shorter legs, and tail that is usually coiled; has finer, thinner, less shaggy fur; lacks mane and long curved tusks. Hybrids between Wild Boar and feral pig have intergrading characteristics; are sometimes spotted black and tan, and are frequently less hairy.

Breeding: Mates at any time of year, but there are usually 2 peaks about 6 months apart: one January–February, and one in early summer (as with domestic pig). After gestation of about 16 weeks, 3–12 young are born, brown with 9 or 10 pale longitudinal body stripes; length at birth 6–8″ (15–20 cm).

Sign: Rooted-up earth; tree rubs from ground level to 36″ (900 mm) high, with clinging hair or mud; muddy wallows.
Nest: Small, grass-lined depression that sow hollows out in a pile of grass and branches in a secluded thicket.
Scat: Massed pellets or sausage-like segments, usually found with other sign.
Trail: Narrower than domestic pig's, almost a single line.
Tracks: Cloven but more rounded and splayed than deer tracks, 2½″ (65 mm) long; hindprints often half covering foreprints; stride 18″ (460 mm). In soft earth, front dewclaws (low, long, and pointed) almost always print as crescents outside and behind main print; hind dewclaws print as dots.

Habitat: Variable: densely forested mountainous terrain, brushlands, dry ridges, and swamps.

Range: Chiefly w North Carolina; e Tennessee (especially Nantahala and Cherokee national forests); West Virginia; Santa Cruz Island (off California); Monterey and San Luis Obispo counties, California. Small numbers, often in preserves, in New Hampshire,

Pennsylvania, Vermont, and perhaps other states. Small feral pigs in Alabama, Arizona, Arkansas, Georgia, Florida, Louisiana, Mississippi, Missouri, Oklahoma, Oregon, South Carolina, and Texas.

Native to Europe and Asia, the feral pig first appeared in North America in 1893, when a herd of 50 animals was brought from Germany's Black Forest to a hunting preserve in New Hampshire's Blue Mountains. Russian Wild Boars were released in 1910 and 1912 on a North Carolina preserve near the Tennessee border; in 1925, near Monterey, California; and a few years later on Santa Cruz Island, off the California coast. Some of these animals escaped from preserves, and many of their progeny bred with feral descendants of domestic pigs. The pure-blooded feral pig is still found in the wild in North Carolina, in Tennessee, in parts of California, and in preserves in other states. Elsewhere North America's wild swine are hybrids or pigs descended from purely domestic stock.
Especially active at dawn and dusk, the feral pig is a fast runner and a good swimmer. It usually trots from one foraging area to another, then slows to a walk. Its wanderings seldom exceed beyond an area of 10 square miles (26 sq km) if food is abundant, but may extend to as far as 50 square miles (130 sq km) when forage is poor. Where oaks are prevalent, acorns are a staple; this animal also favors beech, hickory, and pecan nuts. In late fall and winter, when the nuts accumulated on the forest floor have all been eaten, the feral pig leaves to forage in swamps and marshes. It relies on a wide variety of vegetation, including roots, tubers, grasses, fruit, and berries, but also eats crayfish, frogs, snakes, salamanders, mice, the eggs and young of ground-nesting birds, young rabbits, and any other easy prey or

carrion encountered while foraging; it has even been known to kill and eat fawns. The sow and her young forage in a family group, usually of about half a dozen animals, but they sometimes join other groups in herds of up to 50 individuals. Except during the breeding season, mature males are solitary or band in small groups. During the two breeding seasons, males gather as females enter estrus; the males fight for dominance, slashing at each other's shoulders. Dominant males mate first, allowing other males to mate later. At one week old, piglets follow the sow. They are weaned at three months, and at six weeks lose the pale, longitudinal body stripes they had at birth. Young disperse the following year, are sexually mature at one and a half years, and are fully grown at five to six years.

In America, the feral pig does not grow as large as the boars in some parts of Europe, probably because it must compete for ground food—especially nuts—with so many squirrels and deer. While some European boars weigh over 500 pounds (225 kg), few American boars reach 350 pounds (160 kg). Bears, Bobcats, and feral dogs occasionally kill a baby boar, but predation is light on older individuals because the tusks of a mature pig are as effective for fighting as for rooting; the chief predators are humans. Life span is generally 15 to 25 years. In some states, there are hunting seasons for feral pigs; in others, where they have become agricultural pests, they can be killed at any time. Feral pig populations are neither endangered nor monitored by game departments as a valuable resource.

PECCARIES
Family Dicotylidae

Of the three living species in this family worldwide, one, the Collared Peccary, occurs in our range. Peccaries are relatively small animals descended from large pigs that lived about 25 million years ago. They are distinguished from their distant relatives, the wild and domestic pigs (family Suidae), by their smaller size, straight tusks, single dewclaw, and strong musk gland on the back. They have fewer teeth than pigs, and two-part stomachs, features that seem to be transitional toward the suborder of ruminants (Ruminantia).

299 Collared Peccary
"Javelina"
Tayassu tajacu

Description: *Pig-like.* Head and shoulders large; legs and hindquarters small. Grizzled grayish or blackish above and below, with yellowish tinge on cheeks; *whitish to yellowish irregular collar* from shoulder to shoulder. Heavy, bristly hair from head to back erectable into a mane. Inconspicuous tail. Pig-like snout. Tusks (canines) about 1–1½" (3–3.5 cm); only tips protrude beyond lips. 4 toes on forefeet, 3 on hindfeet; all feet have 2 hooves, on the third and fourth toes. Juvenile brownish with black stripe down back. Ht 20–24" (50–60 cm); L 35–40" (87–102 cm); T ¾–2⅛" (1.9–5.5 cm); HF 7⅛–7⅞" (18–20 cm); Wt 30–65 lb (13.6–30 kg).

Similar Species: Feral pig is much larger, less uniformly and thickly coated, with finer, thinner, less shaggy fur; lacks mane and whitish collar; upper tusks curve up instead of down; tail is moderately long.

Breeding: Breeds at any time of year, but most births in summer; gestation 21 weeks. 2–6 young, generally twins, born with yellowish or reddish hair. 1 litter per

year, although a second can be produced
if first is lost.

Sign: Rooted-up ground; chewed cactus,
especially prickly pear, and other low
vegetation. Sometimes a strong, musky
or cheesy odor.
Scat: Usually large, irregular segments;
flattened disks when feeding on very
succulent vegetation.
Tracks: Similar to pig's but smaller.
Cloven, rounded oblongs, generally
about ¾–1½" (20–38 mm) long, with
hindprint slightly smaller than
foreprint. Stride short, usually 6–10"
(150–250 mm) between pairs of
overlapping foreprints and hindprints.

Habitat: Brushy desert, rocky canyons, and
wastelands.

Range: Southeastern Arizona, extreme se and sw
New Mexico, and sw Texas; south to
Argentina.

The Collared Peccary's activity patterns
are strongly related to temperature.
Active at twilight and at night in
summer, it is much more diurnal in
winter. It often beds down in a hole it
roots in the earth or takes shelter in a
cave during the midday heat; on winter
nights, it may huddle together with
others of its species. The Collared
Peccary is territorial and travels in a
herd that ranges from 6 to 30 animals,
grunting softly while feeding; peccaries
other than the dominant male will help
defend the territory. The herd is quite
stable, but may break up into temporary
feeding subherds for up to several days
at a time. As the animals move about, a
musk gland on the back exudes a strong
skunky or cheesy scent, which probably
serves as a bonding mechanism, helping
to keep members of the group together.
The musk gland also serves as an alarm
signal: When a Collared Peccary is
agitated, the hairs on its back become
erect, uncovering the gland, which then
involuntarily discharges scent. This
animal's alarm call is a barking cough;

the Collared Peccary can squeal, but does so only if terrified or injured. It prefers to flee danger, galloping away at speeds of up to 25 mph (40 km/h) and swimming through streams if necessary, but it will fight viciously if cornered. Primarily herbivorous, this peccary feeds on prickly pear (which provides water as well as food), mesquite fruit, sotol, and agave, as well as other plant species. It roots for tubers but feeds on animal materials only occasionally, eating insects, worms, reptiles, and amphibians. In parts of New Mexico and Arizona, the Collared Peccary sometimes forages at elevations as high as 6,000 feet (1,800 m), moving along the gentler mountain slopes and eating quantities of scrub oak acorns.

The dominant male mates with most of the females in the herd as they come into heat. For a few hours to a few days, the male forms a bond with an estrous female and keeps other males away. Subordinate males normally mate only when more than one female is in heat at one time. Breeding occurs at any time of the year because food is abundant year-round in this species' mild habitat. The peccary has four pairs of nipples, but only the rear two pairs produce milk; while most young mammals nurse alongside their mother, peccaries stand behind to nurse. Peccaries often engage in reciprocal grooming: Two individuals stand side by side, head to tail, each rubbing its head along the rump, legs, and scent gland of the other. This is another behavior thought to reinforce social bonding within the herd.

Life span is 15 to 20 years. Once ranging as far north as Arkansas, the Collared Peccary was extirpated in the northern part of its range by settlers who killed it for its meat and its hide. Now it is hunted as game. The alternate common name, "Javelina," comes from the Spanish *jabalina* ("spear") and refers to the spear-like tusks.

DEER AND THEIR KIN
Family Cervidae

There are 43 species of cervids, as members of this family are known, eight of which occur in North America north of Mexico. Five are native: the Elk, Moose, Caribou, White-tailed Deer, and Mule Deer. Three are introduced: the Sambar, Fallow, and Sika deer. All male members of North American cervid species, as well as female Caribou, are distinguished by having antlers—bony outgrowths of the frontal bone that normally are shed annually. The antlers begin growing in the early summer, at which time they are soft and tender, and covered with a skin that grows as the antlers grow. The skin has short fine hairs, the "velvet"—whence the phrase "in velvet." The velvet contains a network of blood vessels that nourish the growing bone beneath. By late summer, the antlers reach full size and the blood supply diminishes, causing the velvety skin to dry up, loosen, and peel off. The bare horns then serve as sexual ornaments, and rival males may use them as weapons in courtship battles during the rut. After the mating season, decalcification occurs at the point where the antlers attach to the frontal bone, and the antlers are shed, usually both branches within two to three days of each other. Antlers are seldom found on the ground because they are rapidly eaten by rodents craving calcium. An animal's first set of antlers consists of short, straight spikes ("spikehorns"). As long as an individual's diet is adequate, its antlers become larger and have more points each year until it reaches maturity; inadequate diet may result in stunted antlers. Besides antlers, cervids are characterized by a lack of canine teeth in the upper jaw. Deer tend to feed on herbaceous vegetation and thus are grazers in summer; they also feed on

woody vegetation and are browsers in winter. They cut off food items between the lower incisors and a calloused upper pad. In the temperate zone, the winter coats of most deer tend to be darker than their summer coats. Does are generally smaller than bucks and more delicately built. Fawns are usually spotted or striped; when they rest motionless in foliage, the pattern of the coat provides excellent protective coloration. In a few species the juvenile pattern is retained through adulthood. Also enhancing their chance of survival is the fact that the young of many members of this family give off very little odor during the first few days or weeks of their lives.

303 Fallow Deer
Cervus dama

Description: A small deer. *Brown with white spots;* on lower sides and haunches, spots may fuse into white line; individuals also may be black, white, pale yellow, cream, silver-gray, or piebald. *White below.* Black stripe from nape down back onto relatively long tail. Hindlegs slightly longer than forelegs, so that rump is held high. Short neck has prominent larynx. Dewclaws reduced; high on legs. *Buck's antlers have flattened terminal tines;* antlers measure about 12–30″ (30.5–76 cm) from tip to tip. Fawn spotted. Ht about 3′3″ (1 m); L 4′7″–5′11″ (1.4–1.8 m); T 6⅜–7½″ (16–19 cm); Wt 88–176 lb (40–80 kg).

Similar Species: Adult Mule Deer and White-tailed Deer lack spotting in summer; antlers lack flattened tines.

Breeding: Breeds October–November; after gestation of 6–7 months, 1 (rarely 2) spotted young born.

Sign: Browse marks, bed, scat, and tracks similar to those of White-tailed Deer, but tracks never show prints of dewclaw behind main prints.

Habitat: Brushy hills with grassy fields.
Range: First introduced to James Island, British
Columbia, and to Land Between the
Lakes, a 172,000-acre (70,000 ha)
national recreation area in Kentucky;
now occurs also in Maryland, on Saint
Simon's and Jekyll islands off the coast
of Georgia, and in Alabama, Oklahoma,
Texas, and California.

Native to the Mediterranean region and
Asia Minor and introduced in our
region, the Fallow Deer runs in a
distinctive stiff-legged fashion, bouncing
along as if on a pogo stick. It grazes on
grasses and herbaceous plants in
summer, and browses on the woody
parts of deciduous trees and conifers in
winter. The Fallow Deer utters a sound
similar to a dog's bark when it is
nervous, such as when a buck engages in
battle or a doe has lost a fawn. Other
vocalizations include: bleating by a
pregnant female or a female with fawns;
mewing, a submissive call; the peeping
of a fawn to contact or alert its mother;
wailing by a fawn in intense distress;
and groaning, a belch-like sound that
lasts about one second but that may
occur in a series, with four to five
seconds between groans.
The unusually gregarious Fallow Deer
forms herds of up to 150 to 175
members, including young bucks, does,
and fawns. The adult male is solitary.
The buck is polygamous and fights with
other bucks during the rutting season.
At Kentucky's Land Between the Lakes,
during the November rutting season the
male makes scrapes on the ground,
clearing patches about 18 to 24 inches
by 24 to 36 inches (45–60 cm by 60–90
cm), onto which it urinates. A doe in
heat frequents these scrapes, and mating
often occurs on a pathway between
them. After gestation, the female leaves
the herd to give birth. She leaves the
fawn alone most of the day, returning
only for nursing; a fawn learns to

recognize its mother by her bleat. The fawn runs with the mother at about one month, and is weaned at about four months, in October, during the rutting season. Antlers are shed in late winter. The Fallow Deer has an "alerting" behavior during which members of a herd assume a rigid, upright stance and stiff walk with neck extended vertically. The tail elevation indicates the degree of disturbance. If truly wild, the Fallow Deer is wary of humans, but unlike most other deer species, it easily becomes semi-domesticated and has been established as a "park deer" in many parts of the world.

315–317 **Elk**
"Wapiti"
Cervus elaphus

Description: *A very large cervid,* with thick neck and slender legs. Brown or tan above; underparts darker. *Rump patch and tail yellowish brown.* Buck has dark brown mane on throat and *large, many-tined antlers:* 6 tines on each side when mature, with main beam up to 5′ (1.5 m) long. Juvenile spotted until 3 months of age. Ht 4′6″–5′ (1.37–1.5 m); L 6′8″–9′9″ (2.03–2.97 m); T 3⅛–8⅜″ (8–21 cm); HF 18–26″ (46–66 cm); Wt male 600–1,089 lb (272–494 kg), female 450–650 lb (204–295 kg).

Similar Species: Moose has huge, ponderous muzzle and dewlap; lacks yellowish rump and tail.

Breeding: Breeds late August–November, peaking October–November. 1 or 2 young born after gestation of around 9 months; Newborn weighs 25–40 lb (11–18 kg).

Sign: During the rut, thrashed saplings and large shrubs; "rubs" on saplings and small trees made as the male polishes his antlers.
Wallows: Depressions dug in ground by hooves and antlers, where copious urine and feces give a strong, musky odor.

Scat: When feeding in lush pastures in summer and early fall, flattened chips similar to dung of domestic cattle; in winter, when chief foods are dried grasses and browse, dark pellets similar to deer scat but larger, sometimes more than 1″ (25 mm) long.

Tracks: Cloven hearts, much larger and rounder than those of White-tailed Deer or Mule Deer; somewhat smaller and rounder than those of Moose; 4–4½″ (100–115 mm) long. When walking, hindprints slightly ahead of and partly overlapping foreprints; stride 30–60″ (750–1,500 mm). When running and bounding, foreprints and hindprints are separate; stride up to 14′ (4.25 m). In snow or mud, dewclaws often print behind lobed main prints.

Habitat: Variable: in summer, chiefly high, open mountain pastures; in winter, lower wooded slopes, often dense woods.

Range: From e British Columbia, c Alberta, c Saskatchewan, and s Manitoba south to c New Mexico and Arizona, with great numbers in Washington, Montana, Wyoming, and Colorado. Also along coast from Vancouver Island to n California; isolated populations elsewhere in California, Nevada, Utah, Arizona, New Mexico, Oklahoma, South Dakota, Minnesota, and Michigan. Small numbers in several eastern states, notably Pennsylvania.

The Elk is primarily nocturnal, but is especially active at dusk and dawn. Unlike the much smaller White-tailed Deer, which is often heard crashing through the brush, the Elk moves through the forest rapidly and almost silently. The bull can run up to 35 mph (55 km/h), and both bull and cow are strong swimmers. This animal marks the areas it frequents by stripping the bark from seedlings, the cow using her lower incisors and the bull the base of his antlers; they then rub the seedlings

with the sides of the chin and muzzle. These posts may serve as territorial markers, warning other Elk to keep out. The Elk feeds on many kinds of plants, but is primarily a grazer. East of the Continental Divide, it feeds more heavily on woody vegetation, owing to the scarcity of grasses and forbs; it also consumes lichen. The availability of food appears to influence the time of mating, the percentage of cows that become pregnant, and the age of puberty. The Elk vocalizes in several ways. A young Elk squeals, an adult snorts and grunts, and a cow neighs to her calves. The alarm call is a sharp, barking snort. The "bugle," or "whistle," of a bull is a challenge to other bulls and a call of domination to cows; this vocalization begins as a bellow, changes almost immediately to a loud, shrill whistle or scream, and ends with a series of grunts. Only the whistle carries over long distances. A good imitation of this call, usually made with a commercial or homemade whistle about 1 foot (30 cm) long, may be answered by a bull, the most vocal of the American cervids. A cow also whistles, but not as loudly as a bull and chiefly in spring rather than fall.

The Elk is very gregarious. The species' main social unit is the cow/calf band or herd. The size of the herd varies greatly; it is sometimes composed of up to 400 individuals, depending on the terrain, cover, and amount of resources. The larger herds occur in open areas; smaller groups are found in woods. Bulls herd separately, remaining on the outskirts of a cow-dominated herd. During the rutting season, adult bulls join the cow/calf herd. At this time, the bull gives his bugling call, rolls in wallows of stagnant water and mud, and urinates on vegetation, which he then catches in his antlers and tosses over his back. Bulls clash their racks of antlers in

mating jousts. They are seldom hurt, though occasionally there is a major injury or even a death. The most polygamous deer in America and perhaps the world, the bull Elk assembles a harem of up to 60 cows. After a lengthy gestation, the cow leaves the herd to give birth. A week later, she rejoins the herd with her calf, which is entirely dependent on milk for one month, and may suckle for up to nine months. The calf joins a nursery herd for a few weeks after birth; otherwise cows and calves herd together through the summer. As it approaches maturity, the juvenile bull spends less and less time with the cow-dominated herd.

The Elk's main predator is the Mountain Lion, although bears also take some calves. "Elk," the British name for the Moose, was misapplied to the "Wapiti" by early settlers. *Wapiti,* a Shawnee word meaning "white (or pale) deer," alluded to the sides and flanks of the Rocky Mountain subspecies *(C. e. nelsoni),* which are often very pale. The Roosevelt subspecies *(C. e. roosevelti),* shown in plate 317 in its rain forest habitat in Washington's Olympic National Park, is found in the Pacific Northwest. Elk once ranged through most of what is now the U.S. and southern Canada, but their number dwindled as settlements and farming took over their habitats and also as a result of hunting, both for the market and for subsistence. In the 19th century, many Elk were primarily plains animals and were shot by ranchers to reduce grazing competition with domestic livestock. Thousands were also killed solely for two of their upper teeth, which were popular as watch-fob charms. Today Elk herds appear to be stable. In winter, large numbers can be observed as they gather at a refuge outside Jackson Hole, Wyoming, to receive supplemental feed.

307 Sika Deer
Cervus nippon

Description: A medium-size deer. Various shades of brown; *many white spots on back and sides in 7 or 8 rows;* spots more pronounced in summer than in winter. Center of back darker than sides, forming a dark band from head to rump. *Large white rump patch.* Underparts whitish or gray. Buck has narrow, backward-directed antlers, with 2–5 points per antler. Antlers 12–26" (30–65 cm) high, 1" (2.5 cm) in diameter at base. Juvenile is spotted. Ht 3'–6'3" (88–190 cm); L 3'11"–5' (1.2–1.5 m); T 3–5⅛" (7.6–13 cm); HF 12½–14½" (31–36 cm); Wt male 110–309 lb (50–140 kg), female 88–132 lb (40–60 kg).

Similar Species: White-tailed Deer somewhat larger and not spotted as adult.

Breeding: Breeds late September–December; after gestation of about 7 months, 1 young born May–August.

Sign: *Scat:* Dark pellets similar to those of White-tailed Deer.
Tracks: Cloven hearts, slightly smaller than those of White-tailed Deer.

Habitat: Forested areas with dense undercover, but adaptable to many habitats. Does well on Assateague Island in marshes and thickets.

Range: Native in Japan, Manchuria, Taiwan, Korea, and adjacent China, but widely introduced in the U.S.: eastern shore of Maryland; Assateague Island, off coast of Maryland and Virginia; small numbers in Kansas, Oklahoma, and Wisconsin; large herds in Texas.

The Sika Deer is mainly nocturnal, but may be seen during the day. On Assateague Island, it often grazes while standing in water. This deer uses a stiff-legged gallop at lower speeds and a stiff-legged quadrupedal hopping at higher speeds. At lower speeds, it can make jumps up to 30 feet (9 m) long. At higher speeds, it has all four feet 1 foot

(.3 m) off the ground at once; in one stride it can leap up to 20 feet (6 m) in length, and up to 5 feet (1.5 m) in height. As with other deer, the Sika Deer eats many species of plants, but primarily feeds on grasses in summer and on woody plants in winter. It has at least 10 vocalizations, including soft whistles between females, goat-like bleats from doe to fawn, horse-like neighs from fawn to doe, loud screams from males during the rut, and alarm calls from both sexes.

The Sika Deer is not very gregarious; single individuals are seen about as often as small herds. Territorial behavior is initiated by the adult male in summer. He digs holes up to 4 feet (1.2 m) wide and 1 foot (.3 m) deep with his forefeet and antlers, then urinates in the holes; these, plus thrashed ground, designate territorial boundaries. Fierce fighting may occur between rival bucks, and males may attempt to drive females into their territories, where mating occurs. A successful male may mate with up to 12 females. The male does not feed until late in the rutting season; the female moves between male territories and does feed during the breeding period.

308 Sambar Deer
Cervus unicolor

Description: *A large deer, with shaggy mane on throat.* Mostly uniform brown, with *light yellowish brown under chin, inside limbs, between buttocks, and under tail.* Upper jaw has rudimentary canine teeth. On buck's antlers, 1 basal tine projects forward; another projects backward, then forks. Juvenile not spotted. L 6′1″–7′1″ (1.85–2.15 m); T average 10–12″ (25–30 cm); Wt 500–600 lb (227–272 kg).

Sign: *Scat:* Dark pellets similar to those of White-tailed Deer, but larger.

Tracks: Cloven hearts, larger than those of White-tailed Deer.

Habitat: Open woodland in its native land. Little known of habitat in our region.

Range: Native to India, Sri Lanka, Malay Peninsula, East Indies, and the Philippines. Introduced on Saint Vincent Island, Gulf County, Florida.

The Sambar Deer was introduced on Saint Vincent Island, Florida, in 1908 and increased to about 50 individuals by the 1950s. Approximately 50 to 90 apparently are still present. They do not herd, but occur in groups of four or five animals, possibly family groups. Little is known about them in Florida.

309, 310 Mule Deer
"Black-tailed Deer"
Odocoileus hemionus

Description: A medium-size deer. Stocky body, with long, slim, sturdy legs. In summer, reddish brown or yellowish brown above; in winter, grayish above. Throat patch, rump patch, insides of ears, and insides of legs are white; lower parts cream to tan. *Large ears. Buck's antlers are branched equally, each a separate beam forking into 2 tines;* antler spread to 4' (1.2 m). 2 major types: Mule Deer, the more common, has tail white above and tipped with black. "Black-tailed Deer," found only along Pacific Coast, has tail blackish or brown above. Juvenile spotted. Ht 3'–3'5" (90–105 cm); L 3'10"–7'6" (1.16–1.99 m); T 4½–9" (11.4–23 cm); E 4¾–6" (12–15 cm); Wt male 110–475 lb (50–215 kg), female 70–160 lb (32–73 kg).

Similar Species: White-tailed Deer has antlers with 1 main beam; tail is not tipped with black.

Breeding: Breeds from autumn through early winter; after gestation of 6–7 months, 1–2 young born June–August. Newborn weighs about 8 lb (3.6 kg).

Sign: Browse marks, buck rubs, scrapes, bed, and droppings similar to those of White-tailed Deer.
Bed: Examination often reveals sex: Both urinate upon rising, but doe first steps to one side; buck urinates in middle of bed.

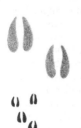

Tracks: Foreprint and hindprint like narrow split hearts, with pointed end forward. Male prints 3¼" (80 mm) long, female 2⅜" (60 mm) long; walking stride 22–24" (550–600 mm). Distinctive bounding gait ("stotting"), with all 4 feet coming down together, forefeet printing ahead of hindfeet.

Habitat: Mixed habitats: forest edges, mountains, and foothills.
Range: Southern Yukon and w Northwest Territories (Mackenzie district) south through w U.S. to Wisconsin and w Texas.

The Mule Deer has large ears that move independently and almost constantly, like a mule's. Primarily active in morning, evening, and on moonlit nights, this deer may also be seen at midday in winter. The Mule Deer has a stiff-legged bounding gait, with back legs and front legs, moving together. It is also a good swimmer. In mountainous areas, the animal migrates up and down seasonally to avoid heavy snows. Its summer forage is chiefly herbaceous plants, but also blackberry, huckleberry, salal, and thimbleberry; winter browse includes twigs of Douglas fir, cedar, yew, aspen, willow, dogwood, serviceberry, juniper, and sage. This deer also eats acorns and apples.
Mule Deer often form herds of both sexes in winter, but herds are seldom large. The usual social group consists of a doe with her fawn or a doe with twin fawns and a pair of yearlings. When does encounter each other, they often fight, so family groups space themselves widely, thereby helping to ensure food and cover for all. Many bucks are

solitary, but some band together before and after the rutting season. The buck has a larger home range than the doe; during the rutting season both buck and doe may leave their home range. The buck is polygamous and seeks out does in estrus, sometimes trying to herd them. A male may breed with most does in his area, and a doe probably breeds with several males. Displays and threats often prevent actual conflict between bucks, but vigorous fights do occur, in which each tries, with antlers enmeshed, to force down the other's head. Even in such battles, injuries are rare; usually the loser withdraws. If antlers become locked, both animals perish through starvation. A first-year doe produces a single fawn, while an older doe usually has twins. For their first month, the young are kept concealed; their mother visits them regularly to nurse. The Mule Deer has glands on the hindlegs above the hooves. A fawn seems able to recognize its mother by the odor from these glands, and when deer are in groups, they frequently sniff these glands. The long hairs around the glands usually become erect when aggressive confrontations between bucks begin. Mountain Lions and wolves are the major natural predators. Bobcats and bears take a few, and Coyotes take juveniles; others are killed by trains and automobiles. Prized as a trophy and for its flesh, this deer is hunted by humans. Mule Deer can cause damage to crops and timber.

304–306   White-tailed Deer
"Whitetail," "Virginia Deer"
Odocoileus virginianus

Description: Size varies greatly; a small to medium-size deer. Tan or reddish brown above in summer; grayish brown in winter. Belly, throat, nose band, eye ring, and inside of ears are white. *Tail brown, edged with*

white above, often with dark stripe down center; *white below.* Black spots on sides of chin. *Buck's antlers have main beam forward, several unbranched tines behind,* and a small brow tine; antler spread to 3′ (90 cm). Doe rarely has antlers. Fawn spotted. Ht 27–45″ (68–114 cm); L 6′2″–7′ (1.88–2.13 m); T 6–13″ (15–33 cm); HF 19–20″ (47.5–51.2 cm); Wt male 150–310 lb (68–141 kg), female 90–211 lb (41–96 kg).

Similar Species: Mule Deer has antlers with both main beams branching; tail tipped with black.

Breeding: Reproductive season varies: first 2 weeks in November in north, January or February in south. 1–3 young born after gestation of about 6½ months.

Sign: Raggedly browsed vegetation, ripped rather than neatly snipped due to lack of upper incisors.
"Buck rubs": Polished scars or oblong sections where bark has been removed from bushes, saplings, or small trees, usually close to the ground; made when a buck lowers his head and rubs antlers against a tree to mark territory; trees chosen to fit antlers (e.g., a rub on a tree with a diameter of 4–5″/100–125 mm would have been made by a very large buck).
"Buck scrapes": Pawed depressions with broken branches about 3–6′ (1–2 m) above the ground.
Bed: Shallow, oval, body-size depression in leaves or snow.
Scat: When browsing, almost always hard, dark, cylindrical pellets, about ¾″ (19 mm) long; sometimes round. When grazing on succulent vegetation, cylindrical and segmented, evenly massed.

Tracks: Like narrow split hearts, pointed end forward, about 2–3″ (50–75 mm) long; dewclaws may print as twin dots behind main prints in snow or soft mud. In shallow snow (1″/25 mm deep), buck may drag its feet, leaving drag marks ahead of prints; in deeper snow, both

buck and doe drag feet. Straddle 5–6″
(125–150 mm) wide. Stride, when
walking, 1′ (.3 m); when running, 6′
(2 m) or more, and hindprints sometimes
register ahead of foreprints; when
leaping, 20′ (6 m). Well-used trails are
very noticeable, with numerous prints
and damaged vegetation.

Habitat: Farmlands, brushy areas, woods, and
suburbs and gardens.

Range: Southern half of southern tier of
Canadian provinces; most of U.S., except
far Southwest.

Although primarily nocturnal, the
White-tailed Deer may be active at any
time. It often moves to feeding areas
along established trails, then spreads out
to feed. The animal usually beds down
near dawn, seeking concealing cover.
This species is a good swimmer. The
winter coat of the northern deer has
hollow hair shafts, which fill with air,
making the coat so buoyant that it
would be difficult for the animal to sink
should it become exhausted while
swimming. The White-tailed Deer is
also a graceful runner, with top speeds
to 36 mph (58 km/h), although it flees
to nearby cover rather than run great
distances. This deer can make vertical
leaps of 8½ feet (2.6 m) and horizontal
leaps of 30 feet (9 m). The White-tailed
Deer grazes on green plants, including
aquatic ones in the summer; eats acorns,
beechnuts, and other nuts and corn in
the fall; and in winter browses on woody
vegetation, including the twigs and
buds of viburnum, birch, maple, and
many conifers. The four-part stomach
allows the deer to feed on items that
most other mammals cannot eat. It can
obtain nutrients directly from the food,
as well as nutrients synthesized by
microbes in its digestive system. This
deer eats 5 to 9 pounds (2.25–4 kg) of
food per day and drinks water from rain,
snow, dew, or a water source. When
nervous, the White-tailed Deer snorts

through its nose and stamps its hooves, a telegraphic signal that alerts other nearby deer to danger. If alarmed, the deer raises, or "flags," its tail, exhibiting a large, bright flash of white; this communicates danger to other deer and helps a fawn follow its mother in flight. There are two types of social grouping: the family group of a doe and her young, which remain together for nearly a year (and sometimes longer), and the buck group. The family group usually disbands just before the next birth, though occasionally two sets of offspring are present for short periods. Bucks are more social than does for most of the year, forming buck groups of three to five individuals; the buck group, which constantly changes and disbands shortly before the fall rut, is structured as a dominance hierarchy. Threat displays include stares, lowered ears, and head-up and head-down postures. Attacks involve kicking and, less commonly, rearing and flailing with the forefeet. Bucks and does herd separately most of the year, but in winter they may gather together, or "yard up." As many as 150 deer may herd in a yard. Yarding keeps the trails open through the movement of large groups of animals, and provides protection from predators. The leadership of the yards is matriarchal. Deer may occupy the same home range year after year, and may defend bedding sites, but otherwise are not territorial. The White-tailed Deer is less polygamous than other deer, and a few bucks mate with only one doe. The extended rutting season begins at about the time the male is losing his velvet, which varies with latitude. At this time, bucks are still in buck groups, and sparring for dominance increases. (Sparring consists of two deer trying to push each other backward.) The buck group then breaks up, and several bucks begin following a doe at a distance of 150 feet (50 m) or so. They follow the

doe's scent; the largest buck stays closest to the female. A buck attempts to dominate other bucks and may mate with several does over the breeding season. He produces "buck rubs" and also "scrapes," revisiting them regularly during the rut; glandular secretions are left on the rubs. Does visit the scrapes and urinate in them; bucks then follow the trails of the does. After the mating season, the doe returns to the subherd until spring (May or June in the North; January to March in the deep South). A young doe bred for the first time usually produces one fawn, but thereafter has twins and occasionally triplets if food is abundant. The female remains near the fawns, returning to feed them only once or twice a day. Twin fawns are separated, which serves to protect them. Weaning occurs between one and two and a half months. Fawns stay with the mother into the fall or winter, sometimes for up to two years, but the doe generally drives off her young of the previous year shortly before giving birth. The Whitetail's first antlers are usually a single spike (the "spikehorn"). A three-year-old would be expected to have eight points, but there can be more or less, as the number of tines is influenced greatly by nutritional factors. A Whitetail's age is determined not by the number of tines on its horns but by the wear on its teeth.

Once nearly exterminated in much of the Northeast and Midwest, this deer is now more abundant than ever—it has become the most plentiful game animal in eastern North America—owing to hunting restrictions and the decline in number of its predators, wolves and the Mountain Lion. Its population has become a public-health concern with the onset of Lyme disease, first identified in 1975 in Lyme, Connecticut; this bacterial disease, transmitted by ticks carried by Whitetails, can be fatal to humans. Thinning the deer population

is best done by hunting both does and bucks, as hunting bucks only alters the herd rather than reducing it.

There are two dwarf subspecies of White-tailed Deer: the Coues' Deer, or Arizona Whitetail *(O. v. couesi)*, of the Arizona desert, and the Key Deer *(O. v. clavium)* of the Big Pine Key area in the Florida Keys. The Coues' Deer, which has somewhat enlarged ears and tail relative to the other Whitetails, reaches a maximum of about 100 pounds (45 kg). Considered endangered, the tiny, dog-size Key Deer, weighing 50 pounds (23 kg) or less, is now fully protected thanks to the establishment in 1961 of the National Key Deer Refuge in Florida. Some mammalogists classify the Key Deer as a separate species.

313, 314 **Moose**
Alces alces

Description: *The largest cervid in the world;* horse-size. Long, dark brown hair. High, *humped shoulders;* long, slender legs; tail inconspicuous. *Huge pendulous muzzle; large dewlap under chin;* large ears. Male much larger than female, with *massive palmate antlers,* broadly flattened. Antler spread usually 4–5′ (1.2–1.5 m); record 6′9″ (2.06 m). Calf light-colored but not spotted. Ht 6′5″–7′5″ (1.95–2.25 m); L 6′9″–9′2″ (2.06–2.79 m); T 6¾″ (17 cm); HF 28–30″ (73–83.5 cm); Wt male 900–1,400 lb (400–635 kg), female 700–1,100 lb (315–500 kg).

Similar Species: Elk has yellowish rump patches and tail, and lacks huge, pendulous muzzle and dewlap.

Breeding: Mates mid-September through late October; after gestation of 8 months, 1 or 2 calves born late May–early June. Newborn weighs 24–35 lb (11–16 kg).

Sign: Browse raggedly torn; thrashed shrubs and trees stripped of bark. *Bed:* Similar to that of other cervids; marked by tracks and droppings.

Wallows: Cleared depressions in ground,
4′ (1.2 m) wide, 4′ (1.2 m) long, 3–4″
(75–100 mm) deep; muddy, smelling of
urine, marked with tracks.

Scat: Chips or masses when feeding on
aquatic plants and lush grasses. Pellets,
more oblong than Elk's, 1½–1¾″
(40–45 mm) long, sometimes round,
when feeding on woody browse. Pale,
resembling compressed sawdust, when
feeding on woody winter browse.

Trails: Wider and deeper than those of
smaller cervids; Moose are more likely
to detour around obstructions.

Tracks: Cloven prints similar to those
of Elk but larger, more pointed; usually
more than 5″ (125 mm) long; 6″ (150
mm) long and 4½″ (115 mm) wide in a
large bull. Lobes somewhat splayed in
snow or mud, or when running.
Dewclaws often print behind main
prints in snow or mud, or when
running, lengthening print to 10″ (250
mm). Stride 3′9″–5′5″ (1.15–1.65 m)
when walking, more than 8′ (2.4 m)
when trotting or running.

Habitat: Spruce forest, swamps, and aspen and
willow thickets.

Range: Most of Canada; in the East south to
Maine, Minnesota, and Isle Royale in
Lake Superior; in the West, Alaska, n
British Columbia, and southeast
through Rocky Mountains to ne Utah
and nw Colorado.

Migrating seasonally up and down
mountain slopes, the Moose is solitary
in summer, but several may gather near
streams and lakes to feed. A good
swimmer, the Moose can move in the
water at a speed of 6 mph (10 km/h) for
a period of up to two hours. At times,
the animal may be completely
submerged for many seconds. When
black flies and mosquitoes torment it,
the Moose may nearly submerge itself or
roll in a wallow to acquire a protective
coating of mud. In winter, the Moose
may herd, packing down snow to

facilitate movement. Winter herding is not social behavior; rather, the Moose are congregating in favorable habitat. Despite its ungainly appearance, this animal can run through the forest quietly at speeds up to 35 mph (55 km/h). A bull's antlers begin growing in March, attain full growth by August, and are shed by breaking or falling off at the pedicel between December and February. The male uses his antlers to thrash brush (probably to mark territory), to threaten and fight for mates, and to root plants from the pond floor. The shedding of the velvet from its antlers, often described as "dripping velvet," is a spectacular sight. The summer diet of the Moose is willows and aquatic vegetation, including the leaves of water lilies. In winter, it browses on woody plants, including the twigs, buds, and bark of willow, balsam, aspen, dogwood, birch, cherry, maple, and viburnum. The Moose loses weight in winter and gains in summer. Vocalizations include the bull's tremendous bellow, and also "croaks" and "barks" during the rut. The cow has a long, quavering moan, which ends in a cough-like *moo-agh,* and also a grunt used in gathering the young. The bull rushes through the forest looking for grunting cows and challenging rival bulls with bellows. It does not gather a harem, but vies for females; it stages mock fights, circling and threatening another male. As with most cervids, either bull can avoid a fight by withdrawing. Occasionally bulls battle, but generally, threat displays prompt one animal to withdraw; if horns interlock, both may perish. Fights include antler-pushing back and forth. If one male falls, he may be hit in the ribs or the flank. The cow is passive during all this activity, until only one bull remains. He will then mate with her over a one- to two-day period, then move on to find another cow. During

mating season, a bull in rut urinates and then rolls in the wallow he creates; cows also roll in it. The newborn calf can stand up the first day; within a couple of weeks, it can swim. It is weaned at about six months, and just before the birth of new calves, the mother drives it off. The life span of the Moose is up to 20 years. Wolves are the main predator, but are extirpated from much of the Moose's range. The Moose is unpredictable and can be dangerous. It is normally a retiring animal and avoids human contact, but a cow with calves is irritable and fiercely protective, and rutting bulls occasionally have charged people, horses, cars, and even trains.

311, 312 **Caribou**
"Barren Ground Caribou"
Rangifer tarandus

Description: A medium-size cervid. Coloration variable; generally *brown shaggy fur, with whitish neck and mane;* belly, rump, and underside of tail white. On Arctic islands, animals are nearly white; tundra, taiga, and forest individuals are more brownish. Large snout; short, furry ears; short, well-furred tail. Foot pads large and soft in summer, shrunken in winter; hooves rounded. *Male and most females have antlers;* flattened brow tine projects vertically over snout. Bull antlers branched, semi-palmated, with flattened brow tines, 21–62″ (52–158 cm) long; cow antlers relatively small and spindly, 9–20″ (23–50 cm) long. Antler spread to 5′ (1.5 m). Fawn unspotted, resembles adult. Ht 27–55″ (68–140 cm); L 4′6″–8′4″ (1.37–2.54 m); T 4–8½″ (10.2–21.8 cm); HF 15–28″ (38–70 cm); Wt male 275–660 lb (125–299 kg), female 150–300 lb (68–136 kg).

Similar Species: Elk lacks flattened brow tines and white throat; has large yellowish-brown rump patch. Deer are smaller, lack brow tines.

Breeding: Breeds October–November; after
gestation of 7½–8 months, 1 or 2 calves
born mid-May through early July; birth
weight about 11 lb (5 kg).

Sign: Distinguishable mainly by tracks and
locale; deeply worn trails made during
migrations; rubs on saplings, thrashed
bushes.
Bed: Depression similar to that of other
cervids.
Scat: Usually small, bell-shaped pellets
similar to White-tailed Deer's;
occasionally massed when animals are
feeding on succulent summer
vegetation.

Tracks: Widely separated crescents, 5″
(125 mm) wide, slightly shorter in
length, almost always followed by
dewclaw marks; on thin, crusted snow,
only round outlines of hooves may print;
hindprints usually overlap foreprints,
leaving double impressions 8″ (200 mm)
long.

Habitat: Tundra and taiga; farther south, where
lichens abound in coniferous forests in
mountains.

Range: Alaska and much of Canada south
through British Columbia to e
Washington and n Idaho; also n Alberta
and northern two-thirds of Manitoba
and Saskatchewan; in the East, most of
Canada south to Lake Superior and east
to Newfoundland.

The Caribou of North America, now
considered to be the same species as the
Reindeer of Europe and Asia, is among
the most migratory of all mammals. It is
the only cervid that lives year-round
north of the tree line in some of the
harshest habitat in North America. The
gregarious Caribou usually forms a
homogeneous band of bulls, or of cows
with calves and yearlings, but may also
gather in groups numbering up to
100,000 of both sexes and all ages in
late winter before the spring migration.
As spring proceeds, herds begin to move
northward. Females move more rapidly,

and soon some of the juveniles drop back, especially if the snow is deep; they will join the bulls, who travel more slowly. The cows spread out as they reach the area for calving, which takes place in mid-May through early July. The newborn calf is well developed, able to stand in about 30 minutes, run some distance after 90 minutes, and keep up with the herd within 24 hours. It begins to eat solid foods at two weeks, but may continue to nurse into the winter.

In October and November, the rut begins; the bulls join the cow/juvenile groups, where they remain until cows become receptive. Mating occurs either at that time, in the early stages of the southward migration, which varies with location, or immediately after fawning. The polygamous bull chases the female, who flees ahead of him. Pursuit is often interrupted by fights with other males. A male may rush about among several cows, thrashing bushes with his antlers and battling other bulls. However, a male actually pursues only one female at a time. After the rut, the animals move south to the winter range; adult bulls often separate at this time from the cow/juvenile group. Different herds move in different ways in order to reach summer, winter, calving, and rutting grounds with adequate food, water, and protection from predators. The most impressive migrations are by the Caribou living on the tundra in the northwest, often called the "Barren Ground Caribou."

Especially active in the morning and the evening, the Caribou can run at speeds of nearly 50 mph (80 km/h), but cannot maintain such a pace for very long. The animal's spongy footpads provide traction and good weight distribution on boggy summer tundra; in winter, when the pads have shrunk and hardened, and are covered with tufts of hair, the hoof rim bites into ice or crusted snow to prevent slipping. The

Caribou is also a good swimmer. It swims with nearly a third of its body above water, the air-filled hollow hairs of its coat giving it great buoyancy. In summer, to avoid heat and insects, the Caribou often lies on snowbanks on the north side of hills; in winter, it suns on frozen lakes. In early spring, the antlers begin to grow; they are lost shortly after rutting. The female retains her antlers through the winter and loses them about the time the calves arrive. In summer, the Caribou feeds on lichens, mushrooms, grasses, sedges, and many other green plants, twigs of birches and willows, and fruit; it also competes with rodents for dropped antlers, a source of calcium. In winter, lichens are the chief food, supplemented by horsetails, sedges, and willow and birch twigs. Food intake is much reduced in winter, and the animal loses weight then. The Caribou needs high-quality forage in summer to supply the energy necessary for reproduction, growth, and winter survival. Cows with insufficient energy reserves will probably not breed, but will build reserves and breed the following year. In the fall, the bull Caribou fattens up to sustain himself through the rigors of the rut, when he seldom eats. Usually quiet, the Caribou may give a loud snort, and herds of snorting animals may sound like pigs. Biting flies and other insects can be a major problem for Caribou in some areas. In years of major outbreaks, the Caribou will seek snowdrifts, windy ridges, water, or other areas with few insects. Sometimes there is nothing the animal can do but run around wildly in an attempt to avoid them. Chief predators are humans and wolves, although Grizzly Bears, Wolverines, Lynx, and golden eagles may take a few Caribou, particularly the young. The Caribou has been a major source of food and clothing for native people of the far north.

PRONGHORN
Family Antilocapridae

This family consists of a single species, which occurs only in western North America. Until fairly recently, all ruminants with horns that were not sheep, goats, or oxen were called antelopes. But despite its genus name (*Antilocapra,* meaning "antelope goat"), the Pronghorn is neither goat nor antelope, or even very closely related to either; instead, it is the sole remnant of an ancient family that dates back 20 million years. Horns, unlike antlers, are not branched; they are sheathed with keratin (the same substance as in human fingernails), and they grow throughout an animal's life. The unique horn that gives the Pronghorn its common name is pronged, and its sheath of keratin (but not its bony core) is shed each year.

318, 319 **Pronghorn**
"American Antelope"
Antilocapra americana

Description: A medium-size, deer-like mammal with long legs. Upper body and outside of legs pale tan or reddish tan; chest, belly, inner legs, *cheeks and lower jaw, sides, and rump patch are white. 2 broad white blazes across tan throat.* Short erectile mane, about 2¾–4″ (7–10 cm) long. Buck has broad black band from eyes down snout to black nose and black neck patch. *Horns black:* buck's 12–20″ (30–50 cm) long when fully grown; lyre-shaped, curving back and slightly inward near conical tips, each with *1 broad, short prong* jutting forward and slightly upward, usually about halfway from base; doe's horns seldom more than 3–4″ (7.5–10 cm) long, usually without prongs. Juvenile grayer and paler than adult; acquires adult coloration at 3 weeks. Ht 35–41″ (88–103 cm); L 4′1″–4′9″ (1.25–1.45 m);

T 2⅜–6¾″ (6–17 cm); HF 15⅜–16⅞″ (39–43 cm); Wt male 90–140 lb (41–64 kg), female 75–105 lb (34–48 kg).

Similar Species: Cervids in its range are larger, lack large white rump patch, and have antlers rather than horns.

Breeding: Breeds September–October, earlier in the South; implantation delayed 1 month; after gestation of 7 months, 1 or 2 (rarely 3) young born May–June; birth weight 2¼–13 lb (1–6 kg).

Sign: *Scat:* Similar to Mule Deer's; segmented masses when grazing on succulent grasses; small pellets when browsing; pellet shape variable, bluntly oval or elongate, bell- or acorn-shaped.
Tracks: Shaped like split hearts, about 3″ (75 mm) long; hindprints slightly shorter than foreprints. Tracking usually relatively unimportant for field observer since Pronghorn inhabits open terrain and can often be seen at a great distance.

Habitat: Grasslands; also grassy brushlands; and bunchgrass-sagebrush areas.

Range: Southeastern Oregon, s Idaho, s Alberta, s Saskatchewan, Montana, and w North Dakota south to Arizona and w Texas.

The fastest animal in the Western Hemisphere and among the fastest in the world, the Pronghorn, often making 20-foot (6 m) bounds, has been clocked at 70 mph (110 km/h) for three to four minutes at a time. Speeds of 45 mph (70 km/h) are not unusual, and the animal can maintain an easy cruising speed of 30 mph (50 km/h) for about 15 miles (25 km). The Pronghorn runs with its mouth open, not from exhaustion but to gasp extra oxygen. Active night and day, it alternates snatches of sleep with watchful feeding. Because it inhabits open terrain, it relies on spotting enemies at a distance and on its ability to flee speedily. The animal's large, protruding eyes have a wide arc of vision and can detect movement 4 miles (6.5 km) away. If the Pronghorn is alarmed, its rump hairs, which are

about twice as long as other body hairs, become erect, almost doubling the size of the white rump rosette and producing a "flash" visible for great distances. When a herd flees, a buck usually serves as rear guard. If the terrain, presence of young, or a surprise attack forces a Pronghorn to fight rather than flee, it uses as weapons only its sharp hooves, which are effective enough to drive off a Coyote. The Pronghorn avoids muddy ground but is a good swimmer. In summer, it grazes on a number of plant species, including grasses, various forbs, and cacti, and drinks little water when moist green vegetation is available; in winter, it browses on many different plants, favoring sagebrush.

The Pronghorn roams in scattered bands in summer, with does and fawns gathering in groups of a dozen or fewer; yearling and two-year-old males form bachelor herds of about the same size. Older males start establishing territories in March or April, and defend them through the end of the rut. Defense is much more vigorous in the center of a territory than on the periphery. Especially during the rut, males defend territories by staring down rivals, giving loud snorts, approaching and interacting with the intruders, chasing them away, and fighting if necessary, battling fiercely with their horns. Most breeding takes place on the most desirable territories. However, there is much variation in breeding systems in this species. A doe's first breeding usually produces one fawn; subsequent breedings produce twins or, rarely, triplets. The doe spaces twins (or triplets) several hundred feet apart. Nearly odorless for their first few days of life, fawns lie quietly in high grass or brush while their mother grazes at some distance to avoid attracting predators. For about one week she returns frequently to nurse, and then does and fawns join the herd. About a month after breeding,

horns are shed. In winter, herds may include 100 animals or more, of both sexes and all ages. Migration from summer to winter range is variable, depending on altitude, latitude, and range conditions. Cold is no deterrent in itself, for the Pronghorn's coat keeps the animal warm even in severe weather; the air that fills the long, hollow outer hairs provides insulation, and the hairs flatten against the body to seal in warmth. (In summer, the animal molts to a thinner coat, and the hairs ruffle up to provide cooling ventilation.) While a Pronghorn can scratch through light snow for food, deep snow forces it to areas where browse is uncovered, including higher elevations where winds have swept away expanses of snow.

The life span of the Pronghorn is 7 to 10 years. Once abundant in its range, this animal declined in numbers almost as precipitously as the American Bison; in 1924, owing to the enmity of ranchers, and the fencing of rangeland, which hampered migration and foraging (Pronghorns cannot leap fences like deer—they crawl under them instead), there were fewer than 20,000 animals. Now, as a result of efforts at transplantation and management of herds by game departments, the Pronghorn's range is expanding and its numbers have increased to about half a million. An Arizona subspecies, the Sonoran Pronghorn *(A. a. sonoriensis),* is classified as endangered by the U.S. government.

GOATS, SHEEP, AND CATTLE
Family Bovidae

The bovids are a big family of mostly larger hoofed mammals, including about 42 genera and 100 species, 72 of them comprising the African antelopes and their allies, and also including domestic cattle, sheep, and goats. There are five native species in our region: the American Bison, Mountain Goat, Muskox, Bighorn Sheep, and Dall's Sheep. We also include here one introduced species, the Barbary Sheep.

Unlike deer, bovids have true horns, which are permanent bony outgrowths of the frontal bone. Present always in males and often in females, the horns are hollow, are never branched, and grow throughout the life of the animal. Males use them as weapons when competing for females, and both males and females may use them as defensive weapons against predators. Most bovids favor grasslands or open areas, depending for protection on their size and speed, as well as their horns and hooves. The domestication of bovids began in Asia about 8,000 years ago; today the domestic cow, goat, and sheep are of great economic importance.

329–333 **American Bison**
"Buffalo"
Bos bison

Description: *The largest terrestrial animal in North America.* Dark brown, with shaggy mane and beard. Long tail with tuft at tip. Broad, massive head; *humped shoulders;* short legs clothed with shaggy hair; large hooves. *Both sexes have short black horns with pointed tips* that protrude from the top of the head, above and behind the eyes, curving outward, then in. Horn spread to 3′ (90 cm). Juvenile reddish brown; acquires adult coloration at 2–3

months of age. Ht male to 6′ (1.8 m),
female to 5′ (1.5 m); L male 10′–12′6″
(3–3.8 m), female 7–8′ (2.1–2.4 m);
T male 17–19″ (43–48 cm), female
12–18″ (30–45 cm); HF 20–26″ (51–
66 cm); Wt male 991–2,000 lb
(450–900 kg), female 793–1,013 lb
(360–460 kg).

Breeding: Varies, but most often June–September;
1 (occasionally 2) young born after
gestation of 9–9½ months.

Sign: In wooded habitat, trees ringed with
pale "horn rubs" or "head rubs" where
bark is worn away; trampled ground
underneath such trees or around rubbed
boulders.
Wallows: Especially in plains habitat,
dusty saucer-like depressions, 8–10′
(2.4–3 m) wide, 1′ (.3 m) deep, where
bison has rolled and rubbed repeatedly,
dust-bathing to relieve itching and to
rid coat of insects. Bull may urinate in
dry wallows, then cake himself with
mud as protection against insect pests.

Scat: Similar to that of domestic cow;
round flat pads, about 10–14″ (25–35
cm) in diameter.
Tracks: Cloven hearts, similar to those
of domestic cow, but rounder and
somewhat larger, about 5″ (130 mm)
wide for mature bull. On hard ground,
cleft between facing crescent lobes may
not show; tracks may then resemble
horse's hoofprints.

Habitat: Varied; primarily plains, prairies, and
river valleys; sometimes forests.

Range: Historically ranged from s Northwest
Territories to nw Mexico, Texas, and
Mississippi, and east to sw New York,
South Carolina, and Georgia. Now
large, free-ranging herds only at Wood
Buffalo National Park, Mackenzie Bison
Sanctuary, and Slave River Lowlands in
Northwest Territories, Canada, and in
Yellowstone National Park, Wyoming.
Small free-ranging herds in Alaska, ne
British Columbia, nw Saskatchewan,
and Northwest Territories. Many
smaller herds in fenced areas.

The American Bison is most active in early morning and late afternoon, but sometimes also on moonlit nights. In the midday heat, it rests, chewing its cud or dust-bathing. This animal commonly rubs its horns on trees, thrashes saplings, and wallows in the dirt. A good swimmer, it is so buoyant that head, hump, and tail remain above water. American Bison will stampede if frightened, galloping at speeds up to 32 mph (50 km/h). Formerly undertaking annual migrations of 200 miles (320 km) or more between winter and summer ranges, some bison in Canada still travel up to 150 miles (240 km) between wooded hills and valleys. The American Bison feeds on many grasses, sedges, and forbs, and sometimes on berries, lichens, and horsetails; in winter, it clears snow from vegetation with its hooves and head. Vocalizations include the bull's bellow during rutting, the cow's snort, and the calf's bawl. Usually between 4 and 20 bison herd together, with sexes separate except during breeding season, when the herds combine and increase greatly in size; occasionally such herds gather into bands of several thousand. There are three kinds of bison groups: matriarchal (cows, calves, yearlings, and sometimes a few bulls), bull (though some bulls are solitary), and breeding (a combination of matriarchal and bull groups). A matriarchal group is relatively stable and often ranges from about 11 to 20 individuals. A bull group is smaller, and the male bison seems to become more solitary with age. The time and length of the rut varies. The bull enters the matriarchal herd and checks for estrous females. He then displays flehmen and tending behavior. Flehmen consists of curling the lip back and extending the neck; it lasts for several seconds and is thought to enhance the sense of smell. The male "tends" a female by remaining between her and

the herd in an attempt to keep the cow
isolated. Tending can last from several
seconds to several days, and may or may
not end in copulation. A female will not
always tolerate tending; thus she has a
choice of mate. Copulation may be
preceded by mutual licking and
butting. The male threatens and battles
other contenders in his attempt to tend
and mate with a cow. Threats usually
ward off fights, but if a rival male
perseveres, fighting may ensue,
involving butting, horn-locking,
shoving, and hooking. When butting,
males walk to within 20 feet (6 m) of
each other, lower their heads, raise their
tails, and charge. Their massive
foreheads, including much hair but not
the horns, collide without apparent
injury; they charge repeatedly until one
animal gives up. Hooking can be very
dangerous, often resulting in injury or
death; it consists of using the horns to
gore the opponent in the side or belly.
During the 24-hour period that a cow is
in heat, a bull may mate with her
repeatedly. The reddish newborn stands
to nurse in 30 minutes, walks within
hours, and in one or two days joins the
herd with its mother. At two months,
hump and horns start to develop. Most
young are weaned by late summer; some
nurse up to seven months.
Life span in the wild averages 25 years.
In the 15th century, millions of American
Bison grazed from the Atlantic Ocean
almost to the Pacific and from Mexico
and Florida into Canada. Probably no
other animal has been as central to a
people's way of life as was the bison to
the Native American, who ate its meat,
used the skins for clothing and shelter,
fashioned thread and rope from sinew,
made glue and tools from the hooves
and bones, and burned the droppings
as fuel. Although Native Americans
occasionally killed more bison than they
could use, stampeding thousands over
cliffs, they had no significant effect on

bison populations. The destruction of the American Bison began about 1830, when U.S. government policy advocated the animal's extermination in order to subdue hostile tribes through starvation, equating bison carcasses with "discouraged Indians." Railroad construction crews often subsisted on bison meat, as did some army posts, and the railroad provided a means of shipping hides to eastern markets. Ultimately millions of pounds of bison bones were ground into fertilizer or used for the manufacture of bone china. By 1900, fewer than 1,000 American Bison remained, and a crusade of rescue and restoration was begun. Estimates of the number of bison in North America before European settlers arrived range from 30 to 70 million. Today more than 65,000 bison roam U.S. and Canadian national parks and ranges, and privately owned rangelands; few are wild and free-ranging.

320–323   Mountain Goat
Oreamnos americanus

Description: A relatively small bovid. Compact, short-legged body. *Yellowish-white fur,* long and shaggy in winter, shorter in summer; "beard," about 5″ (12 cm) long, retained year-round. Eyes, nose, hooves, and horns black. Both sexes have backward-curving, *dagger-like horns,* up to 12″ (30 cm) long in male, 9″ (23 cm) in female. Juvenile similar to adult, but with brown hairs along back. Female approximately 15 percent smaller than male. Ht 35–47″ (90–120 cm); L 4′–5′10″ (1.25–1.8 m); T 3¼–8″ (8.4–20 cm); HF 11¾–14½″ (30–37 cm); Wt average: male 154–180 lb (70–82 kg), female 117–156 lb (53–71 kg).

Similar Species: Dall's and Bighorn sheep lack beard, have much larger, curved horns.

Breeding: Mates mid-November through mid-December; gestation 6 months. 1–3 young born mid-May through mid-June. Newborn weighs about 6½ lb (3 kg).

Sign: *Bed:* Shallow depression scraped out in shale or dirt on ledge at base of cliff. In vicinity, white hair snagged on vegetation and rocks, or blown to ground.

Scat: Similar to that of sheep and deer; massed when feeding on lush grasses; separate, compacted pellets, usually bell-shaped, sometimes oblong or nearly round, when feeding on drier, brushier foods.

Tracks: Double-lobed, widely splayed at front; 2½–3½″ (64–89 mm) long.

Habitat: Rocky, mountainous areas above timberline.

Range: Natural range: extreme s Alaska, s Yukon, British Columbia, sw Alberta, parts of Washington, n Idaho, and nw Montana. Introduced successfully in Oregon, Nevada, Utah, Colorado, Wyoming, and South Dakota.

The Mountain Goat is not a true goat, but belongs to a group known as goat-antelopes, which includes the Chamois (*Rupicapra rupicapra*) of Europe and Asia Minor. Whereas the horns of true goats sweep up and back, and are transversely ridged or tightly spiraled like a corkscrew, the Mountain Goat's horns curve back only slightly and are nearly smooth. Its "beard" is not the true chin beard of male goats, but an extension of a throat mane. The Mountain Goat is active in morning and evening and sometimes during moonlit nights. Its hooves are well adapted for rocky peaks, with a sharp outer rim that grips and a rubbery sole that provides traction on steep or smooth surfaces. Traversing peaks and narrow ledges at a stately walk or trot, a Mountain Goat may seem to move across the face of an almost sheer cliff. However, individuals

have been known to miss their footing and fall to their deaths. On warm days, the animal will bed on a patch of snow, in a shady spot, or on a mountain ledge. Goats spend much time taking dust baths in dry wallows, particularly in May and June. The sexes herd apart until rutting season, usually in November or December. Mountain Goats are polygamous. Females remain on the nursery ranges during rut, and males move from range to range in search of females in heat. Females will allow males on their ranges in October, but will not permit mating until November. During the rutting season, a male often marks a female with a musky oil from glands at the base of his horns by rubbing his head against her body, and digs a pit from which he paws dirt onto his flanks and belly. While rival males frequently threaten each other, breeding battles are uncommon, as skulls and horns are relatively fragile. The kid, usually born on a mountain ledge, can stand and climb shortly after birth. It starts feeding within a few days of birth, but weaning is not complete until August or September. The kid remains with its mother until the next year's young is born. Avalanches and rock slides are the greatest killers of Mountain Goats, accounting for many more deaths than predation. Only the golden eagle can attack this species in high mountains; it may try to drive a kid over a cliff. Carnivores such as the Mountain Lion may attack the Mountain Goat as it descends into a valley, but the goat's sharp hooves make it dangerous prey.

334, 335   Muskox
Ovibos moschatus

Description: A large hoofed mammal. Dark brown *shaggy hair hangs nearly to feet;* lighter "saddle" on back; whitish lower legs.

Broad head; slightly humped shoulders. Male and female have *massive light or dark brown horns that curve down close to sides of head,* then out and up near pointed tips; spread to 30″ (75 cm). Female's horns shorter, more slender, and more curved than male's. Ht 3–5′ (91–152 cm); L 6′4″–8′1″ (1.94–2.46 m); T 2½–7″ (6–17 cm); Wt male 579–900 lb (263–408 kg), female 370–670 lb (168–304 kg).

Breeding: Breeds every other year in late summer; season peaks in September. 1 calf born late April–May after gestation of 8–9 months.

Sign: *Scat:* Flat, round mass similar to that of domestic cow, but often drier and harder.

Tracks: Pair of facing crescents, about 5″ (130 mm) long; front and rear contours so similar that direction of travel may be difficult to ascertain, but front usually prints deeper than rear; similar to tracks of domestic cattle and American Bison. On very hard ground, cleft may not print clearly and then may look similar to horse hoofprints.

Habitat: Arctic tundra; in summer, grassy river valleys, lakeshores, and meadows with willows and heath plants; in winter, windswept hilltops and slopes where vegetation is exposed.

Range: Northern part of Northwest Territories and islands to, but not including, Greenland. Small, introduced semi-domesticated herds in Unalakleet, Alaska, southeast of Nome.

The genus name combines the Latin terms for sheep *(ovis)* and oxen *(bos).* The Muskox is more or less intermediate between the two groups: It is ox-like in size, but its profile, hairy muzzle, thin lips, and relatively small, pointed ears might belong to a giant, long-haired ram. Muskoxen give off a stronger odor than American Bison or domestic cattle; during breeding season, the bull exudes an especially strong scent. The animal's

common and scientific names both refer to this musky odor, which, however, is from urine, not musk, as this animal has no musk glands. Muskoxen have good sight and hearing, and can run rapidly when necessary. They usually travel in a closely packed herd of 15 to 20 individuals in the winter and 10 in the summer, though occasionally a herd may contain up to 100 individuals. Herds include males and females; during the rutting season, the dominant male attempts to take control, driving the other males away by charging and clashing heads with them. The outcast males become solitary or form small groups of their own. The dominant bull continually tests and tends the females, and probably mates with most or all as they come into estrus. The newborn is able to stand shortly after birth, and its horns begin to appear at six months. Undertaking no major migrations, Muskoxen travel only about 50 miles (80 km) between summer and winter ranges. In summer, they feed on green vegetation such as sedges, grasses, and willows; in winter, the diet is primarily woody plants. Both sexes snort and stamp their feet, and excited males give a deep, throaty bellow. When threatened, adults form a fortress-like ring or line with the young inside. If a wolf, the Muskox's chief predator, comes too close, it may be thrown into the air on a bull's horns, then crushed with the hooves; occasionally a bull will leave the ring to attack. Life span is about 20 years. In the 19th century, the Muskox was hunted for meat and hide, and for a time herds were drastically reduced on the mainland. Recently the population was estimated at about 40,000. A herd of about 170 has been maintained at Unalakleet, Alaska, for their valuable underwool, called *quiviut,* which is used by Native American village weavers to fashion luxurious garments.

326–328 Bighorn Sheep
"Mountain Sheep," "Rocky Mountain
Bighorn Sheep"
Ovis canadensis

Description: A medium-size bovid. Muscular body,
with thick neck. Color varies from *dark
brown above* in northern mountains to
pale tan in desert; belly, rump patch,
back of legs, muzzle, and eye patch are
white. Short, dark brown tail. Coat
sheds in patches June–July. *Ram has
massive brown horns* that curve up and
back over ears, then down, around, and
up past cheeks in C-shaped "curl";
spread to 33" (83 cm). Ewe has short,
slender horns that never form more than
half-curl. Juvenile has soft, woolly,
creamy-fawn coat. Ht male 3'–3'5"
(90–105 cm), female 30–36" (75–90 cm);
L male 5'3"–6'1" (1.6–1.85 m), female
4'2"–5'2" (1.28–1.58 m); T male 4–6"
(10–15 cm), female 3½–5" (9–13 cm);
HF 12–22" (27.6–48.2 cm); Wt male
127–316 lb (58–143 kg), female
74–200 lb (34–91 kg).

Similar Species: Dall's Sheep is white, gray, or blackish;
has smaller, more slender horns, and is
found farther north. Mountain Goat is
white and has much smaller horns.

Breeding: Breeds fall–early winter, depending
upon geographical latitude; after
gestation of nearly 6 months, 1 lamb
born April–late June.

Sign: *Bed:* Depression about 4' (1.2 m) wide,
up to 1' (.3 m) deep, usually smelling of
urine, almost always edged with
droppings.
Scat: Dark pellets, usually bell-shaped,
sometimes massed if vegetation is
succulent.
Trails: Similar to that of Mountain Goat
or deer, but at elevations higher than
deer trails and on slopes less precipitous
than Mountain Goat trails. Shed or
snagged hair, longer and generally
darker than deer and Mountain Goat
hair, may help identify trail.

Tracks: Double-lobed prints, 3–3½"
(75–90 mm) long, with hindprints
slightly smaller than foreprints; similar
to deer's but with straighter edges—less
pointed and often more splayed at front,
and less heart-shaped. When walking
downhill on soft ground, dewclaws may
print 2 dots behind hoofprint. Walking
gait about 18" (450 mm); bounding gait
on level ground 15' (4.5 m), down steep
incline 30' (9 m).

Habitat: Semi-open, precipitous terrain with
rocky slopes, ridges, and cliffs or
canyons; from alpine meadow to hot
desert. Grassy vegetation necessary with
scattered shrubby plants; water essential
in desert regions.

Range: Disjunct: from s British Columbia, sw
Alberta, Idaho, and Montana south to se
California, Arizona, and New Mexico.

The Bighorn Sheep inhabits areas
around rocky cliffs rarely disturbed by
humans. It perhaps has adapted to this
rather inhospitable habitat because
there is a lack of competition as well as
protection from predators. A good
swimmer and an excellent rock climber
and jumper, this animal has hooves that
are hard at the outer edge and spongy in
the center, providing good traction even
on sheer rock. The Bighorn is active by
day, feeding in early morning, midday,
and evening; it lies down and chews its
cud at other times and retires to bedding
spots for the night. In heavily congested
areas such as Yellowstone National Park,
this animal has sometimes had to resort
to feeding at night.
The Bighorn has a home range, but not
a territory. It migrates between high
slopes in the summer and valleys in
winter, traveling distances of ½ to 40
miles (.8–64 km); it also makes minor
movements, depending on local
conditions. Most of the year is spent on
the summer range. Highly gregarious,
the Bighorn lives in herds or bands,
usually of about 5 to 15 animals,

including ewes, lambs, yearlings, and two-year-olds; the dominant ewe is the leader. Ram bands usually number two to five. In winter, when ewe herds join, there may be as many as 100 animals, all led by an old ewe. In spring, rams band together and move to separate higher summer ranges. As the fall rutting season approaches, rams have butting contests, which increase in frequency as the season progresses. They charge each other at speeds of more than 20 mph (32 km/h), their foreheads crashing with a crack that can be heard more than a mile away, often prompting other rams to similar contests. Butting battles may continue as long as 20 hours. Horn size determines status; fights occur only between rams with horns of similar size. (A seven- or eight-year-old ram may have a full curl, with tips level with the horn bases; a few old rams exceed a full curl, but often horns are "broomed"— broken off near the tips or deliberately rubbed off on rocks when they begin to block the ram's peripheral vision.) With nose elevated, head cocked to one side, and upper lip curled, the rutting male follows any female in heat; if more than one ram follows the same ewe, they stop occasionally for butting jousts. The species is polygamous; dominant ram does most of the courting and mating. The male moves between herds seeking females in heat. When he finds one, the female will initiate a chase, or he will kick her in an attempt to initiate a chase, which may culminate in mating. If one male tires of mating, another will replace him. Lambing areas are on the most inaccessible cliffs. A single well-developed lamb is born with a soft, woolly, light-colored coat and small horn buds; within a day, it can walk and climb nearly as well as its mother. The lamb remains hidden the first week, then follows its mother about, feeding on grasses, and is weaned at five to six months.

In summer, the Bighorn feeds mainly on grasses and sedges, particularly bluegrass, wheat grass, bromes, and fescues. In winter, it feeds more on woody plants, such as willow, sage, and rabbit brush. Favored forbs are phlox, cinquefoil, and clover. Because of dry conditions, in the desert this animal feeds more on brushy plants, such as desert holly, and on various species of cactus. Like other hoofed mammals in our range, the Bighorn beds down wherever it happens to be each night. Old beds may be reused, but they are pawed out more deeply. Bedding spots are often found along ridges, but sometimes they are in caves or in sites formed by Grizzly Bears digging for ground squirrels. Bighorns respond to disturbance by (1) assuming an attention posture—standing and staring at the source; (2) assuming an alarm posture—snorting, pawing the ground, bowing their heads, or, in the presence of wolves, huddling in a tight circle, facing outward; or (3) running, if startled at close range. Life span is about 15 years. Predators include Mountain Lions, golden eagles, wolves, Coyotes, bears, Bobcats, and Lynx; on cliffs, the Bighorn easily escapes all but the first two, and the eagles attack only lambs. The Bighorn has always been prized for its meat; the horns were used by the Shoshone and Gros Ventre tribes to make powerful bows and are still prized by hunters as trophies.

---

324 Dall's Sheep
"Thinhorn Sheep," "Stone Sheep"
Ovis dalli

Description: A small bovid. *White above and below, often with yellowish or brownish cast;* hooves yellowish brown. Horns of ram massive, light yellow, with well-defined growth rings flaring out and away from head; about 3′ (90 cm) long, with 4′1″

(1.25 m) the largest recorded size; horn spread to 3' (90 cm). Ewe's slender spikes less than 15" (38 cm) long. In southernmost part of range, *black phase* individuals, known as "Stone Sheep," vary from charcoal gray verging on black to light gray or gray brown; in Yukon, where Dall's and Stone phases intergrade, *gray phase* individuals are usually darker on back, occasionally with dark "saddle"; these sheep have white belly, rump patch, back of legs, and facial blaze; dark hooves; and are often slightly bigger than Dall's Sheep, with slightly heavier horns. Ht male 33–41" (83–105 cm), female 30–36" (75–90 cm); L male 4'5"–5' (1.34–1.53 m), female 3'5"–4'5" (1.05–1.35 m); T male 3½–4½" (8.9–11.5 cm), female 3–3½" (7.5–9 cm); HF 15–20" (38–51 cm); Wt male 174–200 lb (79–91 kg), female 100–125 lb (45–57 kg).

Similar Species: Bighorn Sheep is brown, has larger, thicker horns, and occurs to the south.

Breeding: Breeds late fall; after gestation of slightly less than 6 months, 1 (rarely 2) young born mid-May.

Sign: Tufts of white hair, snagged or shed, sometimes left on bushes and rocks. Tracks, beds, and droppings similar to those of Bighorn Sheep, but when Dall's Sheep has been licking a salt lick, scat forms round pellets.

Habitat: Rocky, mountainous areas.

Range: Disjunct populations in Alaska, Yukon, w Mackenzie district (Northwest Territories), and n British Columbia.

Dall's Sheep is diurnal. Its habits are similar to those of the Bighorn Sheep, but it seems more wary and agile. In winter, the entire herd feeds together on such woody plants as willow, sage, crowberry, and cranberry. In summer, the animal grazes on grasses, sedges, and forbs. Staple foods are fescue and sedges, particularly the seed heads. Willow, horseweed, and alpine fireweed are also eaten, and horsetail and Richardson's

saxifrage are highly relished. Dall's Sheep seldom eats lichens and mosses. In spring, the herd splits into two groups, with ewes, lambs, and yearling rams in one group, older rams in the other, though older rams sometimes remain solitary; the oldest member in each group is its leader. In late fall, when rams try to gather a harem of ewes, butting contests among rivals are common. After walking apart 15 to 20 yards (14–18 m), rams turn, rise up on their hindfeet, then drop to all fours and race toward each other. The clash of their horns can be heard more than a mile away. Collisions may be repeated, but ultimately the contest becomes a matter of pushing and shoving, until the stronger, heavier ram drives off the weaker one. The newborn lamb walks when three to four hours old, and begins eating grasses at about 10 days. Life span is up to 15 years. Wolves are the chief predators; occasionally a Lynx, Wolverine, Mountain Lion, or bear takes a sheep, and the golden eagle sometimes seizes a lamb.

325　Barbary Sheep
Ammotragus lervia

Description: A small to medium-size bovid. Coat uniformly tawny, except for white on chin, insides of ears, and line on underparts. *Long mane* hangs from throat, chest, and upper forelegs. Both sexes have heavy, wrinkled, *goat-like horns* that bend outward, then backward and in; measure to 33″ (84 cm) long. Ht 3′–3′5″ (90–105 cm); L 4′3″–6′3″ (1.3–1.9 m); T 10″ (25 cm); Wt male 220–320 lb (100–145 kg), female 88–121 lb (40–55 kg).

Similar Species: Dall's and Bighorn sheep lack mane and have coiled horns. Mountain Goat is white and has much smaller horns.

Breeding: Mates September–November, occasionally throughout the year.

Gestation about 5½ months; usually
1–3 young born March–May.

Sign: Tracks and droppings similar to those of
Bighorn Sheep.

Habitat: Rocky, mountainous terrain in arid or
semi-arid areas.

Range: Native to North Africa, where it is
hunted for meat. Introduced into
southwestern U.S. Largest populations
in Palo Duro Canyon of Texas
Panhandle; the Canadian River Gorge,
Canyon Largo, and Hondo Valley of
New Mexico; and the Santa Lucia
Mountains of wc California.

The single species in this genus was
introduced as a game animal in the
southwestern U.S. in 1950. A gland
beneath the tail gives the Barbary Sheep
a goat-like odor. The animal gets its
water primarily from dew and from the
green vegetation upon which it chiefly
feeds. In behavior, the Barbary Sheep
is sheep-like, following its leader,
usually an adult female. Families of
male, female, and young remain
together. During breeding season,
males stand apart, run toward each
other, and clash horns. They also stand
side by side, lock horns, and try to pull
each other down in a display of strength
similar to arm wrestling. The newborn
lamb nurses for about six months.

PART III
APPENDICES

GLOSSARY

Blastula   Early stage of developing embryo that consists of a layer of cells around a central cavity.

Buff   Dull brownish yellow.

Cache   A place in which food stores are hidden; also, food hidden in such a place.

Calcar   In bats, a small bone or piece of cartilage that projects from the inner side of the hindfoot into the interfemoral membrane.

Carnivorous   Flesh-eating.

Cephalopod   One of the higher mollusks, such as octopus, squid, and cuttlefish.

Chaparral   Shrubby thicket or thorny shrubs, cacti, or evergreen oak.

Colonial   Living in a colony or group, in which individuals usually maintain close association (e.g., prairie dogs).

Coprophagy   Feeding on fecal pellets that have passed through the digestive tract in relatively undigested condition; common in rabbits and hares.

Crepuscular   Appearing or becoming active at dusk and/or dawn.

Cud   Food brought up into the mouth of a ruminant animal from the stomach to be chewed again. See *Ruminant*.

Delayed fertilization   A variation of the reproductive process exhibited in some mammals in which mating takes place, but the sperm remain free in the female reproductive tract for an extended period before fertilization occurs.

Delayed implantation   A variation of the reproductive process exhibited in some mammals in which fertilization occurs normally, but development stops shortly thereafter, and the blastula remains free for an extended period before it implants in the uterus.

Dewclaw   A functionless digit or "toe," usually on the upper part of a mammal's foot; on deer, it is located above the hoof.

Dewlap   A loose fold of skin hanging from the throat of an animal, such as the Moose.

Diurnal   Active by day.

Dusky   Dull grayish brown.

Echolocation   A method with which certain animals orient themselves by emitting high-frequency sounds and interpreting the reflected sound waves; many bats navigate and locate prey by this means.

Endangered   Used to refer to an animal species or subspecies whose prospects for survival and reproduction are in immediate jeopardy.

Estivation   A state of dormancy similar to hibernation that occurs during hot or dry periods.

Estrus   The period during which a female mammal is sexually receptive to males and capable of conceiving; also called heat.

Extinction   Total extermination of a species.

Extirpation   Extermination of a species in a particular geographical area.

Feral   Having escaped from domestication and become wild.

Forb   An herbaceous plant other than grass.

Fossorial   Adapted for a burrowing existence (e.g., moles).

Gregarious   Living harmoniously with other individuals of the same species; sociable.

Grizzled   Sprinkled or streaked with gray or another (usually contrasting) color.

Guard hairs   Long, coarse hairs that form a protective coating over the underfur of a mammal.

Habitat   The environment in which an organism lives.

Harem   A group of two or more breeding females, plus their young, monopolized and sometimes protected by one male.

Heat   The period during which a female mammal is sexually receptive to males and capable of conceiving; also called estrus.

Herbivorous   Plant-eating.

Hibernation   A state of dormancy during winter in which temperature and all bodily processes are greatly reduced, thereby conserving energy and enabling certain mammals to sleep through much of the winter.

Home range   Area in which an animal moves during its normal day-to-day travels; differs from territory in not being defended.

Hybrid   Offspring born of two different species.

Intergrade   To merge gradually with one another by interbreeding, usually through a continuous series of intermediate forms.

Interfemoral membrane   The thin skin that stretches between the hindlegs of bats; also called tail membrane.

Keeled calcar   In bats, a calcar with a flat projection protruding along part of its length toward the interfemoral membrane.

Lanugo   Soft, usually white, woolly coat with which many seals are born.

Melanistic   Refers to an individual with an increased development of black pigment in the skin, fur, or other body coverings.

Midden   A dung hill or refuse heap, often heaped with discarded feeding debris, such as nutshells.

Molariform teeth   Teeth usually adapted for grinding; all teeth behind the canines, including premolars and molars.

Montane   Pertaining to or inhabiting mountainous country.

Nocturnal   Active by night.

Olivaceous   Olive-tinted or slightly olive.

Omnivorous   Feeding on both plant and animal material.

Palmate   Branching like fingers on a hand.

Pedicle   Projection on frontal bone of skull of cervids from which the antlers grow.

Plantar tubercles   Raised areas on soles of feet.

Polygamous   Having more than one mate at a time.

Ruminant   An even-toed hoofed mammal, such as a sheep or a deer, having a complex, usually four-chambered stomach from which stored food, the "cud," is regurgitated into the mouth for chewing to aid more complete digestion.

Runway   A beaten path made by animals.

Rut   The periodic sexual excitement of male deer, sheep, and goats (corresponding to estrus or heat in females); also the period when this occurs.

Scat   Fecal pellet or dropping; feces.

Sedge   Family of marsh plants related to grasses.

Semi-colonial   Living in loose association with other individuals, usually without very close contact.

Straddle   The width of a set of animal tracks.

Stride   The distance between sets of animal tracks.

Subalpine   Mountain regions below timberline.

Subspecies   As used here, a population isolated geographically from other populations of a species, and evolving in its own direction.

Talus   A slope formed by an accumulation of rock debris; rock debris at the base of a cliff.

Territory   A defended area, within the home range, in which an animal lives permanently or temporarily.

Threatened   Used here to refer to a species or subspecies likely to become endangered in the foreseeable future.

Torpor   Temporary loss of all or part of the power of sensation or motion, resulting from the reduction of body temperature and the slowing of bodily processes.

Tragus   A lobe projecting upward from inside the base of the ear, as in bats.

Underfur   A thick undercoat of fur.

Ungulates   Hoofed mammals.

Unicuspid   In shrews, any of the small teeth between the two front teeth and the large molariform teeth.

Velvet   The soft, furry covering on a cervid's growing antlers, which contains a network of capillaries to supply nourishment; when antlers are so covered, the animal is said to be "in velvet."

Yard   A place where hoofed mammals, such as deer or Moose, herd together to feed during the winter.

ENDANGERED AND
THREATENED MAMMALS

Some species of North American mammals are listed by
the U.S. and Canadian governments as "endangered" or
"threatened." An endangered species is one whose existence
is threatened with imminent extinction through all or a
significant portion of its range. A threatened species is
one that is likely to become endangered if the factors
affecting its vulnerability are not reversed. Both federal
and state governments give full protection to these species.
With some species, a certain subspecies or geographic
population is at risk, not the entire species. Many
conservation organizations, like the National Audubon
Society, are working on restoring animal populations and
their corresponding habitats; we urge all concerned about
diminishing habitats and endangered species to join one
of these groups.

The following list of endangered and threatened mammals
of North America north of Mexico is based on official lists
compiled by federal agencies of the United States and
Canada. The animals are grouped by order, and arranged
in the same sequence as in the species accounts. Species
and subspecies are listed by their common names, followed
by the scientific name, and the location in which they are
endangered. For a listing of the abbreviations of state and
province names, see the range charts in the appendices.

ENDANGERED

Order Chiroptera: Bats
Lesser (Sanborn's) Long-nosed Bat
  Leptonycteris curasoae yerbabuenae
  (L. sanborni)   AZ, NM
Mexican Long-nosed Bat
  Leptonycteris nivalis   TX
Gray Myotis
  Myotis grisescens   c and se U.S.
Indiana Myotis
  Myotis sodalis   e and c U.S.
Hawaiian Hoary Bat
  Lasiurus cinereus semotus   HI
Ozark Big-eared Bat
  Plecotus townsendii ingens   AR, MO, OK
Virginia Big-eared Bat
  Plecotus townsendii virginianus   KY, NC, VA, WV

Order Lagomorpha: Pikas, Rabbits, and Hares
Lower Keys Rabbit
  Sylvilagus palustris hefneri   FL

Order Rodentia: Rodents
Point Arena Mountain Beaver
  Aplodontia rufa nigra   CA
Vancouver Marmot
  Marmota vancouverensis   Vancouver I., Can.
Delmarva Peninsula Fox Squirrel
  Sciurus niger cinereus   DE, MD, PA, VA
Mount Graham Red Squirrel
  Tamiasciurus hudsonicus grahamensis AZ
Carolina Northern Flying Squirrel
  Glaucomys sabrinus coloratus   NC, TN
Virginia Northern Flying Squirrel
  Glaucomys sabrinus fuscus   VA, WV
Pacific Pocket Mouse
  Perognathus longimembris pacificus   CA
Morro Bay Kangaroo Rat
  Dipodomys heermanni morroensis   CA
Giant Kangaroo Rat
  Dipodomys ingens   CA
Fresno Kangaroo Rat
  Dipodomys nitratoides exilis   CA
Tipton Kangaroo Rat
  Dipodomys nitratoides nitratoides   CA
Stephen's Kangaroo Rat
  Dipodomys stephensi   CA

Key Rice Rat
  Oryzomys argentatus   FL
Salt-marsh Harvest Mouse
  Reithrodontomys raviventris   CA
Key Largo Cotton Mouse
  Peromyscus gossypinus allapaticola   FL
Choctawahatchee Beach Mouse
  Peromyscus polionotus allophrys   FL
Alabama Beach Mouse
  Peromyscus polionotus ammobates   AL
Anastasia Island Beach Mouse
  Peromyscus polionotus phasma   FL
Perdido Key Beach Mouse
  Peromyscus polionotus trissyllepsis   AL, FL
Key Largo Woodrat
  Neotoma floridana smalli   FL
Amargosa Vole
  Microtus californicus scirpensis   CA
Hualapai Mexican Vole
  Microtus mexicanus hualpaiensis   AZ
Florida Salt-marsh Vole
  *Microtus pennsylvanicus
  dukecampbelli*   FL

Order Carnivora: Carnivores

Gray Wolf   lower 48 states
  Canis lupus   except MN
Red Wolf
  Canis rufus   se U.S. to c TX
Northern Swift Fox
  Vulpes velox hebes   n U.S., Can.
San Joaquin Kit Fox
  Vulpes velox mutica   CA
West Indian Monk Seal
  Monachus tropicalis   Gulf of Mexico
Black-footed Ferret
  Mustela nigripes   w U.S., w Can.
Florida Panther
  Felis concolor coryi   se U.S.
Eastern Cougar
  Felis concolor couguar   e N.A.
Ocelot
  Felis pardalis   AZ, TX
Margay
  Felis wiedii   TX
Jaguarundi
  Felis yagouaroundi   AZ, TX
Jaguar
  Panthera onca   AZ, NM, TX

Order Sirenia: Manatees and Sea Cows
Manatee
    Trichechus manatus   se U.S.

Order Artiodactyla: Even-toed Hoofed Mammals
Key Deer
    Odocoileus virginianus clavium   FL
Columbian White-tailed Deer
    Odocoileus virginianus leucurus   OR, WA
Woodland Caribou
    Rangifer tarandus caribou   ID, WA, sw Can.
Sonoran Pronghorn
    Antilocapra americana sonoriensis   AZ
Wood Bison
    Bos bison athabascae   Can.

THREATENED

Order Insectivora: Insectivores
Dismal Swamp Shrew
    Sorex longirostris fisheri   NC, VA

Order Rodentia: Rodents
Utah Prairie Dog
    Cynomys parvidens   UT
Southeastern Beach Mouse
    Peromyscus polionotus niveiventris   FL

Order Carnivora: Carnivores
Gray Wolf
    Canis lupus   MN
Black Bear
    Ursus americanus   LA, MS, TX
Grizzly Bear
    Ursus arctos   lower 48 states
Guadalupe Fur Seal
    Arctocephalus townsendi   CA
Northern Sea Lion
    Eumetopias jubatus   AK, CA, OR, WA
Southern Sea Otter
    Enhydra lutris nereis   CA, OR, WA
Mountain Lion
    Felis concolor   FL

RANGE CHARTS

Many of the smaller mammals, such as shrews and pocket gophers, are so similar in appearance and so reclusive in their habits that often one gets only a glimpse of them in the field. However, since most species occupy a specific geographical area, it is often possible to distinguish between species simply on the basis of where the animal is seen. For example, if you are in Florida and see a pocket gopher poke its head out of the ground, you will find, on consulting the range chart, that only one species of pocket gopher occurs in Florida—the Southeastern Pocket Gopher *(Geomys pinetis).* Reading the species account confirms your identification.

Groups
Range charts are provided for the following groups of smaller mammals:

Shrews	Pocket Mice
Bats	Kangaroo Rats
Rabbits and Hares	Harvest Mice
Chipmunks	Deer Mice
Ground Squirrels	Woodrats
Pocket Gophers	Voles

Regions
The range charts for each group are divided into the following geographical regions:

Eastern	*Western*
Northeast	North Central
Southeast	South Central
Midwest	West
	Northwest

Each region is further divided into states and provinces. Because of space restrictions, we have combined North and South Dakota as one area with the abbreviation NSD. The range for North Dakota appears at top; the range for South Dakota at bottom.

Key
The following location symbols are used in the charts:

◯	=	throughout
Ⓝ	=	north
Ⓢ	=	south
Ⓔ	=	east
Ⓦ	=	west
Ⓒ	=	central

Abbreviations
The following abbreviations of states and provinces are used in the charts:

Northeast	NB	New Brunswick
	NE	New England
	NJ	New Jersey
	NY	New York
	NF	Newfoundland
	NS	Nova Scotia
	PA	Pennsylvania
	QU	Quebec
Southeast	DE	Delaware
	FL	Florida
	GA	Georgia
	MD	Maryland
	NC	North Carolina
	SC	South Carolina
	VA	Virginia
	WV	West Virginia
Midwest	IL	Illinois
	IN	Indiana
	IA	Iowa
	KS	Kansas
	MI	Michigan
	MN	Minnesota
	MO	Missouri
	NE	Nebraska
	OH	Ohio
	ON	Ontario
	WI	Wisconsin

North Central	AB	Alberta
	MB	Manitoba
	MT	Montana
	ND	North Dakota
	NSD	North and South Dakota
	SK	Saskatchewan
	SD	South Dakota
	WY	Wyoming
South Central	AL	Alabama
	AR	Arkansas
	KY	Kentucky
	LA	Louisiana
	MS	Mississippi
	OK	Oklahoma
	TN	Tennessee
	TX	Texas
West	AZ	Arizona
	BC	British Columbia
	CA	California
	CO	Colorado
	ID	Idaho
	NV	Nevada
	NM	New Mexico
	OR	Oregon
	UT	Utah
	WA	Washington
Northwest	AK	Alaska
	NW	Northwest Territories
	YK	Yukon

BERING SEA

BEAUFORT SEA

AK

YK

GULF OF ALASKA

N

BC

AB

SK

WA

MT

OR

ID

WY

NV

UT

CA

CO

PACIFIC OCEAN

AZ

NM

MEXICO

	NB	NE	NJ	NY	NF	NS	PA	QU	DE	FL	GA
Northeast									**Southeast**		

SHREWS

	NB	NE	NJ	NY	NF	NS	PA	QU	DE	FL	GA
Arctic Shrew (*Sorex arcticus*)	○					○		(S)			
Masked Shrew (*S. cinereus*)	○	○	○	○	○	○	○	○	○		(NE)
Long-tailed Shrew (*S. dispar*)	(W)	○	(NW)	(E)			○				
Maryland Shrew (*S. fontinalis*)							(SE)	○			
Smoky Shrew (*S. fumeus*)	○	○	(NW)	○		○	○	(S)			(NE)
Gaspé Shrew (*S. gaspensis*)	(C)										
Hayden's Shrew (*S. haydeni*)											
Pygmy Shrew (*S. hoyi*)	○	○		○	(S)	○	(NW)				
Southeastern Shrew (*S. longirostris*)										(N)	○
Water Shrew (*S. palustris*)	○	○		(E)	(S)	○	(NE)	(S)			
Northern Short-tailed Shrew (*Blarina brevicauda*)	○	○	○	○		○	○	(S)		(WC)	(N)
Southern Short-tailed Shrew (*B. carolinensis*)										○	○
Elliot's Short-tailed Shrew (*B. hylophaga*)											
Least Shrew (*Cryptotis parva*)		(SW)	○	(S)			○		○	○	○
Desert Shrew (*Notiosorex crawfordi*)											

BATS

	NB	NE	NJ	NY	NF	NS	PA	QU	DE	FL	GA
Southeastern Myotis (*Myotis austroriparius*)										(N)	(S) (C)
Gray Myotis (*M. grisescens*)											
Eastern Small-footed Myotis (*M. leibii*)		(S) (C)		○				(S)			(N)

Midwest

MD	NC	SC	VA	WV	IL	IN	IA	KS	MI	MN	MO	NE	OH	ON	WI
									N	◯				◯	◯
◯	W		W	E	N E	◯			◯	N C		SW	◯	◯	◯
W	W		W	E											
◯				N											
W	W	NW	W	◯		SC							E	S	
							◯			S W		◯			
	W		◯		N	SC	N		◯	◯			E	◯	◯
S	SE	◯	E	S	E	S C									
	W		W	◯					N	N C				W S	N
◯	W	W	W N	◯	◯	◯	◯		◯	◯	NE	◯	◯	S	◯
◯	◯	◯	E		S		SW	◯				◯	S		
						S	◯					◯	S		
◯	◯	◯	◯	◯	◯	◯	◯	◯	◯	S	S	◯	◯	◯	S
							S								
					S	S									
					S	S		SE				◯			
◯	W		W	◯				NW			S	W		SE	

| | Northeast | | | | | | | | | | Southeast |
	NB	NE	NJ	NY	NF	NS	PA	QU	DE	FL	GA
BATS (continued)											
Little Brown Myotis (*Myotis lucifugus*)	○	○	○	○	○	○	○	Ⓢ	○		Ⓝ Ⓒ
Northern Myotis (*M. septentrionalis*)	○	○	○	○	○	○	○	Ⓢ	○	Ⓝ🄲	Ⓝ Ⓦ
Indiana Myotis (*M. sodalis*)		Ⓢ	Ⓝ	🅂🄴			○				Ⓝ🅆
Cave Myotis (*M. velifer*)											
Long-legged Myotis (*M. volans*)											
RABBITS AND HARES											
Swamp Rabbit (*Sylvilagus aquaticus*)											Ⓝ Ⓦ
Desert Cottontail (*S. audubonii*)											
Eastern Cottontail (*S. floridanus*)		Ⓢ	○	Ⓢ Ⓒ			○	Ⓢ	○	○	○
Appalachian Cottontail (*S. obscurus*)							Ⓒ 🅂🅆				Ⓝ🄴
Marsh Rabbit (*S. palustris*)										○	Ⓢ
New England Cottontail (*S. transitionalis*)		Ⓢ		Ⓔ			Ⓔ				
Snowshoe Hare (*Lepus americanus*)	○	○	○	○	○	○	○	○			
Arctic Hare (*L. arcticus*)					Ⓝ Ⓔ			Ⓝ			
Black-tailed Jackrabbit (*L. californicus*)		○									
European Hare (*L. europaeus*)		Ⓦ	Ⓦ	○			Ⓔ	Ⓢ			
White-tailed Jackrabbit (*L. townsendii*)											
CHIPMUNKS											
Least Chipmunk (*Tamias minimus*)								Ⓦ			
Eastern Chipmunk (*T. striatus*)	○	○	○	○		○	○	Ⓢ	○		Ⓝ

MD	NC	SC	VA	WV	IL	IN	IA	KS	MI	MN	MO	NE	OH	ON	WI
○	W	N	○	○	○	○	○	E	○	○	○	E/W	○	○	○
○	N/W	W	○	○	○	○	○	E	○	○	○	N/E	○	S	○
N	W		W	○	○	○	E		S		○		○		S
								SC							
												W			
		W			S	SW		SE			S				
								W				W			
○	○	○	○	○	○	○	○	○	○	○	○	○	○	SE	○
W	W		W	E											
	E	E/C	SE												
○	E		E	E											
W	W		W	E					○	N			NE	○	N
								○				○			
									N/E	NE				S	N
							N	NW		SW		○			
									N	N		NW		○	N/C
○	W	NW	○	○	○	○	○	E	○	○	○	E	○	S	○

	Northeast									Southeast	
	NB	NE	NJ	NY	NF	NS	PA	QU	DE	FL	GA
GROUND SQUIRRELS											
Wyoming Ground Squirrel *(Spermophilis elegans)*											
Franklin's Ground Squirrel *(S. franklinii)*											
Richardson's Ground Squirrel *(S. richardsonii)*											
Spotted Ground Squirrel *(S. spilosoma)*											
Thirteen-lined Ground Squirrel *(S. tridecemlineatus)*											
POCKET GOPHERS											
Northern Pocket Gopher *(Thomomys talpoides)*											
Plains Pocket Gopher *(Geomys bursarius)*											
Southeastern Pocket Gopher *(G. pinetis)*										N	S
Yellow-faced Pocket Gopher *(Cratogeomys castanaps)*											
POCKET MICE											
Olive-backed Pocket Mouse *(Perognathus fasciatus)*											
Plains Pocket Mouse *(P. flavescens)*											
Silky Pocket Mouse *(P. flavus)*											
Hispid Pocket Mouse *(Chaetodipus hispidus)*											
KANGAROO RATS											
Ord's Kangaroo Rat *(Dipodomys ordii)*											

MD	NC	SC	VA	WV	IL	IN	IA	KS	MI	MN	MO	NE	OH	ON	WI
												(SW)			
					(N/C)	(NW)	○	(N/E)		○	(N/C)	○		(SW)	(SW)
										(W)					
								(W)				(W/C)			
					(N/C)	(N/C)	○	○	○	○	(N/W)	○	(W)		○
										(NW)		(W)			
					(C)	(NW)	○	○		○	(N/E)	○			(W)
								(SW)							
												(NW)			
							(N/W)	(W/C)		(S/W)		○			
								(W)				(W)			
								○				○			
								(W/C)				(W/C)			

| | Northeast | | | | | | | | Southeast | | |
	NB	NE	NJ	NY	NF	NS	PA	QU	DE	FL	GA
HARVEST MICE											
Fulvous Harvest Mouse (*Reithrodontomys fulvescens*)											
Eastern Harvest Mouse (*R. humulis*)										○	○
Western Harvest Mouse (*R. megalotis*)											
Plains Harvest Mouse (*R. montanus*)											
DEER MICE											
Texas Mouse (*Peromyscus attwateri*)											
Brush Mouse (*P. boylii*)											
Cotton Mouse (*P. gossypinus*)										○	S, C
White-footed Mouse (*P. leucopus*)	SE	○	○	○		○	○				N, C
Deer Mouse (*P. maniculatus*)	○	○		○	W	○	○	○			NE
Oldfield Mouse (*P. polionotus*)										N, W	○
Florida Mouse (*Podomys floridanus*)										S, C	
Golden Mouse (*Ochrotomys nuttalli*)										N, W	○
WOODRATS											
Bushy-tailed Woodrat (*Neotoma cinerea*)											
Florida Woodrat (*N. floridana*)										N, C	○
Allegheny Woodrat (*N. magister*)		SW	N	SE			○				
Southern Plains Woodrat (*N. micropus*)											

Midwest

ID	NC	SC	VA	WV	IL	IN	IA	KS	MI	MN	MO	NE	OH	ON	WI
								SE			SW				
W	○	○	○	○									S		
					N	NW	○	○		S	○	○			SW
								○				○			
								SE			SW				
								SE			SW				
	E	E/C	SE		S						SE				
○	W/C	W/C	○	○	○	○	○	○	S	○	○	S/E	○	SE	○
W	W	NW	W	N/W	○	○	○	○	○	○	N/C	○	○	○	○
		W													
	○	○	S	S	S						S				
													W		
	S	S			S			○			S/C				
○	W	W	○	○		SC							S		
								SW							

	Northeast								Southeast		
VOLES	NB	NE	NJ	NY	NF	NS	PA	QU	DE	FL	GA
Southern Red-backed Vole *(Clethrionomys gapperi)*	◯	◯	◯	◯	(w)	◯	◯	◯	◯		
Heather Vole *(Phenacomys intermedius)*	◯				◯	◯		◯			
Rock Vole *(Microtus chrotorrhinus)*	(w)	(N)		(E)	(E)		(NE)	(s)			
Prairie Vole *(M. ochrogaster)*											
Meadow Vole *(M. pennsylvanicus)*	◯	◯	◯	◯	◯	◯	◯	◯	◯		(NE)
Woodland Vole *(M. pinetorum)*		◯	◯	◯			◯		◯	(NC)	◯

Midwest

ID	NC	SC	VA	WV	IL	IN	IA	KS	MI	MN	MO	NE	OH	ON	WI
○	w		○	○					c	○				○	○
										NE			○		
NE	w		w	E						NE				SE	
				w	○	○	○	○	SW	S / w	○	○	SW		SW
○	w / c	○	○	○	N	○	○		○	○	N	NE / SW	○	○	○
○	○	○	○	○	○	○	○	E	S / c	SE	○	SE	○	SE	S

	North Central						South Central				
	AB	MB	MT	NSD	SK	WY	AL	AR	KY	LA	MS
SHREWS											
Arctic Shrew *(Sorex arcticus)*	○	○		(N-E)(NE)	○						
Arizona Shrew *(S. arizonae)*											
Baird's Shrew *(S. bairdii)*											
Pacific Water Shrew *(S. bendirii)*											
Masked Shrew *(S. cinereus)*	○	○	(W)		(N)	○			(N)(E)		
Long-tailed Shrew *(S. dispar)*											
Smoky Shrew *(S. fumeus)*									(E)		
Hayden's Shrew *(S. haydeni)*	(SE)	(SW)	(E)	○	(S)						
Pygmy Shrew *(S. hoyi)*	○	○	(W)	(NE)(NE)		(NW)(SC)					
Pribilof Island Shrew *(S. hydrodromus)*											
St. Lawrence Island Shrew *(S. jacksoni)*											
Southeastern Shrew *(S. longirostris)*							○	(N)	(W)(E)	(NE)	○
Mt. Lyell Shrew *(S. lyelli)*											
Merriam's Shrew *(S. merriami)*			(E)	(SW)(NW)		○					
Dusky Shrew *(S. monticolus)*	○	(WC)	(W)		○	○					
Dwarf Shrew *(S. nanus)*				○ (SW)		○					
Ornate Shrew *(S. ornatus)*											
Pacific Shrew *(S. pacificus)*											
Water Shrew *(S. palustris)*	○	○	(W)(C)	(NE)(NE)	○	○					
Preble's Shrew *(S. preblei)*				○							

OK	TN	TX	AZ	BC	CA	CO	ID	NV	NM	OR	UT	WA	AK	NW	YK
				NE										SW	SE
			SE						SW						
										NW					
				SW	NW					W		W			
	E			○		○	○		NC		○	○	○	○	○
	E														
	E														
					○		C	N				E	○	S	S
													W		
													NW		
	○														
					EC										
			NE		E	W	S	○	NW	E		E			
				○	N	W	○	WC	C/NC	W	○	○	○	SW	○
			NE		W/C	SE			○		E				
					S										
					NW					SW					
	E	N	NE	○	NE	○	○	○	NC	○	○	○	SE	SC	S
							SW			E		SE			

	North Central						South Central				
	AB	MB	MT	NSD	SK	WY	AL	AR	KY	LA	MS
SHREWS (continued)											
Fog Shrew (*Sorex sonomae*)											
Inyo Shrew (*S. tenellus*)											
Trowbridge's Shrew (*S. trowbridgii*)											
Tundra Shrew (*S. tundrensis*)											
Vagrant Shrew (*S. vagrans*)			W		◯						
Northern Short-tailed Shrew (*Blarina brevicauda*)		S		E-C / E	SE				E / C		
Southern Short-tailed Shrew (*B. carolinensis*)							◯	◯	W	◯	◯
Elliot's Short-tailed Shrew (*B. hylophaga*)								NW			
Least Shrew (*Cryptotis parva*)				SC			◯	◯	◯	◯	◯
Desert Shrew (*Notiosorex crawfordi*)								W			
BATS											
Southwestern Myotis (*Myotis auriculus*)											
Southeastern Myotis (*M. austroriparius*)							S / NW	S	W	◯	◯
California Myotis (*M. californicus*)											
Long-eared Myotis (*M. evotis*)	S		◯	W / W	SW	◯					
Gray Myotis (*M. grisescens*)							◯	N	◯		
Keen's Myotis (*M. keenii*)											
Small-footed Myotis (*M. leibii*)	S		◯	SW		◯	NE	NW	NC		
Little Brown Myotis (*M. lucifugus*)	◯	◯	◯	◯	◯	◯	N / C	N			N / C
Northern Myotis (*M. septentrionalis*)		S	E	◯	S / C	NE	◯	◯	◯		

OK	TN	TX	AZ	BC	CA	CO	ID	NV	NM	OR	UT	WA	AK	NW	YK
					(NW)					(SW)					
					(EC)			(WC)							
				(SW)	(N)			(WC)		(W)		(W)			
													○	(NW)	(N/W)
			(E/N)	(SW)	(N)	(W)	○	(N)	(W)	○	○	○			
	(E)														
(E)	○	(E)				(NE)									
(E)						(NE)									
(NE)	○	(E)				(NE)									
○		(W/C)	○		(S)	(S)		(S)	○		(S)				
			(SE)							(SW)					
(SE)	(SW)	(E)													
		(W)	○	(W)	○	(W/C)	(N)	(W/S)	(W)	○	(SE)	○			
			(NE)	(S)	○	(W/C)	○	○	(NW)	○	○	○			
(NE)	○														
				(W)								(NW)			
(SW)	(E)	(W)	(E/C)	(E/SE)	○	○	○	○	○	○	(E)	(E/C)			
(E)	○	(NC)	○	○	(N/C)	○	○	○	(N/SW)	○	○	○	(S)	(S)	(S)
(E)	○														

	North Central						South Central				
	AB	MB	MT	NSD	SK	WY	AL	AR	KY	LA	MS
BATS (continued)											
Indiana Myotis (*Myotis sodalis*)							(E)	(N)	(O)		
Fringed Myotis (*M. thysanodes*)				(SW)		(E)					
Cave Myotis (*M. velifer*)											
Long-legged Myotis (*M. volans*)	(SW)		(W)	(W)		(O)					
Yuma Myotis (*M. yumanensis*)			(W)			(NW)					
RABBITS AND HARES											
Pygmy Rabbit (*Brachylagus idahoensis*)			(SW)								
Swamp Rabbit (*Sylvilagus aquaticus*)							(O)	(O)	(W)	(O)	(O)
Desert Cottontail (*S. audubonii*)			(SE) (C)	(SW) (W)		(O)					
Brush Rabbit (*S. bachmani*)											
Eastern Cottontail (*S. floridanus*)		(S)	(SE)	(O)		(E)	(O)	(O)	(O)	(O)	(O)
Mountain Cottontail (*S. nuttallii*)	(S)		(O)	(W) (W)	(S)	(O)					
Marsh Rabbit (*S. palustris*)							(S)				
Allegheny Cottontail (*S. obscurus*)							(N)		(SE)		
Antelope Jackrabbit (*Lepus alleni*)											
Snowshoe Hare (*L. americanus*)	(O)	(O)	(W)	(NE)	(O)	(W)					
Arctic Hare (*L. arcticus*)		(NE)									
Black-tailed Jackrabbit (*L. californicus*)				(S)		(W) (SE)		(NW)	(O)		
White-sided Jackrabbit (*L. callotis*)											
Alaska Hare (*L. othus*)											
White-tailed Jackrabbit (*L. townsendii*)	(S)	(S)	(O)	(O) (O)	(S)	(O)					

OK	TN	TX	AZ	BC	CA	CO	ID	NV	NM	OR	UT	WA	AK	NW	YK
NE	○														
		W	○	SC	○	SW	W	○	○	○	○	○			
W		W/C	○		SE				SE/SW						
		W	E	W/S	○	W/C	○	○	○	○	○	○			
		W	○	S	○	S	W	S/W	○	○	S/E	○			
					NE		S	N/C		E/C	NW	SE			
E	W	E													
W		W/C	○		S/C	○		S	○						
					W/C					W					
○	○	○	SW/NE		E				E-C/SW						
		NE	SC	E	W/C	○	○	N	E/C	○	E				
	E														
		SC						SW							
	E	NE	○	NE	W/C	○	W	NC	○	NE	○	○	W/C	○	
													N/C		
○		○	○		○	○	S	○	○	○	W/S	SC/SE			
									SW						
													W/N		
			SC	NE	○	○	N/C	NC	E/C	○	E/C				

CHIPMUNKS	North Central						South Central				
	AB	MB	MT	NSD	SK	WY	AL	AR	KY	LA	MS
Alpine Chimpunk *(Tamias alpinus)*											
Yellow-pine Chipmunk *(T. amoenus)*	(SW)		(W)			(NW)					
Gray-footed Chipmunk *(T. canipes)*											
Gray-collared Chipmunk *(T. cinereicollis)*											
Cliff Chipmunk *(T. dorsalis)*						(SW)					
Merriam's Chipmunk *(T. merriami)*											
Least Chipmunk *(T. minimus)*	◯	◯	◯	(W)(W)	◯	◯					
California Chipmunk *(T. obscurus)*											
Yellow-cheeked Chipmunk *(T. ochrogenys)*											
Palmer's Chipmunk *(T. palmeri)*											
Panamint Chipmunk *(T. panamintinus)*			·								
Long-eared Chipmunk *(T. quadrimaculatus)*											
Colorado Chipmunk *(T. quadrivittatus)*											
Red-tailed Chipmunk *(T. ruficaudus)*	(SW)		(W)								
Hopi Chipmunk *(T. rufus)*											
Allen's Chipmunk *(T. senex)*											
Siskiyou Chipmunk *(T. siskiyou)*											
Sonoma Chipmunk *(T. sonomae)*											
Lodgepole Chipmunk *(T. speciosus)*											
Eastern Chipmunk *(T. striatus)*		(S)		(E)(E)							

)K	TN	TX	AZ	BC	CA	CO	ID	NV	NM	OR	UT	WA	AK	NW	YK
					(c)										
				(S)	(N)		(○)	(NE)(WC)		(E)(C)	(NW)	(○)			
		(NW)							(SE)						
			(E)						(SW)						
			(○)			(NW)	(SE)	(E)	(W)		(○)				
					(S)										
			(EC)(N)	(N)	(NE)	(W)(C)	(S)	(○)	(N)(SC)	(E)(C)	(○)	(SC)		(S)	(S)
					(SC)										
					(NW)										
								(S)							
					(SC)			(SW)							
					(EC)			(W)							
			(NE)			(W)(C)			(N)		(SE)				
				(SE)			(N)					(NE)			
			(NE)			(W)					(SE)				
					(N)			(WC)		(WC)					
					(NW)					(SW)					
					(NW)										
					(E)(C)			(W)							
(E)															

	North Central						South Central				
	AB	MB	MT	NSD	SK	WY	AL	AR	KY	LA	MS

CHIPMUNKS
(continued)

	AB	MB	MT	NSD	SK	WY	AL	AR	KY	LA	MS
Townsend's Chipmunk *(Tamias townsendii)*											
Uinta Chipmunk *(T. umbrinus)*			(SC)			(W)(S)					

GROUND SQUIRRELS

	AB	MB	MT	NSD	SK	WY	AL	AR	KY	LA	MS
Uinta Ground Squirrel *(Spermophilus armatus)*			(SW)			(W)					
California Ground Squirrel *(S. beecheyi)*											
Belding's Ground Squirrel *(S. beldingi)*											
Idaho Ground Squirrel *(S. brunneus)*											
Columbian Ground Squirrel *(S. columbianus)*	(SW)		(W)								
Wyoming Ground Squirrel *(S. elegans)*			(SW)			(NW)(S)					
Franklin's Ground Squirrel *(S. franklinii)*	(EC)	(S)		(E-C)(E)	(S)						
Golden-mantled Ground Squirrel *(S. lateralis)*	(SW)		(W)			(W)(S)					
Mexican Ground Squirrel *(S. mexicanus)*											
Mohave Ground Squirrel *(S. mohavensis)*											
Arctic Ground Squirrel *(S. parryii)*	(NW)										
Richardson's Ground Squirrel *(S. richardsonii)*	(S)	(S)	(◯)	(NE)	(S)	(SW)					
Cascade Golden-mantled Ground Squirrel *(S. saturatus)*											
Spotted Ground Squirrel *(S. spilosoma)*				(SW)		(SE)					
Round-tailed Ground Squirrel *(S. tereticaudus)*											

OK	TN	TX	AZ	BC	CA	CO	ID	NV	NM	OR	UT	WA	AK	NW	YK
				(SW)						(W)		(W)			
			(NC)		(EC)	(NW)	(E)	(E)(S)			(○)				
							(E)				(NC)				
					(○)			(WC)		(W)		(SC)			
					(NE)		(SW)	(N)		(E)	(NW)				
							(SW)								
				(SE)			(N)			(NE)		(E)			
						(N)	(SW)(E)	(NE)		(NE)	(SW)				
			(NE)(C)	(SE)	(○)	(W)(C)	(○)	(○)	(W)(N)	(E)	(○)	(NC)(SE)			
		(W)(S)							(SE)						
					(S)										
				(N)									(○)	(○)	(○)
											(○)				
(W)		(S)(W)	(N)(SE)			(E)(SW)				(○)	(SE)				
			(SW)		(SE)			(S)							

	North Central						South Central				
	AB	MB	MT	NSD	SK	WY	AL	AR	KY	LA	MS
GROUND SQUIRRELS (continued)											
Townsend's Ground Squirrel (*Spermophilus townsendii*)											
Thirteen-lined Ground Squirrel (*S. tridecemlineatus*)	SE	S	E C	◯ ◯	S	◯					
Rock Squirrel (*S. variegatus*)											
Washington Ground Squirrel (*S. washingtoni*)											
POCKET GOPHERS											
Botta's Pocket Gopher (*Thomomys bottae*)											
Camas Pocket Gopher (*T. bulbivorus*)											
Wyoming Pocket Gopher (*T. clusius*)						SC					
Idaho Pocket Gopher (*T. idahoensis*)			SW								
Western Pocket Gopher (*T. mazama*)											
Mountain Pocket Gopher (*T. monticola*)											
Northern Pocket Gopher (*T. talpoides*)	S	SW	◯	◯ ◯	S	◯					
Townsend's Pocket Gopher (*T. townsendii*)			SE								
Southern Pocket Gopher (*T. umbrinus*)											
Desert Pocket Gopher (*Geomys arenarius*)											
Attwater's Pocket Gopher (*G. attwateri*)											
Baird's Pocket Gopher (*G. breviceps*)								W		W	
Plains Pocket Gopher (*G. bursarius*)		SC		E S-E		SE		W S		W	
Jones' Pocket Gopher (*G. knoxjonesi*)											

OK	TN	TX	AZ	BC	CA	CO	ID	NV	NM	OR	UT	WA	AK	NW	YK
					NE		S	○		E	W	SC			
○		N E				○			N E		NE				
W		W C	○		SE	○		S	○	○					
										NC		EC			
		W	○		○	SW		S E	○	SW	○				
										NW					
							SE								
					NC					NW W		W			
					NE			WC							
			N	SE	NE	○	○	N	N	E C	○	E C			
					NE			N		SE					
			SE						SW						
			W						SC						
		SE													
E		E													
○		N E				E			EC						
		W							SE						

	North Central						South Central				
	AB	MB	MT	NSD	SK	WY	AL	AR	KY	LA	MS
POCKET GOPHERS (continued)											
Texas Pocket Gopher (*Geomys personatus*)											
Southeastern Pocket Gopher (*G. pinetis*)							SE				
Llano Pocket Gopher (*G. texensis*)											
Yellow-faced Pocket Gopher (*Cratogeomys castanops*)											
POCKET MICE											
White-eared Pocket Mouse (*Perognathus alticolus*)											
Arizona Pocket Mouse (*P. amplus*)											
Olive-backed Pocket Mouse (*P. fasciatus*)	SE	SW	E	W-C / W-C	S	◯					
Plains Pocket Mouse (*P. flavescens*)				SE / E-C		SE					
Silky Pocket Mouse (*P. flavus*)						SE					
San Joaquin Pocket Mouse (*P. inornatus*)											
Little Pocket Mouse (*P. longimembris*)											
Merriam's Pocket Mouse (*P. merriami*)											
Great Basin Pocket Mouse (*P. parvus*)			SW			SW					
Yellow-eared Pocket Mouse (*P. xanthonotus*)											
Bailey's Pocket Mouse (*Chaetodipus baileyi*)											
California Pocket Mouse (*C. californicus*)											
San Diego Pocket Mouse (*C. fallax*)											
Long-tailed Pocket Mouse (*C. formosus*)											

OK	TN	TX	AZ	BC	CA	CO	ID	NV	NM	OR	UT	WA	AK	NW	YK
		S													
			C												
W		W				SE			E						
					S										
			◯												
						N					NE				
W		N	NE		◯			◯			SE				
W		W	N / SE		E / SC			◯			SE				
					C										
			N / SW		S / C			◯		SE	W / S				
		S													
			NW	SC	NE		W / S	◯		E / C	◯	E / C			
					S										
			S		S				SW						
					SW / C										
					SW										
			NW		S			S / W				W			

	North Central						South Central				
	AB	MB	MT	NSD	SK	WY	AL	AR	KY	LA	MS

POCKET MICE
(continued)

	AB	MB	MT	NSD	SK	WY	AL	AR	KY	LA	MS
Hispid Pocket Mouse (*Chaetodipus hispidus*)			SE	SC / WC		E				W	
Rock Pocket Mouse (*C. intermedius*)											
Nelson's Pocket Mouse (*C. nelsoni*)											
Desert Pocket Mouse (*C. penicillatus*)											
Spiny Pocket Mouse (*C. spinatus*)											
Mexican Spiny Pocket Mouse (*Lyomis irroratus*)											
KANGAROO RATS											
Dark Kangaroo Mouse (*Microdipodops megacephalus*)											
Pale Kangaroo Mouse (*M. pallidus*)											
Agile Kangaroo Rat (*Dipodomys agilis*)											
California Kangaroo Rat (*D. californicus*)											
Desert Kangaroo Rat (*D. deserti*)											
Texas Kangaroo Rat (*D. elator*)											
Big-eared Kangaroo Rat (*D. elephantinus*)											
Heermann's Kangaroo Rat (*D. heermanni*)											
Giant Kangaroo Rat (*D. ingens*)											
Merriam's Kangaroo Rat (*D. merriami*)											
Chisel-toothed Kangaroo Rat (*D. microps*)											
Fresno Kangaroo Rat (*D. nitratoides*)											
Ord's Kangaroo Rat (*D. ordii*)	SE		E	SW / C	SW	◯					

	West									Northwest					
OK	TN	TX	AZ	BC	CA	CO	ID	NV	NM	OR	UT	WA	AK	NW	YK
○		○	SE			E			E / S						
		W	NC / S						SW		SC				
		W							SE						
		W	NW / S		S			S	S		SW				
					S			S							
		S													
					NE			○		SE	NW				
					NE			W							
					SW										
					N					SC					
			W / S		S			SW							
SW		NC													
					W										
					WC										
					SW										
		W	S / W		S			W / S	S / C		SW				
			NW		E		SW	○		SE	W				
					WC										
W		W / S	N / E		NE	○	S	○	○	E / C	○	SC			

	North Central						South Central				
	AB	MB	MT	NSD	SK	WY	AL	AR	KY	LA	MS
KANGAROO RATS (continued)											
Panamint Kangaroo Rat (*Dipodomys panamintinus*)											
Banner-tailed Kangaroo Rat (*D. spectabilis*)											
Stephens' Kangaroo Rat (*D. stephensi*)											
Narrow-faced Kangaroo Rat (*D. venustus*)											
HARVEST MICE											
Fulvous Harvest Mouse (*Reithrodontomys fulvescens*)								W C		○	SW
Eastern Harvest Mouse (*R. humulis*)							○	N	○	S	○
Western Harvest Mouse (*R. megalotis*)	SE		E	S ○		N W		NE			
Plains Harvest Mouse (*R. montanus*)				SW		E					
Salt-marsh Harvest Mouse (*R. raviventris*)											
DEER MICE											
Texas Mouse (*Peromyscus attwateri*)								NW			
Brush Mouse (*P. boylii*)											
California Mouse (*P. californicus*)											
Canyon Mouse (*P. crinitus*)						SW					
Cactus Mouse (*P. eremicus*)											
Cotton Mouse (*P. gossypinus*)							○	S E	SW	○	○
Osgood's Mouse (*P. gratus*)											
White-footed Mouse (*P. leucopus*)	SE		E	○	S	NE	N C	○	○	○	○
Deer Mouse (*P. maniculatus*)	○	S C	○	○	○	○		C	E		
Black-eared Mouse (*P. melanotis*)											

K	TN	TX	AZ	BC	CA	CO	ID	NV	NM	OR	UT	WA	AK	NW	YK
					C / E			W / S							
		W	SE / NE						◯						
					S										
					W										
E		E / SW	SE						SW						
	◯	SE		SE											
◯		W	◯	SC	◯	E / S-C	◯	◯	◯	S-C / E	◯	E			
◯		W / N	SE		SE				◯						
					WC										
E / C		NC													
		W	◯		◯	S / W		S / W	◯	S	◯				
					SW										
			N / W		E	W	SW	◯	NW	SE	◯				
		SW	◯		S			S	S		SW				
SE	W / C	E													
			SE						SW						
◯	◯	◯	SE / C			SE				◯					
E / C	NE	W / C	◯	◯	◯	◯	◯	◯	◯	◯	◯	◯	◯	SW	◯
			SE												

	North Central						South Central				
	AB	MB	MT	NSD	SK	WY	AL	AR	KY	LA	MS
DEER MICE (continued)											
Merriam's Mouse (*Peromyscus merriami*)											
Northern Rock Mouse (*P. nasutus*)											
Columbian Mouse (*P. oreas*)											
White-ankled Mouse (*P. pectoralis*)											
Oldfield Mouse (*P. polionotus*)							E / C				
Sitka Mouse (*P. sitkensis*)											
Piñon Mouse (*P. truei*)						SW					
Golden Mouse (*Ochrotomys nuttalli*)							◯	◯	◯	◯	◯
WOODRATS											
White-throated Woodrat (*Neotoma albigula*)											
Bushy-tailed Woodrat (*N. cinerea*)	SW		◯	SW / W	SW	◯					
Arizona Woodrat (*N. devia*)											
Florida Woodrat (*N. floridana*)				S			◯	◯	W	◯	◯
Dusky-footed Woodrat (*N. fuscipes*)											
Desert Woodrat (*N. lepida*)											
Allegheny Woodrat (*N. magister*)									◯		
Mexican Woodrat (*N. mexicana*)											
Southern Plains Woodrat (*N. micropus*)											
Stephens' Woodrat (*N. stephensi*)											
VOLES											
Western Red-backed Vole (*Clethrionomys californicus*)											
Southern Red-backed Vole (*C. gapperi*)	◯	◯	◯	◯◯	◯	◯					

OK	TN	TX	AZ	BC	CA	CO	ID	NV	NM	OR	UT	WA	AK	NW	YK
			SC												
W		W	E			W / C			○		SE				
				SW								W			
S		W / C							SE						
					W								SW		
W		N	N / E		○	W / S		○	○	SW / C	○				
E	○	E													
		W	○		SE	SE / SW			○		SE				
			N	○	N / EC	W / C	○	○	NW	○	○	○		SW	SE
			W												
E	W	E				EC									
					W					W					
			W		NE / S		SW	○		SE	W / S				
	○														
		W	E			NC / S			○		SE				
W		W / C-S			SE				○						
			N						W		SE				
					NW					W					
			EC	○		○	○		○	NE	○	○		S	

	North Central						South Central				
	AB	MB	MT	NSD	SK	WY	AL	AR	KY	LA	MS
VOLES (continued)											
Northern Red-backed Vole (*Clethrionomys rutilus*)											
White-footed Vole (*Phenacomys albipes*)											
Heather Vole (*P. intermedius*)	○	○	(W)		○	○					
Red Tree Vole (*P. longicaudus*)											
California Vole (*Microtus californicus*)											
Gray-tailed Vole (*M. canicaudus*)											
Rock Vole (*M. chrotorrhinus*)											
Long-tailed Vole (*M. longicaudus*)	(SW)		(W)(C)	(SW)		○					
Mexican Vole (*M. mexicanus*)											
Singing Vole (*M. miurus*)											
Montane Vole (*M. montanus*)			(SW)			(W)(C)					
Prairie Vole (*M. ochrogaster*)	(SE)	(SW)	(E)	○	(S)	(E)(C)		(N)	○	(SW)	
Tundra Vole (*M. oeconomus*)											
Creeping Vole (*M. oregoni*)											
Meadow Vole (*M. pennsylvanicus*)	○	○	○	○	○	(E)(W)			(E)(C)		
Woodland Vole (*M. pinetorum*)							(N)(C)	○	○	(N)(C)	○
Water Vole (*M. richardsoni*)	(SW)		(W)			(W)					
Townsend's Vole (*M. townsendii*)											
Yellow-cheeked Vole (*M. xanthognathus*)	(N)	(N)			(N)						
Sagebrush Vole (*Lemmiscus curtatus*)	(S)	(SW)	(E)(C)	(W)(NW)		(S)(NW)					

OK	TN	TX	AZ	BC	CA	CO	ID	NV	NM	OR	UT	WA	AK	NW	YK
				(N)									○	○	○
					(NW)					(W)					
				○	(N)		○		(N)	(SE)	○	○		(S)	(S)
					(NW)					(W)					
					○					(SW)					
										(NC)		(SC)			
	(E)														
			(NE)	○	(N)(SC)	(W)(C)	○	○	(W)(C)	○	○	○	(SE)	(SW)	(S)
		(W)	(E)(WC)			(SW)			(W)(S)		(SC)				
				(NW)									○	(W)	○
			(EC)	(SC)	(NE)	(W)	○	○	(NW)	○	○	(E)(C)			
(N)	○	(SE)				(NE)									
				(NW)									○	(N)	(W)
				(S)	(NW)					(W)		(W)			
	(E)			○		(C)	(N)(E)		(NW)		(N)	(NE)	(C)(S)	○	○
(E)(C)	○	(NE)													
				(SE)(SW)			(N)			(C)(E)	(NC)	(C)(E)			
				(SW)	(NW)					(W)		(W)			
				(NE)									(EC)	(W)	(N)(C)
					(NE)	(NW)	(S)	○		(E)	○	(E)			

PHOTO CREDITS

The photo credits are listed by plate number. Those followed by the letter "a" refer to the inset photograph.

1 Ronn Altig
2 James F. Parnell
3 Roger W. Barbour
4 James F. Parnell
5 Rob & Ann Simpson
6 Roger W. Barbour/
   Morehead State Univ.
7 B. Moose Peterson/WRP
8 B. Moose Peterson/WRP
9 Victor B. Scheffer
10 Jan L. Wassink
11 Rod Planck
12 Dwight R. Kuhn
13 James F. Parnell
14 Rob & Ann Simpson
15 Dale & Marian
   Zimmerman
16 Ronn Altig
17 Edmund Brodie
18 John Serrao
19 Gary Meszaros/Dembinsky
   Photo Associates
20 Dwight R. Kuhn
21 Rob & Ann Simpson
22 Gerry Ellis/Ellis Nature
   Photography
23 Merlin D. Tuttle/Bat
   Conservation Int'l
23a Merlin D. Tuttle/Bat
   Conservation Int'l
24 Merlin D. Tuttle/Bat
   Conservation Int'l
24a Merlin D. Tuttle/Bat
   Conservation Int'l
25 Charles W. Melton
25a Merlin D. Tuttle/Bat
   Conservation Int'l
26 Merlin D. Tuttle/Bat
   Conservation Int'l
26a Merlin D. Tuttle/Bat
   Conservation Int'l
27 Merlin D. Tuttle/Bat
   Conservation Int'l
28 Merlin D. Tuttle/Bat
   Conservation Int'l
29 Roger W. Barbour/
   Smithsonian
30 Roger W. Barbour/
   Smithsonian
31 Merlin D. Tuttle/Bat
   Conservation Int'l
31a Merlin D. Tuttle/Bat
   Conservation Int'l
32 B. Moose Peterson/WRP
33 Merlin D. Tuttle/Bat
   Conservation Int'l
33a Merlin D. Tuttle/Bat
   Conservation Int'l
34 Merlin D. Tuttle/Bat
   Conservation Int'l

99 James F. Parnell
100 James F. Parnell
101 Roger W. Barbour/
    Morehead State Univ.
102 Jeff Foott
103 B. Moose Peterson/WRP
104 B. Moose Peterson/WRP
105 C. Allan Morgan
106 B. Moose Peterson/WRP
107 Dale & Marian
    Zimmerman
108 B. Moose Peterson/WRP
109 C. Allan Morgan
110 B. Moose Peterson/WRP
111 C. Allan Morgan
112 Betty Randall
113 James F. Parnell
114 B. Moose Peterson/WRP
115 Tom McHugh/Photo
    Researchers, Inc.
116 Anthony Mercieca
117 Herbert Clarke
118 Roger W. Barbour/
    Smithsonian
119 Ron Austing
120 Tom McHugh/Photo
    Researchers, Inc.
121 Tom J. Ulrich
122 Rick & Nora Bowers
123 B. Moose Peterson/WRP
124 B. Moose Peterson/WRP
125 Tom McHugh/Photo
    Researchers, Inc.
126 Roger W. Barbour/
    Morehead State Univ.
127 Tom McHugh/Photo
    Researchers, Inc.
128 Dwight R. Kuhn
129 James F. Parnell
130 Charlie Ott/Photo
    Researchers, Inc.
131 Dwight R. Kuhn
132 Rob & Ann Simpson
133 G. C. Kelley
134 Tom McHugh/Photo
    Researchers, Inc.
135 Karl H. Maslowski/Photo
    Researchers, Inc.
136 Roger W. Barbour
137 Robert W. Murphy
138 Harry M. Walker
139 Stan Osolinski
140 Robert Pollock
141 Tom McHugh/Photo
    Researchers, Inc.
142 Ronn Altig
143 Jeff Foott
144 Jeff Foott
145 Roger W. Barbour
146 Roger W. Barbour
147 Roger W. Barbour
148 Tom J. Ulrich
149 James F. Parnell
150 G. C. Kelley
151 Mike Danzenbaker
152 Robert Pollock
153 Herbert Clarke
154 Herbert Clarke
155 Charles W. Melton
156 Phillip C. Roullard
157 Rob & Ann Simpson
158 Roger W. Barbour/
    Morehead State Univ.
159 Carl R. Sams, II/
    Dembinsky Photo
    Associates
160 G. C. Kelley
161 Charlie Ott/Photo
    Researchers, Inc.
162 Stephen Krasemann/
    Nature Conservancy/
    Photo Researchers, Inc.
163 Larry Sansone
164 Tom J. Ulrich
165 Mike Danzenbaker
166 Mike Danzenbaker
167 Anthony Mercieca/Root
    Resources
168 Michael H. Francis
169 Rick & Nora Bowers
170 G. C. Kelley
171 Jeff Foott
172 Charles W. Melton
173 Herbert Clarke
174 Leonard Lee Rue, Jr.
175 Carroll W. Perkins
176 Herbert Clarke
177 Mark F. Wallner
178 Stan Osolinski
179 Rick & Nora Bowers

337 Art Wolfe
338 Daniel J. Cox/
     naturalexposures.com
339 Daniel J. Cox/
     naturalexposures.com
340 Charles G. Summers, Jr.
341 Daniel J. Cox/
     naturalexposures.com
342 Bud Nielson/Images
     International
343 Harry M. Walker
344 Karen McClymonds
345 Daniel J. Cox/
     naturalexposures.com
346 Stan Osolinski
347 Walt Enders/Ellis Nature
     Photography
348 Walt Enders/Ellis Nature
     Photography
349 Fritz Polking/Dembinsky
     Photo Associates
350 Stan Osolinski
351 Jeff Foott

352 Michael H. Francis
353 C. Allan Morgan
354 Tom J. Ulrich
355 Rod Planck
356 Betty Randall
357 Tom J. Ulrich
358 Karen McClymonds
359 Stefan Lundgren/
     The Wildlife
     Collection
360 Kevin Schafer
361 Carleton Ray/Photo
     Researchers, Inc.
362 Bud Lehnhausen/Photo
     Researchers, Inc.
363 Fred Bruemmer
364 Tom McHugh/Photo
     Researchers, Inc.
365 C. Allan Morgan
366 Douglas Faulkner/Photo
     Researchers, Inc.
367 Daniel J. Cox/
     naturalexposures.com

INDEX

Numbers in boldface type refer to color-plate numbers. Numbers in italic type refer to page numbers.

N

O

NOTES